CONTENTS

Exploring Ecology and its Applications

Exploring Ecology and its Applications

Readings from American Scientist

EDITED BY
Peter M. Kareiva
The University of Washington

Sinauer Associates, Inc. *Publishers*
Sunderland, Massachusetts U.S.A.

The Cover

Migrating western sandpipers forage on an intertidal flat. The predation of such migratory shorebirds on soft-bodied intertidal invertebrates is intense but brief; their influence on the ecology of the intertidal zone is usually temporary. See "Intertidal Zonation of Marine Invertebrates in Sand and Mud" by Charles H. Peterson, page 142. (Photograph by T. Leeson/Photo Researchers.)

Exploring Ecology and Its Applications:
Readings from *American Scientist*

Sinauer Associates, Inc.
23 Plumtree Road
Sunderland, MA 01375 U.S.A.

FAX: 413-549-1118
Internet: publish@sinauer.com
World Wide Web: www.sinauer.com

Sigma Xi, The Scientific Research Society: www.sigmaxi.org

Photographs in these articles appear with permission; the photographers are credited in the captions. The book's artists are listed at the end of this book. Sigma Xi and Sinauer Associates thank the authors, artists, and photographers for their kind permission to reprint their work.

Library of Congress Cataloging-in-Publication Data
Exploring ecology and its applications : readings from American scientist / edited by Peter M. Kareiva.
p. cm.
Includes bibliographical references.
ISBN 0-87893-414-6 (pbk.)
1. Ecology. I. Kareiva, Peter M., 1951- . II. American scientist.
QH541.145.E97 1998
577—dc21
6053558X
98-14698
CIP

Printed in U.S.A.

5 4 3 2 1

PREFACE

Ecology has never been a more exciting science than it is today. Challenged with everything from guiding society toward a sustainable biosphere to passing judgment on conservation plans for spotted owls, ecology is attracting the best and brightest students. To meet its challenges, ecology draws inspiration from economics, chemistry, physics, and atmospheric sciences, while remaining firmly grounded in the natural history of organisms. And professional scientists are not the only ones who must struggle with ecological questions and think in interdisciplinary terms; environmental debates commonly find their way into courtrooms, legislatures, and voting booths. The task of nurturing our biosphere demands economic sacrifices and hard political choices; this means we all need to have some familiarity with ecology, if only so that we can distinguish false prophets of ecological doom from those justified calls to political action in the service of our environment. The twentieth century was marked by the industrial and computer revolution; the next century will likely see an environmental revolution. We need sound, yet provocative, information—such as that developed by the contributors to this volume—that adds to our basic understanding of both general ecological principles and specific environmental crises.

Exploring Ecology and its Applications is a collection of articles in which leading researchers explain their personal approaches and points of view regarding ecological problem-solving. It is not intended to be a comprehensive textbook. Too often textbooks homogenize science into a bland soup of principles and examples, as though all participants in a field have agreed on one common pathway to the truth. Such is never the case in any science, and it is less the case in ecology than in most other sciences. Ecology is a fractious field. Approaches to the subject range from the painstaking observation of organisms in their natural environments, to clever manipulations of populations in small experimental settings, to computerized interpretation of satellite images. The articles in *Exploring Ecology and Its Applications* expose students to the many different ways of doing ecology by encompassing the major questions that ecologists are currently tackling.

This book is designed to provide vivid supplementary readings for courses in ecology or environmental sciences. Because the authors have been selected for their skill at communicating with a wide audience, this collection could profitably be used by students who are not science majors. At the same time, the research reported here is sufficiently topical and "cutting edge" that the articles can introduce the novice scientist to a richly rewarding technical literature.

Seattle, January 1998

PART I
The Ecology of Individuals: Adaptations and Strategies of Organisms

The study of ecology begins with the organism and makes little progress without a solid understanding of plant and animal adaptations to their environment. Although understanding such adaptation has long been a goal of ethologists, physiologists, and morphologists, recent developments in optimization theory have revitalized this branch of evolutionary ecology. By studying adaptations with an aim toward predicting how organisms "should behave or should be designed," we gain much more than knowledge about specific plants or animals; we often uncover patterns that allow us to predict what we might expect to see in different environments, or we identify the constraints that press most severely on an organism's evolutionary success. The articles in this section display the range of insights delivered by studies of adaptations and life histories, as well as the diversity of research approaches that deliver these insights.

We lead off with what might at first seem like "old-fashioned" physiology: an article by French that discusses hibernation in mammals. However, instead of simply recounting adaptations to cold, French explores the thermoregulatory implications of how metabolism scales nonlinearly with body size, whereas the capacity for fat storage scales linearly with body size. The result is that large hibernating mammals face a strikingly different set of constraints than do small hibernating mammals. Whereas small animals are precariously balanced between the energy savings provided by hibernation and the possibility their fat reserves may run out, large hibernators are constrained primarily by the possibility of too little time for growth and reproduction should they hibernate for too long. This is just one of many cases of evolutionary physiology identifying those aspects of the environment that present the greatest risks to organisms.

Being "too cold" is something countless organisms must face, not just hibernating mammals. As the article by Seeley shows, the risk associated with freezing temperatures often translates into a question of whether an animal can get enough food to stay warm. Seeley makes this risk vivid by pointing out that in our North American temperate climate, as many as 90% of wild honeybee colonies die each winter from starvation; therefore, much attention has been given to the thermoregulatory abilities of honeybees in temperate regions. But overlooked in our fascination with honeybee heating systems is the fact that honeybees evolved in the tropics, which remain their center of diversity. And in the tropics, cold winters are not a problem—rather, predation on nests is the greatest problem the bees face. The major adaptations of tropical honeybees often take the form of antipredator defenses that range from concealment, to sticky traps, to aggressive stinging behavior. Seeley synthesizes physiology and ecology to obtain predictions about broad patterns of social and nesting traits in honeybees as a function of the climate in which the bees live.

A second venue for studies of individual ecology is behavior. The past 20 years have been golden ones in behavioral ecology, with enormous advances in our understanding of foraging behavior, mating behavior, and fighting behavior. Three papers in this section deal with the fundamental decisions individual animals must make about when to disperse, when to fight, where to lay their eggs, and how to hunt for food. Reichert examines encounters among territorial grass spiders and asks what determines whether spiders back off from their disputes as opposed to engaging in escalating and potentially lethal battles. A modern development in behavioral ecology, called evolutionary game theory, is used to successfully predict frequencies of "retreats" versus "engaged battles" as a function of environmental circumstances. Game theory, with its concept of evolutionary stable strategies (ESS), has captured the imagination of behavioral ecologists because it focuses on the fact that the outcome of an organism's behavior depends on what other animals in the population are doing—a realization that is probably a simple fundamental truth of interactions among organisms.

Of course, "game theory" is not the only way to understand behavior, as Holekamp and Sherman eloquently show with their analysis of dispersal in male ground squirrels. The beauty of Holekamp and Sherman's research is that it addresses both proximate (physiological) and ultimate (evolutionary) explanations for dispersal behavior in squirrels. Finally, studies of individual behavior lend themselves to elegant experiments, something exemplified in the research reported by Prokopy and Roitberg. Ingenious arrays of "model lures" are used to isolate the cues flies use when laying their eggs on fruit such as apples. This represents a case in which fundamental understanding of a pest's behavior yields practical, environmentally sound methods for pest control.

The final two papers in this section explore the life history strategies of organisms. Cook analyzes the demography of ramets (genetically identical modules of the same genetic unit) to understand why clonal growth in plants is so extraordinarily successful, leading in some cases to single clones that occupy almost half a million square meters. Cook's perspective introduces a way of looking at clones that raises questions of "clonal foraging behavior" and "developmental decisions." Jackson and Hughes meld Cook's perspective with quantitative data on lifetime success to contrast clonal versus sexual reproduction among invertebrates, especially corals. The result is striking evidence for the great advantage of rapid clonal expansion in disturbed environments. The central idea in Jackson and Hughes' analysis is that fundamental "investment strategies"—trade-offs in the allocation of resources to growth, maintenance, and sexual reproduction—underlie the life history patterns we observe among invertebrates. This notion of critical trade-offs is one of the unifying ideas in evolutionary ecology, and underlies most predictions about how organisms should develop, behave, or be designed in different environments.

The Patterns of Mammalian Hibernation

Alan R. French

Most of the food eaten by mammals is used for generating heat to maintain their relatively high and constant body temperatures. Obviously the advantages of this internal heat production offset its energetic costs—mammals would not be so successful if it were otherwise. However, strongly seasonal environments may present formidable thermoregulatory challenges to small mammals. Energy demands increase during cold winters at the same time the productivity of the environment declines, and snow cover frequently prevents foraging well into the spring (Fig. 1). In such habitats, many mammals that do not migrate must temporarily rely on stored energy supplies. Should these energy stores be inadequate for the continuous maintenance of high body temperatures — a thermoregulatory state known as euthermy— a mammal's survival depends on its ability to reduce energy consumption by allowing itself to cool and hence become torpid. Some of the smaller hibernators are capable of reducing their rates of metabolism as much as a hundredfold.

Thermoregulatory strategies of hibernating mammals balance the need to conserve energy over the winter and the demands of springtime reproductive success

Even though the energy savings of torpor can be substantial, no mammalian hibernator remains continuously at low body temperatures all winter. All hibernators rewarm and then briefly maintain high body temperatures at periodic intervals throughout the dormant season. The universal occurrence of these arousal episodes suggests that, even within the confines of a secluded cave or underground burrow, high body temperatures are somehow adaptive.

It seems reasonable to assume that natural selection would favor a compromise between these opposing selection pressures, between the presumed advantages of the euthermy common to all mammals and the necessity for torpor in the face of energy shortages. Accordingly, hibernators would be expected to remain torpid no more than necessary to ensure survival. If so, then the total amount of time spent in torpor should be correlated with the energetic constraints on an individual during the winter. As will be elaborated below, these energetic constraints are a function of the size of the hibernator and its method of storing energy.

Hibernators can be classified as either those that store energy in the form of body fat or those that rely predominantly on stored food, a fundamental distinction that is usually unambiguous. Although some hibernators that become obese also occasionally cache food, such stores do not appear to be necessary for successful hibernation, are usually small when found in natural areas away from agricultural development, and may not be consumed until hibernation is terminated in the spring (Shaw 1925). Likewise, species that rely on stored food also may deposit some fat, but such internal stores usually represent a minor component of the total energy available.

The physical limitations of fat storage permit a theoretical analysis of the energetic problems facing hibernators that store fat. This in turn provides a framework for understanding the observed differences among their patterns of thermoregulation during dormancy.

Energetic consequences of size

In general, large mammals can fast for a longer period of time than can small ones, a phenomenon governed by the fact that the capacity to store fat and metabolic rate are related differently to body size (Morrison 1960). The capacity to store fat is directly related to mass ($mass^1$). In contrast, the rate at which energy is used to maintain euthermy is proportional to $mass^{3/4}$ at warm temperatures where metabolic rates are at basal or minimal levels (Kleiber 1947); the rate varies approximately with $mass^{1/2}$ at colder temperatures where resting metabolism is elevated above basal levels (Herreid and Kessel 1967). Therefore, as body size increases, the ability to store fat increases faster than the rate at which fat is metabolized. Small mammals will starve within a few days unless they become torpid, but large mammals can remain warm for many months without eating. In fact, although many large mammals have long dormant seasons, no mammal over approximately 5 kg in mass has evolved the capacity to reduce its body temperature more than a few degrees. This observation reinforces the idea that it must be very important for mammals to maintain high body temperatures if they can possibly do so.

Figure 1. (Next page) This adult male ground squirrel (*Spermophilus beldingi*) has emerged from hibernation through the snowpack at a high elevation in the Sierra Nevada well before green vegetation is available. Such an early emergence in anticipation of the appearance of the females is an energetic gamble, which requires that the males have ample post-hibernation stores of fat. (Photo courtesy of P. Sherman.)

These relationships demonstrate that even among those mammals that must hibernate, large species are less constrained by energy requirements than are small ones. The deficit between the energy available and the energy needed to maintain high body temperatures during a dormant season of fixed duration is inversely proportional to body size (Fig. 2). This theoretically presents large hibernators with several options. In comparison with small individuals, they potentially can have longer hibernation seasons, deposit proportionately less fat prior to hibernation, spend more time at high body temperatures, or finish hibernation with proportionately more fat unused.

There is no evidence that large hibernators either have longer dormant seasons or put on proportionately less fat than small hibernators. Variations in the duration of dormancy appear to be related to climatic conditions and not to body size. For example, jumping mice and marmots differ by two orders of magnitude in mass, but they frequently live together in montane meadows, where they may go without eating for eight months or more out of the year (Cranford 1978; Andersen et al. 1976). Fat content is also variable but unrelated to body size. Small jumping mice may enter hibernation with

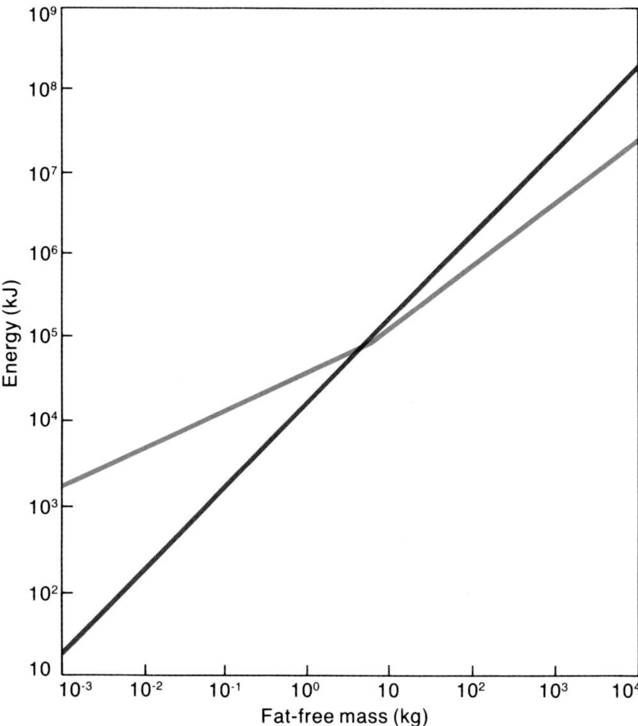

Figure 2. The energy available to mammals when fat is 50% of body weight (*gray line*) and the energy necessary to maintain high body temperatures (*red line*) are related differently to body size. Energy requirements were calculated for mammals resting for a period of six months at a temperature of 5°C; the slope of the line changes because large mammals metabolize at basal levels at this temperature but small mammals do not. These relationships indicate that although large mammals consume more food each day, they can also go without eating for much longer periods of time than can small mammals. As lean body weight falls below about 5 kg, the energy available from body fat becomes increasingly inadequate for the continuous maintainance of high body temperatures, and the selection pressures for the use of torpor increase commensurately.

50% of their weight as fat (Cranford 1978), but equally small or smaller bats often deposit only half that percentage (Krulin and Sealander 1972). This range of values is similar to that found in some of the largest marmots (Andersen et al. 1976).

In contrast, the amount of time spent at high body temperatures during dormancy is directly related to size. Large hibernators arouse only slightly more frequently in midwinter than do small ones, but the duration of their euthermic intervals is much greater (French 1985). Small bats usually remain euthermic for only one or two hours, whereas large marmots maintain high body temperatures for about a day at a time (Fig. 3). The duration of these euthermic episodes increases with size at about the same rate that mass-specific rates of metabolism decrease (French 1985). This means that the longer arousal episode of a large hibernator costs proportionately the same amount as the shorter one of a small hibernator. It also means that large and small hibernators starting with the same percentage of body fat can have hibernating seasons of about the same length, despite their differing rates of metabolism.

In addition, large hibernators are more likely than small ones to terminate their hibernation seasons spontaneously before food becomes available and, as a result of this early termination, to start their active seasons with relatively large fat reserves. Often such individuals remain euthermic for several weeks before feeding (Snyder et al. 1961); their energy stores are necessary for survival after hibernation and, in some cases, for reproductive activities during the spring. This thermoregulatory strategy is associated with an early onset of breeding, the selective advantage of which is also related to body size.

Reproductive consequences of size

Large hibernators should resume activity and begin breeding earlier in the spring than small hibernators, because both the need for an early emergence and the ability to overcome the energetic risks involved with such behavior increase with increasing size. Time, like energy, is at a premium for hibernators. The longer they are forced to remain dormant, the less time they have for reproduction and preparation for the next hibernation season. These time constraints increase with increasing body size, because gestation and juvenile growth are slower in large species than in small ones (Western 1979). However, there are energetic risks associated with an early emergence, because environmental conditions in the spring can be unpredictable. Obviously if hibernators become active before food is available, they run the risk of starving, especially in years when springtime conditions are delayed. Even the synchronization of emergence with the first appearance of food is perilous, because those resources may prove to be ephemeral in the face of late-season storms (Morton and Sherman 1978). The longer a hibernator waits to emerge, the more it will be assured that the environment will be hospitable, but the less time it will have to breed.

The timing of emergence should represent a balance between these constraints of time and energy. Large species need more time for reproduction, and by virtue of their large size they can afford to gamble more with their energy reserves to attain it than can small species.

On the other hand, small hibernators must employ a conservative thermoregulatory strategy in the spring, but they can afford to do so, and thereby to delay breeding, because their young reach adult size more quickly.

An accurately timed emergence is of value to hibernators of all sizes, and there are numerous examples to indicate that the resumption of activity in the spring is synchronized with relatively unpredictable changes in weather conditions (Michener 1984). In the most extreme case, members of a high-altitude population of Belding's ground squirrels (*Spermophilus beldingi*) emerged six weeks earlier in the year with the lowest snowfall and earliest snowmelt on record than they did in the year with the heaviest snowfall and latest snowmelt (Morton and Sherman 1978). The synchrony between emergence dates and climatic changes occurs despite the fact that hibernators spend the winter underground or in caves where they are insulated from most changes in the weather. Apparently many species assess environmental conditions at or near the entrances to their dens, and it stands to reason that the more frequently this is done, the more accurately their emergence can be timed. Such assessment requires hibernators to be alert and mobile, and thus there are selective advantages in increasing the time spent at high body temperatures at the time of year when appropriate conditions for emergence are likely. In other words, the frequency and duration of arousal episodes should change during the course of the dormant season, because the benefits of euthermy change relative to the costs.

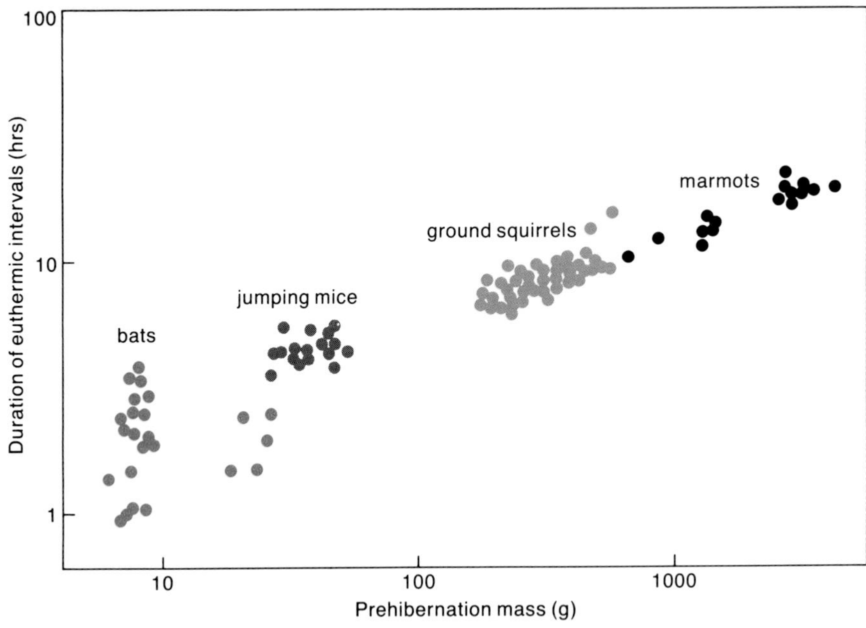

Figure 3. Large hibernators kept at 5°C spend more time at high body temperatures during each arousal episode in midwinter than do small hibernators. The duration of the arousal increases or decreases with mass$^{0.38}$, whereas the basal metabolic rate is proportional to mass$^{-0.38}$ (French 1985; Kayser 1950). This match suggests that hibernators remain awake until they have completed a fixed amount of metabolism. Thus small and large hibernators with the same percentage of body fat can have hibernating seasons of equal length.

The magnitude of this springtime increase in euthermy should increase as body size increases, because large species accrue more reproductive benefits from an early and accurately timed emergence than do small species. A spontaneous termination of hibernation in the absence of food or environmental change can be viewed as the culmination of this trend. Furthermore, intraspecific differences should occur whenever there are large differences in size among individuals in a population, or when not all individuals breed.

Long-term hibernators

The ramifications of body size can be seen clearly in a comparison of the patterns of thermoregulation exhibited by three hibernators that live sympatrically in many mountainous areas of the western United States. Yellow-bellied marmots (*Marmota flaviventris*), Belding's ground squirrels, and western jumping mice (*Zapus princeps*) span two orders of magnitude in size, yet all face approximately the same climatic conditions and often must rely exclusively on stored fat for eight or more months out of the year. The intermediate-sized Belding's squirrels are one of the most thoroughly studied hibernators from both physiological and ecological perspectives, and they provide a good basis of comparison for other species.

All Belding's squirrels increase the time they spend at high body temperatures toward the end of the dormant season, but the extent of this energetic gamble varies as a function of sex and age (French 1982). These animals are sexually dimorphic with respect to body size at maturity; large adult males spontaneously terminate hibernation, whereas females and small, nonbreeding males (yearlings) hedge their bets. These latter groups increase the frequency at which they arouse in the spring but do not stop hibernating until they are fed (Fig. 4).

Because Belding's squirrels do not reach adult size in their first year, it is beneficial for females to breed as soon as possible and thereby to maximize the time their offspring have to grow and deposit fat before their first hibernation season. At high elevations in the Sierra Nevada, females hibernate in areas of the meadows that first become free of snow in the spring, and they synchronize their emergence with the snowmelt and the availability of green vegetation near their burrow entrances (Sherman and Morton 1979). Presumably the consumption of this food triggers the physiological termination of hibernation and the continued presence of the females above ground. The increasing frequency of arousal in the springtime most likely improves the accuracy with which they can time their emergence, but the benefits of a commitment to continuous euthermy before food becomes available appear to be slight compared to the risks involved.

Adult males are active in the early spring, and at high elevations they regularly tunnel through the snowpack to reach the surface (Morton and Sherman 1978). It appears that the endogenously timed termination of their hibernation allows them to anticipate the emergence of females. This is advantageous because female ground squirrels become sexually receptive a few days after they come above ground, and a male that is late to emerge would find females that had previously emerged already impregnated. However, because females time their emergence to coincide with the first availability of food, males are forced to fast while waiting for their potential mates to appear. This is a substantial energetic gamble, and the large size and concomitant fat storage capabilities of adult males are clearly adaptive. Nevertheless, males rarely survive as long as females (Morton and Parmer 1975).

Small yearling males do not have the energy stores necessary to undergo a prolonged fast at euthermic body temperatures; nor would they gain much by doing so, because sexual maturation and breeding do not usually occur until males are two years old (Morton and Gallup 1975). Their open-ended hibernation seasons and delayed emergence thus facilitate survival over the winter.

The differences between adult and immature males appear to be related to differences in their size and fat content rather than their age. Juveniles kept under ideal laboratory conditions attain adult weight in their first summer, and they also spontaneously terminate hibernation and become reproductively active as yearlings the following spring. Conversely, males two years and older that fail to deposit normal quantities of fat have open-ended hibernation seasons and do not mature sexually (French 1982).

Intraspecific differences in the patterns of hibernation also occur in yellow-bellied marmots, which as adults are an order of magnitude larger than Belding's squirrels. Adult marmots of both sexes spontaneously stop hibernating, but the smallest juveniles do not (Fig. 4). Marmots grow more slowly than ground squirrels, and winter mortality of juveniles is inversely related to their weight at the start of dormancy (Armitage et al. 1976). To help compensate for this slow growth, marmots emerge and begin breeding early in the spring, well before there is food to eat (Andersen et al. 1976). In contrast to the smaller ground squirrels, the larger size of female marmots enables them to remain at high body temperatures and undergo the energy drains of early pregnancy while fasting. The smallest juveniles do not stop hibernating until they are fed, even though they are over twice as large as adult male ground squirrels. Apparently there is little value for these large but immature mammals to make such an extreme energetic gamble, because neither sex breeds as a yearling (Andersen et al. 1976). In fact, yearlings usually emerge above ground a month after breeding adults (Carey 1985).

Western jumping mice are an order of magnitude smaller than Belding's squirrels, and all individuals have open-ended hibernation seasons (Fig. 4). However, males do arouse more frequently than females

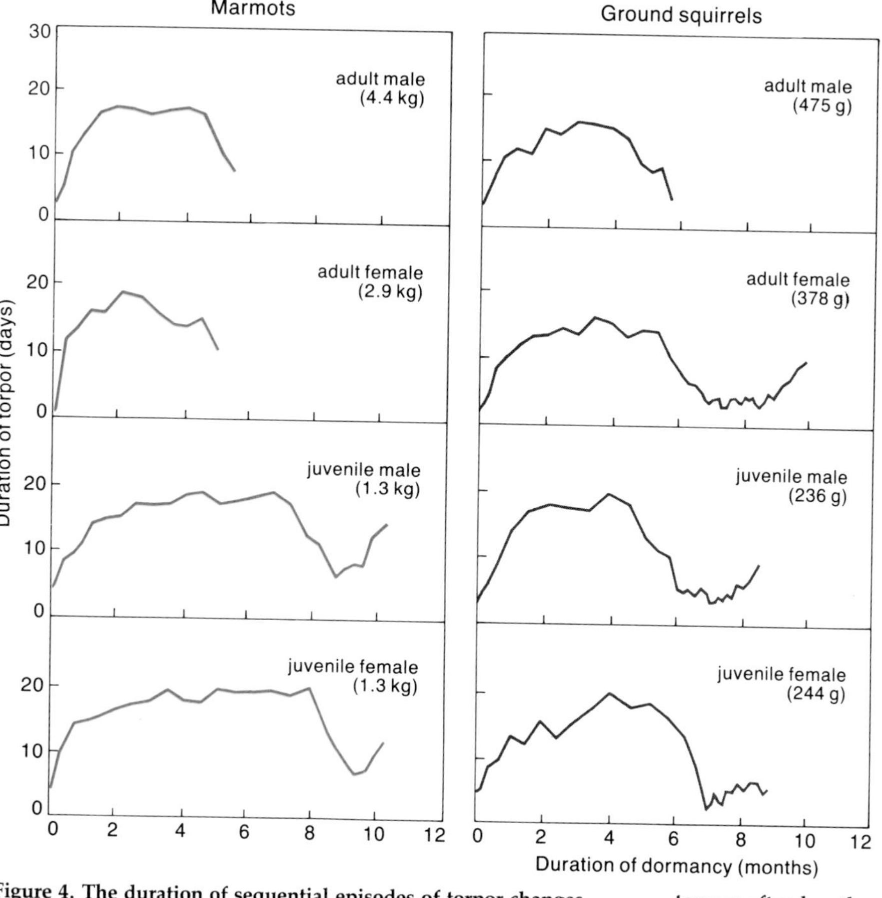

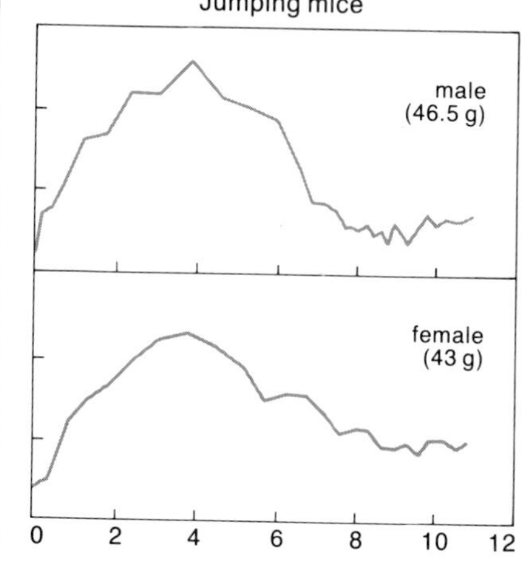

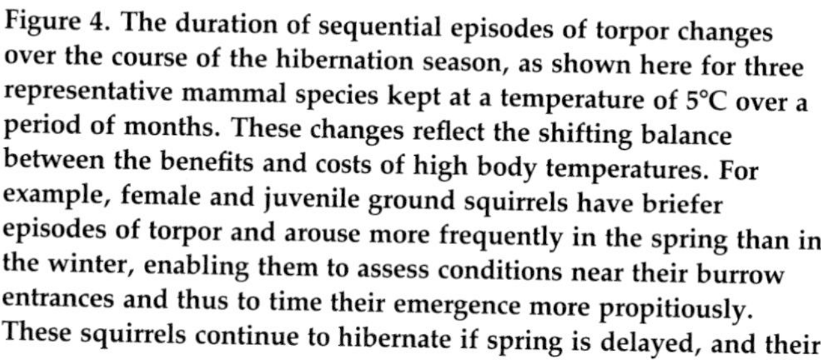

Figure 4. The duration of sequential episodes of torpor changes over the course of the hibernation season, as shown here for three representative mammal species kept at a temperature of 5°C over a period of months. These changes reflect the shifting balance between the benefits and costs of high body temperatures. For example, female and juvenile ground squirrels have briefer episodes of torpor and arouse more frequently in the spring than in the winter, enabling them to assess conditions near their burrow entrances and thus to time their emergence more propitiously. These squirrels continue to hibernate if spring is delayed, and their torpors often lengthen as they begin to run out of fat reserves and energy conservation becomes more important than an accurately timed emergence. By contrast, adult male squirrels spontaneously terminate hibernation before their stores of fat are depleted in anticipation of the emergence of the smaller females, which become sexually receptive as soon as they appear above ground. In the experiment, bouts of torpor ended spontaneously in adult marmots of both sexes and in adult male ground squirrels; torpor ended in the other animals only when they were fed.

during the spring phase of hibernation, and like most other rodent hibernators, they emerge one to two weeks earlier as well (Brown 1967). Therefore, even in a species strongly constrained by energy, males gamble more with their fat stores than females. Nevertheless, the mice are conservative, emerging only after the snow has melted and the soil at the depth of their winter nests has begun to warm (Cranford 1978). The relatively rapid growth rates of jumping mice more than compensate for their late dates of emergence, and even at high altitudes youngsters grow to near adult size in the summer of their birth and breed the next spring.

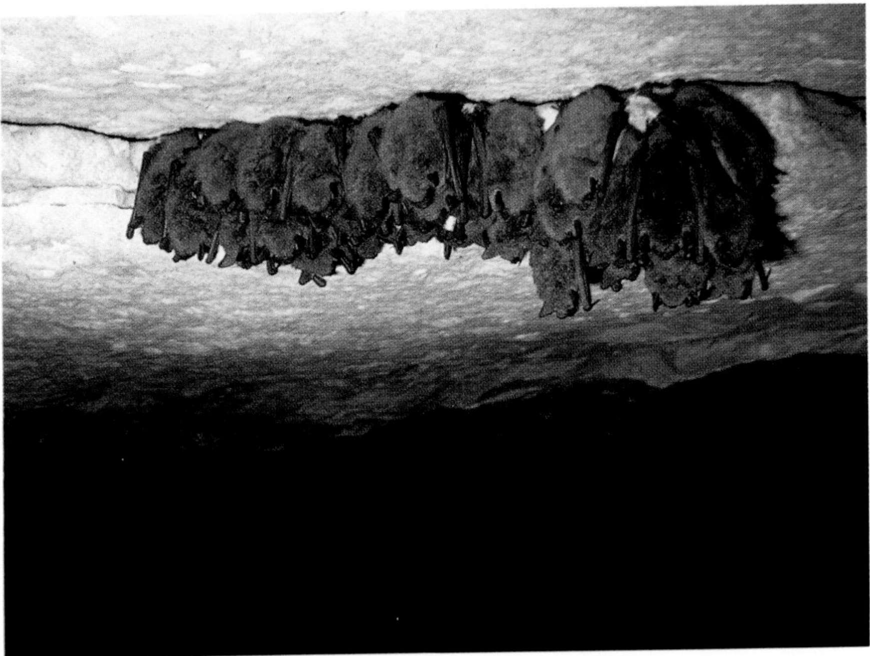

Figure 5. Many bats hibernate communally, like these cave bats (*Myotis velifer*), and winter aggregations may number in the thousands. This clustering may help tiny bats to conserve heat during their arousal episodes in much the way a nest helps insulate a rodent. Most females are inseminated before hibernation. Males store previously produced sperm, however, and during their brief arousals they will occasionally copulate with torpid females nearby. (Photo courtesy of T. H. Kunz.)

Long-term hibernators at the extremes of body size mate prior to dormancy. For example, bears (family Ursidae) need to do so because the development time of their offspring is so long. They complete gestation and lactation during dormancy, and their young are mobile at the time of spring emergence and are ready to feed soon thereafter. Bears can do this because they are so large that they need not reduce their body temperatures during their winter fast (Hock 1960).

At the other extreme are many species of insectivorous bats, the smallest and most energetically constrained hibernators, which also are the only mammals that undergo delayed fertilization (Fig. 5). Mating occurs prior to or occasionally during dormancy, sperm is stored in the female reproductive tract over the winter, and ovulation and fertilization take place as soon as the females resume activity in the spring (Wimsatt 1969). This remarkable reproductive strategy helps to alleviate the delay in offspring development imposed by an insectivorous diet. Flying insects can be an ephemeral food resource early in the spring, and pregnant bats lower their body temperatures following nights of poor foraging, slowing gestation as a result. Pregnancy has been known to last over three months when the prevalence of torpor is high (Pearson et al. 1952).

The energetic constraints of their small size dictate that bats have a conservative hibernation strategy, but their prolonged gestation favors early conception in the spring. Bats cannot afford to resume activity until flying insects first appear. If copulation occurred in the spring, females would have to delay ovulation until males emerged and became reproductively competent. Autumn mating provides females with a source of sperm early in the spring, when the availability of competent mates would otherwise be at best unpredictable.

Patterns of thermoregulation have not yet been well documented for hibernating bats, but available data indicate that some species have relatively short torpors in the spring (Twente et al. 1985). It may turn out that female bats have more frequent arousals than males. Females must initiate pregnancy as soon as possible, but males, with females already inseminated, need not gamble as much with their limited energy reserves in order to achieve an early emergence. In fact, among little brown bats (*Myotis lucifugus*), males have been known to remain hibernating in caves well after females have departed (Fenton 1970).

Short-term hibernators

The analysis so far has been restricted to long-term hibernators in order to illustrate the effects of body size. However, the costs and benefits of torpor also vary in response to environmental conditions. Species of similar size may have dissimilar reproductive strategies and patterns of thermoregulation if their dormant seasons are of substantially different durations.

This point is well illustrated by a comparison between thirteen-lined ground squirrels (*Spermophilus tridecemlineatus*), which remain underground for at least eight months out of the year (McCarley 1966), and round-tailed ground squirrels (*S. tereticaudus*), which are dormant for only half as long (Dunsford 1977). Both species are about half the size of Belding's squirrels, and thus one might predict that they would be energetically more conservative during dormancy than their larger cousins. This is true for the long-term hibernator *S. tridecemlineatus;* both males and females have open-ended hibernation seasons when kept at 5°C without food. In the desert-dwelling *S. tereticaudus*, however, females have open-ended hibernation seasons but males of all sizes and ages spontaneously terminate hibernation. A short dormant period allows these small male squirrels to start their active seasons with relatively large fat reserves, and thereby makes the risk of starvation before food becomes available relatively small.

Figure 6. An energetic strategy based on stored food rather than fat deposits is exemplified by this chipmunk (*Tamias striatus*), which has stuffed its cheek pouches with 93 ragweed seeds. The energy content of this animal's winter store is not limited by its size, and thus it need not spend as much time in torpor over the winter as a similar-sized hibernator that stores energy as fat. (Photo courtesy of K. Maslowski/Photo Researchers.)

Similarly, American badgers (*Taxidea taxus*) often live in environments with long winters, but because of their carnivorous diet, food is unavailable for a relatively short period and activity ceases for only two and a half to three months during the coldest years. Although badgers are not much different in size from the larger marmots, their patterns of thermoregulation during dormancy differ greatly; badgers allow their body temperatures to fall from 38°C to 29°C only during irregular torpors of about a day in duration (Harlow 1981). Again, we see that need and ability to reduce body temperatures are closely matched.

Hibernators that store food

The constraints of body size on energy availability during hibernation have been circumvented by species that store food (Fig. 6). For example, a bat weighing 8 g must survive all winter on less than 8 g of fat, but a pocket mouse weighing 8 g can easily store 25 to 50 times that amount of energy in the form of seeds. If high body temperatures are advantageous, then species that store food should not spend as much time in torpor as similar-sized species that deposit fat. Furthermore, if the energy content of food caches increases with body size in a fashion similar to that of fat deposits, then large species that rely on stored food should be euthermic more than small species.

Data from both the smallest and largest hibernators that store food indicate again that high body temperatures tend to be maximized during dormancy. Many of the tiny pocket mice in the genus *Perognathus* are seasonal hibernators, and some remain underground ten months out of each year (O'Farrell et al. 1975). These mice may experience over a hundred arousal episodes, each lasting several hours, whereas a similar-sized bat may awaken for only an hour or two on fewer than two dozen occasions throughout the winter. The longest torpors in pocket mice are less than a week (French 1977); in bats they are over a month (Menaker 1959). Comparatively large chipmunks, *Tamias striatus*, also remain torpid for less than a week at a time, but torpor is infrequent in many individuals and may not occur at all in some, even though they remain underground for several months (Maclean 1981).

Differences between large and small species do exist and are similar to those observed in hibernators that store fat. Both male and female *Tamias* usually stop hibernating spontaneously after several months in the cold, but small pocket mice continue to hibernate until they run out of food or experience an increase in environmental temperature. However, pocket mice do increase the time they spend at high body temperatures in the spring, and the magnitude of this increase is greater in males than in females. Furthermore, the more food these mice store, the less time they spend in torpor (French 1976); this indicates both the importance of high body temperatures and the ability of the animals to adjust their physiology to match energetic demands.

It is tempting to speculate that the energetic principles derived for species that store fat also hold for those that rely on cached food. If, for ecological reasons, external energy stores do in fact increase with increasing body size, then energy availability should match the energy necessary for continuous euthermy at a much smaller body mass for species that store food than for species that store fat. This does not appear to be an unreasonable assumption. The largest mammals that rely exclusively on stored food and also reduce their body temperatures during dormancy are chipmunks and hamsters, which weigh between 100 and 140 g and are considerably smaller than the largest mammals, 5-kg marmots, that depend on stored fat and also reduce their body temperatures.

Alan R. French received an A.B. in zoology from the University of California at Berkeley and a Ph.D. in biology from the University of California at Los Angeles. He did postdoctoral research at Stanford University and taught at the University of California at Riverside before joining the faculty of the State University of New York at Binghamton in 1984. His interests lie in the adaptations of organisms to harsh environments, with particular emphasis on the timing and energetics of reproduction and dormancy. Address: Department of Biological Sciences, State University of New York, Binghamton, NY 13901.

References

Andersen, D. C., K. B. Armitage, and R. S. Hoffmann. 1976. Socioecology of marmots: Female reproductive strategies. *Ecology* 57: 552–60.

Armitage, K. B., J. F. Downhower, and G. E. Svendsen. 1976. Seasonal changes in weights of marmots. *Am. Midl. Nat.* 96:36–51.

Brown, L. N. 1967. Seasonal activity patterns and breeding of the western jumping mouse (*Zapus princeps*) in Wyoming. *Am. Midl. Nat.* 78:460–70.

Carey, H. V. 1985. The use of foraging areas by yellow-bellied marmots. *Oikos* 44:273–79.

Cranford, J. A. 1978. Hibernation in the western jumping mouse (*Zapus princeps*). *J. Mammal.* 59:496–509.

Dunsford, C. 1977. Behavioral limitation of round-tailed ground squirrel density. *Ecology* 58:1254–68.

Fenton, M. B. 1970. Population studies of *Myotis lucifugus* (Chiroptera: Vespertilionidae) in Ontario. *Roy. Ontario Mus. Life Sci. Contrib.* 77: 1–34.

French, A. R. 1976. Selection of high temperatures for hibernation by the pocket mouse, *Perognathus longimembris*. *Ecology* 57:185–91.

———. 1977. Periodicity of recurrent hypothermia during hibernation in the pocket mouse, *Perognathus longimembris*. *J. Comp. Physiol.* 115: 87–100.

———. 1982. Intraspecific differences in the pattern of hibernation in the ground squirrel *Spermophilus beldingi*. *J. Comp. Physiol.* 148:83–91.

———. 1985. Allometries of the durations of torpid and euthermic intervals during mammalian hibernation: A test of the theory of metabolic control of the timing of changes in body temperature. *J. Comp. Physiol. B.* 156:13–19.

Harlow, H. J. 1981. Torpor and other physiological adaptations of the badger (*Taxidea taxus*) to cold environments. *Physiol. Zool.* 54:267–75.

Herreid, C. F., II, and B. Kessel. 1967. Thermal conductance in birds and mammals. *Comp. Biochem. Physiol.* 21:405–14.

Hock, R. J. 1960. Seasonal variations in physiologic functions of Arctic ground squirrels and black bears. *Bull. Mus. Comp. Zool.* 124:155–71.

Kayser, C. 1950. Le problème de la loi des tailles et de la loi des surfaces tel qu'il apparait dans l'étude de la calorification des Batraciens et Reptiles et des Mammifères hibernants. *Arch. Sci. Physiol.* 4:361–78.

Kleiber, M. 1947. Body size and metabolic rate. *Physiol. Rev.* 27:511–41.

Krulin, G. S., and J. A. Sealander. 1972. Annual lipid cycle of the grey bat, *Myotis grisescens*. *Comp. Biochem. Physiol.* 42A:537–49.

Maclean, G. S. 1981. Torpor patterns and microenvironment of the eastern chipmunk, *Tamias striatus*. *J. Mammal.* 62:64–73.

McCarley, H. 1966. Annual cycle, population dynamics and adaptive behavior of *Citellus tridecemlineatus*. *J. Mammal.* 47:294–316.

Menaker, M. 1959. The frequency of spontaneous arousal from hibernation in bats. *Nature* 203:540–41.

Michener, G. R. 1984. Age, sex, and species differences in the annual cycles of ground-dwelling sciurids: Implications for sociality. In *Biology of Ground-Dwelling Squirrels*, ed. J. O. Murie and G. R. Michener, pp. 81–107. Univ. of Nebraska Press.

Morrison, P. 1960. Some interrelations between weight and hibernation function. *Bull. Mus. Comp. Zool.* 124:75–91.

Morton, M. L., and J. S. Gallup. 1975. Reproductive cycle of the Belding ground squirrel (*Spermophilus beldingi beldingi*): Seasonal and age differences. *Great Basin Nat.* 35:427–33.

Morton, M. L., and R. J. Parmer. 1975. Body size, organ size, and sex ratios in adult and yearling Belding ground squirrels. *Great Basin Nat.* 35:305–09.

Morton, M. L., and P. W. Sherman. 1978. Effects of a spring snowstorm on behavior, reproduction, and survival of Belding's ground squirrels. *Can. J. Zool.* 56:2578–90.

O'Farrell, T. P., R. J. Olson, R. O. Gilbert, and J. D. Hedlund. 1975. A population of Great Basin pocket mice, *Perognathus parvus*, in the shrub-steppe of south-central Washington. *Ecol. Monogr.* 45:1–28.

Pearson, O. P., M. R. Koford, and A. K. Pearson. 1952. Reproduction of the lump-nosed bat *Corynorhinus rafinesquei* in California. *J. Mammal.* 33:273–320.

Shaw, W. T. 1925. The hibernation of the Columbia ground squirrel. *Can. Field Nat.* 39:56–61, 79–82.

Sherman, P. W., and M. L. Morton. 1979. Four months of the ground squirrel. *Nat. Hist.* 88:50–57.

Snyder, R. L., D. E. Davis, and J. J. Christian. 1961. Seasonal changes in the weights of woodchucks. *J. Mammal.* 42:297–312.

Twente, J. W., J. Twente, and V. Brack, Jr. 1985. The duration of the period of hibernation of three species of vespertilionid bats. II. Laboratory studies. *Can. J. Zool.* 63:2955–61.

Western, D. 1979. Size, life history and ecology in mammals. *Afr. J. Ecol.* 17:185–204.

Wimsatt, W. A. 1969. Some interrelations of reproduction and hibernation in mammals. *Symp. Soc. Exp. Biol.* 23:511–49.

Thomas D. Seeley

The Ecology of Temperate and Tropical Honeybee Societies

Ecological studies complement physiological ones, offering a new perspective on patterns of honeybee adaptation

In his masterful synthesis of insect sociology, Wilson (1971) identified three overlapping stages in the history of research on social insects: a natural history phase, a physiology phase, and a population-biology or ecology phase. Until the mid-1970s, most students of the social insects focused on the first two stages, either describing the diverse societies of termites, ants, wasps, and bees or experimentally analyzing the physiological bases of their social systems. Quite recently, ecological studies of the social insects have also flourished. As a result, we are beginning to understand the properties of social insect colonies as adaptations fixed by natural selection.

Throughout the history of insect sociology the premier object of study has been a single species of social bee: *Apis mellifera*, one of four living honeybee species. Its preeminence stems from several factors. First, honeybee societies rank among the most complex of all insect societies, with such advanced features as strong dimorphism of queen and worker, elaborate division of labor by age, precise control of nest temperature, and a remarkable system of communication based on a dance language (Michener 1974). Ease of study also favors research on honeybees. Not only are their colonies easily maintained in man-made hives of the sort used by beekeepers, but they will even live in glass-walled observation hives, enabling humans to peer into the heart of their society.

Thomas D. Seeley is Assistant Professor of Biology at Yale University. He is a graduate of Dartmouth College and Harvard University. His scientific interests center on the social behavior of insects, especially the honeybee. The area of research described in this article will be reviewed more fully in his book Honeybee Ecology, *to be published by Princeton University Press. Address: Department of Biology, Yale University, New Haven, CT 06511.*

Still another factor is surely man's ancient fascination with the honeybee. The great insect sociologist William Morton Wheeler (1923) expressed this accurately when he wrote:

> Its sustained flight, its powerful sting, its intimacy with flowers and avoidance of all unwholesome things, the attachment of the workers to the queen—regarded throughout antiquity as a king—its singular swarming habits and its astonishing industry in collecting and storing honey and skill in making wax, two unique substances of great value to man, but of mysterious origin, made it a divine being, a prime favorite of the gods, that had somehow survived from the golden age or had voluntarily escaped from the garden of Eden with poor fallen man for the purpose of sweetening his bitter lot. [p. 91]

When the human fascination with the honeybee turned scientific, man began to describe and experimentally analyze the interwoven phenomena of colony life cycle, caste structure, and communication codes that make up the social organization of bees. This approach gained strong impetus from the highly crafted studies of the Nobel Laureate Karl von Frisch (1967, 1971) and continues apace today. We now understand in fair detail such topics as honeybee caste determination, sensory physiology, nest micrometeorology, and communication among colony members. In stark contrast, the ecology of honeybee societies remains a largely uncharted area of study. In short, we know a great deal about how honeybee societies work, but remarkably little about the pressures of natural selection that have shaped them.

The honeybee's dance language provides a clear example of this imbalance between mechanistic and functional knowledge. Over the past 60 years, aided by more than 40 graduate students, von Frisch (1967) has assembled a detailed picture of the physiological mechanisms of the behaviors that unfold when a scout bee discovers a rich patch of flowers, flies back to her nest, and recruits nestmates to gather the food by using the dance language. However, the precise ecological significance of this finely tuned system of communication has until recently remained a matter of speculation. Only in the past three years have researchers begun to analyze in a systematic way how the dance language helps colonies living in nature collect their food (Visscher and Seeley 1982; Seeley, in press).

In this essay I will describe some recent developments in the ecological study of honeybee social behavior, drawing in particular on a general program of behavioral-ecological research I have conducted over the past eight years. Most of my research deals with colonies of *A. mellifera* living in the northeast region of the United States. *A. mellifera* occurs as a native in Europe, western Asia, and Africa, but has been introduced throughout the world by man. A smaller portion of the research program considers the ecologies of the three other honeybee species: *A. florea*, *A. cerana*, and *A. dorsata*. Except for *A. cerana*, whose range extends north into China and Japan in eastern Asia, these three species are found only in southern Asia.

Colonies living in nature

Wild colonies of honeybees have been hunted by the peoples of Asia, Africa, and Europe for hundreds if not thousands of years (Crane 1975).

Thus man has long had contact with honeybee colonies in nature, but with the intention of robbing them of their beeswax and honey, not studying them. When, in the 1500s, Europeans turned from exclusively plundering bees to also examining the fundamental facts of honeybee life, they observed bees living in man-made hives. To the present day, virtually all scientific research on honeybee biology has been conducted with colonies occupying beehives placed in locations that are convenient for scientists. For example, much of von Frisch's pioneering research on honeybee communication was carried out in the courtyard of the Munich Zoological Institute, a converted monastery in the heart of a city.

An important first step in studying honeybee ecology, therefore, was to describe the nests and life history of wild colonies of *A. mellifera.* This required turning away from colonies living in man-made hives in ecologically disturbed habitats and instead studying colonies inhabiting hollow trees in forests—in this case colonies found in the countryside near Ithaca, New York (Fig. 1). We located wild colonies either through information from local residents or by beelining, the old bee hunter's technique of inducing bees to forage from a comb filled with sugar water and then tracing the bees back along their flight lines until the bee tree is reached (Edgell 1949). Some of the bee trees were felled to collect the colonies and dissect their nests. Others were left standing for long-term observations of colony mortality. To determine the distinctive reproductive patterns of wild colonies, we simultaneously studied reproduction in colonies living in man-made hives the same size as the tree cavities occupied by wild colonies.

When allowed to remain in hollow trees in the forest, honeybee colonies live quite differently than they do when they inhabit a beekeeper's or bee researcher's standard beehives. Whereas a beekeeper desires a large, nonreproducing colony capable of stockpiling a vast quantity of honey (much more honey, in fact, than the colony would ever need), wild colonies grow to only one-third to one-half the population size, sequester only as much honey as they need, and devote their remaining

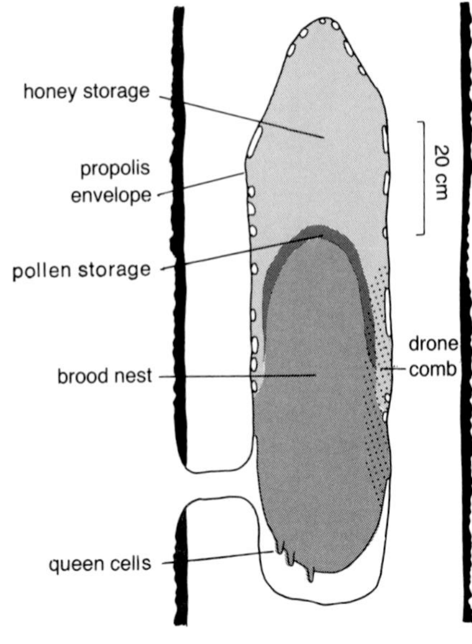

Figure 1. A knothole that serves as the entrance to an *Apis mellifera* nest is visible high up in the left fork of this intact bee tree in central New York State (*above*). A cross-section through a typical nest (*right*) shows what lies beyond the entrance. Layers of vertical comb nearly fill the tree hollow. Honey is stored in the upper region, pollen is packed in a narrow band directly below, and brood is reared in the lowermost portion of each comb. Queen cells house the new queens before swarming; drones occupy special cells at the edge of the nest. A polished layer of propolis, or tree resins, seals the nest cavity. Before a swarm occupies such a cavity, scouts carefully check its suitability by measuring the volume of the cavity and noting the size of the entrance opening, its compass orientation, and its proximity to the floor of the cavity. (All photographs by the author; sketch after Seeley and Morse 1976.)

energies to colony reproduction, swarming nearly every year (Seeley and Morse 1976; Seeley 1978; Winston 1980). These differences between wild colonies and beekeeper's colonies do not reflect genetic differences between the two colony types. Rather, they reflect the beekeeper's practice of providing abundant hive space (usually more than 160 liters), so that colonies rarely become overcrowded and thus are rarely stimulated to swarm. Wild colonies, in contrast, live in cavities that are usually smaller than 80 liters in volume; they soon outgrow this space and so swarm annually (Seeley 1977).

The two types of colonies also differ strikingly in their patterns of survival. Because beekeepers provide their colonies with snug hives, preconstructured combs, and supplementary food in lean times, only about 10% of domestic colonies perish annually. Wild colonies face a much harsher existence. Each colony must search the forest for a tree cavity, construct energetically expensive beeswax combs, and laboriously collect from millions of flowers the tens of kilograms of pollen and honey consumed by a colony yearly (Ribbands 1953). If in a given summer the plants yield abundant food, wild colonies thrive and multiply; if not, many starve. In their first year fully 75% of wild colonies starve, but if a colony survives this critical period, mortality drops to about 20% per year (Seeley 1978). Evidently selection is most intense, and hence evolution potentially fastest, at the stage of colony founding. It is therefore perhaps not surprising that honeybee colonies possess the complex adaptations for colony founding discussed below.

The challenge of winter

The finding of greatest ecological meaning from our initial descriptive studies is that there is strong seasonality in colony mortality. The vast majority of wild colonies perish during the winter; 77% and 90% of the deaths of first-year and established colonies, respectively, occur during the winter (Seeley 1978). Death is almost always due to starvation. It appears that the paramount ecological challenge faced by honeybee colonies in cold temperate regions of the world is winter survival. Much of the honeybee's biology can be understood as adaptations to this particular problem.

Unlike other social insect species in cold climates, honeybee colonies do not overwinter either as dormant, solitary queens or as inactive colonies in refuge deep underground, but as full, active colonies in self-heated nests. To achieve this, each colony contracts in autumn into a tight cluster of bees whose surface temperature is maintained above about 10°C. Heat is generated within the cluster by microvibrations produced by the bees' massive flight muscles. The colony's stored honey provides the fuel (Johansson and Johansson 1979; Seeley and Heinrich 1981).

The reason that honeybees practice such an exotic technique of overwintering is largely historical. Honeybees apparently arose in the Old World tropics and only later penetrated temperate regions (Wilson 1971; Michener 1974). All honeybee species living in the tropics today show a basic ability to control the central, nursery region of the nest at 30 to 36°C year round (Morse and Laigo 1969; Darchen 1973; Akratanakul 1977 diss.). When *A. mellifera* expanded out of the tropics into regions with harsher climates, its colonies adapted by simply refining their preexisting methods of controlling nest microclimate, rather than evolving wholly new physiological techniques for surviving periods of low body temperature.

Success in maintaining a warm microenvironment inside the nest and thus in overwintering requires that a colony occupy a well-sheltered tree cavity which tightly encloses the bees and their honey-filled combs, and that it possess the 20 or more kilograms of honey needed for winter heating fuel. Bees fulfill the first requirement through their elegant process of nest site selection (Lindauer 1955, 1961; Seeley et al. 1980; Seeley 1982). After leaving the parent nest and settling nearby, a honeybee swarm sends out a few hundred scout bees who search the surrounding forest for potential nest sites; from among the twenty or so sites found they select the single best site for their colony's future home.

A number of investigators (Seeley and Morse 1978a,b; Jaycox and Parise 1980; Gould 1982; Rinderer et al. 1982) have analyzed the preferences of house-hunting bees by setting out series of paired nest-boxes differing in a single variable such as entrance height or cavity volume and noting the patterns of occupation by wild swarms. Scouts reject cavities that have a volume of less than about 15 liters, undoubtedly because such small cavities cannot enclose enough honey-filled combs to fuel a colony through a winter. They measure the size of the cavity by walking about inside it and integrating the sensations thus experienced into a perception of volume (Seeley 1977).

Scouts also reject cavities whose entrance holes are larger than about 70 cm^2, probably because such sites are excessively drafty. Small openings in the walls of the cavity do not disqualify a site, because bees can plug these with tree resins. The bees' preference for an entrance opening that lies at the floor of the cavity and that faces south is probably also an adaptation for overwintering (Avitabile et al. 1978). An entrance located at the bottom of the cavity may help minimize convectional heat loss, and a southern exposure ensures a sun-warmed platform from which bees can make the critical cleansing flights to eliminate accumulated body wastes on occasional mild days in midwinter.

Fulfilling the second requirement for overwintering, sufficient honey stores, is fostered by several unique elements of honeybee social organization. One is the dance language, a system of communication whereby a scout that has discovered a rich food source recruits nestmates to help gather her find by performing inside the nest a miniature reenactment of a flight out to the flowers (von Frisch 1967; Michener 1974; Gould 1976). To understand exactly how this ability to communicate helps colonies collect the many kilograms of honey needed for winter, Visscher and I (1982) recently monitored the foraging behavior of a honeybee colony living in a deciduous forest in central New York State (Fig. 2). We envisaged the colony as a gigantic amoeba fixed on its nest site but able to send pseudopods—i.e., groups of foragers—out across the forest to patches of flowers rich in nectar and pollen. Of course, it is utterly impossible to follow a colony's entire force of 10,000 or more foragers by observing them directly; they are spread across some 25 km^2 of forest. However, we suspected that one could infer where a colony is foraging by tapping into the bees' own communication system—that is, by reading the dances by which foragers were recruited.

The heart of our experimental procedure, therefore, was to establish a normal-sized colony in a glass-walled observation hive and to sample, record, and translate the colony's dances. In each dance a bee walks briskly and repeatedly across one of the nest's vertical combs, shaking her abdomen back and forth to attract the attention of the surrounding bees. The duration of these so-called "waggle runs" is proportional to the distance to the patch of flowers being advertised. Nearby sites are indicated by waggle runs lasting less than one second, distant sites by runs of up to several seconds. The direction of the patch of flowers is encoded in the angle of the dance. Waggle runs in which the bee walks straight up the vertical comb indicate flower patches directly in line with the sun, whereas waggle runs performed 90° to the right of this indicate patches 90° to the right of the sun, and so forth. Whether the patch being advertised yields pollen or nectar is indicated by the presence or absence of loads of pollen on the dancer's hind legs.

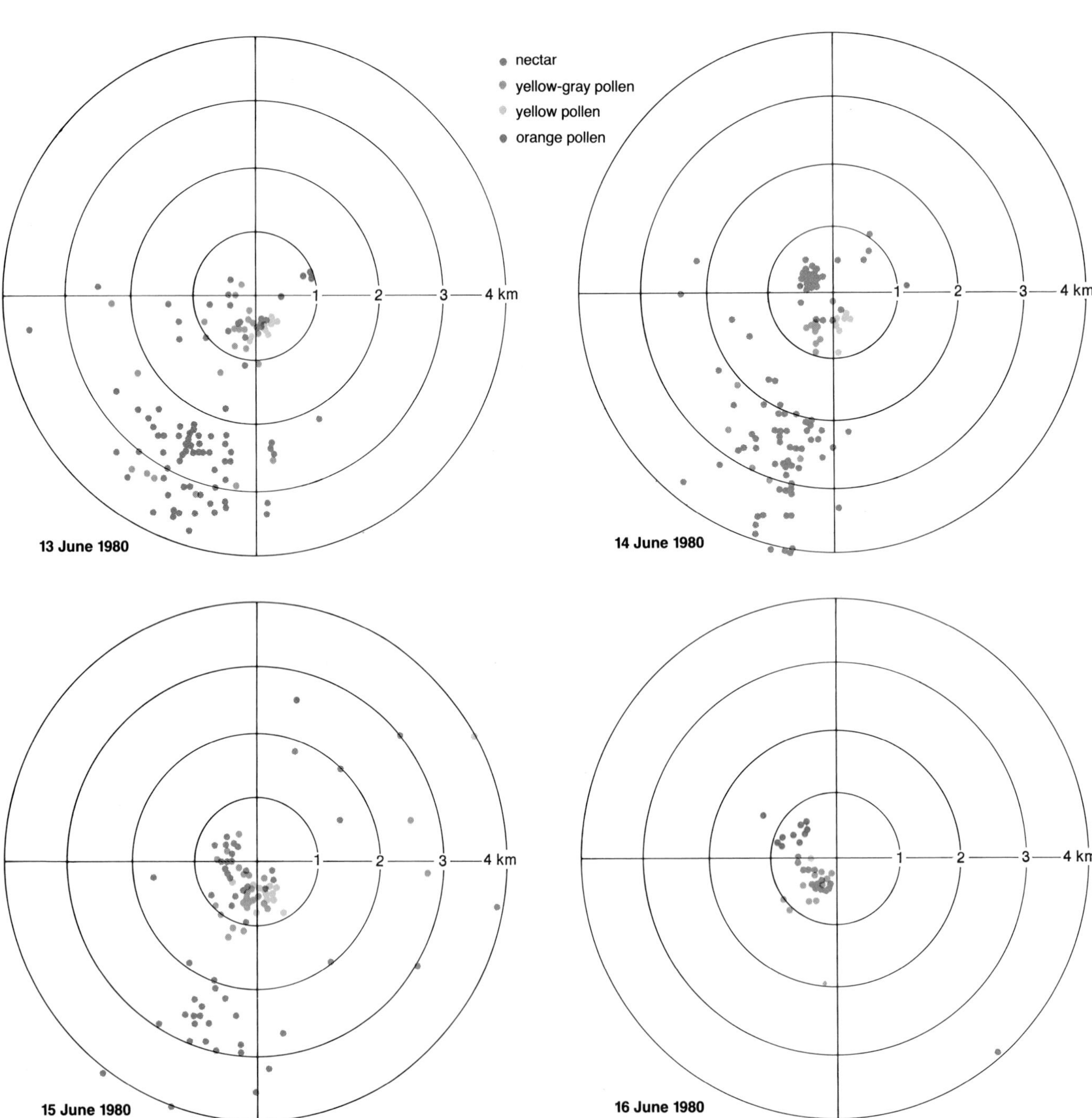

Figure 2. By reading the recruitment dances of foragers, it was possible to map the location of food sources on consecutive days and thus to study the overall dynamics of a colony's foraging activity. The nature of the food source was determined by observing whether returning foragers unloaded nectar (*gray dots*) or pollen (*colored dots*); recording the color of the pollen brought back in each case gave a further key to the distribution of the foragers. The pattern that emerges is one of a shifting mosaic of food sources, as the honeybee colony uses information gained from the searches of thousands of foragers to help direct its forces to the most rewarding flower patches. (After Visscher and Seeley 1982.)

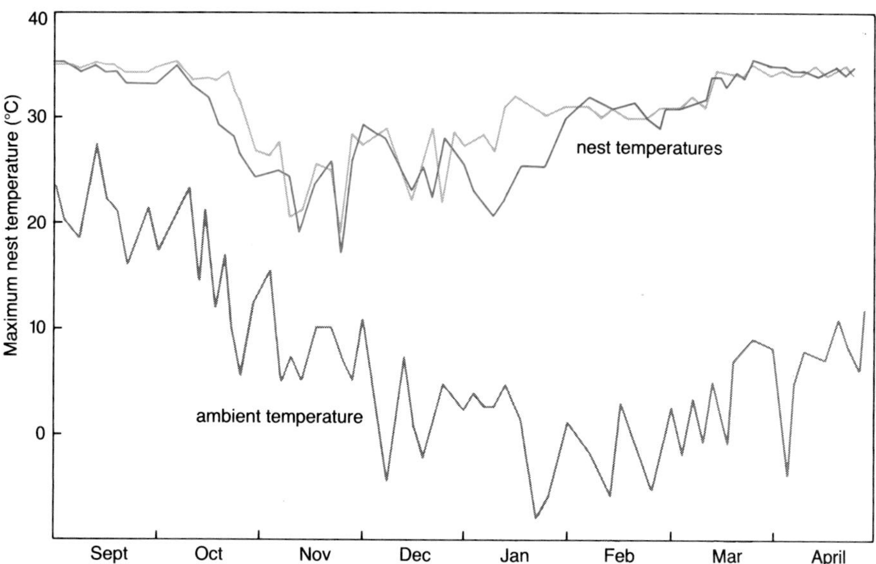

Figure 3. Records of the maximum temperatures inside two *A. mellifera* nests in the course of one winter show a remarkable ability to control the climate within the nest. When colonies cease brood rearing in October, they lower the nest temperature from about 34°C to around 20°C to conserve fuel. The return to temperatures of about 30°C in December and January marks the midwinter resumption of brood rearing.

By carefully timing the duration and measuring the angles of over 100 dances each day, we were able to plot the colony's forage sites as shown in Figure 2. Comparison of the maps for consecutive days revealed the dynamics of the colony's foraging. We discovered that the colony routinely foraged several kilometers from its nest; 95% of the foraging took place within 6 km, and the median distance traveled was 1.7 km. The colony frequently—at least once daily—adjusted the distribution of its foragers among the patches, and it worked relatively few patches each day. On average, 9.7 patches accounted for 90% of the colony's foraging in a given day. In short, the overall pattern was one of exploitation of a vast, rapidly changing mosaic of flower patches. Evidently the dance language is one component of a strategy that enables the colony to forage efficiently on patches of flowers which are widely dispersed and which vary greatly in profitability. By pooling information gained from the reconaissance of thousands of foragers, the colony is able to monitor the flower patches around the nest continuously. By also somehow identifying which patches are best (Boch 1956) and using the dance language to direct foragers to these patches, colonies appear to be able to keep their forces consistently focused on a few top-quality patches of flowers. The result is highly efficient food collection based on what might be called an "information center" strategy of foraging.

To stockpile sufficient winter stores, colonies need not only efficient foraging techniques but also adequate time to collect food. Time is especially valuable for newly founded colonies. Within the short summer season of temperate regions, each first-year, or "founder," colony must find a proper homesite, build from scratch a nest of costly beeswax combs, rear young that can survive the winter, and gather provisions for winter. All that a swarm takes from the parent colony to help establish the daughter colony, besides the workers themselves, is one stomachful of honey per worker (Combs 1972).

Apparently colonies have been selected to swarm as early as possible in late spring or early summer to allow maximum time for a founder colony to become established. Around Ithaca, swarming starts by mid-May, peaks in early June, and is largely completed by early July (Fell et al. 1977). A similar peak period for swarming in late spring and early summer has been observed in other north temperate locations (Simpson 1959; Martin 1963). The importance of the swarming date was demonstrated by experiments conducted in Ithaca over three years in which standard-sized swarms were placed in empty hives on May 20 and June 30 (20 days before and after the median date for swarming, respectively) and patterns of survival through the following winter were noted. Although, as expected, the probability of surviving to the next spring was low for both groups, it was significantly higher for colonies that swarmed early than for those that swarmed late: 0.55 and 0.33, respectively (Seeley and Visscher, unpubl.).

Colonies cannot divide by swarming until they have grown large enough to ensure that both parent and daughter colonies will have a sufficiently large worker force. Thus selection favors rapid growth of the colony in the spring. This may explain why colonies, having ceased rearing brood in October, resume this activity in mid-December, shortly after the winter solstice (Jeffree 1956; Avitabile 1978). What enables colonies to perform the remarkable feat of raising young in midwinter is again their outstanding capacity to regulate nest temperature. As shown in Figure 3, a colony can maintain the core of its winter cluster at a cozy 32 to 34°C, the brood-nest temperature, even when the temperature outside the nest cavity is 0°C or colder (Simpson 1961; Seeley and Heinrich 1981).

The theme that emerges from these ecological studies of *A. mellifera* in North America is one of adaptation to a harsh physical environment, especially the long, freezing winters of north temperate regions. Adaptations include the choice of nest sites providing adequate shelter, the ability to regulate nest temperature well above the ambient temperature for months on end, storage of tens of kilograms of honey for heating fuel, and even the timing of the basic annual cycle of rearing brood and swarming, which helps founder colonies become established during the short summers of the temperate zone.

Predation in the tropics

Paradoxically, although the studies described so far reveal a pattern of adaptation to a challenging physical environment, honeybees are essentially tropical insects, most of whose evolution occurred in mild climates. Thus most of the features of honeybee biology discussed above (the foraging strategy being an exception)

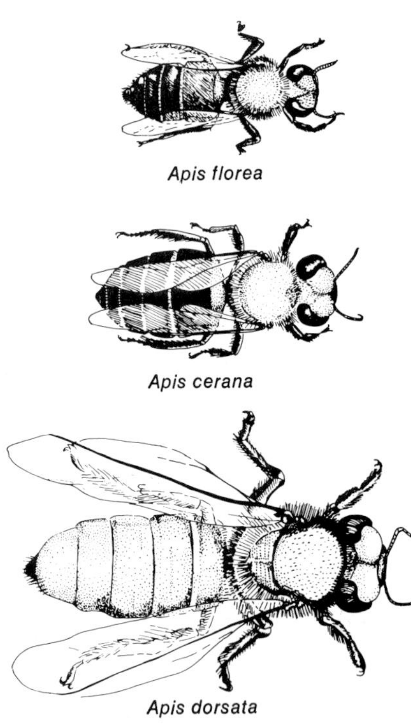

Figure 4. Workers of the three honeybee species found in tropical Asia show substantial differences in size. The smallest worker, that of *A. florea* (*top*), differs by approximately 30% in head width from the *A. cerana* worker (*center*), which in turn differs 40% from the largest worker, *A. dorsata* (*bottom*). An important related difference is stinger size; larger stingers such as those possessed by *A. dorsata* workers produce several times the pain of smaller stingers.

are secondary adaptations evolved by *A. mellifera* in expanding out of the tropics. To understand the primary evolution of the honeybee societies, therefore, one must study these bees in the tropics.

I chose to study the tropical honeybees living in Southeast Asia (Seeley et al. 1982). Here one finds three sympatric honeybee species—*A. florea*, *A. cerana*, and *A. dorsata*—a situation which provides a wonderful opportunity to analyze the ecological forces shaping insect societies through a comparison of species. The logic of this approach is that different species have evolved in relation to different ecological pressures, and that comparisons among closely related species can reveal how differences in their traits reflect differences in their ecologies (Lack 1968; Clutton-Brock and Harvey 1979; Krebs and Davies 1981). Because the species being compared are closely related phylogenetically, one can assume that their differences reflect recent adaptive divergence within a monophyletic group rather than long-held historical differences among phylogenetically distant species. One special value of comparative studies is that they can reveal how entire constellations of traits in a given species—morphological, behavioral, and physiological—are interrelated and are influenced by the same ecological pressures. In other words, the comparative approach can reveal the overall adaptive strategy of a species.

The three Asian honeybee species are ideally suited to this type of study. Their close phylogenetic relationship is indicated by the numerous anatomical and behavioral traits (such as the dance language) by which insect systematists have grouped them in the genus *Apis* (Michener 1974; Winston and Michener 1977). Equally important, however, is the fact that despite this close relationship the three species exhibit a curious array of differences in behavior and ecology and so provide rich material for comparison.

Although little was known about these bees, a few papers (Tokuda 1924; Roepke 1930; Lindauer 1956; Morse and Laigo 1969) reported enough to allow a preliminary comparison of the three species before fieldwork was begun (Table 1). The available information suggested that each species possesses a different strategy of colony defense. Specifically, it appeared that *A. florea* colonies defend themselves against visual predators by building low, widely dispersed, hidden nests, and that they use sticky barriers for protection against ants; that *A. cerana* colonies rely on a strategy of nesting inside cavities for protection against

Figure 5. Nest sites of the three Asian honeybee species differ widely. This typical *A. florea* nest (*top*) is built low on a slender branch of a small tree; a protective blanket of bees covers the nest comb. About 20 cm wide, the nest was originally hidden by vegetation. *A. cerana* colonies characteristically nest in cavities with small entrance openings (*center*). Here the nest was constructed in the cavity of a tree, which has been cut open to expose the interior. The bees have been killed, and most have dropped off the combs. The two taller combs at the right each measure about 30 cm. Colonies of *A. dorsata* typically nest in tall trees (*bottom*). The two nests shown here, each about 100 cm wide, are located about 20 m high in a dipterocarp tree. As in *A. florea* nests, bees cover the nest comb.

Figure 6. An *A. florea* colony salvages wax from a nest it was forced to vacate by a predator's attack. The loads of wax visible on each bee's hindlegs will be used to build comb at a new site. All three species of Asian honeybees suffer from intense predation that causes frequent shifts in nest sites.

large predators, and on direct attacks by their relatively large workers to repel smaller enemies such as ants; and that *A. dorsata* colonies counter predators by nesting in lofty aggregations in the treetops and by launching powerful stinging attacks against intruders capable of penetrating to such heights. It seemed likely that two factors set the stage for this radiation in defense strategies: differences in the size of the workers (Fig. 4), which could strongly influence the effectiveness of defense by direct fighting, and differences in the type of nest site (tree cavity or open air), which might profoundly influence a colony's basic defense situation. Presumably these differences could be traced to interspecific competition for food and nest sites.

To test this hypothesis, Robin Hadlock Seeley and I assembled basic information on the natural history of the three species by observing them living under natural conditions. Our study was conducted in the mountains of northeast Thailand, in and around the Khao Yai National Park. This park comprises 2,000 km^2 of tropical semievergreen rain forest—apparently an ideal habitat for honeybees, since it contains abundant colonies of all three species (Fig. 5).

We found that the bees in this forest suffer severe predation, probably because their brood- and honey-filled nests are bonanzas for numerous species. To quantify the intensity of this predation, we monitored groups of colonies for several months, inspecting each colony twice monthly. We found that approximately 25% of all *A. florea* and 10% of all *A. cerana* colonies are forced each month to abandon their nests because of a predator's attack (Fig. 6). These figures imply that on average each colony's nest is destroyed every four months in the case of *A. florea*, and every ten months for *A. cerana*. Comparable data could not be collected for *A. dorsata* colonies because they migrate between the highlands and the lowlands, following the rains and the flowers (Koeniger and Koeniger 1980), and so were absent from the area during much of our seven-month stay. However, it is clear from attacks we witnessed on *A. dorsata* colonies and destroyed nests we encountered that colonies of this species are also constantly threatened by predators.

What predators cause these high rates of nest destruction? A number of vertebrates—tree shrews (*Tupaia glis*), rhesus monkeys (*Macaca mulatta*), Eurasian honey buzzards (*Pernis apivorus*), and Malayan honey bears (*Helarctos malyanus*)—together with giant social wasps (*Vespa* spp.) are the principal predators capable of mounting massive, overwhelming attacks which can force colonies to abandon their nests. Other predators, such as agamid lizards and weaver ants (*Oecophylla smaragdina*), impose a chronic stress on colonies by extracting a few workers daily but rarely drive colonies from their nests.

As we came to know the three Asian honeybee species during seven months of fieldwork, we found that our initial hypothesis neatly meshed with our observations. Most of the differences in colony design among the three species do seem to reflect three different strategies of colony defense.

A. florea colonies build their defense around avoiding detection by predators. To this end, the bees build small, widely dispersed nests usually set deep in dense, shrubby vegetation along forest margins. One dramatic demonstration of this emphasis on nest concealment came in the dry season, when most *A. florea* colonies lost their cover as plants shed their leaves. Once exposed, colonies waited about two weeks,

Table 1. Preliminary comparison of the three honeybee species in tropical Asia

Property	*A. florea*	*A. cerana*	*A. dorsata*
Worker size (length)	small	medium	large
Nest site	open	cavity	open
Nest height	low (<5 m)	?[a]	high (>10 m)
Nest visibility	hidden?	?	conspicuous?
Nest dispersion	solitary	solitary	aggregated
Sticky barrier	present	absent	absent
Colony population	small (<5,000)	?	large (>20,000)
Colony aggressiveness	timid?	?	fierce

SOURCES: Tokuda 1924; Roepke 1930; Lindauer 1956; Morse and Laigo 1969
[a] Question marks indicate that the characteristic is uncertain or unknown.

until their brood had matured, and then moved a hundred meters or so to a new, well-hidden site. To help minimize the cost of switching nest sites, each colony transported all its stored honey to the new nest and even salvaged much of the old nest's wax for use in building the new comb.

We experimentally tested whether these shifts in nest site served to conceal nests from predators or simply to keep them out of the burning tropical sun by clipping away the vegetation surrounding several nests but leaving vegetation overhead to provide shade. Other nests that were similarly disturbed but left surrounded by vegetation served as controls. All colonies whose nests were exposed abandoned their nests within two weeks, whereas after eight weeks only one control colony had shifted nest sites. Evidently loss of concealment does trigger a shift in nest site by *A. florea* colonies. Reliance on nest concealment makes good sense for these bees. Being built in the open, their nests are basically vulnerable, and because their workers are small and so deliver only relatively painless stings, colonies usually cannot repel large invaders. Their best option seems to be to prevent discovery of the nest in the first place.

Figure 7. *A. florea* workers protect their colony from the weaver ant (*Oecophylla smaragdina*), an important predator, by plastering a sticky band of plant resin around the branch leading to their nest. Here four trapped weaver ants – one partially hidden by the branch – show the effectiveness of this strategy. *A. florea* evades larger predators by concealing its nest.

Nest concealment is not the sole line of defense mounted by *A. florea* colonies. In fact, one of their most important predators, the weaver ant, whose colonies contain hundreds of thousands of workers (Hölldobler 1970), easily locates *A. florea* nests. Because of their diminutive size, *A. florea* workers are no match for this large ant in direct fighting, and colonies rely instead on sticky bands of plant resin plastered around the branch supporting the nest (Fig. 7). The vulnerability of colonies without such sticky bands was easily demonstrated by bending a leaf bearing a weaver ant onto the nest. The bees would immediately retreat, but the ant always managed to yank a bee off the nest and kill it.

A. cerana colonies have evolved a different strategy of colony defense, one that centers on their habit of nesting inside cavities such as caves and hollow trees. These colonies are quite conspicuous (at least to humans, but probably also to most other vertebrates), because their medium-sized workers stream in and out of low, clearly visible entrance holes. Their defense thus relies little upon nest concealment. Instead, *A. cerana* colonies make it difficult for large predators to gain access to their nests by selecting cavities whose largest entrance opening is usually a narrow crack or hole no bigger than 40 cm^2.

When we simulated attacks on these colonies by scraping at the entrances of nests, the guard bees simply retreated to safety inside their nest cavities. In our long-term observations of these colonies we frequently encountered evidence of attacks by predators in the form of signs of digging and scratch marks around the nest entrance, but apparently the predators could not reach the walled-in colonies and the bees were unharmed. Adaptations related to this defense include multiple combs (rather than a single comb, as in the nests of *A. florea* and *A. dorsata*), to fit within the limited space of the cavity, small and nonaggressive colonies (since massive counterattacks are not needed), and probably special techniques of ventilation to prevent asphyxiation within the enclosed nest cavity (Seeley 1974). Small predators such as ants can enter these nests easily, but *A. cerana* workers are large enough to fend off such enemies by direct fighting.

The defense strategy of *A. dorsata* colonies follows a still different design: they nest high—usually above 15 m—in the tallest trees in the forest, thereby evading all predators not skilled in climbing or flying, and launch massive stinging attacks with their sizable colonies of relatively large, ferocious workers against whatever predators can reach their nests. No attempt is made to conceal their huge nests, which can measure a meter or more across and which appear as conspicuous dark objects hanging in the treetops. When we simulated an attack by the Eurasian honey buzzard, a specialist feeder on brood of social wasps and bees, by swinging a black binocular case near an *A. dorsata* colony without striking it, over 100 guards immediately chased the case in a comet's-tail formation while it "flew" back and forth within 5 m of the nest. When we retrieved the case after three minutes of attack we counted 72 stings deeply embedded in its leather cover.

One striking feature of *A. dorsata* is their habit of nesting in groups: a single tree can contain 10 or more *A. dorsata* colonies. At first we thought that this was also an adaptation to colony defense, since grouped colonies could pool their forces. However, it seems probable that this phenomenon is merely a by-product of the fact that colonies converge on a few especially tall trees in choosing nest sites. When I climbed up into an aggregation of nests and attacked one, the grouped colonies did not unleash a concerted counterattack. Only the colony attacked fought back with hundreds of stinging bees.

In general, our observations in the forests of Thailand support the hypothesis that the honeybees of tropical Asia have strongly diverged in their strategies of colony defense, and that this divergence underlies the initially puzzling array of differences among colonies of the three species. In each species the colony's system of defense involves numerous interwoven lines of adaptation, including choice of nest site, nest architecture, population size, and worker morphology, physiology, and behavior. It seems clear that predation has been a pervasive and powerful force in the evolution of these bee societies.

In this discussion I have tried to show how an ecological approach to the honeybee societies reveals the adaptive theme of each species. I have also tried to demonstrate that a synthesis of physiological and ecological approaches provides a broad framework for understanding the social behavior of honeybees, and that, when combined, the two approaches are mutually inspiring. A clear example of this synergism comes from the study of the foraging strategy of honeybee colonies. Here we rely on being able to interpret the bees' dance language, but at the same time find that understanding how this communication functions in nature raises new physiological questions, such as how colonies identify the most profitable flower patches. In essence, physiological studies have paved the way for precise ecological investigations, and in turn, the fledgling ecological approach illuminates pathways into new areas of physiological investigation.

References

Akratanakul, P. 1977. The natural history of the dwarf honey bee, *Apis florea* F., in Thailand. Ph.D. diss., Cornell Univ.

Avitabile, A. 1978. Brood rearing in honeybee colonies from late autumn to early spring. *J. Apic. Res.* 17:69–73.

Avitabile, A., D. P. Stafstrom, and K. J. Donovan. 1978. Natural nest sites of honeybee colonies in trees in Connecticut, USA. *J. Apic. Res.* 17:222–26.

Boch, R. 1956. Die Tänze der Bienen bei nahen und fernen Trachtquellen. *Z. vergl. Physiol.* 38:136–67.

Clutton-Brock, T. H., and P. H. Harvey. 1979. Comparison and adaptation. *Proc. Roy. Soc. London B* 205:547–65.

Combs, G. F. 1972. The engorgement of swarming worker honeybees. *J. Apic. Res.* 11:121–28.

Crane, E. 1975. *Honey: A Comprehensive Survey.* Heinemann.

Darchen, R. 1973. La thermorégulation et l'ecologie de quelques espèces d'abeilles sociales d'Afrique (Apidae, Trigonini et *Apis mellifica* var. *adansonii*). *Apidologie* 4:341–70.

Edgell, G. H. 1949. *The Bee Hunter.* Harvard Univ. Press.

Fell, R. D., J. T. Ambrose, D. M. Burgett, D. DeJong, R. A. Morse, and T. D. Seeley. 1977. The seasonal cycle of swarming in honeybees. *J. Apic. Res.* 16:170–73.

Gould, J. L. 1976. The dance-language controversy. *Quart. Rev. Biol.* 51:211–44.

———. 1982. Why do honey bees have dialects? *Beh. Ecol. Sociobiol.* 10:53–56.

Hölldobler, B. 1979. Territories in the weaver ant (*Oecophylla longinoda* [Latreille]): A field study. *Z. Tierpsychol.* 51:201–13.

Jaycox, E. R., and S. G. Parise. 1980. Homesite selection by Italian honeybee swarms, *Apis mellifera ligustica* (Hymenoptera: Apidae). *J. Kansas Ent. Soc.* 53:171–78.

Jeffree, E. P. 1956. Winter brood and pollen in honeybee colonies. *Insectes Soc.* 3:417–22.

Johansson, T. S. K., and M. P. Johansson. 1979. The honey bee colony in winter. *Bee World* 60:155–70.

Koeniger, N., and G. Koeniger. 1980. Observation and experiments on migration and dance communication of *Apis dorsata* in Sri Lanka. *J. Apic Res.* 19:21–34.

Krebs, J. R., and N. B. Davies. 1981. *An Introduction to Behavioral Ecology.* Sinauer.

Lack, D. M. 1968. *Adaptations for Breeding in Birds.* Methuen.

Lindauer, M. 1955. Schwarmbienen auf Wohnungssuche. *Z. vergl. Physiol.* 37:263–324.

———. 1956. Über die Verständigung bei indischen Bienen. *Z. vergl. Physiol.* 38:521–57.

———. 1961. *Communication among Social Bees.* Harvard Univ. Press.

Martin, P. 1963. Die Steuerung der Volksteilung beim Schwärmen der Bienen. Zugleich ein Beitrag zum Problem der Wanderschwärme. *Insectes Soc.* 10:13–42.

Michener, C. D. 1974. *The Social Behavior of the Bees: A Comparative Study.* Harvard Univ. Press.

Morse, R. A., and F. M. Laigo. 1969. *Apis dorsata* in the Philippines. *Mono. Philipp. Assoc. Entomol.* 1:1–96.

Ribbands, C. R. 1953. *The Behaviour and Social Life of Honeybees.* London: Bee Research Association.

Rinderer, T. E., K. W. Tucker, and A. M. Collins. 1982. Nest cavity selection by swarms of European and Africanized honeybees. *J. Apic. Res.* 21:98–103.

Roepke, W. 1930. Beobachtungen an indischen Honigbienen, insbesondere an *Apis dorsata* F. *Meded. Landbouw. Wageningen* 34:1–28.

Seeley, T. D. 1974. Atmospheric carbon dioxide regulation in honey-bee (*Apis mellifera*) colonies. *J. Insect Physiol.* 20:2301–05.

———. 1977. Measurement of nest cavity volume by the honeybee (*Apis mellifera*). *Beh. Ecol. Sociobiol.* 2:201–27.

———. 1978. Life history strategy of the honey bee, *Apis mellifera. Oecologia* 32:109–18.

———. 1982. How honeybees find a home. *Sci. Am.* 247:158–68.

———. In press. Division of labor between scouts and recruits in honeybee foraging. *Beh. Ecol. Sociobiol.*

Seeley, T. D., and B. Heinrich. 1981. Regulation of temperature in the nests of social insects. In *Insect Thermoregulation*, ed. B. Heinrich, pp. 159–234. Wiley.

Seeley, T. D., and R. A. Morse. 1976. The nest of the honey bee (*Apis mellifera* L.). *Insectes Soc.* 23:495–512.

———. 1978a. Nest site selection by the honey bee, *Apis mellifera. Insectes Soc.* 25:323–37.

———. 1978b. Dispersal behavior of honeybee swarms. *Psyche* 84:199–209.

Seeley, T. D., R. A. Morse, and P. K. Visscher. 1980. The natural history of the flight of honeybee swarms. *Psyche* 86:103–13.

Seeley, T. D., R. H. Seeley, and P. Akratanakul. 1982. Colony defense strategies of the honeybees in Thailand. *Ecol. Mono.* 52:43–63.

Seeley, T. D., and P. K. Visscher. Unpubl. Survival of honeybee colonies in cold temperate climates: The critical timing of brood rearing and colony reproduction.

Simpson, J. 1959. Variation in the incidence of swarming among colonies of *Apis mellifera* throughout the summer. *Insectes Soc.* 6: 85–99.

———. 1961. Nest climate regulation in honeybee colonies. *Science* 133:1327–33.

Tokuda, Y. 1924. Studies on the honeybee, with special reference to the Japanese honeybee. *Trans. Sapporo Nat. Hist. Soc.* 9:1–27.

Visscher, P. K., and T. D. Seeley. 1982. Foraging strategy of honeybee colonies in a temperate deciduous forest. *Ecology* 63:1790–1801.

von Frisch, K. 1967. *The Dance Language and Orientation of Bees.* Harvard Univ. Press.

———. 1971. *Bees: Their Vision, Chemical Senses, and Language.* Cornell Univ. Press.

Wheeler, W. M. 1923. *Social Life among the Insects.* Harcourt, Brace and Co.

Wilson, E. O. 1971. *The Insect Societies.* Harvard Univ. Press.

Winston, M. L. 1980. Swarming, afterswarming, and reproductive rate of unmanaged honeybee colonies (*Apis mellifera*). *Insectes Soc.* 27:391–98.

Winston, M. L., and C. D. Michener. 1977. Dual origin of highly social behavior among bees. *PNAS* 74:1135–37.

Spider Fights as a Test of Evolutionary Game Theory

Susan E. Riechert

The behavioral differences between two populations of the same species can be largely predicted and explained by game theory

The extent to which the behavior of a species changes from one local environment to another is an area of current ecological and evolutionary theory for which the spider has made an excellent test subject. Differences in traits between populations of the same species have been termed ecotypic variation (Turesson 1922). Although most studies of ecotypic variation deal with morphological and physiological characteristics, such as body color, the timing of reproduction, and temperature tolerances, it is well known that in animal evolution behavior tends to change before morphology (von Wahlert 1965; Krebs and Davies 1981). Mayr (1963) even states that for animals "a shift into a new niche or adaptive zone is almost without exception initiated by a change in behavior."

I have been working on a territorial system in spiders which lends itself to the use of game theory to analyze how behavior may adjust populations to local conditions. I share my findings here to demonstrate both the value of spiders as tools in ecological research and the importance of such studies in understanding evolutionary processes. Readers might wish to consult papers by Mitchell and his colleagues (1977) and Brockmann and Dawkins (1979), which also consider the variation in behavioral traits between populations.

The genus *Agelenopsis* belongs to the funnel-web spider family (Araneae, Agelenidae) and consists of the "grass spiders," so called because they frequent grassy areas throughout North America. They are, in fact, common inhabitants of lawns and hedgerows. I chose to work with the western species, *A. aperta* (Gertsch), because it has the widest geographic distribution and resides in the broadest range of habitats of any member of the genus: its populations are distributed from northern Wyoming to southern Mexico and from California to central Texas and are abundant in such divergent habitats as wet woodland and cactus scrub (Fig. 1).

Like all spiders, *A. aperta* is a predator that feeds on a diverse array of insects, other spiders, and their allies. Its web-traps are nonsticky sheets horizontal to the ground. Because the sheet is not sticky, it is used more for the detection of prey than for their actual capture, and it does not need to be replaced each day, being instead merely repaired and added to. The spider itself spends most of its time within the protection of a silk-lined funnel that is attached to the sheet and extends under a rock or into leaf litter, grasses, or shrub branches.

A. aperta overwinters in the juvenile stage and matures in the early summer just prior to the onset of seasonal rains (Riechert 1974). Spiderlings emerge from egg cases in August and September, generally weeks after the deaths of their mothers.

Susan Riechert, professor in the Department of Zoology and in the Graduate Program in Ecology at the University of Tennessee, received her B.S., M.S., and Ph.D. (1973) from the University of Wisconsin. She is an evolutionary ecologist and, using the spider as her experimental tool, specializes in social structure, competition, and predator-prey relationships. The research reported here was supported by grants from the NSF and by a fellowship from the Fogarty Foundation. Address: Department of Zoology, University of Tennessee, Knoxville, TN 37996.

Levels of competition in different habitats

Food, water, shelter, and potential mates are often scarce in natural systems. For *A. aperta*, the availability of suitable locations for building webs is a major problem. Such sites must provide the structural support needed for attachment of the web. Their temperatures must be favorable; body temperature in spiders is determined by external factors that vary with local topography—factors such as solar and thermal radiation, air temperature, wind speed, and humidity—and *A. aperta* can capture prey only when its body temperature is between 21 and 35°C (Riechert and Tracy 1975). Finally, a web site must provide an abundance of prey, because the web-building spider does not actively search for prey but relies on prey coming to it.

I have studied populations of *A. aperta* at two habitats that differ markedly in the number of suitable web sites: a desert grassland in south-central New Mexico and a desert riparian habitat, a woodland bordering a stream, in the Chiricahua mountains of southeastern Arizona (Fig. 2). The grassland habitat is characterized by daily temperature extremes and strong winds. As a result, only 12% of the available space is capable of providing both shelter and the levels of prey necessary to support the growth and reproduction of *A. aperta* (Riechert and Tracy 1975). In this habitat, spiders settle in the limited number of sites that provide enough shade to decrease solar radiation, enough ground litter to decrease the amount of heat that would otherwise be radiated if the web were placed over dirt or rock substrates, and enough insect attractants, such as flowering herbs and shrubs and cow, deer, and rabbit feces (Riechert 1976).

Figure 1. The funnel-web spider *Agelenopsis aperta* at the entrance to the funnel that extends from its web. (Photo by P. Riechert.)

The riparian habitat offers a much more favorable environment, with over 90% of the available space suitable for occupation by *A. aperta*. The tree canopy provides protection from wind and more constant temperatures. The permanent spring-fed stream both attracts insects from the surrounding woods and fields and offers a rich supply of aquatic insects on which spiders can feed; even in drought years, spiders anywhere in this habitat are assured an abundant supply of prey. And the structurally complex substrate of grass sod, fallen tree branches, and low shrubs is well suited for building webs.

A. aperta is capable of using visual, chemical, temperature, and vibratory senses to discriminate among habitats of different quality in selecting a suitable location for building a web (Riechert 1985). The spider seeks shade, and once in shade, it moves along a temperature gradient, seeking a location that offers a temperature of approximately 30°C. At the same time, the spider uses vibratory and olfactory cues from insects to locate patches where prey is abundant. In the case of the grassland population, this behavior leads to the construction of webs near animal feces, which attract flies, and near flowering plants, which are visited by a variety of insect pollinators.

The location of a web in the grassland habitat is strongly correlated with the number of healthy offspring a female spider produces (Riechert and Tracy 1975). Protected sites afford both more prey and more time for capturing them, and females that consume more prey produce more and heavier eggs. Male spiders in the grassland also benefit from the increased consumption of prey at protected sites, because heavier males can wander farther in search of potential mates and, furthermore, have a much higher probability of being accepted by females than lighter males, who may even be eaten rather than mated with. In the riparian habitat, there is no correlation between the site of a web and reproductive success, because prey levels are generally high, and the entire habitat is protected from unfavorable winds, temperatures, and levels of humidity by a tree canopy.

By itself, a difference in the availability of web sites between grassland and riparian habitats does not prove that different levels of competition for these sites exist. The grassland spiders may merely cluster in the more favorable patches, and the riparian spiders may disperse more evenly throughout the habitat. However, *A. aperta* is a territorial species, and territorial behavior can limit the number of individuals a given area will support (Whitham 1979; Riechert 1981). With the exceptions of newly hatched juveniles and males searching for mates, all *A. aperta* defend an area around the web against intrusion by other members of their population (Riechert 1978a, 1981). Unlike a person defending his yard against tresspassers, however, no specific boundaries are patrolled. Rather, an intolerance is exhibited toward the construction of webs within distances that are genetically determined for the respective populations (Riechert 1978a, in press).

The territories defended by *A. aperta* are much larger

Figure 2. The desert grassland of south-central New Mexico (*top*) can support growth and reproduction among *A. aperta* on only 12% of its surface at best, whereas nearly all of the riparian habitat in southeastern Arizona (*bottom*) is inhabitable. Competition for suitable web sites is therefore expected to be much greater among grassland than among riparian spiders. (Photos by S. Riechert.)

in the grassland than in the riparian, with a mean size of 3.7 m^2 versus 0.6 m^2. This difference is determined by the accessibility of food. By having sole occupancy of high-quality grassland territory or of any riparian territory, a spider is assured of the food needed for survival and for reproduction, even during times of food shortage (Riechert 1981). Using artificial sticky webs to assess the difference in the amount of prey that would be available in shared versus unshared territories, I found that a spider would lose approximately 40% of its food supply if it did not defend a territory of the size specified for each of these habitats, a reduction that would place the spider under food stress and would ultimately limit its potential to produce offspring (Riechert 1978a).

The system of fixed territory sizes can limit population size, something that is realized in the grassland habitat. Since *A. aperta* produces approximately 125 eggs per clutch, it is obvious that there might be more individuals than can be supported by the grassland habitat, where only 3% of the available space is adequate to ensure survival to reproduction in drought years, and only 12% in most years (Riechert 1981). Spiders in this habitat compete for precious few sites. In contrast, most riparian spiders obtain territories and survive to reproduction, regardless of the year's weather. Here, other factors, such as predation by birds, spider wasps, and other spiders, appear to maintain population size below the threshold of competition for available web sites.

Territorial disputes and game theory

Because the two populations of *A. aperta* exist under markedly different levels of competition for web sites, it is reasonable to expect differences in the fighting behavior they exhibit in settling territorial disputes. This hypothesis deviates from classical ethology's view of animal conflict as being highly ritualized and constant within a species. The species-specific model supported by classical ethology assumes that the repetitive use of the same visual and vocal displays and the lack of escalated fighting evolved to prevent injury to competing members of a species (e.g., Lorenz 1966). By convention, the contestant that exhibits the "best" display wins the contested resource. Male Galapagos tortoises, for instance, settle contests over mates on the basis of head height; the ritualized display exhibited by competing tortoises, then, consists of two animals facing one another and stretching their necks skyward, with the tortoise perceived as being "taller" by this means winning the mating opportunity (Fritts 1984).

In *Agelenopsis*, however, fights are not highly ritualized, and the sequence of events varies greatly from contest to contest. For example, the number of distinct behavior patterns might be as few as 2 (approach and retreat) and as many as 28, and contests may last anywhere from a few seconds to 21 hours (Riechert 1978b). Fighting among these spiders is not limited to displays: biting, shoving, dragging, and tumbling are common in the disputes (Fig. 3).

A recently developed model, evolutionary game theory, provides a closer fit to *Agelenopsis* territorial disputes than does the species-specific model of classical ethology, because it allows for changes in conflict behavior with varying contexts, such as differences in size, experience, and age. Evolutionary game theory is based on the costs and benefits to individuals of particular strategies. By this theory, if a spider, say, fails to engage in biting and other physical forms of interaction during the course of a dispute, it is because such direct fighting is not in the spider's best interest in that instance. There may be a risk of retaliation, or perhaps the value of the disputed resource is so low that the energy expenditures and the risk of injury do not warrant escalated fighting. Consider, for example, a woman whose purse is grabbed while she is walking down a street; one would expect that she would put up far more resistance to the theft if she had her life savings in the purse than if it contained but a few coins.

Evolutionary game theory deviates from the simple calculation of the maximum ratio of benefit to cost by the addition of a factor called frequency dependence. For example, a student's performance on an exam that is graded on a curve is frequency dependent; his grade depends not only on his but also on his classmates' performances. Thus, evolutionary game theory fits those cases in which the success of an individual using a particular behavioral strategy is not independent of the performances of other members of its population. An example from biology is sex ratio allocation by parents: the optimum proportion of males to females a mother should produce depends on the proportion present in the population; if there are, say, more females than males, then an individual parent will achieve a higher reproductive success by producing a higher proportion of males to mate with the existing females. Other examples from biology to which this theory might be applied include animal contests, parent-offspring conflict, optimal growth patterns and other life-history traits, social structure (e.g., when to cooperate), and local patterns of distribution.

Evolutionary game theory was adapted from classical game theory developed by von Neumann and Morgenstern (1944) to explain human behavior in conflict situations. It is called game theory because the mathematics underlying it is similar in structure to that which forms the basis of such parlor games as chess and checkers. In both classical and evolutionary games, "decisions" are made as to what strategy or action is to be taken in a given context. Each strategy has associated with it a set of payoffs which specify what an individual's level of success will be against contestants exhibiting the same and other strategies. That set of actions which provides the best average payoff is expected to prevail in the system. Classical game theory has been widely applied in economics and business (e.g., McDonald 1975; Schotter and Schwodiauer 1980), the social sciences and politics (Brams 1975), the military (Aumann and Maschler 1966), and social psychology (Bartos 1967).

Evolutionary game theory is a relatively new introduction. John Maynard Smith is largely responsible for the application of classical game theory to evolutionary contexts (e.g., Maynard Smith and Price 1973; Maynard Smith 1983; reviewed in Riechert and Hammerstein 1983; Parker 1984). There are two major differences between the classical and evolutionary theories. First, whereas in classical game theory it is assumed that rational thought is used in determining which action to take, evolutionary game theory assumes that natural selection ultimately determines the strategies that are exhibited, individuals being merely the performers of inherited programs. The other difference is in the characterization of the payoffs: in classical game theory, payoffs are indicated by an individual's personal value judgment of what constitutes success; in evolutionary game theory, the payoffs are expressed in terms of individual fitness, which is defined as the reproductive success of one's offspring but is often estimated by individual weight gain or reproductive rate.

The solution to the evolutionary game is referred to as the evolutionarily stable strategy, or ESS, and thus the theory is frequently referred to as the ESS theory. The ESS specifies the set of strategies that will be exhibited in all roles and situations an individual may find itself in; if most members of a population utilize this set of strategies, which provides the best average payoff in fitness, then mutants exhibiting a different set of strategies will have little success in the population. Hence, we say the ESS is stable against invasion by a mutant strategy set.

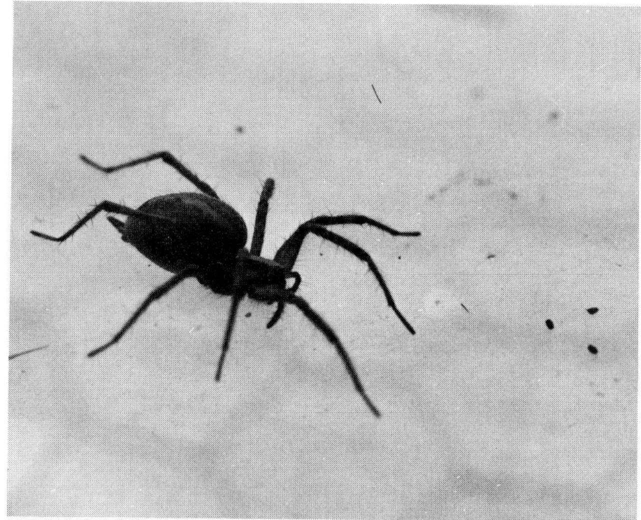

Figure 3. A large variety of characteristic behaviors are used in territorial disputes among *A. aperta*. A number of these fall under the category of assessment (*top*), which is indicated by subtle moves, such as an adjustment of the spider's position or a spreading of the front legs, made while the opponent is on the web but beyond visual contact. Display (*middle*) involves both vibratory and visual signals to the opponent, such as waving the front legs. Fighting (*bottom*) comprises all acts of physical contact, including shoving, biting, and tumbling. (Photos by S. Riechert.)

The number of biological systems to which game theory might be applied is enormous. However, because detailed data are required to determine the payoffs associated with different sets of strategies in natural contexts, few empirical tests of the theory have been completed. Behavioral applications include work on competition among dung flies for mates at cowpats; an ESS was obtained specifying how long a male should hover at a cowpat, given that the rates at which females visit a cowpat decrease with its age (Parker 1970). In another study involving behavioral strategies, an ESS model was developed to explain the choice exhibited by digger wasps between digging a new burrow and using an existing one (Brockmann et al. 1979). This particular study fits into the more general problem of whether animals should be producers or should steal the products of other individuals' labor, a line of investigation that has been pursued in the case of foraging house sparrows (Barnard and Sibly 1981). The ESS theory has also been applied to cooperative foraging systems, specifically to foraging bird flocks in which individual birds spend a proportion of their time watching for predators instead of feeding (Pulliam et al. 1982).

Evolutionarily stable strategies and ecotypic variation

Because the fighting behavior of *A. aperta* is frequency dependent and because my long-term studies of its population biology included estimates of the fitness of individuals from both the riparian and grassland populations occupying sites of different quality, I was interested in applying ESS theory to the ecotypic variation observed in the levels of competition for web sites. The aim was to determine whether each of the populations is at the expected ESS—whether the predicted strategy is the prevalent one in the population.

Although spider contests vary greatly and can involve dozens of different behaviors, they do show a definite structure, which consists of four basic categories of behaviors that range from assessment and display to threat and, finally, to actual fighting—from lower to higher levels of energy expenditure and risk of injury. The shorter contests rarely go beyond the display phase; most often, a retreat signifies a withdrawal from the territory, and a contest consists only of one bout. Many contests, however, consist of multiple bouts separated by retreats from the web.

John Maynard Smith, Peter Hammerstein, and I undertook game-theory analyses of territorial disputes among *A. aperta* (Maynard Smith 1983; Hammerstein 1981; Riechert 1978b, 1979, 1984). Based on my observations of the territorial system, our current analysis includes the following strategy sets: withdraw if the opponent is larger, otherwise display; display only; display only, if the site quality is low, if the probability of winning is low, or both, otherwise escalate; always escalate to fighting (Hammerstein and Riechert, unpubl.). A large number of parameters were required to describe the payoffs for these complex sets of strategies, including the probability that the territory owner will win a random contest, the encounter rate, the costs of biting and display, the time spent on the web exposed to predators during territorial disputes, the rate of preda-

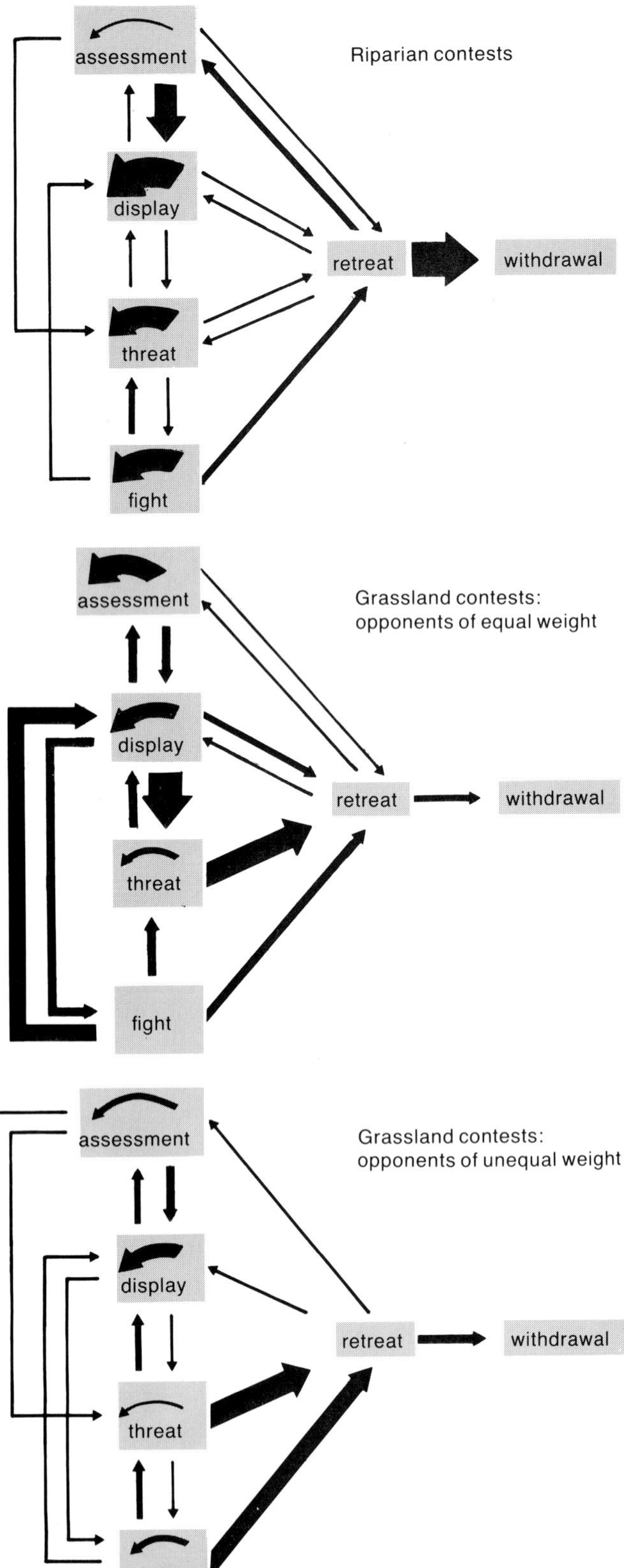

Figure 4. The sequence of events that occur during territorial disputes among *A. aperta* are summarized in these flow diagrams, in which the different widths of the arrows represent differences in the probabilities of transition between various levels of the disputes. For example, riparian contests often escalate from assessment to display, but are far less likely to escalate further, presumably because an abundance of suitable territories reduces the potential benefits of such risky behavior. In contrast, the scarcity of suitable grassland territories means that disputes over them (between equal opponents) are much more likely to escalate.

tion on *A. aperta* in the habitat, the daily increment to reproductive success associated with gaining or keeping the territory in dispute, and the probability of winning a subsequent site if this dispute is lost. It was possible to collect these data because *A. aperta* is locally abundant and can be readily observed and manipulated with no discernible disruption of its natural behavior.

The results of the analyses indicate that members of the grassland population should exhibit the following strategy: display only, if site quality is low, if the probability of winning is low, or both, otherwise escalate. For the riparian population, the optimum strategy is to withdraw if the opponent is larger, otherwise display.

It is apparent from Figure 4 that the contest behavior observed in riparian spiders deviates from the predicted ESS for that population. Contrary to what the model predicts, the majority of riparian spiders do not withdraw from occupied territories when the owner is encountered. Rather, both engage in disputes and escalate to potentially injurious behavior. Indeed, if escalation is reached in fights involving riparian spiders, the opponents tend to stay there rather than returning to less risky behavior as grassland spiders typically do.

Conflicts among grassland spiders apparently do exhibit the ESS predicted by evolutionary game theory. As Figure 4 and Table 1 show, the time and energy expended by these spiders in fights varies with their probabilities of winning them, which for these animals are functions of the opponents' relative weights. The intensity of the contests also corresponds to the potential reproductive value of the different sites under dispute, with far greater energy expended and risk of injury taken in disputes over high-quality sites.

Why is the behavior for settling territorial disputes at its predicted ESS within the grassland population but not within the riparian? To answer this question, it is helpful to compare some of the characteristic differences between the two populations outlined in Table 1. First, there is a much greater variability in the parameters associated with the contests in the grassland disputes than in the riparian (Riechert 1979). This variability is largely attributable to the greater disparity in quality among different locations in the grassland habitat. Furthermore, there is an owner bias in the grassland disputes: owners tend to win disputes with intruders of similar weight. This bias is not observed in the riparian disputes, where, even if there is a sizable weight difference between opponents, the outcomes of the disputes are not as predictable as in the grassland (Maynard Smith and Riechert 1984). There thus appears to be a relaxation of the rules of decision that govern fighting in the riparian population; that is, the riparian spider can sometimes get away with failing to implement the optimum strategy. I noted one other interesting but puzzling difference. Biting and other escalated behaviors in the riparian disputes are ritualized in that no injury has been observed to result from them. This is despite the fact that such fighting is more lengthy when it does occur than in the grassland disputes. In the grassland contests, 50% of these escalated contests result in some form of injury to one of the opponents.

These differences in behavior between populations, especially the level of aggression individual spiders exhibit in the contests, are to some extent under genetic control (Maynard Smith and Riechert 1984). The ritualization of the fighting behavior, however, is more flexible; when I pitted riparian spiders against grassland spiders at quality sites in the grassland habitat, the behavior of the riparian spiders changed and they bit and injured their grassland opponents in response to being bitten by them.

What, then, can be concluded from these observations? Historically, *A. aperta* is a desert species, and most populations exist under higher levels of competition for limited sites than observed for the riparian population. Therefore, the fighting behavior of the riparian spiders, in many respects similar to that of the grassland spiders but with some significant modifications, is probably derived from the type of contest observed in the grassland population. The derived riparian ecotype has resulted, in part, from a relaxation of the rules of decision governing the outcome of disputes and from the ritualization of potentially injurious behavior.

But why have riparian contests not reached the level of change predicted by our ESS analysis? If one assumes that the model is correct—that it has taken into account all the important parameters and includes all possible sets of strategies—then there must be some biological explanation for the observed deviation. A number of hypotheses have been proposed as to why some populations fail to reach a predicted ESS, and several of these are applicable to the riparian ecotype of *A. aperta*. One possibility is that the release from strong competition is a recent event and that there just has not been sufficient time for natural selection to operate on the behavioral traits to complete the expected change. Another possibil-

Table 1. Comparison of contests between opponents of differing weights and for sites of varying quality

Site quality (opponents equal in weight, within 10%)	*Cost* *Estimated energy expended*	*Length* *Number of acts*	*Number of bouts*	*Outcome* *Probability that owner will win*
Grassland				
Poor: surface	123.3	11.9	1.9	0.56
Average: holes	344.0	31.4	3.0	0.76
Excellent: holes +	556.7	51.6	3.7	0.92
Riparian				
Excellent: rocks	126.2	13.5	1.8	0.92
Excellent: grasses	146.2	14.5	2.1	0.60
Excellent: leaf litter	185.5	16.1	2.1	0.89
Weight of opponents (site quality excellent)				
Grassland				
Owner heavier	493.7	40.8	2.9	0.96
Opponents equal (within 10%)	556.7	51.6	3.7	0.92
Intruder heavier	1,012.5	75.9	4.0	0.28
Riparian				
Owner heavier	160.4	16.9	2.0	0.85
Opponents equal (within 10%)	215.4	19.9	2.1	0.70
Intruder heavier	145.5	15.6	2.1	0.17

NOTE: A minimum of 25 contests were used in each context; the values shown represent averages. Estimated energy expended is given as a rank-score based on intensity (after Riechert 1979). All contests took place in natural habitats.

ity is that the expected or "best" set of strategies has been vitiated by interbreeding with individuals from populations existing under stronger levels of competition for sites. In the case of the riparian habitat, there might be an influx of individuals from nearby cactus-scrub and mesquite habitats, where the levels of competition for limited web sites are similar to the grassland habitat. Finally, a major change in the wiring of *A. aperta*'s nervous system might be required to achieve the new ESS, and such a mutant may simply not have arisen yet, a delay referred to as phylogenetic inertia. It is not yet clear which if any of these possibilities is operating in the case of *A. aperta*. These are ongoing studies.

References

Aumann, R. J., and M. Maschler. 1966. Game theoretic aspects of gradual disarmament. In *Development of Utility Theory for Arms Control and Disarmament, Report to the US Arms Control and Disarmament Agency*. Contract 80:1–55.

Barnard, C. J., and R. M. Sibly. 1981. Producers and scroungers: A general model and its application to captive flocks of house sparrows. *Anim. Behav.* 29:543–50.

Bartos, O. J. 1967. *Simple Models of Group Behavior*. Columbia Univ. Press.

Brams, S. J. 1975. *Game Theory and Politics*. Macmillan.

Brockmann, H. J., and R. Dawkins. 1979. Joint nesting in a digger wasp as an evolutionarily stable preadaptation to social life. *Behaviour* 71:203–45.

Brockmann, H. J., A. Grafen, and R. Dawkins. 1979. Evolutionarily stable nesting strategy in a digger wasp. *J. Theor. Biol.* 77:473–96.

Fritts, T. H. 1984. Evolutionary divergence of giant tortoises in the Galapagos. *Biol. J. Linn. Soc.* 21:165–76.

Hammerstein, P. 1981. The role of asymmetries in animal contests. *Anim. Behav.* 29:193–205.

Hammerstein, P., and S. Riechert. Unpubl. Payoffs and strategies in spider territorial contests.

Krebs, J. R., and N. B. Davies. 1981. *An Introduction to Behavioral Ecology*. Sinauer.

Lorenz, K. 1966. *On Aggression*. Methuen.

Maynard Smith, J. 1983. *Evolution and the Theory of Games*. Cambridge Univ. Press.

Maynard Smith, J., and G. R. Price. 1973. The logic of animal conflict. *Nature* 246:15–18.

Maynard Smith, J., and S. E. Riechert. 1984. A conflicting-tendency model of spider agonistic behavior: Hybrid-pure population line comparisons. *Anim. Behav.* 32:564–78.

Mayr, E. 1963. *Animal Species and Evolution*. Harvard Univ. Press.

McDonald, J. D. 1975. *The Game of Business*. Doubleday.

Mitchell, D. E. T., E. T. Beatty, and P. K. Cox. 1977. Behavioral differences between two populations of wild rats: Implications for domestication research. *Behav. Biol.* 19:206–17.

Parker, G. A. 1970. The reproductive behavior and the nature of sexual selection in *Scatophaga stercoraria* (Diptera: Scatophagidae). II. The fertilization rate and the spatial and temporal relationships of each sex around the site of mating and oviposition. *J. Anim. Ecol.* 39: 205–23.

———. 1984. Evolutionarily stable strategies. In *Behavioral Ecology*, ed. J. R. Krebs and N. B. Davies, pp. 30–61. Sinauer.

Pulliam, H. R., G. H. Pyke, and T. Caraco. 1982. The scanning behavior of juncos: A game theoretical approach. *J. Theor. Biol.* 95:89–103.

Riechert, S. E. 1974. The pattern of local web distribution in a desert spider: Mechanisms and seasonal variation. *J. Anim. Ecol.* 43:733–46.

———. 1976. Web-site selection in a desert spider, *Agelenopsis aperta* (Gertsch). *Oikos* 27:311–15.

———. 1978a. Energy-based territoriality in populations of the desert spider *Agelenopsis aperta* (Gertsch). *Symp. Zool. Soc.* 42:211–22.

———. 1978b. Games spiders play: Behavioral variability in territorial disputes. *Behav. Ecol. Sociobiol.* 4:1–28.

———. 1979. Games spiders play. II. Resource assessment strategies. *Behav. Ecol. Sociobiol.* 6:121–28.

———. 1981. The consequences of being territorial: Spiders, a case study. *Am. Nat.* 117:871–92.

———. 1984. Games spiders play. III. Cues underlying context-associated changes in agonistic behavior. *Anim. Behav.* 32:1–5.

———. 1985. Decisions in multiple goal contexts: Habitat selection of the spider, *Agelenopsis aperta* (Gertsch). *Zeitschrift fur Tierpsychol.* 70:53–69.

———. In press. Between population variation in spider territorial behavior: Hybrid-pure population line comparisons. In *Colloquium on Behavioral Genetics*, ed. M. Huettel. Plenum.

Riechert, S. E., and P. Hammerstein. 1983. Game theory in an ecological context. *Ann. Rev. Ecol. Syst.* 14:317–409.

Riechert, S. E., and C. R. Tracy. 1975. Thermal balance and prey availability: Bases for a model relating web-site characteristics to spider reproductive success. *Ecology* 56:265–84.

Schotter, A., and G. Schwodiauer. 1980. Economics and the theory of games: A survey. *J. Econ. Lit.* 18:479–527.

Turesson, G. 1922. The genotypic response of the plant species to the habitat. *Hereditas* 3:211–350.

von Neumann, J., and O. Morgenstern. 1944. *Theory of Games and Economic Behavior*. Princeton Univ. Press.

Wahlert, G. von. 1965. The role of ecological factors in the origin of higher levels of organization. *Syst. Zool.* 14:288–300.

Whitham, T. G. 1979. Territorial defence in a gall aphid. *Nature* 279:324–25.

Why Male Ground Squirrels Disperse

Kay E. Holekamp
Paul W. Sherman

When they are about two months old, male Belding's ground squirrels *(Spermophilus beldingi)* leave the burrow where they were born, never to return. Their sisters behave quite differently, remaining near home throughout their lives. Why do juvenile males, and only males, disperse? This deceptively simple question, which has intrigued us for more than a decade *(1, 2)*, has led us to investigate evolutionary, ecological, ontogenetic, and mechanistic explanations. Only recently have answers begun to emerge.

A multilevel analysis helps us to understand why male and not female Belding's ground squirrels leave the area where they were born

Dispersal, defined as a complete and permanent emigration from an individual's home range, occurs sometime in the life cycle of nearly all organisms. There are two major types: breeding dispersal, the movement of adults between reproductive episodes, and natal dispersal, the emigration of young from their birthplace *(3, 4)*. Natal dispersal occurs in virtually all birds and mammals prior to first reproduction. In most mammals, young males emigrate while their sisters remain near home (the females are said to be philopatric); in birds, the reverse occurs *(4–6)*. Although naturalists have long been aware of these patterns, attempts to understand their causal bases have been hindered by both practical and theoretical problems. The former stem from difficulties of monitoring dispersal by free-living animals, and of quantifying the advantages and disadvantages of emigration *(6)*. The latter stem from failure to distinguish the two types of dispersal, and from confusion among immediate and long-term explanations for each type.

We begin with a discussion of the latter point and develop the idea that natal dispersal, like other behaviors and phenotypic attributes, can be understood from multiple, complementary perspectives. Separating these levels of analysis helps organize hypotheses about cause and effect in biology *(7)*. In the case of natal dispersal, this approach can minimize misunderstandings in terminology and allow for clearer focus on the issues of interest.

Questions of the general form "Why does animal A exhibit trait X?" have always caused confusion among biologists. And even today, the literature is full of examples. The nature-nurture controversy, which arose over the question of whether behaviors are innate or acquired through experience, is a classic case *(8)*. After two decades of spirited but inconclusive argument in the nature-nurture debate, it became apparent to Mayr *(9)* and Tinbergen *(10)* that a lack of consensus was caused by the failure to realize that such questions could be analyzed from multiple perspectives.

In 1961, Mayr proposed that causal explanations in biology be grouped into proximate and ultimate categories. Proximate factors operate in the day-to-day lives of individuals, whereas ultimate factors encompass births and deaths of many generations or even entire taxa. Pursuing this theme in 1963, Tinbergen further subdivided each of Mayr's categories. He noted that complete proximate explanations of any behavior involve elucidating both its ontogeny in individuals and its underlying physiological mechanisms. Ultimate explanations require understanding both the evolutionary origins of the behavior and the behavior's effects on reproduction. The former involves inferring the phylogenetic history of the behavior, and the latter requires comparing the fitness consequences of present-day behavioral variants.

There are two key implications of the Mayr-Tinbergen framework. First, competition among alternative hypotheses occurs within and not between the four analytical levels. Second, at least four "correct" answers to any question about causality are possible, because explanations at one level of analysis complement rather than supersede those at another. Deciding which explanations are most interesting or satisfying is largely a matter of training and taste; debating the issue is usually fruitless *(7)*.

With the Mayr-Tinbergen framework in mind, let us turn to the question of natal dispersal in ground squir-

Kay Holekamp is a research scientist in the Department of Ornithology and Mammalogy at the California Academy of Sciences. She received a B. A. in psychology in 1973 from Smith College, and a Ph.D. in 1983 from the University of California, Berkeley. From 1983 to 1985 she studied reproductive endocrinology as a postdoctoral fellow at the University of California, Santa Cruz. She is currently observing mother-infant interactions and the development of social behaviors in hyenas in Kenya. Paul Sherman is an associate professor of animal behavior at Cornell. He received a B. A. in biology in 1971 from Stanford, and a Ph.D. in zoology in 1976 from the University of Michigan. Following a postdoctoral appointment at the University of California, Berkeley (1976–78), he joined the psychology faculty there. He moved to his present position at Cornell in 1980, and is currently studying the behaviors of naked mole-rats, Idaho ground squirrels, and wood ducks. Address for Dr. Sherman: Section of Neurobiology and Behavior, Seeley G. Mudd Hall, Cornell University, Ithaca, NY 14853.

Figure 1. A female Belding's ground squirrel *(Spermophilus beldingi)* sits with two of her pups in the central Sierra Nevada of California. The pups are about four weeks old, and have recently emerged above ground. At about six or seven weeks of age, male ground squirrels begin to disperse; young females always remain near home. The causes of male dispersal in ground squirrels and many other mammals are complex, but can be explained by using a multilevel analytical approach in which four categories of causal factors are considered separately. (Photo by Cynthia Kagarise Sherman.)

rels. Following analyses of why natal dispersal occurs from each of the four analytical perspectives, we attempt an integration and a synthesis. Our studies reaffirm the usefulness of levels of analysis in determining biological causality.

From 1974 through 1985 we studied three populations of *S. beldingi* near Yosemite National Park in the Sierra Nevada of California (Figs. 1 and 2). In each population, the animals were above ground for only four or five months during the spring and summer; during the rest of the year they hibernated *(1, 2)*. Females bore a single litter of five to seven young per season, and reared them without assistance from males. Most females began to breed as one-year-olds, but males did not mate until they were at least two. Females lived about twice as long as males, both on average (four versus two years) and at the maximum (thirteen versus seven years) *(11)*.

During each field season ground squirrels were trapped alive, weighed, and examined every two to three weeks. About 5,300 different ground squirrels were handled. The animals were marked individually and observed unobtrusively through binoculars for nearly 6,000 hours. Natal dispersal behavior was measured by a combination of direct observations, livetrapping, radio telemetry, and identification of animals killed on nearby roads *(12)*. The day on which each emigrant was last seen within its mother's home range was defined as its date of dispersal. Only those juveniles that were actually seen after leaving their birthplace were classified as dispersers.

Observations of marked pups revealed that natal dispersal was a gradual process, visually resembling the fissioning of an amoeba (see Fig. 3). Young first emerged from their natal burrow and ceased nursing when they were about four weeks old. Two or three weeks later some youngsters began making daily excursions away from, and evening returns to, the natal burrow. Eventually these young stopped returning, restricting their activities entirely to the new home range; by definition, dispersal had occurred.

As shown in Figure 4, natal dispersal is clearly a sexually dimorphic behavior. In our studies, every one of over 300 surviving males dispersed by the end of its second summer; a large majority (92%) dispersed before their first hibernation, by the age of about 16 weeks. In contrast, only 5% of over 250 females recaptured as two-year-olds had dispersed from their mother's home range. The universality of natal dispersal by males suggested no plasticity in its occurrence; however, there was variation among individuals in the age at which dispersal occurred.

During the summer following their birth, males that

Figure 2. ***S. beldingi*** **in the central Sierra Nevada are found above ground only four or five months of the year, during the spring and summer; they hibernate during the rest of the year. The group above is emerging from an underground burrow. Female adults bear litters of five to seven young each year and rear them in underground burrows without assistance from males. (Photo by George D. Lepp)**

had dispersed as juveniles often moved again, always farther from their birthplace (Fig. 4). Yearling males were last found before hibernation an average of 170 m from their natal burrow, whereas yearling females moved on average only 25 m from home in the same time period. As two-year-olds, males mated at locations that were on average ten times farther from their natal burrows than the mating locations of females *(13)*.

By the time they were two years old, male *S. beldingi* had attained adult body size. In the early spring they collected on low ridges beneath which females typically hibernated. As snow melted and females emerged, the males established small mating territories. Only the most physically dominant males—especially the old, heavy ones—retained territories throughout the three-week mating period. Although dominant males usually copulated with multiple females, the majority of males rarely mated. After mating, the most polygynous males again dispersed. They typically settled far from the places where they had mated; indeed, their new home ranges usually did not include their mating territories. Less successful males tended not to move, and they attempted to mate the following season in the same area where they were previously unsuccessful.

Females were all quite sedentary. After mating on a ridge top close to her hibernation burrow, each female dug a new nest burrow or refurbished an old one—sometimes her own natal burrow. There she reared her pups. As a result of philopatry, females spent their lives surrounded by and interacting with female relatives. Close kin cooperated to maintain and defend nesting territories and to warn each other when predators approached *(13, 14)*. Natal philopatry has facilitated similar nepotism, or favoring of kin, among females in many other species of ground-dwelling sciurid rodents *(15)*.

Physiological mechanisms

We began our analysis of natal dispersal in *S. beldingi* by considering physiological mechanisms. Of the two broad categories of such mechanisms, neuronal and hormonal, we were most interested in the latter. Gonadal steroids can influence the development of a specific behavior in two general ways: through organizational effects, which are the result of hormone action, in utero or immediately postpartum, on tissues destined to control the behavior, and through activational effects, which result from the direct actions of hormones on target tissues at the time the behavior is expressed *(16)*. We suspected that gonadal steroids might mediate natal dispersal, and so we tested for organizational versus activational effects of androgens.

Under the activational hypothesis, levels of circulating androgens should be elevated in juvenile males at the time of natal dispersal. Conversely, in the absence of androgens, males should not disperse. To test this, we studied male pups born and reared in the laboratory. Blood samples were drawn every few weeks for four months *(17)*. We also conducted a field experiment: soon after weaning but prior to natal dispersal, a number of juvenile males and females were gonadectomized; sham operations were performed on a smaller sample of each sex. After surgery, these juveniles were released into their natal burrow and subsequent dispersal behavior was monitored.

Castration was found to have little effect on natal dispersal. Although castrated males and those subjected to sham operations dispersed a few days later than untreated males, probably because of the trauma of surgery, castration did not significantly reduce the fraction that dispersed. Likewise, removal of ovaries did not increase the likelihood of dispersal by juvenile females. Finally, radioimmunoassays revealed only traces of testosterone in the blood of lab-reared juvenile males throughout their first four months, and no increase in circulating androgens was detected at the age when natal dispersal typically occurs (7–10 weeks).

Sex and body mass together were the most consistent predictors of dispersal status

---

Under the organizational hypothesis, exposing perinatal or neonatal females to androgens should masculinize subsequent behavior, including natal dispersal. We tested this idea by capturing pregnant females and housing them at a field camp until they gave birth. Soon after parturition, female pups were injected with a small amount of testosterone propionate dissolved in oil; a control group was given oil only. After treatment, the pups and their mothers were taken back to the field, where the mothers found suitable empty burrows and successfully reared their young.

Twelve of the female pups treated with androgens were located when they were at least 60 days old, and

75% of them had dispersed *(17)*. The distances they had traveled and their dispersal paths closely resembled those of juvenile males. By comparison, only 8% of untreated juvenile females in the same study area had dispersed by day 60, whereas 60% of juvenile males from the transplanted litters and 74% of males from unmanipulated litters born in the same area had dispersed by day 60.

It is possible that transplantation and not treatment with androgens caused the juvenile females in our experiment to disperse; unfortunately, we were unable to test this because none of the transplanted females treated with only oil were recovered. However, transplantation did not seem to affect the behavior of the juvenile males in the experiment. Also, other behavioral evidence linked natal dispersal in the females with androgen treatment. For example, treated juvenile females did not differ significantly from untreated juvenile males of the same age, but did differ from control females with respect to several indices of locomotor and social behavior. Androgen treatment masculinized much of the behavior of juvenile females, apparently including the propensity to disperse.

These results, which suggest an organizational role for steroids in sexual differentiation of *S. beldingi,* are consistent with those from studies of many other vertebrates *(18)*. In mammals, females are homogametic (XX) and males are heterogametic (XY), whereas in birds the situation is reversed. In each taxon, natal dispersal occurs primarily in the heterogametic sex. In both birds and mammals, sex-typical adult behavior in the homogametic sex can often be reversed by perinatal exposure to the gonadal steroid normally secreted at a particular developmental stage by the heterogametic sex. These considerations suggest that natal dispersal in mammals and birds has a common underlying mechanism, namely the organizational effects of gonadal steroids on the heterogametic sex.

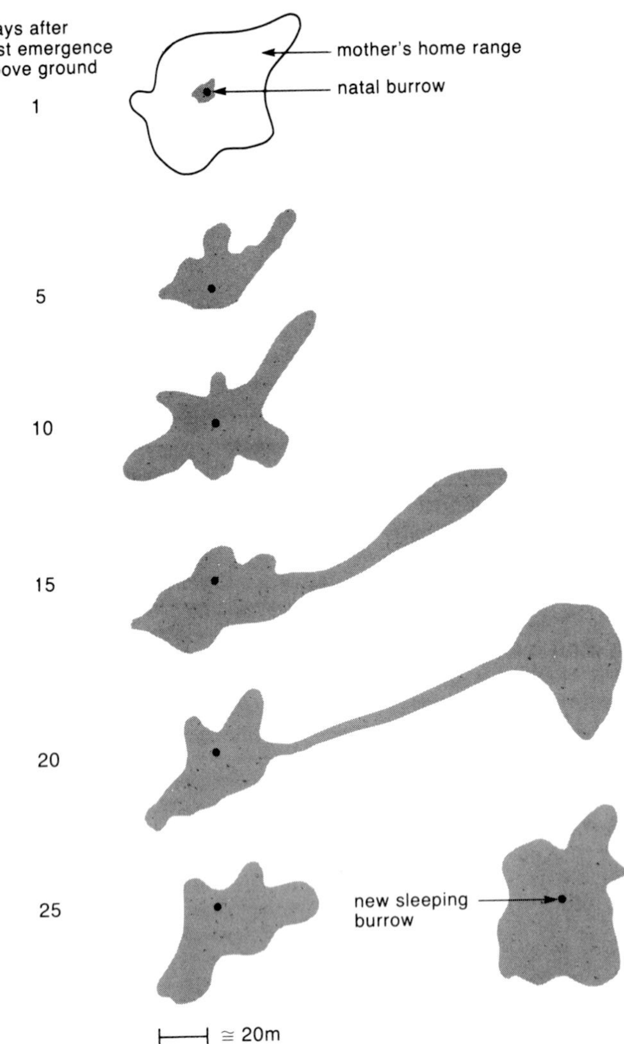

Figure 3. The process by which *S. beldingi* males disperse visually resembles the fissioning of an amoeba. When a male first emerges from the natal burrow, at an age of about four weeks, his daily range of movement is restricted to the immediate vicinity of the burrow. He soon enlarges that range into an amorphous shape, the boundaries of which are established by topographic features or the presence of other animals. By about the 15th day above ground, his range has surpassed the scope of his mother's home range. At this time he may spend long periods far from the natal burrow, yet he will return home at nightfall. Near the 25th day, when he is roughly seven weeks old, he will cease returning at dusk, thereby accomplishing dispersal. (After ref. *35*.)

Ontogenetic processes

Natal dispersal might be triggered during development by changes in either the animal's internal or external environment. We tested two hypotheses about external factors. First, natal dispersal might be caused by aggression directed at juveniles by members of their own species. Under this hypothesis, prior to or at the time of dispersal, the frequency or severity of agonistic behavior between adults and juvenile males should increase. However, observations revealed that adults neither attacked nor chased juvenile males more frequently or vigorously than juvenile females *(19)*, and there was no increase in aggression toward juvenile males at the time of dispersal. Moreover, there were no differences between juvenile males and females in the number and severity of wounds inflicted by other ground squirrels. Thus the data offered no support for the social aggression hypothesis.

A second hypothesis is that natal dispersal occurred because juvenile males attempted to avoid their littermates (current and future competitors) or their mother *(20)*. For a large number of litters, we found no significant relationship between litter size or sex ratio and dispersal behavior *(2, 19)*. Males who dispersed during their natal summer were not from especially large or small litters, or predominantly male or female litters. Also, the timing of juvenile male dispersal depended neither on the mother's age nor on whether the mother was present or deceased. Thus the ontogeny of natal dispersal was apparently not linked to either of the exogenous (external) influences usually invoked to account for it.

In view of these results, we suspected that natal dispersal was triggered by endogenous (internal) factors. In particular, we hypothesized that males might stay home until they attained sufficient size or energy reserves to permit survival during the rigors of emigration. This ontogenetic-switch hypothesis predicts that juvenile

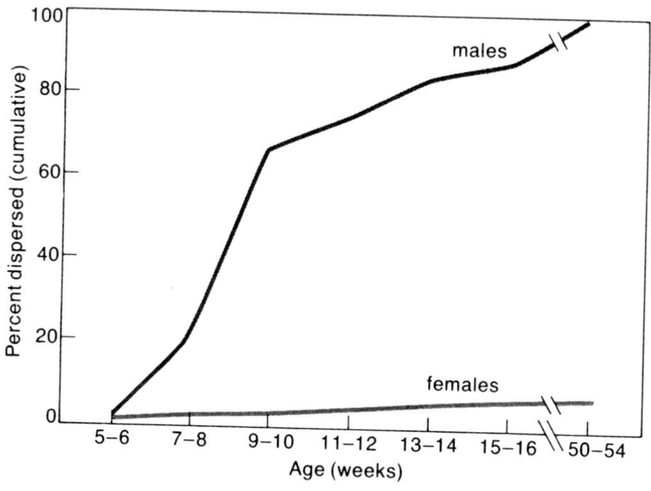

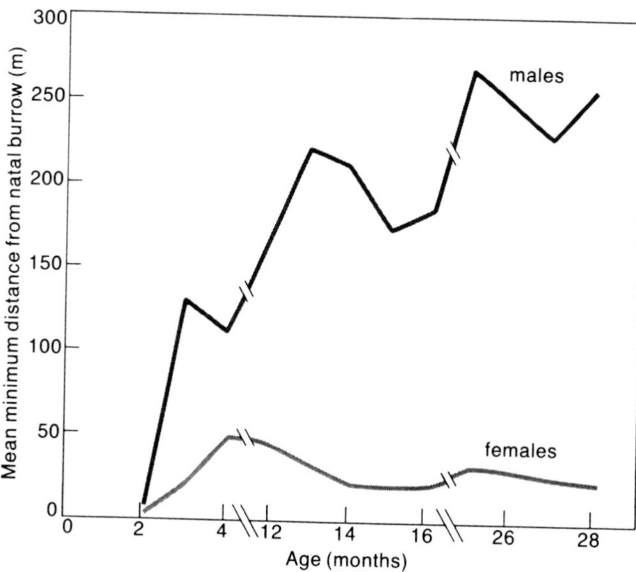

Figure 4. Although a small fraction of female *S. beldingi* disperse, the behavior is very evidently male-biased. The majority of male pups disperse by the 10th week; by about the 54th week, all males have dispersed *(above)*. Although many other mammals exhibit male-biased natal dispersal, *S. beldingi* is unusual in that all males eventually disperse. Males also move considerably farther from their natal burrows than do females, and they continue to move away from home throughout their first three years *(below)*. (After refs. *2* and *12*.)

males will disperse when they attain a threshold body mass and that dispersers should be heavier, or exhibit different patterns of weight gain, than predispersal males of equivalent ages.

Our data were consistent with the ontogenetic-switch hypothesis. Emigration dates were correlated with the time at which males reached a minimum body weight of about 125 g, as shown in Figure 5. Emigrant juveniles were significantly heavier than male pups that had not yet dispersed. Most males attained the threshold weight during their natal summer, and dispersed then. Only the smallest males, who did not put on sufficient weight in the first summer, overwintered in their natal area. All these males dispersed the following season once they had become heavy enough.

Sex and body mass together were the most consistent predictors of dispersal status. Occasionally, however, predispersal and immigrant juvenile males with body weights exceeding the threshold were captured in the same area. This observation suggested that something closely associated with body weight, such as fat stores, may be the actual dispersal trigger.

Behavioral changes also accompanied natal dispersal. The frequencies of movement and distances moved per unit time by juvenile males were found to be greater than those of females, and these behaviors peaked at the time of dispersal. Relative to juvenile females, juvenile males also spent significantly more time climbing and digging and exploring nonfamilial burrows and novel objects—for example, a folding footstool; they also reemerged from a burrow into which they had been frightened much sooner than did females. These observations of spontaneous ontogenetic changes in the behavior of young males reinforced the hypothesis that endogenous factors triggered natal dispersal.

Effects on fitness

Natal dispersal might enable juvenile males to avoid fitness costs associated with life in the natal area and might allow them to obtain benefits elsewhere *(6)*. Possible disadvantages of remaining at home include shortages of food or burrows *(21)*, ectoparasite infestations or diseases, competition with older males for mates *(5, 22)*, and nuclear family incest *(4, 23, 24)*. We examined each of these hypotheses as functional explanations for natal dispersal in *S. beldingi*.

If natal dispersal occurs because of food shortages, then juveniles whose natal burrow is surrounded by abundant food should be more philopatric than those from food-poor areas; immigration to food-rich areas should exceed emigration from them; dispersing individuals should be in poorer condition (perhaps weigh less) than males of the same age residing at home; and, based on the strong sexual dimorphism in natal dispersal, food requirements of young males and females should differ.

Detailed observations revealed that juvenile males and females ate similar amounts of the same plants and at similar rates. Juvenile males spent only slightly more time foraging than did juvenile females. The diets and foraging behaviors of males that had not yet dispersed and males that had immigrated to that same area were indistinguishable. As discussed previously, dispersing males were significantly heavier than predispersal males, a result contrary to that predicted in the scenario of emigration because of lack of food. Finally, juvenile male immigration equaled emigration every year. This is important because preferred foods were unevenly distributed within and among populations *(1, 2)*. Evidence consistently suggested no link between immediate food shortages and natal dispersal.

A second reason for natal dispersal might be to locate a nest burrow. Ground squirrels depend on burrows for safety from predators, as places to spend the night, and as nests in which to hibernate *(25)*. Given the sexual dimorphism in natal dispersal, this hypothesis predicts differences between males and females in the type or location of habitable burrows and implies that dispersers should emigrate from areas of high population density or low burrow quality to areas where unoccupied holes of high quality are available. To test this idea we monitored population density each week and counted burrow entrances in the territories of lactating females. We found that neither the probability of juvenile male dispersal nor its timing was significantly related to population density or burrow availability near home, and that dispersers did not settle in areas of higher burrow density.

The only unusual aspect is that every male eventually leaves home

Another cause of natal dispersal might be ectoparasite infestation. If parasites build up in the natal nest and if juvenile males are more affected by them than are juvenile females, then males in particular might emigrate to avoid them. We examined this hypothesis indirectly, by counting the number of fleas and ticks on every captured juvenile. We found low levels of ectoparasitism throughout the animals' natal summer, and no consistent differences between infestations in males and females prior to or at the time of dispersal.

Do juvenile males disperse to avoid future competition with older males for sexual access to females? Because males always emigrated, it was not possible to determine if dispersers experienced less severe mate competition than hypothetical nondispersers. However, the mate competition hypothesis was examined indirectly by comparing, at sites where males were born and on ridge tops where those males mated two or more years later, three parameters: the density of breeding adult males, the mean number of fights adults engaged in for each successful copulation, and the mean daily ratio of breeding males to receptive females. We found no significant differences in any of these parameters, suggesting that dispersing males did not find better access to females than they would have if they had remained at home.

Do juvenile males disperse to avoid future nuclear family incest? A test of this hypothesis requires comparing the reproductive consequences of various degrees of inbreeding *(26, 27)*. However, of more than 500 copulations observed, none occurred between close kin; therefore we could not directly test this hypothesis. Nonetheless, the nonrandom movements of males away from the natal area clearly resulted in complete avoidance of kin as mates (Fig. 4). Furthermore, during post-breeding dispersal, the highly polygynous males moved farthest. Under this hypothesis, the polygynous males who had sired many female pups in an area would have the most to gain by emigrating. Under the mate-competition hypothesis, successful males would be expected to stay put, while unsuccessful males might gain by dispersing. The observed pattern is thus most consistent with avoiding inbreeding.

Belding's ground squirrels are not unusual in the rarity of close inbreeding. Consanguineous mating is minimized in most mammals and birds *(23, 24, 28, 29)*, often via the mechanism of sex-specific natal dispersal. But why are males the dispersive sex in mammals generally and ground squirrels particularly? The answer probably relates to a sexual asymmetry in the significance

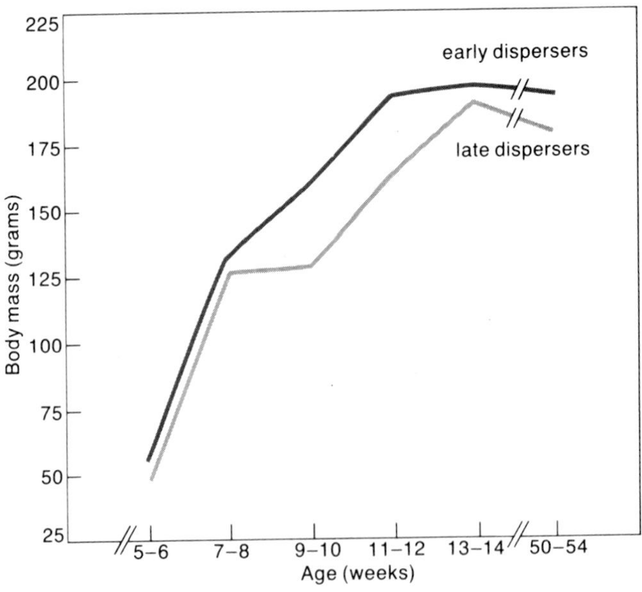

Figure 5. Weight gain among juvenile *S. beldingi* has been positively correlated with the onset of dispersal. Early dispersers (most males) left home at 7–10 weeks of age; late dispersers, in contrast, remained at home until they were 11–14 weeks old. Dispersal seems to occur when a threshold body mass of 125–150 g is attained. (After ref. *19*.)

of the location and quality of burrows for procreation *(6, 30)*. The depth and dryness of nest burrows, their proximity to food, and their degree of protection from both inter- and intraspecific predators are vital to pup survival *(31)*. The significance of the burrow, in turn, favors females who seek out and defend high-quality nest sites and who remain in them from year to year *(25)*. The quality of a nursery burrow is of negligible significance to nonparental males. To avoid predators and inclement weather, and to forage, males can move frequently without jeopardizing the survival of their young. Thus the sexual bias in natal dispersal might occur because inbreeding is harmful to both sexes and males incur lower procreative costs by leaving home.

Sexual selection could reinforce a sex-bias in natal dispersal generated by incest avoidance. If consanguineous mating is indeed harmful, then the philopatric females should prefer to mate with unrelated (unfamiliar) males. A reproductive advantage should therefore accrue to males that seek and locate unfamiliar females *(32)*.

Evolutionary origins

The fourth component of our investigation of natal dispersal was an attempt to infer evolutionary origins. A first hypothesis was that the male bias in natal dispersal arose in an evolutionary ancestor of *S. beldingi* as a developmental error (for example, in the timing of hormone secretion) or as a by-product of natural selection on males for the high levels of activity associated with finding mates and defending mating territories. Alternatively, perhaps natal dispersal was favored directly by selection, for example, as a mechanism to avoid inbreeding, throughout the evolutionary history of *S. beldingi*.

One way to evaluate these alternatives is to consider

Why do juvenile male Belding's ground squirrels disperse? Answers have been found at each of four levels of analysis.

Level of analysis	Summary of findings
Physiological mechanisms	Dispersal by juvenile males is apparently caused by organizational effects of male gonadal steroid hormones. As a result, juvenile males are more curious, less fearful, and more active than juvenile females.
Ontogenetic processes	Dispersal is triggered by attainment of a particular body mass (or amount of stored fat). Attainment of this mass or composition apparently also initiates a suite of locomotory and investigative behaviors among males.
Effects on fitness	Juvenile males probably disperse to reduce chances of nuclear family incest.
Evolutionary origins	Strong male biases in natal dispersal characterize all ground squirrel species, other ground-dwelling sciurid rodents, and mammals in general. The consistency and ubiquity of the behavior suggest that it has been selected for directly across mammalian lineages.

the taxonomic distribution of male-biased natal dispersal. If selection has consistently and directly favored dispersal by juvenile males, then phylogenetic relatives of *S. beldingi* should share this trait to a greater degree than if it were a hormonally mediated side effect or developmental error. This is because any hormonal link between adult male sexual activities and dispersal by juveniles two years previously could presumably be broken by mutation in some species through evolutionary time. This, in turn, would lead to a spotty taxonomic distribution of the behavior if it were neutral for fitness.

Members of the squirrel family first appeared in the fossil record 35 to 40 million years ago; thus they are one of the most ancient of extant rodent families *(33)*. Belding's ground squirrel is one of 32 species in the genus *Spermophilus*; this genus is more closely related to marmots and prairie dogs than to tree squirrels *(34)*. Strongly male-biased natal dispersal occurs in all 12 *Spermophilus* species that have been studied in this regard *(5, 15, 35)*. Male-biased natal dispersal patterns are also the rule in marmots *(35)* and prairie dogs *(36)*. The dispersal behavior of *S. beldingi* is therefore probably a conservative rather than a derived trait; in other words, it is likely quite ancient.

As far as we know, the only unusual aspect of natal dispersal in *S. beldingi* is that every male eventually leaves home, whereas in a few other species a tiny fraction of males are philopatric. Male-biased natal dispersal is widespread among mammals *(4–6, 30, 32, 37)*, suggesting that this behavior may predate the appearance of the squirrel family. The ubiquity of natal dispersal seems more consistent with the hypothesis that it has been favored directly by natural selection in various lineages than that it originated as a mistake or a correlated response to selection for some other male attribute and is maintained by phylogenetic inertia rather than adaptive value.

Synthesis

Our data reveal that there are at least four types of answers to the question of why juvenile male Belding's ground squirrels disperse (see the box). These answers complement rather than supersede each other. Clearly, however, the causal variables we have identified within each analytical level do not operate in isolation, and it seems appropriate to consider how they may interrelate.

During embryogenesis, sex chromosomes cause the formation of testes in male *S. beldingi*. The gonads secrete a pulse of androgens before birth, which, we hypothesize, sets up an ontogenetic switch, presumably by modifying the morphology or behavior of neurons or nuclei in the brain. When juvenile males have accumulated sufficient weight or fat stores, the switch turns on. The young males then boldly explore their environment, making increasingly longer forays away from home. The timing of dispersal by each individual may be influenced by any environmental factor that accelerates or delays arrival at the dispersal threshold (for example, food abundance or scarcity). The main cost of natal dispersal is probably mortality during emigration; the main benefits are likely related to reduced inbreeding and optimal outbreeding. Male biases in natal dispersal occur consistently across modern mammalian taxa *(37)*, suggesting an evolutionary history of natural selection favoring such behavior directly, and a taxon-wide consistency of function.

By employing the levels-of-analysis framework for developing and testing hypotheses, we have come to appreciate the complexity of what at first appeared to be a simple behavior. We suspect that our explanations for the proximate and ultimate causes of natal dispersal in *S. beldingi* will be applicable to other species. Perhaps equally important, our study illustrates that there can be multiple correct answers to questions of causality in behavioral biology *(38)*. The usefulness of the levels-of-analysis approach is thereby reemphasized.

References

1. P. W. Sherman. 1976. Natural selection among some group-living organisms. Ph.D. thesis, Univ. of Michigan.
2. K. E. Holekamp. 1983. Proximal mechanisms of natal dispersal in Belding's ground squirrels *(Spermophilus beldingi beldingi)*. Ph.D. thesis, Univ. of California.
3. W. Z. Lidicker, Jr. 1975. The role of dispersal in the demography of small mammals. In *Small Mammals: Their Productivity and Population Dynamics*, ed. F. B. Golley, K. Petruscewicz, and C. Ryszkowski, pp. 103–28. Cambridge Univ. Press.
4. P. J. Greenwood. 1980. Mating systems, philopatry and dispersal in birds and mammals. *Animal Behav.* 28:1140–62.
5. F. S. Dobson. 1982. Competition for mates and predominant juvenile male dispersal in mammals. *Animal Behav.* 30:1183–92.
6. A. E. Pusey. 1987. Sex-biased dispersal and inbreeding avoidance in birds and mammals. *Trends in Ecol. and Evol.* 2:295–99.
7. P. W. Sherman. 1988. The levels of analysis. *Animal Behav.* 36: 616–19.
8. D. S. Lehrman. 1970. Semantic and conceptual issues in the

nature-nurture problem. In *Development and Evolution of Behavior*, ed. L. R. Aronson, E. Tobach, D. S. Lehrman, and J. S. Rosenblatt, pp. 17–52. W. H. Freeman.

9. E. Mayr. 1961. Cause and effect in biology. *Science* 134:1501–06.
10. N. Tinbergen. 1963. On aims and methods of ethology. *Zeitschrift für Tierpsychologie* 20:410–33.
11. P. W. Sherman and M. L. Morton. 1984. Demography of Belding's ground squirrels. *Ecology* 65:1617–28.
12. K. E. Holekamp. 1984a. Natal dispersal in Belding's ground squirrels *(Spermophilus beldingi). Behav. Ecol. Sociobiol.* 16:21–30.
13. P. W. Sherman. 1977. Nepotism and the evolution of alarm calls. *Science* 197:1246–53.
14. P. W. Sherman. 1981a. Kinship, demography, and Belding's ground squirrel nepotism. *Behav. Ecol. Sociobiol.* 8:251–59.
15. G. R. Michener. 1983. Kin identification, matriarchies, and the evolution of sociality in ground-dwelling sciurids. In *Recent Advances in the Study of Mammalian Behavior*, ed. J. F. Eisenberg and D. G. Kleiman, pp. 528–72. Am. Soc. Mammal.
16. C. H. Phoenix, R. W. Goy, A. A. Gerall, and W. C. Young. 1959. Organizing action of prenatally administered testosterone propionate on the tissues mediating mating behavior in the female guinea pig. *Endocrinology* 65: 369–82.
17. K. E. Holekamp, L. Smale, H. B. Simpson, and N. A. Holekamp. 1984. Hormonal influences on natal dispersal in free-living Belding's ground squirrels *(Spermophilus beldingi). Hormones and Behavior* 18:465–83.
18. E. Adkins-Regan. 1981. Early organizational effects of hormones: An evolutionary perspective. In *Neuroendocrinology of Reproduction*, ed. N. T. Adler, pp. 159–228. Plenum Press.
19. K. E. Holekamp. 1986. Proximal causes of natal dispersal in Belding's ground squirrels *(Spermophilus beldingi). Ecol. Monogr.* 56: 365–91.
20. S. Pfeifer. 1982. Disappearance and dispersal of *Spermophilus elegans* juveniles in relation to behavior. *Behav. Ecol. Sociobiol.* 10:237–43.
21. F. S. Dobson. 1979. An experimental study of dispersal in the California ground squirrel. *Ecology* 60:1103–09.
22. J. Moore and R. Ali. 1984. Are dispersal and inbreeding avoidance related? *Animal Behav.* 32:94–112.
23. A. E. Pusey and C. Packer. 1987. The evolution of sex-biased dispersal in lions. *Behaviour* 101:275–310.
24. A. Cockburn, M. P. Scott, and D. J. Scotts. 1985. Inbreeding avoidance and male-biased natal dispersal in *Antechinus* spp. (Marsupialia: Dasyuridae). *Animal Behav.* 33:908–15.
25. J. A. King. 1984. Historical ventilations on a prairie dog town. In *The Biology of Ground-dwelling Squirrels*, ed. J. O. Murie and G. R. Michener, pp. 447–56. Univ. of Nebraska Press.
26. W. M. Shields. 1982. *Philopatry, Inbreeding, and the Evolution of Sex.* State Univ. of New York Press.
27. P. J. Greenwood, P. H. Harvey, and C. M. Perrins. 1978. Inbreeding and dispersal in the great tit. *Nature* 271:52–54.
28. J. L. Hoogland. 1982. Prairie dogs avoid extreme inbreeding. *Science* 215:1639–41.
29. K. Ralls, P. H. Harvey, and A. M. Lyles. 1986. Inbreeding in natural populations of birds and mammals. In *Conservation Biology: The Science of Scarcity and Diversity*, ed. M. E. Soulé, pp. 35–56. Sinauer.
30. P. M. Waser and W. T. Jones. 1983. Natal philopatry among solitary mammals. *Q. Rev. Biol.* 58:355–90.
31. P. W. Sherman. 1981b. Reproductive competition and infanticide in Belding's ground squirrels and other animals. In *Natural Selection and Social Behavior*, ed. R. D. Alexander and D. W. Tinkle, pp. 311–31. Chiron Press.
32. A. E. Pusey and C. Packer. 1986. Dispersal and philopatry. In *Primate Societies*, ed. B. B. Smuts, D. L. Cheney, R. M. Seyfarth, R. W. Wrangham, and T. T. Struhsaker, pp. 250–66. Univ. of Chicago Press.
33. W. P. Luckett and L. J. Hartenberger, eds. 1985. *Evolutionary Relationships among Rodents.* Plenum Press.
34. D. J. Hafner. 1984. Evolutionary relationships of the nearctic Sciuridae. In *The Biology of Ground-dwelling Squirrels*, ed. J. O. Murie and G. R. Michener, pp. 3–23. Univ. of Nebraska Press.
35. K. E. Holekamp. 1984b. Dispersal in ground-dwelling sciurids. In *The Biology of Ground-dwelling Squirrels*, ed. J. O. Murie and G. R. Michener, pp. 297–320. Univ. of Nebraska Press.
36. M. G. Garrett and W. L. Franklin. 1988. Behavioral ecology of dispersal in the black-tailed prairie dog. *J. Mammal.* 69:236–50.
37. B. D. Chepko-Sade and Z. T. Halpin, eds. 1987. *Mammalian Dispersal Patterns.* Univ. of Chicago Press.
38. P. W. Sherman. 1989. The clitoris debate and the levels of analysis. *Animal Behav.* 37:697–98.

Ronald J. Prokopy
Bernard D. Roitberg

Foraging Behavior of True Fruit Flies

Concepts of foraging can be used to determine how tephritids search for food, mates, and egg-laying sites and to help control these pests

Until early this century, most of what we knew about the behavior of insects in nature was based on observation and description by naturalists. Then, in the 1930s, a systematic experimental approach to the analysis of insect behavior was launched. The next two decades saw the beginning of a concerted effort both to elucidate mechanisms by which insects perceive and orient themselves to environmental stimuli and to determine how their physiological state influences their response to such stimuli. The emphasis of research in recent years has been on detailed quantitative analysis of the behavior of individual insects, using an ecological framework. This approach is especially appealing to many field biologists, because it calls attention to the manner in which ecological factors affect the contribution of behavior to survival and reproductive success.

One aspect of insect behavioral ecology currently of widespread interest is resource-foraging behavior. The central issue in foraging behavior, as explicated in Hassel and Southwood (1978), Krebs and Davies (1978), Kamil and Sargent (1981), among others, concerns how a creature adjusts its activities in response to the nature and distribution of potential resources. Changes in behavior as a result of changes in the spatial and temporal distribution of resources may affect foraging efficiency and, ultimately, reproductive success. A thorough understanding of foraging requires integration of mechanistic approaches to the analysis of behavior, which accentuate immediate causation, with evolutionary, ecological approaches, which stress the adaptive significance of a behavior.

In the past, most investigations of foraging behavior have been aimed at determining how an animal goes about "deciding" which type of resource "patch" to visit, which particular resource unit within the chosen patch to utilize, how long to remain in that location, and how to move efficiently between patches (Pyke et al. 1977). At present, several additional questions are receiving increased attention: How does a forager go about sampling resources to arrive at an average value for a locale? How does it make trade-offs in attempting to satisfy all its resource requirements? Should a forager aim at making the most efficient use of time and energy in acquiring the benefits of a resource? Should it be more concerned with minimizing variation in the amount or quality of that resource? Or should it concentrate on reducing risks during foraging by minimizing exposure to potential hazards such as predators and harmful inanimate factors at a possible cost of reducing foraging efficiency?

In attempting to provide meaningful answers to these and other questions that arise in the course of studying foraging behavior, much of the challenge lies in choosing a species of animal whose activities are amenable to thorough analysis and in gaining sufficient background information on the physiology, behavior, and ecology of that species. Thus, it must be possible to identify each type of essential resource utilized by that animal in nature and to estimate the quantity, quality, and distribution of each resource in the normal habitat. The activity pattern of the forager and the stimuli to which it responds when seeking a particular resource must be known, as must the sorts of competitors, natural enemies, and nonbiological factors that might influence its behavior. Also important is knowledge of the role of genetic background and physiological state in a forager's behavior and of whether the animal is capable of learning. Moreover, the researcher must be able to construct a variety of realistic experimental resource patches and distributions under natural or seminatural conditions and to devise a reliable way of tracking individual foragers as they move within or among patches.

Among vertebrates, birds have proved to be particularly valuable for investigating foraging behavior, and indeed, much current foraging theory derives from studies of birds. Among invertebrates, bees have been the object of quantitative analysis of food-foraging behavior, dung flies of mate foraging, and parasitoids of foraging for egg-laying sites.

True fruit flies (Tephritidae) would also seem to be highly suitable organisms for this kind of research under natural, as well as seminatural and laboratory, conditions. Because certain species of fruit-infesting te-

Ronald J. Prokopy is Professor of Entomology at the University of Massachusetts. Since receiving his Ph.D. from Cornell University in 1964, he has conducted studies on various aspects of fruit pest biology and control in several regions of the United States and Europe, in Central and South America, and in Japan. His present research interests lie in the application of behavioral ecological theory and investigations to fruit pest management. Bernard D. Roitberg received his Ph.D. from the University of Massachusetts in 1981 and is now Assistant Professor of Biology at Simon Fraser University in British Columbia. The principal focus of his research is foraging behavior of parasitic insects. *The research reported here was supported in part by a Guggenheim Fellowship, by grants from NSF and the Science and Education Administration of the US Department of Agriculture, and by the Massachusetts Agricultural Experiment Station.* *Address for Dr. Prokopy: Department of Entomology, University of Massachusetts, Amherst, MA 01003.*

phritids are among the world's most damaging agricultural pests, there have been many studies of their physiology, behavior, and ecology (reviewed in Bateman 1972, 1976; Boller and Chambers 1977; Burk 1981; Bush 1974; Drew et al. 1978; Prokopy 1977, 1980; and Zwölfer, in press). The knowledge thereby obtained, coupled with the comparative ease with which tephritid resources can be manipulated and adult flies tracked as they move within resource patches, thus goes a long way toward satisfying the criteria for selecting an animal for research into foraging behavior.

From an evolutionary standpoint, the Tephritidae are a younger family than their distant relatives the pomace flies (Drosophilidae), which are also commonly referred to as fruit flies. Tephritids may have arisen in the middle Tertiary, some 40–50 million years ago. So far, about 5,000 species have been described, the vast majority of which infest living plants. Adults of nearly all species must ingest food to attain sexual maturity, and mating is required for egg viability. Non-fruit-infesting tephritids lay their eggs in roots, stems, buds, leaves, seeds, or flowers, and the larvae consume these organs, whereas the fruit-infesting species deposit their eggs in growing fruits, and the larvae develop in the flesh of the fruit. On reaching maturity, the larvae leave the fruit and form pupae in the soil beneath the host plant.

In this article, we will discuss elements of our present knowledge of the biology of fruit-infesting tephritids that bear on adult foraging for food, mates, and egg-laying sites. We will also point out instances in which knowledge of foraging behavior has given rise to methods of monitoring and controlling tephritids that are pests. Finally, we will describe some of our own recent quantitative studies of foraging for egg-laying sites.

We will focus on the four tephritid species listed in Table 1. These were chosen in part because more is known about their foraging behavior than is perhaps the case for any other tephritids, in part because they are all agricultural pests. *Rhagoletis pomonella* attacks apples in northeastern and central North America, and, within the past few years, has spread to apple-growing regions on the West Coast. *Ceratitis capitata*, the Mediterranean fruit fly, or medfly, has invaded the continental United States on at least eight occasions, the most recent of which was in California from 1980 to 1982. *Anastrepha suspensa* was first observed in Florida in 1965 and has since gained a strong foothold there. *Dacus tryoni* is native to Australia and has not spread to the United States. If any of these species or other *Anastrepha* or *Dacus* species should ever become established in California, they would pose a significant threat to the multibillion-dollar fruit industry of that state.

Furthermore, as a group, the four species listed in Table 1 exhibit rather contrasting biological characteristics. For example, the temperate *R. pomonella* infests only a few different kinds of fruits, breeds only one generation per year, spends most of its life in a state of pupal dormancy (adults live only ~30 days), has comparatively low mobility and fecundity, and tends to form permanent and rather stable local populations which inhabit the same locales in successive years. By contrast, the tropical *D. tryoni* utilizes many kinds of fruits, produces several generations per year, spends most of its life in the adult stage (adults can live for many months), and tends to form local populations that are distinctly transient. The other two species, *C. capitata* and *A. suspensa*, lie somewhere between these two extremes but closer to the *D. tryoni* type.

Table 1. Species of fruit-infesting tephritids discussed in the article

Species	*Hosts*	*Location*
Rhagoletis pomonella (apple maggot fly)	hawthorn, apple, sour cherry, rose hip	North America
Ceratitis capitata (medfly)	>250 species of fruits, nuts, and vegetables	Has spread from central Africa to many other parts of the world
Anastrepha suspensa (Caribbean fruit fly)	~100 species of fruits	North, Central, and South America
Dacus tryoni (Queensland fruit fly)	>100 species of fruits and vegetables	Australia and other islands of the South Pacific

Foraging for food and water

Adults of the *Rhagoletis*, *Ceratitis*, *Anastrepha*, and *Dacus* genera are unable to sustain high fecundity without ingesting water and such nutrients as carbohydrates, amino acids, B vitamins, and salts. This they do during periodic diurnal grazing bouts, more or less on a daily basis, throughout life.

A particularly favored nutrient source is honeydew, which is excreted by aphids, scales, leafhoppers, and some other plant-sucking insects. In temperate regions inhabited by tephritids, honeydew usually accumulates in substantial quantities and is likely to be available most of the time in most places during periods when flies are active. Where rainfall is frequent or daily, however, as during the wet season in subtropical and tropical areas, accumulations may not be appreciable, and fruit flies may then feed on juices from bird-pecked, insect-bitten, or overripe fruit.

When deprived of food or water, tephritids kept in cages in the laboratory may initiate flights with exceptional frequency. This suggests that in nature, hungry or thirsty individuals may emigrate from locales that are lacking in these resources. Ripley and co-workers (1940) and various other researchers have reported large numbers of *C. capitata* adults moving downwind when warm, dry, windy weather follows a dewless night. Lack of sufficient water could, in fact, have been an important factor leading to the emigration of *C. capitata* flies to new locales in the vicinity of streams, lakes, and other sources of water during the hot, dry, windy summer days of the outbreak of this pest in California in 1981.

In searching for food, tephritids respond to both odors and visual cues. Curiously, even at present, efforts to characterize olfactory stimuli rarely include a deliberate attempt to identify volatile substances in the air surrounding a natural food source.

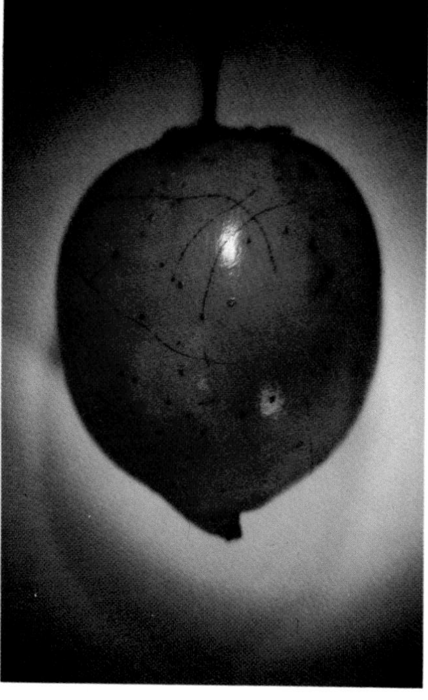

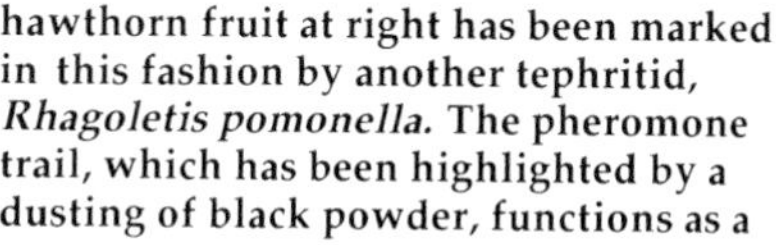

Figure 1. In a number of tephritid species, such as *Ceratitis capitata* shown above, the female drags its ovipositor over the surface of the fruit after egg-laying, leaving a trail of pheromone contained in the feces. The hawthorn fruit at right has been marked in this fashion by another tephritid, *Rhagoletis pomonella.* The pheromone trail, which has been highlighted by a dusting of black powder, functions as a sexual stimulant to males and, more importantly, as a deterrent to the laying of more eggs than can be supported by a single fruit. (Photo at left by G. K. Uchida; photo at right by A. L. Averill.)

Rather, beginning about the turn of the century, they have taken a somewhat circuitous route involving testing of a scattered array of thousands of substances for potential attractiveness. Some of the more enticing of these, in chronological order of investigation, were found to be molasses, wheat bran, ammonium salts, McPhail's mixture (sugar and brewer's yeast), and a variety of protein hydrolysates. Recent work on *D. tryoni* suggests that ammonia may be the principal attractant emanating from these substances or mixtures, and possibly from natural food sources as well (Morton and Bateman 1981). Ammonium salts or protein hydrolysates have been widely used in traps to monitor populations of food-seeking *Rhagoletis*, *Ceratitis*, *Anastrepha*, and *Dacus* species.

Attempts to characterize visual stimuli serving as cues to food-seeking tephritids have been undertaken only within the past decade or so. By erecting variously colored rectangular models of trees and coating them with an odorless, sticky substance that captures alighting flies, Moericke and his colleagues (1975) showed that the green color of foliage and the contrast of a plant silhouetted against a background of skylight are significant visual stimuli for *R. pomonella*, with regard not only to feeding sites, but also to potential mating, egg-laying, and shelter sites. Using models of different shapes and sizes, these scientists found that compact shapes, like those of host plants, were more attractive than highly elongated ones, and that, not surprisingly, larger models were more attractive than smaller ones. Similar findings have been reported for *C. capitata* by Sanders (1968) and for *D. tryoni* by Meats (in press).

Interestingly, in these and several other tephritid studies (Agee et al. 1982; Owens 1982), it turned out that the models eliciting the most positive response from the flies were yellow. This seems to be because most yellow pigments, like green leaves, reflect little energy of wavelengths less than 500 nanometers (nm) and a great deal of energy, much more than green leaves, between 500 and 580 nm. Thus, it appears that flies respond to yellow models as if they were masses of very bright leaves. Yellow pigments that mimic foliage have proved to be attractive to a wide range of tephritid species. In fact, a small sticky-coated yellow rectangle baited with some kind of protein hydrolysate is currently the standard trap for monitoring populations of food-seeking *R. pomonella* flies and other temperate tephritids.

Research on the food-foraging behavior of tephritid adults has led not only to the development of effective traps of this sort, but also to an effective approach (albeit somewhat harsh on nontarget organisms) to suppression or elimination of fly populations. During the outbreak of *C. capitata* in Florida in 1929, molasses was used as a bait and, in combination with arsenic, was sprayed over more than 500 km^2, thereby eliminating the flies. At the time of the 1956 outbreak of *C. capitata* in Florida, which affected 3,000 km^2, protein hydrolysate was substituted for molasses, and malathion for arsenic, and again the pest was eliminated. The protein-hydrolysate spray containing malathion was used in the US in all six succeeding campaigns to eliminate this pest. In has also been used to suppress or eliminate *D. tryoni* in certain parts of Australia but has been found ineffective against *R. pomonella*, possibly because natural food sources are more abundant for *R. pomonella* than for subtropical and tropical tephritids.

Another approach to population suppression based on food-foraging behavior was developed some years ago by Monro (1966), who released large numbers of sterile *D. tryoni* flies cultured in the laboratory, in an attempt to overload existing food resources of wild flies. Natural populations declined sharply within a very few days. This approach has been proposed for *D. tryoni* and other tephritids and perhaps deserves more emphasis than it has received to date.

Foraging for mates

As far as is known, females of all tephritid species require mates if viable offspring are to be produced. For *R. pomonella* and related, temperate species, copulation appears necessary on a weekly, or even more frequent, basis to ensure high egg fertility. For *Ceratitis*, *Anastrepha*, and *Dacus* species, however, a single mating usually suffices for several weeks, if not an entire lifetime, and females usually resist further mating attempts following initial copulation. In all species, males, who reach sexual maturity a few days after emerging from the pupal stage and a few days before females of comparable age, are promiscuous.

Before considering where tephritid males and females mate, we would do well to note Parker's (1978) observation that insects of virtually all species come together for mating purposes at specific encounter sites—that is, males rarely, if ever, search randomly for females within all parts of the habitat. Parker further indicates that encounter sites are often locales that have value to females in other respects—as feeding, egg-laying, or shelter sites.

Studies in nature reveal that in *R. pomonella* mating takes place exclusively on trees where egg-laying occurs—that is, on potential host trees. In *A. suspensa* and *D. tryoni*, on the other hand, it may also occur on neighboring nonhost trees. No one has yet published a systematic investigation of where *C. capitata* flies mate, which is remarkable, considering the economic injury that can be inflicted by this species. Recent observations by Arita and Kaneshiro (in press) in Hawaii, however, suggest that at least some copulations are initiated on host trees.

Confinement of mating to host plants may be characteristic not only of *R. pomonella* but also of other temperate fruit-infesting tephritids that breed only one generation per year, and whose populations and hosts are predictable in distribution and abundance from one year to the next. In fact, such confinement may be extremely important in ensuring reproductive isolation. For example, in an area of Wisconsin where *R. pomonella* recently shifted to sour cherry trees that had been introduced, neither sex was readily able to distinguish members of its own species from those of *R. fausta*, which, in an earlier decade, had likewise shifted (from pin cherry) to sour cherry as a new host. Numerous cases of interspecific mating were observed (Prokopy 1980). Findings of this sort suggest that when the range of host plants used by temperate tephritids expands in this way, a previously efficient sexual communication system may be somewhat compromised.

Mating sites seem to be less predictable in subtropical and tropical tephritids, which produce several generations per year, and whose host plants are patchy in distribution in both time and space. Sites at which these flies assemble may shift among host plants, plants furnishing food or shelter, and even plants that have no apparent resource value whatever, depending on the particular habitat.

In searching for mates, as for food, flies are guided by olfactory and visual cues emanating from the host. In addition, males provide cues to females in the form of pheromones—volatile sex odors—which attract virgin females. In *C. capitata*, *A. suspensa*, and *D. tryoni*, the chemicals that compose the male pheromonal blend have been partially identified (Nation 1977). In *A. suspensa*, males have been found to produce a second type of sexual stimulus: a wing vibration sound that is attractive to virgin females (Webb et al., in press). A third type of sexual signal operates in *R. pomonella*, whereby a female, after laying eggs, deposits on the surface of the fruit a trail of feces containing a pheromone that discourages egg-laying, as shown in Figure 1. A male arriving on such a fruit and contacting the trail within an hour or so of its deposition becomes sexually aroused and either waits for the female to reappear or searches for her in the vicinity.

Among *R. pomonella* and other temperate *Rhagoletis* species that infest only one kind of fruit (or did until relatively recently), strict confinement of sexual encounters to the host plant may confer both a savings in energy for males looking for mates and a comparatively high probability of copulatory success. In such instances, host volatiles probably play a larger part than male sex pheromones in long-range attraction of virgin females.

The sequence of events leading to copulation once the male and female have arrived on the same plant differs considerably for the temperate and more tropical species. In *R. pomonella*, whose peak mating time is from mid-morning to mid-afternoon, nearly all sexual encounters early in the fruiting season, when most females are virgin, take place on foliage (Prokopy 1980). Studies suggest that a female flies to the general vicinity of a pheromone-releasing male. Upon visual detection of a nearby moving female of the same species or another insect of similar morphology, the male flies toward it, makes a frontal approach to within 1–2 cm, waves his wings briefly, and then, if the gestalt of visual characteristics suggests that the insect is indeed of the same species, jumps onto its abdomen and attempts to copulate.

Male courtship in *R. pomonella* changes drastically with the onset of egg-laying (Prokopy 1980). At that time, males begin to spend most of their time patrolling or establishing territories on fruit and frequently "fight" one another to maintain a territory, as illustrated in Figure 2. When a female arrives on a fruit occupied by a male or on a nearby fruit within the visual range of a male—up to ~40 cm—the male is alerted by movements made by the female immediately prior to or during egg-laying. In most cases, the male proceeds to approach the female from the rear. Just as the female is about to extend the ovipositor—her pointed egg-laying tube—to initiate egg deposition or while she is laying her eggs or withdrawing her ovipositor from the fruit afterward, the male leaps onto her back and attempts copulation, usually with success. Such forced mating is probably highly advantageous to the male, because much of his sperm takes precedence over the sperm of previous males, which is not the case with many other insects. Males attempt to copulate with other males of their own species just as often as they do with females, indicating an inability to discriminate between sexes prior to direct physical contact.

The quality, density, and distribution of host trees and fruit, as well as the relative population densities of males and virgin females, may well play a major role in determining which mate-foraging strategy a *R. pomonella* male will employ within a given period: release of sex pheromone while foraging among host leaves to attract virgin females, or patrolling host fruit to attempt force-mating with already mated females.

In *C. capitata* and *A. suspensa*, the predominant mode of courtship involves sexual selection by females among males that have come together in a lek (Prokopy and Hendricks 1979; Burk 1983). A lek, in the words of Emlen and Oring (1977), is a "communal display area where males congregate for the sole purpose of attracting and courting females and to which females come for mating." Parts of plants which receive above-average amounts of light are especially likely to function as gathering sites. Attraction to pheromone given out by other males also seems to be important in forming leks. Indeed, in *C. capitata*, the chemical lure (Trimedlure) that is most effective in detecting male populations apparently owes much of its attractiveness to its mimicry of the male lek-forming pheromone.

Figure 2. Two *R. pomonella* males fight to establish possession of an apple as their territory. During egg-laying season, males spend much of their time establishing and defending territories for the purpose of gaining sexual access to females that alight nearby. (Photo by G. L. Bush.)

Far more matings are initiated within a lek than outside one (Arita and Kaneshiro, in press; Burk 1983). The male exposes a conspicuous droplet of pheromone from the anus while taking up prolonged residence—from morning until early afternoon in *C. capitata* and from late afternoon to early evening in *A. suspensa*—on the underside of a leaf. A lek normally consists of from three to ten males gathered on closely adjacent leaves. Each resident male is vigorous in defending its leaf against intruding males, and fights often break out. In *A. suspensa*, resident males, especially the larger ones, win more fights than intruders (Burk 1983). In both species, copulation with a pheromone-emitting male occurs after a female has flown near the male, moved to within 1–2 cm of him, and engaged in a rather elaborate wing-waving display. Peak daily egg-laying occurs before or after the peak mating period. When not in a lek or engaged in secreting pheromone, some males of each species may shift strategies and attempt to court or force-mate apparently unreceptive females who are laying eggs. Compared with *Rhagoletis* males, however, they meet with little success in this regard, since a single mating often provides enough sperm to fertilize egg production for a lifetime, and females resist mating after first copulation (Prokopy and Hendricks 1979; Burk 1983).

The mating behavior pattern in *D. tryoni* is similar to that in *C. capitata* and *A. suspensa*, except that males congregate in leks only during a half-hour period at dusk when they gather on foliage in swarms (Tychsen 1977). Courtship display after initial encounter with a female is not elaborate, and, as far as is known, there are few, if any, attempts by males to court or force-mate females on fruit.

So far, the greatest impact of our knowledge of mate-foraging behavior on tephritid management strategy has been in the release of mass-produced sterile males who then mate with wild females. This approach has been employed with greatest advantage in the case of *C. capitata* and has been especially successful in dealing with small, confined wild populations, where wild males can be easily outnumbered. Some 800 million sterile *C. capitata* flies are now produced each week in Tapachula, Mexico, for release along the Guatemalan-Mexican border, to suppress medfly incursion into Mexico. During spring and summer of 1981, several million such flies were released each week in central California, in an attempt to suppress *C. capitata* there. The technique was not totally successful in this instance, however, possibly because the wild population was too large in part of the affected area, and the number of sterile flies available for release was too small by comparison.

To compete with their wild counterparts, laboratory-produced sterile males must be able to locate mating sites efficiently, and, in the case of *C. capitata* and some other species, to engage effectively in lek formation and in competition for territories within leks. The pioneering work of Boller and his colleagues (1981) and Chambers and his coworkers (1983) on fly-rearing procedures which reduce deleterious genetic effects and on developing biologically realistic quality control measures for evaluating the competency of sterile males has contributed greatly to production of high-grade sterile *C. capitata* males.

Figure 3. Sticky-coated models spaced one meter apart were used to investigate the methods by which *R. pomonella* locates apples. For the purpose of the photograph, these models were brought closer together and flies of other species were removed. The spherical model, similar in shape, size, and color to a large red apple, was preferred to the other shapes of similar size and attracted as many flies as the actual fruit. (Photo by R. J. Prokopy.)

Foraging for egg-laying sites

With the onset of sexual maturity, fruit-infesting tephritids begin making frequent visits to hosts suitable for egg-laying. The quality of the particular fruit in which eggs are laid can have a profound impact on the survival and the reproductive success of the ensuing larvae and adults.

How do females go about locating hosts and then assessing their suitability as egg-laying sites? Before answering, it is important to recognize that the physical and chemical characteristics of fruits may well have evolved principally as advertisements to potential seed dispersers, particularly birds and mammals (Herrera 1982). A major constraint faced by plants in advertising their fruit is the need to conceal or protect the fruit until the seeds are mature. That fruit-infesting insects can significantly reduce the reproductive success of plants in this respect has been shown recently by Manzur and Courtney (in press). It is perhaps testimony to their physiological, behavioral, and ecological flexibility that some of these tephritids—in particular the four species considered here—have been able to cope with or overcome plant defenses against fruit and seed depredation and to utilize increasing numbers of plants as hosts.

The major long-range stimuli drawing both sexes of tephritids to host plants are volatile components of ripening fruit. In only one case, however—in a study involving *R. pomonella*—have the chemicals comprising an attractive fruit-volatile blend been identified and synthesized (Fein et al. 1982). Olfactory components of the foliage appear to play a minor role, if any, in locating host plants. As pointed out earlier, visual properties of plants, such as color, silhouette against background, shape, and size, are used by both sexes of *R. pomonella*, and probably by other species too, as cues for finding plants, but none of these is viewed as being specific to hosts.

To determine how a fly detects an individual fruit after it has arrived on a host plant, we hung wooden mimics of apples on twigs near real apples in trees harboring *R. pomonella* and then counted the number of adults alighting on them. We found that just as many adults arrived on the wooden models as on the real fruits, which strongly suggests that at close range—within a meter—visual characteristics are the predominant or sole cues to finding individual fruits. By employing sticky-coated models of various shapes, sizes, and colors, some of which are shown in Figure 3, we found that *R. pomonella* and *C. capitata* flies were most attracted by spherical shape, above average but not excessive size, and a dark color that contrasts with the background of skylight and the light reflected or transmitted by the foliage.

These visually attractive properties have been incorporated in a sticky-coated trap consisting of a red wooden sphere the size of a croquet ball, which has proved to be very effective in monitoring fruit-seeking *R. pomonella* females and males in commercial apple orchards, and which has even provided excellent, though very labor-intensive, direct control of *R. pomonella* in small orchards. Used in conjunction with synthetic attractive apple volatiles, it is even more effective (Reissig et al. 1982). One major advantage of this sort of trap over a trap incorporating feeding-type stimuli is that comparatively fewer beneficial and other nontarget insects are attracted.

To date, more attention has been focused on the stimuli eliciting egg-laying after arrival at a potential host site than on any other aspect of the host-foraging process. Using wax-covered, agar-filled structures as artificial fruits, various investigators have found that in all four species considered here, the major stimuli affecting egg-laying include convex shape, appropriate size and color, and the chemical and physical composition of both the surface and flesh of the fruit. Chemically speaking, acceptability of a fruit for egg-laying may depend not only on the blend of positive stimuli present—sugars and acids, for example—but equally, or more so, on negative stimuli—phenols, glucosides, or alkaloids—in immature host or nonhost fruit (Szentesi et al. 1979). These negative components, which are toxic to the larvae, may constitute one of the plant's primary defenses against tephritids and other fruit parasites.

When plants or plant patches cease to provide enough high-quality fruit for egg-laying, mature adults are likely to emigrate (Fletcher 1973; Chapman 1982). The distances covered by emigrating flies in search of new host plants may equal or exceed those traveled by emigrants in search of food or water—1–2 km for *R. pomonella*, somewhat more for *C.*

capitata and *A. suspensa*, and as many as 20 km or more for *D. tryoni*.

In the recent campaign to eliminate *C. capitata* from California, host fruits were stripped from currently infested or potentially vulnerable trees. However, this may have stimulated long-distance movement of some flies to previously uninfested areas, an unintended side effect that may have exacerbated the problem. Lack of available host fruit can give rise to extensive attempts at egg-laying in various nonhost fruits. We are reminded of an occasion in Wisconsin when, with Guy L. Bush, we observed *R. pomonella* females laying eggs in tomatoes, a nonhost fruit, that we had hung in birch trees, also nonhosts, near a sour cherry orchard. It appears that the females had emigrated from the sour cherry trees, which at that time had few remaining fruits.

Bush and Diehl (1982) have suggested that permanent new hosts for such plant parasites as tephritid flies probably arise through changes in host recognition and in larval survival genes. Indeed, the expansion in the host range of *R. pomonella* from hawthorn to apple, sour cherry, and rose hip constitutes one of the hypothesized prototypical examples of insects forming new host races without antecedent geographical isolation (Bush 1974; Diehl and Bush, in press). Although it had been postulated that there are behavioral and ecological differences among populations of *R. pomonella* inhabiting various hosts, it is only lately that experimental evidence has come forth to support this hypothesis (Prokopy et al. 1982a; Diehl and Bush, in press). The genetic basis of these differences is yet to be determined, however.

We recently discovered that *R. pomonella* females are capable of changing their behavior as a result of experience. In the laboratory, as well as in nature, we found that after laying eggs in hawthorn or apple, they could learn to accept or reject these hosts in future encounters (Prokopy et al. 1982b). A possible selective advantage of this sort of learning may be that it permits a female foraging for an egg-laying site to concentrate her effort on the host species of greatest local abundance (presuming the most abundant species to be the one encountered first), thereby decreasing the probability that she will waste energy or die searching for a rarer host.

Yet another element influencing foraging behavior for host fruit is the presence or absence of intraspecific competitors. A female olive fly of the species *D. oleae*, a relative of *D. tryoni*, who has just deposited an egg, uses her proboscis to spread the olive juice that exudes from the egg-laying puncture over the surface of the fruit. In an artful series of experiments, Cirio (1971) found that the juice signals the presence of an egg or a developing larva in that fruit, thereby acting as a deterrent to subsequent egg-laying by either the same fly or other olive flies.

In *R. pomonella*, *C. capitata*, *A. suspensa*, and several related tephritids, the female likewise circles around the fruit after laying her eggs. Instead of spreading fruit juice, however, she drags her extended ovipositor across the surface of the fruit and deposits a pheromone that deters further egg-laying, as shown in Figure 1 (Prokopy 1981). There is an upper limit to the number of *R. pomonella* larvae that can develop in a fruit without reducing their survival rate or diminishing adult fitness (see Fig. 4). Both the amount of pheromone deposited and the amount required for deterrence are closely linked to the number of larvae which a fruit can support (Averill and Prokopy, unpubl.). Although we know a fair amount about the physiology of the production, release, and detection of pheromones that inhibit egg-laying in tephritids, the chemical identity of the active components remains somewhat of a mystery, despite concerted efforts at identification. Once the chemistry is determined, however, it may be possible to synthesize such pheromones. Spraying host trees with synthetic pheromone might then become an important new approach to fly management, especially if fruit-mimicking traps are used to capture the females thus deterred from laying eggs.

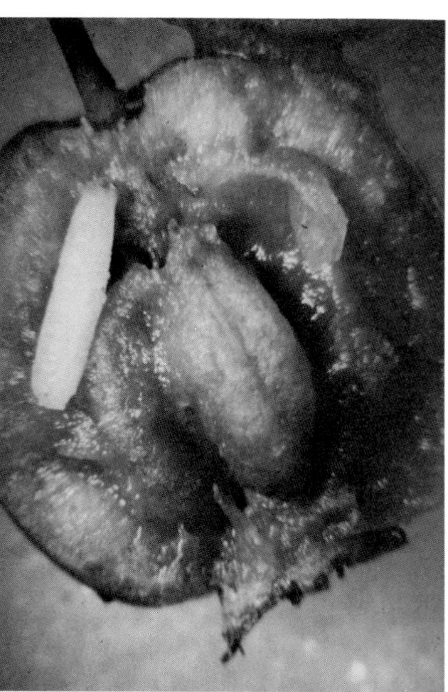

Figure 4. This mature larva feeding inside a hawthorn fruit is a member of *R. pomonella*, one of the tephritid species that deposit their eggs inside growing fruits. Usually, only a single larva can be sustained by such a fruit, and tephritids have evolved ways, as shown in Figure 1, of limiting the number of eggs in a host fruit. (Photo by A. L. Averill.)

D. tryoni females apparently neither spread droplets of fruit juice nor emit a pheromone deterring further egg-laying. However, when present simultaneously on the same fruit, they are known to fight for possession of it. Moreover, they seem to be able to detect living larvae in fruit and to refrain from laying eggs there (Fitt 1983).

Interspecific competition may affect the fruit-foraging behavior of certain tephritid females, at least in part, although the underlying mechanisms are not yet fully understood. The classic case involves *C. capitata*, which was first introduced into the Hawaiian Islands early this century. In parts of some islands, it has since been displaced by the oriental fruit fly *D. dorsalis*, introduced in the 1940s. When found together in the same fruit, *D. dorsalis* larvae appear capable of suppressing the development of *C. capitata* larvae (Keiser et al. 1974), although at present there exists conjecture as to the extent of suppression. Competitive interactions may be further exacerbated by the preference of fruit-foraging *D. dorsalis* females for preexisting punctures made by such insects as *C. capitata*.

Quantitative analysis of foraging

In spite of the comparative wealth of basic knowledge of tephritid biology, little has been done in the way of constructing realistic experimental patches to evaluate quantitatively foraging for food, water, or mates. However, we recently adopted this kind of approach in a study of ways in which *R. pomonella* flies forage for egg-laying sites (Roitberg and Prokopy, in press, and references therein). We released and then monitored and recorded the behavior of individual females in host trees on which we had hung fruit in varying densities and distributions. The trees were divided into several hundred individually numbered sectors, which permitted detailed assessment of fly movements within a single tree.

We employed the information gleaned from our observations to address specific questions in contemporary foraging theory. The primary question with which we wish to concern ourselves here is how females determine the amount of time they allocate to foraging for egg-laying sites within host trees. We will also consider briefly how the criteria used might promote foraging efficiency and reproductive success.

Our research suggests that females arriving on host trees employ at the outset what Murdoch and Oaten (1975) term a fixed-threshold-rate rule to determine search persistence. According to this rule, animals either remain in the patch containing their prey until they discover what they are looking for, or they emigrate after some fixed time has elapsed. Encounters with prey cause them to reset their search allocation "clocks," and the process is repeated. Such behavior has been shown to promote foraging efficiency, though not optimally, when the environment is highly variable (see Cowie and Krebs 1979).

In *R. pomonella*, the discovery of host fruit dramatically influences the allocation of search time. This was clearly demonstrated by releasing flies in trees devoid of fruit (Roitberg et al. 1982). Flies that had not laid eggs recently stayed only about 4 minutes in such trees, whereas flies that had laid eggs immediately after being released searched significantly longer—about 9 minutes—before emigrating. Subsequent visits to uninfested host fruit and egg-laying therein led to an increase in the time allocated for that tree, as well as to an intensive search in the general area in which eggs had been laid. Both these responses promote foraging efficiency in environments where resources are distributed patchily (Hassel and Southwood 1978).

As we have seen already, each host fruit can harbor only a limited number of *R. pomonella* larvae, and intensive foraging immediately following egg-laying may lead to an increase in the numbers of encounters with previously exploited hosts, and, more important, with hosts harboring the forager's own offspring. In view of the relatively few eggs—about 300—that a *R. pomonella* female produces in her lifetime and the apparent intensity of intraspecific competition for egg-laying sites, it is not surprising that a pheromone which, in effect, labels parasitized fruit has evolved.

It is only where fruits are abundant and have not been extensively exploited previously that the response of remaining in the same fruit patch would appear to be a "wise" tactic. In areas already heavily exploited, reducing the search time would seem to be an equally wise decision. Our findings clearly show that females do indeed reduce the time they spend both within trees and within specific regions of trees, following encounters with already parasitized hosts. Thus, two antagonistic mechanisms, one increasing and the other decreasing search allocation times, determine the observed foraging persistence of individual females. It is important to consider each mechanism operating as a separate entity.

Many theoretical studies of foraging suggest that animals determine search times on the basis of information relating to their average rate of success. By contrast, our work suggests that *R. pomonella* flies process and respond to information concerning the richness of a patch in a sequential fashion. Thus, flies visiting parasitized and unparasitized hosts at the same rate, but in different sequence, allocate different search times (Roitberg and Prokopy, in press).

We suggest, therefore, that search persistence of *R. pomonella* within single trees can be predicted by a model similar to, but more complex than, that developed by Waage (1979) for parasitic wasps. In Waage's model, the forager, after entering a patch containing the desired host, is arrested within it. The level at which foragers are arrested declines with time until some threshold is crossed, at which point emigration takes place. However, if a forager encounters and then parasitizes a host, the level increases until it reaches an asymptote, the value of which depends on the time that has elapsed since a host was last exploited. In our model for *R. pomonella*, a similar process occurs, but attempts to exploit already parasitized hosts cause the level of arrest to decline by an amount which shows an inverse dependence on the time that has elapsed since the fly previously encountered an already parasitized host. This model predicts foraging persistence with reasonable accuracy, at least under seminatural conditions (Roitberg and Prokopy, in press).

With our colleague J. S. Elkinton, we are currently evaluating host-foraging and reproductive success in *R. pomonella* as functions of energy allocation, availability of hosts of different quality, previous experience, offspring survival, and risk to predation and parasitism. This multidimensional approach should permit a more precise assessment of host-foraging behavior.

The economic devastation that can be wrought by certain kinds of fruit-infesting tephritids, including the four species on which we have focused in this article, has stimulated much research into their physiology, behavior, and ecology. This work has not only enhanced our knowledge of how tephritids go about finding sustenance, mates, and places to lay their eggs, it has also led to ways of detecting and suppressing fly populations, thereby saving fruit-growers millions of dollars. We fully expect that further research into foraging behavior, along the lines described here, will lead to even more effective strategies for controlling these pests. We also believe that this research will help address questions that have recently arisen in foraging-behavior theory.

References

Agee, H. R., E. Boller, U. Remund, J. C. Davis, and D. L. Chambers. 1982. Spectral sensitivities and visual attractant studies on the Mediterranean fruit fly, *Ceratitis capitata* (Wiedemann), olive fly, *Dacus oleae* (Gmelin), and the European cherry fruit fly, *Rhagoletis cerasi* (L.). *Z. Angew. Entomol.* 93: 403–12.

Arita, L. H., and K. Y. Kaneshiro. In press. Pseudo-male courtship behavior of the female Mediterranean fruit fly, *Ceratitis capitata*. *Proc. Hawaiian Entomol. Soc.*

Averill, A. L., and R. J. Prokopy. Unpubl. Pheromonal mediation of competition in *Rhagoletis pomonella*.

Bateman, M. A. 1972. The ecology of fruit flies. *Ann. Rev. Entomol.* 17:493–518.

———. 1976. Fruit flies. In *Studies in Biological Control*, ed. V. L. Delucchi, pp. 11–49. Cambridge Univ. Press.

Boller, E. F., and D. L. Chambers. 1977. *Quality Control.* Bull. Sect. Reg. Ouest Palearctique.

Boller, E. F., B. I. Katsoyannos, U. Remund, and D. L. Chambers. 1981. Measuring, monitoring, and improving the quality of mass-reared Mediterranean fruit flies, *Ceratitis capitata.* 1. The rapid quality control system for early warning. *Z. Angew. Entomol.* 92: 67–83.

Burk, T. 1981. Signaling and sex in acalypterate flies. *Fla. Entomol.* 64:30–43.

———. 1983. Behavioral ecology of mating in the Caribbean fruit fly, *Anastrepha suspensa* (Loew). *Fla. Entomol.* 66:330–40.

Bush, G. L. 1974. The mechanism of sympatric host race formation in the true fruit flies (Tephritidae). In *Genetic Mechanisms of Speciation in Insects*, ed. M. J. D. White, pp. 3–23. Australian and New Zealand Book Co.

Bush, G. L., and S. R. Diehl. 1982. Host shifts, genetic models of sympatric speciation, and the origin of parasitic insect species. In *Proceedings of the 5th International Symposium on Insect-Plant Relationships*, ed. J. H. Visser and A. K. Minks, pp. 297–305. PUDOC, Netherlands.

Chambers, D. L., C. O. Calkins, E. F. Boller, Y. Ito, and R. T. Cunningham. 1983. Measuring, monitoring, and improving the quality of mass-reared Mediterranean fruit flies, *Ceratitis capitata.* 2. Field tests for confirming and extending laboratory results. *Z. Angew. Entomol.* 95:285–303.

Chapman, M. G. 1982. Experimental analysis of the pattern of tethered flight in the Queensland fruit fly, *Dacus tryoni. Physiol. Entomol.* 7:143–50.

Cirio, U. 1971. Reperti sul meccanismo stimulorisposta nell' ovideposizione del *Dacus oleae* Gmelin. *Redia.* 52:577–600.

Cowie, R. J., and J. R. Krebs. 1979. Optimal foraging in patchy environments. In *Population Dynamics*, ed. R. M. Anderson, B. D. Turner, and L. R. Taylor, pp. 183–205. Blackwell.

Diehl, S. R., and G. L. Bush. In press. An evolutionary and applied perspective of insect biotypes. *Ann. Rev. Entomol.*

Drew, R. A. I., G. H. S. Hooper, and M. A. Bateman. 1978. *Economic Fruit Flies of the South Pacific Region.* Watson Ferguson, Brisbane.

Emlen, S. T., and L. W. Oring. 1977. Ecology, sexual selection, and the evolution of mating systems. *Science* 197:215–23.

Fein, B. L., W. H. Reissig, and W. L. Roelofs. 1982. Identification of apple volatiles attractive to the apple maggot, *Rhagoletis pomonella. J. Chem. Ecol.* 8:1473–87.

Fitt, G. P. 1983. Factors limiting the host range of tephritid fruit flies, with particular emphasis on the influence of *Dacus tryoni* on the distribution and abundance of *Dacus jarvisi.* Ph.D. diss., Univ. of Sydney, Australia.

Fletcher, B. S. 1973. The ecology of a natural population of the Queensland fruit fly, *Dacus tryoni.* IV. The immigration and emigration of the adults. *Australian J. Zool.* 21: 541–65.

Hassel, M. P., and T. R. E. Southwood. 1978. Foraging strategies of insects. *Ann. Rev. Ecol. Syst.* 9:75–98.

Herrera, C. M. 1982. Defense of ripe fruit from pests: Its significance in relation to plant-disperser interactions. *Am. Nat.* 120:218–41.

Kamil, A. C., and T. D. Sargent. 1981. *Foraging Behavior: Ecological, Ethological, and Psychological Approaches.* Garland STPM Press.

Keiser, I., R. M. Kobayashi, D. H. Miyashita, E. J. Harris, E. L. Schneider, and D. L. Chambers. 1974. Suppression of Mediterranean fruit flies by Oriental fruit flies in mixed infestations in Guava. *J. Econ. Entomol.* 67: 355–60.

Krebs, J. C., and N. B. Davies. 1978. *Behavioral Ecology: An Evolutionary Approach.* Blackwell.

Manzur, M. I., and S. P. Courtney. In press. Coevolution of *Crataegus monogyna*, frugivores and seed predators. II. Influence of insect damage on bird foraging and seed dispersal. *Oecologia.*

Meats, A. In press. The response of the Queensland fruit fly, *Dacus tryoni*, to tree models. In *Proceedings of International Symposium on Fruit Flies, Athens.* Commission of European Communities Press.

Moericke, V., R. J. Prokopy, S. Berlocher, and G. L. Bush. 1975. Visual stimuli eliciting attraction of *Rhagoletis pomonella* flies to trees. *Entomol. Exp. Appl.* 18:497–507.

Monro, J. 1966. Population flushing with sexually sterile insects. *Science* 151:1536–38.

Morton, T. C., and M. A. Bateman. 1981. Chemical studies on proteinaceous attractants for fruit flies, including the identification of volatile constituents. *Austral. J. Agric. Res.* 32:905–16.

Murdoch, W. W., and A. Oaten. 1975. Predation and population stability. *Adv. Ecol. Res.* 9:1–131.

Nation, J. L. 1977. Pheromone research in tephritid fruit flies. In *Proceedings of the International Society of Citriculture*, vol. 2, pp. 481–85.

Owens, E. D. 1982. The effects of hue, intensity, and saturation on foliage and fruit finding in the apple maggot, *Rhagoletis pomonella.* Ph.D. diss., Univ. of Massachusetts.

Parker, G. A. 1978. Evolution of competitive mate searching. *Ann. Rev. Entomol.* 23: 173–96.

Prokopy, R. J. 1977. Stimuli influencing trophic relations in Tephritidae. *Colloq. Int. CNRS* 265:305–36.

———. 1980. Mating behavior of frugivorous Tephritidae in nature. In *Proceedings of Symposium on Fruit Fly Problems*, pp. 37–46. Nat. Inst. Agric. Sci. Japan.

———. 1981. Epideictic pheromones that influence spacing patterns of phytophagous insects. In *Semiochemicals: Their Role in Pest Control*, ed. D. A. Nordlund, R. L. Jones, and W. J. Lewis, pp. 181–213. Wiley.

Prokopy, R. J., A. L. Averill, S. S. Cooley, C. A. Roitberg, and C. Kallet. 1982a. Variation in host acceptance pattern in apple maggot flies. In *Proceedings of the 5th International Symposium on Insect-Plant Relationships*, ed. J. H. Visser and A. K. Minks, pp. 123–29. PUDOC, Netherlands.

Prokopy, R. J., A. L. Averill, S. S. Cooley, and C. A. Roitberg. 1982b. Associative learning in egglaying site selection by apple maggot flies. *Science* 218:76–77.

Prokopy, R. J., and J. Hendricks. 1979. Mating behavior of *Ceratitis capitata* on a field-caged host tree. *Ann. Entomol. Soc. Am.* 72:642–48.

Pyke, G. H., H. R. Pulliam, and E. L. Charnov. 1977. Optimal foraging: A selective review of theory and tests. *Quart. Rev. Biol.* 52: 137–54.

Reissig, W. H., B. L. Fein, and W. L. Roelofs. 1982. Field tests of synthetic apple volatiles as apple maggot attractants. *Environ. Entomol.* 11:1294–98.

Ripley, L. B., G. A. Hepburn, and E. E. Anderssen. 1940. *Fruit fly migration in the Kat River valley.* South Africa Dept. Agric. Forestry Plant Ind. Ser. 49.

Roitberg, B. D., and R. J. Prokopy. In press. Host visitation sequence as a determinant of search persistence by fruit-parasitic tephritid flies. *Oecologia.*

Roitberg, B. D., J. C. van Lenteren, J. J. M. van Alphen, F. Galis, and R. J. Prokopy. 1982. Foraging behavior of *Rhagoletis pomonella*, a parasite of hawthorn (*Crataegus viridis*) in nature. *J. Animal Ecol.* 51:307–25.

Sanders, W. 1968. Die Eiablagehandlung der Miltelmeer fruchtfliege, *Ceratitis capitata.* Ihre Abhangigkeit von Farbe und Gliederung des Umfeldes. *Z. Tierpsychol.* 25:588–607.

Szentesi, A., P. D. Greany, and D. L. Chambers. 1979. Oviposition behavior of laboratory-reared and wild Caribbean fruit flies (*Anastrepha suspensa*). I. Selected chemical influences. *Entomol. Exp. Appl.* 26:227–38.

Tychsen, P. H. 1977. Mating behavior of the Queensland fruit fly, *Dacus tryoni*, in field cages. *J. Austral. Entomol. Soc.* 16:459–65.

Waage, J. K. 1979. Foraging for patchily distributed hosts by the parasitoid, *Nemeritis canescens. J. Animal Ecol.* 48:353–71.

Webb, J. C., T. Burk, and J. Sivinski. In press. Attraction of female Caribbean fruit flies (*Anastrepha suspensa*) to males and male-produced stimuli in field cages. *Ann. Entomol. Soc. Am.*

Zwölfer, H. In press. Life systems and strategies of resource exploitation in tephritids. In *Proceedings of International Symposium on Fruit Flies, Athens.* Commission of European Communities Press.

Robert E. Cook

Clonal Plant Populations

A knowledge of clonal structure can affect the interpretation of data in a broad range of ecological and evolutionary studies

Within the temperate woodlands of northeastern America, the starflower (*Trientalis borealis*) lives as one of the common inhabitants of the forest floor (Fig. 1). In Concord, Massachusetts, where I work, this inconspicuous herb emerges through the leaf litter by the middle of May and is surrounded by other vernal species characteristic of second-growth vegetation—Canadian lily (*Maianthemum canadense*), wild sarsaparilla (*Aralia nudicaulis*), bunchberry (*Cornus canadensis*), and blue and white species of violets (*Viola* spp.). The erect shoot of the starflower, which reaches a height of about 15 cm when fully expanded, grows from the tip of an underground tuber and forms a small umbrella of seven or eight leaves at the apex. One to three small white flowers may develop and set seed by the end of July, and the entire plant normally begins to senesce in August without producing any new leaves. As with many occupants of the forest understory, seedlings of *Trientalis* are rarely found.

The modest number of leaves formed by this plant belies a much more significant productivity which is hidden from sight. As the canopy of the forest closes in early June, the buried tuber at the base of the parent shoot initiates one to four horizontally growing stems, known as rhizomes or stolons, which penetrate the complicated matrix of roots, stones, and decomposing litter that constitutes the organic layer of the forest floor. In the course of the summer these stems will extend away from the location of the parent, and each may reach a distance of close to a meter. By September the tips of these rhizomes have swollen into starchy tubers with several buds, some incipient roots, and a potential shoot. Sugars and proteins metabolized in the senescing leaves of the parent plant have been transported to this developing organ and stored as reserves. By late fall, all of the connecting rhizomes have withered and disappeared. Next year's generation of daughter plants, although genetically identical, lie isolated up to two meters away from each other, forming a small clone along with the progeny of earlier generations. The growth and proliferation of tubers, usually referred to as vegetative reproduction, is the dominant process by which the plants of the starflower survive and multiply in the forest habitat (Anderson and Loucks 1973).

Throughout the mountains and forests of much of North America, often in the same habitats occupied by the starflower, a second, quite unrelated species proliferates by means of underground vegetative growth. Quaking aspen (*Populus tremuloides*) is a tree which often appears after fire, cutting, or the abandonment of agricultural land. A germinating seed produces a seedling that will develop into a sapling up to a meter high in its first year of growth. The tap root is surrounded by a shallow system of lateral roots extending about 20 cm from the shoot on all sides. In the second year the distribution of growth above and below ground is reversed: the lateral root system expands up to two meters, while the shoot grows less than a meter. Over the next two or three years, sinker roots initiated along the lateral root system penetrate the deeper layers of the soil. Simultaneously, buds forming along the spreading root system send out shoots that emerge at the surface, surrounding the original sapling.

This proliferation of root suckers continues at ever-increasing distances from the parent plant over the next 50 years, creating a grove of aspens all derived from a single seedling. Functional connections within this clone may survive intact for decades, and the death of individual trees may still leave a root system capable of supplying water and nutrients to other living shoots (Barnes 1966). Eventually, through the accidents of time, parts of the expanding clone become physically separated and proliferate as independent systems. Some aspen clones are very big. In aerial photographs of the Fish Lake Basin in Sevier County, Utah, one individual can be seen to occupy an area of 43 ha, with more than 47,000 trees of an average age of 100 years; the age of the entire clone is unknown (Kemperman and Barnes 1976).

Both the starflower and the quaking aspen grow in such a way that, at some point in the life of a successful individual, more than one plant is involved. Distinctions are critical here: What constitutes an individual? Barring somatic mutation for the moment, we may define the genetic individual as all genetically identical members of a clone derived from a single zygote, and call this entity the genet (Sarukhan and Harper 1973). This is the important unit upon which natural selection operates (Harper 1977).

Robert E. Cook completed his graduate education at Yale University in 1973 and taught plant ecology at Harvard University for eight years. During that time he studied the population ecology of three species of wild violets. He is currently Program Director for Population Biology and Physiological Ecology at the National Science Foundation, and has recently accepted an appointment as Associate Professor at Cornell University, where he will serve as Director of Cornell Plantations. He is editing a book on the biology of clonal organisms. Address: Population Biology/Physiological Ecology, National Science Foundation, Washington, DC 20550.

Equally distinct is the physiological individual, the shoot with its associated root system capable of independent survival and therefore independent death (Cook 1979b). Each such shoot may be called a ramet, though when in the process of vegetative reproduction one ramet becomes two is very difficult to pinpoint. In the starflower, daughter ramets become physiologically independent when the parent shoot senesces. Aspen suckers, however, may remain physiologically integrated—capable of exchanging water, nutrients, metabolites, and hormones—for more than 25 years despite the presence of well-developed root systems of their own. Many plant species, of course, do not grow this way. In most annuals and biennials and in many forest trees the genetic individual and the physiological individual remain identical throughout life, although individual branches in a tree and blades in an annual grass will exhibit some independence from each other. Why, ask plant ecologists, do many species characteristically separate these two individuals while other species do not? What is it that distinguishes the biology of a clonal species such as an aspen from the growth of an oak?

Botanists have known for a long time that plant clones can be very successful in nature. Some clones are very large and therefore probably very old (Table 1). The majority of the plant species growing at high latitudes and altitudes, in aquatic environments, within the temperate and boreal forest understory, or in grazed grasslands and areas of fire-disturbed vegetation display some form of clonal growth and proliferation. Two-thirds of the common perennial species of Great Britain are clonal (Abrahamson 1980), and many of our most pernicious weeds are serious economic pests precisely because they grow from underground roots, rhizomes, or buds (Leakey 1981). Plant clones may also be formed above ground by the repeated basal branching often seen in shrubs, by layering (the rooting of branch tips), by tillering (the growth of multiple shoots, as in grasses), by the rooting of surface runners, and by the formation of bulbils or plantlets on aerial shoots.

All these mechanisms share a common resemblance to a branching process: a region of numerous coordinated cells on the body of the parent plant known as a meristem begins to grow out to form a ramet. Eventually an independent root system is developed which supports the growth of this second plant. The creation of a clone can also occur through a purely genetic process known as agamospermy, in which a single cell within an inflorescence begins to grow and differentiate as an asexual seed. In this process the ramet becomes independent when the seed is dispersed; the success of dandelions (*Taraxacum* spp.) and hawkweeds (*Hieracium* spp.) attests to the efficiency of this means of reproduction.

The clonal propensity of plants derives in part from their form of construction, which is very different from that of most terrestrial animals. Nearly all plants grow through the sequential reiteration of a basic structural unit or module consisting of the leaf with its associated axillary bud and a segment of stem that connects it to other such units (Harper 1981). This modular architecture lends a potential immortality to plants, because active or potentially active meristems are always available to continue the exponential production of structural units. Depending on the patterns of tissue differentiation and branching, a vast degree of variation in the form and size of individuals results.

In many ways, therefore, the formation of a clone represents an extension of this process of growth by a genet, and it is quite distinct from the formation of a new sexual or asexual individual through the development of a single cell in a differentiated organ of reproduction. Harper (1981) has argued that it is misleading to apply the phrase "vegetative reproduction" to a process that is fundamentally one of sequential branching of ramets through the elaboration of modular units. Clones enlarge by a process of growth, not reproduction. In some species the clone fragments into unconnected ramets, while in others the connections endure for many years. The starflower and the aspen clearly represent the extremes of this continuum.

Yet the process of clonal growth remains distinctive, and it seems important, while acknowledging that it is a branching process, to distinguish the proliferation of ramets from the ordinary branching seen in an oak tree with its single trunk. What is it about this cloning process that leads to the formation of ramets instead of branches? Two factors stand out: first, all ramets appear to be capable of developing their own independent root systems through the adventitious initiation of roots from stem tissue; second, the proliferation of ramets in clonal growth results in the occupation of horizontal space by a genet. These factors lead to two questions posed by the

Figure 1. Like many herbaceous species, the starflower (*Trientalis borealis*) reproduces primarily through a process of clonal growth, forming subsurface stems that extend horizontally through the litter of the forest floor to produce genetically identical daughter plants. The duration of such clonal connections varies from species to species; in the starflower, daughter plants become physiologically independent when the parent plant dies at the end of the summer. Although starflowers also produce seeds, the rate of germination is low and seedlings are rare. (Photo by Kitty Kohout, courtesy of Hillstrom Stock Photo.)

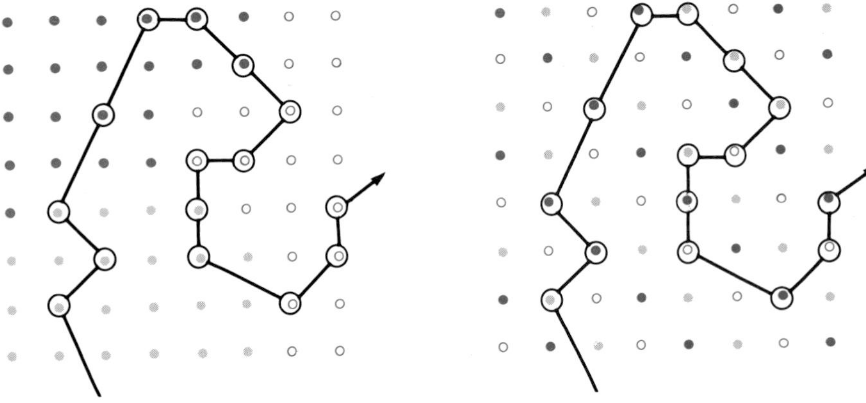

Figure 2. The frequency of outcrossing is plotted for two different hypothetical plant populations, each consisting of three clones with approximately twenty-one ramets. In the population at the left the three clones are separated from one another; in that at the right they are evenly intermingled. The path of the insect pollinator is indicated by the black line, and pollinated ramets are circled. Assuming that the pollinator follows the same path in both populations, the rate of outcrossing would be about twice as high when the clones are intermingled, with nine such events taking place as opposed to only four in the first population.

plant ecologist: What are the ecological advantages of forming a system of independent ramets for plants that can grow horizontally and branch? Is this phenomenon related to the acquisition of territory—to an imperative to control as much ground as possible?

Unfortunately the answers to these questions still elude us, because there are significant practical difficulties in the study of clonal plants. In many species, branching and the formation of ramets occur below ground; tedious excavation is the only means of observation. Existing clones are exceedingly difficult to identify at the surface, although the practiced eye has been able in some instances to identify genetically identical ramets (Harberd 1961; Cahn and Harper 1976). In the case of genets that develop in very tightly organized units, such as bunch grasses, bracken, and aspens (Fig. 3), the form of the clone may suggest a common genetic origin (Oinonen 1967a; Vasek 1980). The self-incompatibility system that makes crossings between ramets sterile in some species of grasses has also been used to identify genets. When this technique was systematically applied to one field population of velvet grass (*Holcus mollis*), only four different genetic individuals could be found, and one stretched over half a mile of rolling hillsides (Harberd 1967). The most promising method of genet identification is electrophoresis, which can distinguish genotypes by detecting the unique constellations of proteins created by highly variable loci (Silander 1979).

Table 1. Large and successful plant clones

Species	*Size*	*Age (yrs)*	*Origin*	*Source*
Ground pine (*Lycopodium complanatum*)	250 m	850	rhizome	Oinonen 1967b
Bracken (*Pteridium aquilinum*)	489 m	1,400	rhizome	Oinonen 1967a
Black spruce (*Picea mariana*)	14 m	330+	layering	Legere and Payette 1981
Red fescue (*Festuca rubra*)	220 m	1,000+	tillering	Harbard 1961
Sheep fescue (*Festuca ovina*)	8.25 m	1,000+	tillering	Harbard 1962
Velvet grass (*Holcus mollis*)	880 m	1,000+	tillering	Harbard 1967
Reed grass (*Calamagrostis epigeios*)	50 m	400+	rhizome	Oinonen 1969
Lily of the valley (*Convallaria majalis*)	83 m	670+	rhizome	Oinonen 1969
Quaking aspen (*Populus tremuloides*)	81 ha	10,000+	root buds	Kemperman and Barnes 1976
Creosote (*Larrea tridentata*)	7.8 m	11,000+	basal branching	Vasek 1980
Huckleberry (*Gaylussacia brachycerium*)	1,980 m	13,000+	rhizome	Wherry 1972

The importance of clones

Based entirely on measurements of ramets alone, plant geneticists have focused on the ways in which the breeding systems of different species may affect estimates of gene flow and the intensity of natural selection. Because each individual shoot is rooted to a particular location for its lifetime, the dispersal of pollen grains and seeds can be very local. When coupled with strong selection, this could rapidly lead to sharp differences in genetic identities over short distances within the population. Any assumption of random mating, although valid for animal species, may be very inappropriate in the case of plants. A further complication is present in dioecious species, where the ratio of male to female individuals and the relative contribution of each sex to reproduction can place strong constraints on the genetically effective size of the population.

Yet correct identification of genets is critical to the evolutionary interpretation of any field data. For instance, in wild sarsaparilla (*Aralia nudicaulis*), both male and female shoots can be found flowering together, and two recent studies have measured the sex ratio of plants in several *A. nudicaulis* populations (Barrett and Helenurm 1981; Bawa et al. 1982). But sarsaparilla grows clonally from underground rhizomes that branch to form perennial ramets over a meter from each other, and populations therefore usually consist of an unknown number of interspersed clones of one sex or another. In both of these studies, the data regarding phenology, sex ratio, and the relative contributions of each sex to reproduction were based entirely on the measurement of ramets whose genetic identity and clonal connections were unknown. As the authors acknowledge, however, both the true sex ratio and the significance of the differences found between the sexes depend entirely on the number of clones present and the degree to which they are physiologically integrated.

Even measurements of gene

flow in bisexual species can be misleading if the populations are composed of an unknown number of clones intermingling to an unknown degree. Estimates of the movement of pollen are often based on the foraging behavior of insect pollinators that tend to fly from the flower of one plant to that of a near neighbor. This has led to measures of gene movement which are very small. For example, it has been estimated on the basis of flight distances of pollinators that the effective size of an interbreeding population of the sweet white violet (*Viola blanda*) is approximately 16 m^2 (Beattie and Culver 1979). Yet *V. blanda* forms extensive clones through the production of stolons. Ramets of different clones may be interspersed, and thus accurate measures of gene flow must distinguish flights between members of the same clone, which do not produce seeds owing to the presence of self-incompatibility, from flights between two different clones. Moreover, the particular pattern of genet interspersion may determine the degree to which gene flow is local or random (Fig. 2).

A knowledge of clonal structure is also potentially important in the study of competition in plant populations. Traditional models of competition assume that interactions among individuals are spread homogeneously through the population. For plants, however, the most intense interference is likely to occur between nearest neighbors. Plants also vary dramatically in size, which greatly affects their competitive ability. Plant ecologists have therefore argued that a better index of potential competition is the position of nearest neighbors combined with some measure of these neighbors' performance. For instance, in populations of dune annuals growing in sand cultures, 69% of the variation in the performance of plants could be accounted for through a knowledge of the position and growth of surrounding individuals (Harper 1977).

This approach has recently been extended to populations of *V. blanda* (Waller 1981; Schellner et al. 1982). Unfortunately, when measurements of ramet growth in eight populations of *V. blanda* were analyzed in the light of the position and performance of neighboring shoots, no indication of interference among shoots could be found; surrounding plants did not appear to influence either the growth or the mortality of individuals. As noted earlier, however, this species forms interspersed clones of connected ramets which remain physiologically integrated for several years, and competitive interactions may well be occurring at the level of the genet. Thus the performance of an individual ramet will depend not only on the constellation of nearest neighbors, but also on the number and performance of other ramets connected with it through underground stolons. The clonal growth habit will tend to alter the nature of interference in cases where interspersed clones function as single physiological organisms, rendering the analysis of competitive interactions among genets very complex indeed.

Figure 3. In the quaking aspen (*Populus tremuloides*), variations in fall coloring and the timing of leaf loss may offer clues to clonal identity. This stand of aspens in the Rocky Mountains appears to be made up of at least four separate genetic individuals, each covering an extensive area. Aspen clones are often very large, and may remain physiologically integrated for more than 25 years. (Photo courtesy of D. W. Johns.)

Clonal identities and demography

Not knowing clonal identities can also be a problem in demographic research. Borrowing concepts and terminology from the demographic study of animal populations, plant ecologists have recently begun mapping and counting the ramet modules in a number of clonal plant species (Harper and Bell 1979). Censuses of all the ramets present in permanent plots have been repeatedly taken to determine the rates at which new shoots appear and older individuals die. As with any demographic study, a census of marked individuals produces birth and death rates from which the growth rate and future size of the population may be predicted.

This research on clonal plants, generally rhizome-forming herbs, has yielded a fairly consistent picture of the dynamics of such populations. The rates of birth and death, when compared with the density of the standing population, reveal a remarkable stability of population levels amid a great flux of ramets. When Noble and his co-workers (1979) conducted monthly censuses of ramet populations in permanent plots of the sand sedge (*Carex arenaria*), they found that population numbers remained surprisingly constant over a period of 20 months despite high rates of birth and death (Fig. 4). One plot contained 160 shoots of *C. arenaria* in January 1975; one year later the number had dropped to 149. Yet during this period 355 new ramets were born and 366 ramets died. Censuses of other plots in the same population revealed a similar pattern. In general, the risk of death is highest for ramets of all species during the season of active growth, and the high rates of mortality are usually preceded by high rates of birth. Similar demographic patterns have been found in the buttercup (*Ranunculus repens*), the hawkweed (*Hieracium pilosella*), and the sweet white violet (*V. blanda*), all of which form clones with stolons or rhizomes (Sarukhan and Harper 1973; Lovett-Doust 1981; Bishop et al. 1978; Newell et al. 1981).

These facts strongly suggest the presence of some regulatory force that determines stable population numbers, and plant demographers have hypothesized that patterns of birth and death may be a response to density mediated by competition for limited resources. This view is stated precisely by Noble and his colleagues (1979): "The seasonal cycles of birth and death rate [are] generally so closely synchronized that it is difficult to believe they are not causally related. It can be surmised that it is the birth of new shoots that generates the density stress responsible for the death of old shoots" (p. 1006).

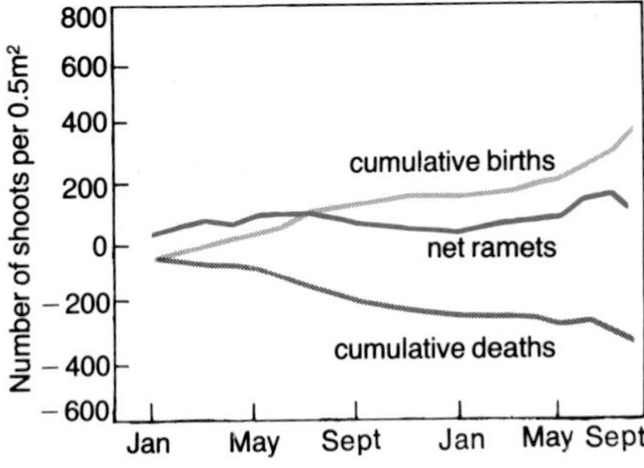

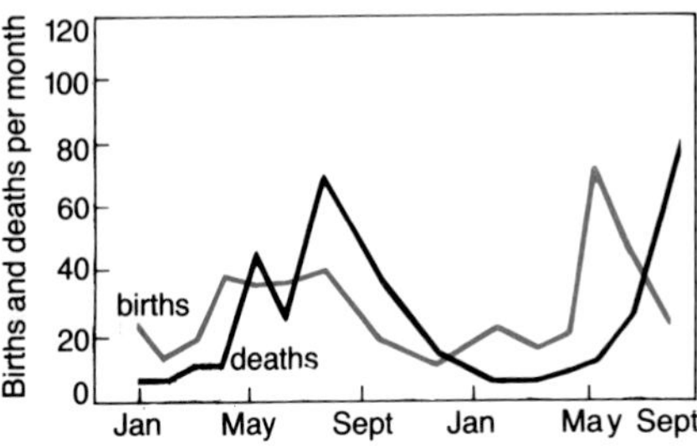

Figure 4. Monthly censuses of a single population of *Carex arenaria* ramets carried out over a 20-month period from January 1975 to August 1976 showed a striking balance between the cumulative number of births and deaths (*above*). High mortality rates for the season of active growth were generally preceded by high birth rates (*below*), suggesting the presence of a regulatory force that keeps population levels stable. (After Noble et al. 1979.)

Such an interpretation rests on the assumption, usually true in animal populations, that the censused individuals are independent units; they are born, compete with each other for resources, and die as a consequence of such competition. Is this assumption valid for clonal species of plants?

Clearly there is a strong dependency among ramets of a clone, especially between parent and daughter generations. What appears at the surface to be a birth is actually the growth of a branching meristem initiating new leaves, either from a rhizome, a root, or a stolon. This new ramet will depend on photosynthates, water, and minerals from the parent until its own root and shoot systems begin to function. The behavior of stolons and rhizomes can also be strongly influenced by hormones produced by the parent plant, as well as by nutritional factors (Kumar and Wareing 1972).

Different species have evolved different patterns of ramet dependency. A study of two forest herbs, Clintonia (*Clintonia borealis*) and whorled wood aster (*Aster acuminatus*), revealed a marked contrast between the two in the physiological relations of mature ramets connected by rhizomes through the base of the senesced parent plant (Ashmun et al. 1982). Photoassimilates identified by labeled carbon moved readily through the entire rhizome system of *C. borealis*, whereas the absence of such circulation in *A. acuminatus* indicated that the ramets were functionally independent at maturity. These physiological relations are believed to be regulated through the interactions between metabolic sinks (storage tissues, growing meristems, and developing fruits) and sources of assimilation (leaves and senescing parent tissues). Additional work with *R. repens* (Ginzo and Lovell 1973), *C. arenaria* (Tietema 1980), and *V. blanda* (Newell 1982) indicates continuing physiological relations among ramets of species that have been studied demographically.

Finally, the conserved mineral and carbon resources in senescing older tissues are known to be redistributed to younger, actively growing tissue (Chapin 1980). In many species, therefore, clones function as integrated developmental systems and the performances of related ramets are not independent.

Thus the regulation of ramet populations in clonal species may be given a different interpretation. Rather than a balanced process of birth and death mediated by competition, the flux of ramets may in part represent the growth and development of daughter ramets followed by the senescence of older parent shoots and the redistribution of metabolic resources through clonal connections to younger growing tissues. Leaf and shoot senescence is part of a genetically determined and hormonally mediated program of development, although it can be influenced by environmental factors such as shortage of light or mineral nutrients. At the level of the genet, then, the regulation of ramets actually represents the movement of the biomass of the clone in space, controlled by a number of developmental processes: branch formation, interactions between metabolic sinks and sources of assimilation, and tissue senescence. If each parent shoot is replaced, on average, by a single daughter, ramet density will appear to be very stable despite high rates of "birth" and "death." Distinguishing internal regulation from density mediated by competition will require a knowledge of the developmental ecology of natural clones.

Clonal development in *V. blanda*

The sweet white violet, a common perennial in the second-growth deciduous forests of eastern Massachusetts, produces clones through the production of several axillary stolons, each of which grows just below the surface of the litter to form a single daughter (Fig. 5). For the last six years I and Thomas Ducker have been studying the biology of genets of this species. We have taken two approaches: first, seedlings are marked at germination, and a census is taken every two weeks to measure growth, reproduction, and mortality; second, entire populations of *V. blanda* have been mapped and excavated to measure the spatial development of clones.

Each spring we identify a new cohort as it germinates during May and early June, and mark individual seedlings with small numbered flags. Although the age of the seeds at the time of germination is unknown (Cook 1980a), marking a set of seedlings from "birth" permits the development of individuals to be charted and the rates of fecundity and mortality in the population to be calculated from a common point. At each later census the size of the survivor is measured by adding together the widths of all leaves that have been produced. This nondestructive measure is closely correlated with dry weight (r = 0.94; n = 69). An estimate of damage from grazing is made and the production of seed capsules is recorded. After growth has ceased each year, the paths of stolons are carefully tracked and the positions of the new ramets are marked. In this way an entire genet can be mapped from the germination of the seed until the last ramet, or the patience of the investigator, succumbs.

Deep in the forest understory, violet seedlings do not fare very well in their first year. Levels of light are very low, the soil dries out rapidly as water is consumed by the roots of surrounding shrubs and trees, and the forest floor is populated by a wide array of herbivores such as slugs and aphids. Even the most vigorous seedling may produce a rosette of only two or three leaves, each no larger than a quarter, and these are frequently grazed to the petiole. Not surprisingly, mortality rates are very high for these small individuals, as is the case with plants in general (Cook 1979a). Out of each cohort of 100 seedlings marked in May, more than 64 will die before the following spring. Even this risk varies greatly over short distances of a meter or two, and it is strongly influenced from year to year by the pattern of local rainfall. Once a violet survives into the second year, its chance of death declines below 50% as it grows larger, and size rather than age determines the risk (Cook 1980b). For instance, mature ramets whose summed leaf width is less than 10 cm (about three leaves) have a 22% chance of death, whereas larger plants have a risk of less than 4%. Larger plants have deeper roots, more nutritional reserves, and a

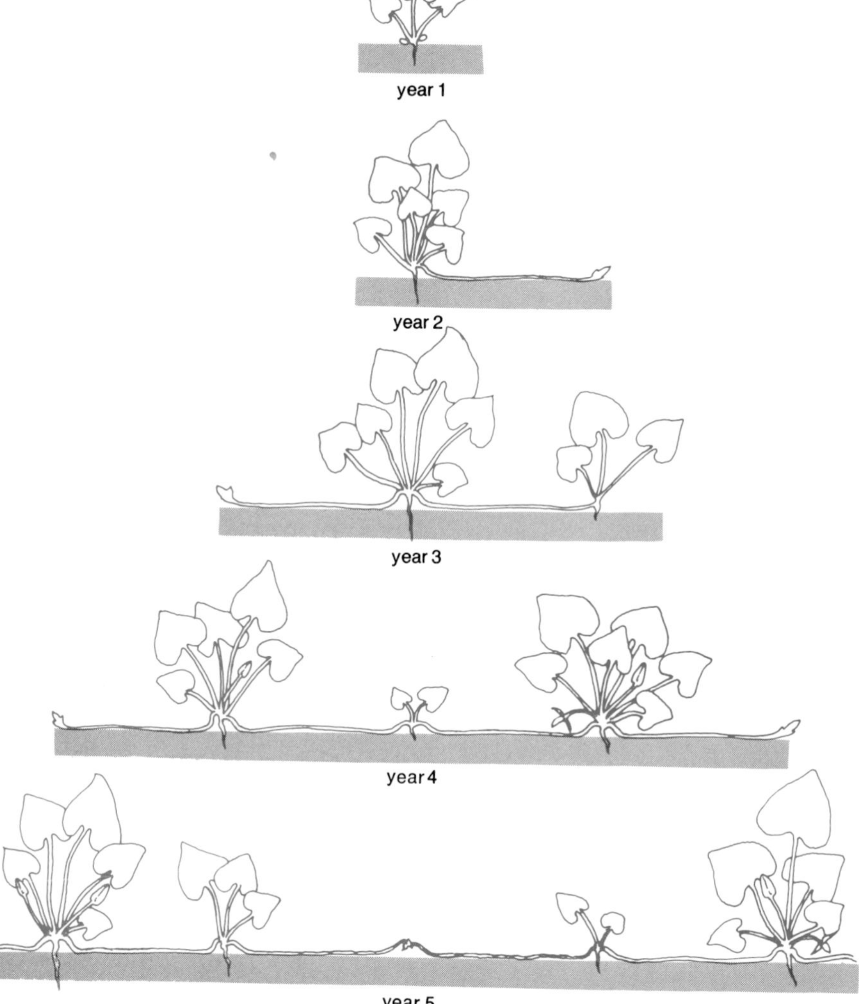

Figure 5. This schematic sketch of the life cycle of a *Viola blanda* ramet shows the way in which the distribution of growth shifts to younger generations as the genet moves slowly across the forest floor, continually acquiring new locations. After germinating, the original ramet may produce three or more leaves in the first year. In the second year, if growth is vigorous, one or more stolons are produced, which form resting buds in September. Daughter ramets emerge in the spring of the third year, and themselves produce stolons, self-fertilized seed capsules (shown both closed and open), and possibly flowers in the fourth year. In the fifth year a second generation of ramets develops, and the original ramet fails to emerge following a year of greatly reduced vigor.

greater potential for regrowth following disturbance.

It is these larger individuals that begin to produce stolons in their second year, long before any flowers are formed. The axillary buds in the first three or four leaves to emerge in spring initiate horizontal growth in June, extending through the complex matrix of litter during the summer months. A very vigorous seedling ramet may produce a seed capsule in this or a later year, but such an occurrence is rare; most seeds will be produced by its daughter ramets. When the overwintering buds of the daughter ramet emerge in the third year, leaf production is often very vigorous, and a second generation of stolons will be produced in the same season. Self-fertilized seed capsules are frequently initiated after stolon production, and in some daughter ramets the axillary buds at the tip of the overwintering shoot will emerge as brilliant white flowers in the spring of the clone's fourth year (*see cover*). These flowers can be cross-fertilized by pollinating insects; flowers produced later in the summer are inconspicuous, without petals, and are the result of self-fertilization.

With this knowledge of both ramet and genet patterns of development, we are able to ask an important question about previous work on clonal species. Will demographic studies of ramets alone yield the same understanding of mortality in the population as demographic studies of genets? To obtain an answer, we pooled data based on over 2,220 seedlings from cohorts that germinated in 1976, 1977, and 1978, and calculated the rate of mortality by age from germination both for ramets alone and for genets (Fig. 6). As might be expected, the mortality rate first declines with age (and thus size) in a parallel fashion for both ramets and genets, primarily because ramets and genets are the same entity until the first daughter stolons are produced in the second or third year. By the fourth year, however, the risk of death for parent ramets has risen dramatically, while the mortality rate for genets remains relatively low. In the fifth year the death rate for ramets is still high, but that for genets continues to decline. Although a ramet may live for only four or five years, the clone survives much longer through the establishment of daughter shoots. Thus demographic studies of ramets alone give a misleading picture of the fate of genets.

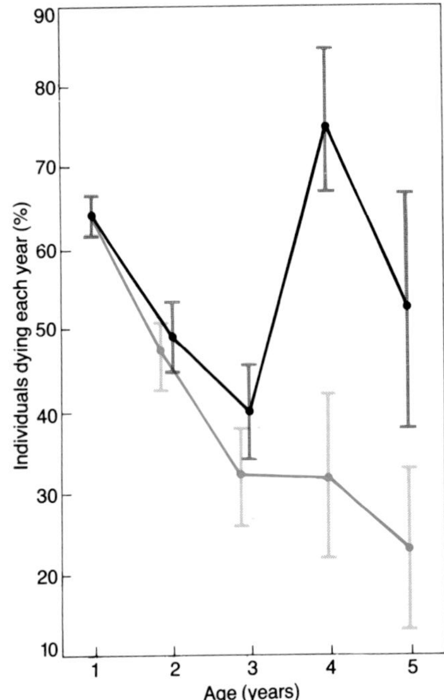

Figure 6. When the mortality rates for *V. blanda* ramets (*black line*) and genets (*colored line*) are analyzed separately, it is apparent that demographic studies of ramets alone may be misleading. Although the mortality rate for both ramets and genets first declines with age as the individuals increase in size, the risk of death for ramets is already slightly higher than that for genets in the second and third years, when stolons begin to be produced, and rises dramatically by the fourth year. By contrast, the death rate for genets continues to decline throughout the five-year period.

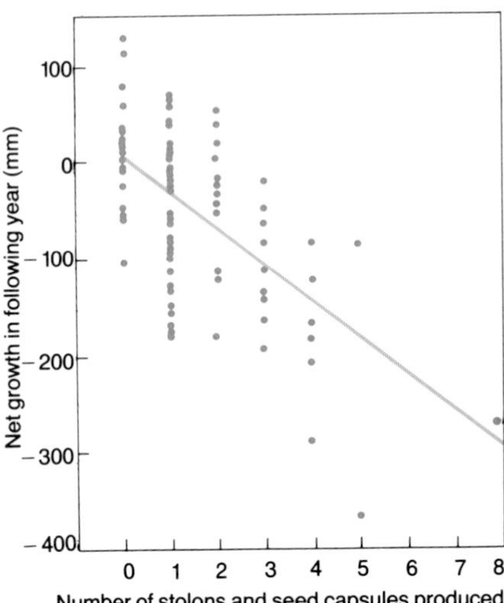

Figure 7. When the net growth of *V. blanda* ramets was calculated by subtracting size in the second year (measured by adding the widths of all leaves produced) from size in the first year, a strong negative correlation was found between the production of stolons and seed capsules in the first year and growth in the following year. Plants that produced no stolons or seeds were about the same size in the next year, whereas any reproduction led to a decline in future growth.

The death of a parent ramet at the age of four or five is a curious phenomenon to observe in the field. Sometimes the individual will thrive luxuriantly in one growing season, producing several stolons and a seed capsule or two, but will simply fail to emerge the following year. Careful excavation reveals the dead husk of the original ramet in the litter. More often, a vigorous shoot one year will initiate growth very late the following spring, producing only a single very small leaf; death often follows the slightest grazing or drought. Thus a weak and declining parent ramet will be surrounded by three or four very healthy daughter plants, which may themselves have produced granddaughter ramets. These observations suggest that reproduction in one year may cost the parent ramet its future survival through a reduction in growth, with death following as a consequence of small size.

To explore this possibility, Ducker (1981 diss.) examined the net growth of ramets from one year to the next and plotted this measure of vigor against the sum of the number of stolons and seed capsules produced by the shoot in the first year. Figure 7 shows the striking negative relation between the production of stolons and seeds in one year and the net growth of the ramet the following year. He also performed a partial correlation analysis to separate the effects of stolon production from those of producing seed capsules. Holding the effects of seed production constant, stolon production is negatively correlated with net production ($r = -0.40$, $p < 0.001$); likewise, holding the effects of stolon production constant, seed production is negatively correlated with net production ($r = -0.29$; $p < 0.001$). There appears to be a cost of reproduction, as predicted by demographic theory, but is the growth of stolons the same kind of "reproduction" as the formation of seeds?

The natural history of a clone

From the life cycle of a violet ramet, we can reconstruct the broader history of a clone. Surviving into its second or third year, the individual ramet begins to produce stolons which themselves become vigorous and prolific (see Fig. 5). In the third or fourth year the focus of growth shifts to younger generations, while the original ramet declines in size. By the fifth year the ramet is no longer producing green tissue, but the clone thrives through the production of daughter ramets. Much of the "cost" of ramet reproduction actually represents the redistribution of genet growth.

This longer ontogeny of the clone can be interpreted in the context of the seasonal cycle of ramet growth and dormancy. During the month of August, when the apical meristems of growing stolons initiate overwintering buds with their stores of nutrients, the leaves of parent ramets begin to senesce, and the resulting metabolic products are transported below ground. The relative allocation of these metabolites between the crown of the parent and the developing buds of the daughter ramets is unknown. In the spring these dormant buds with their differentiated leaves and floral structures begin growing more than a month before the closing of the forest canopy. Remaining stores of starch and mineral nutrients support this early growth, and the actively growing meristems probably compete for these resources. Daughter ramets appear to have the edge.

In effect the clone moves across the forest floor. Any one location is occupied for about five years, but the genet continually acquires new locations by the proliferation of daughter ramets. This spatial development now becomes of considerable interest. How do violet clones explore and occupy habitat?

In an attempt to answer this question, three small populations of *V. blanda* were located and the positions of all ramets were mapped (Ducker 1981 diss.). By careful excavation, all extant clonal connections were identified. The length of each stolon was measured and the angles between the positions of two or more daughter ramets and the parent were obtained from the map. Finally, the degree of dispersion among ramets was calculated using the Hopkins index (Goodall and West 1979), which compares the distribution of distances between a randomly chosen point and the nearest plant with the distribution of distances between a randomly chosen plant and its nearest neighbor.

The length of individual stolons is quite variable and appears to follow no fixed rule. The average stolon is approximately 21 cm long, but stolons may range from 6 to 35 cm in length. One might guess that the angle between daughter ramets would be approximately 140°—the angle between successive axillary buds. In fact, when plants with two, three, or four stolons were tested against this expectation, the angles between daughters were found to be significantly smaller (t-test, $p < 0.01$). In the three populations taken as a whole, over half of all angles were less than 60°. Thus daughter ramets growing from the same parent are much closer together than would be predicted on the basis of the position of the original axillary bud. Stolons appear to grow in such a way that daughter ramets cluster together.

An analysis of ramet dispersion confirmed this suspicion. In two of the three populations, shoots of *V. blanda* were closer together than would be predicted from the Hopkins index, and the closeness of ramets in the third population fell just short of being statistically significant. Figure 8 is a map of one of these populations, showing the tendency of new ramets to cluster together. Two separate processes appear to contribute to the degree of clustering: first, daughter ramets from the same parent are often located close to one another; second, stolons from different parents also converge on the same location, establishing small clumps of ramets. It is difficult not to find the explanation for this behavior in the effects of habitat conditions on the growth of stolons. Over 20 years ago Kershaw

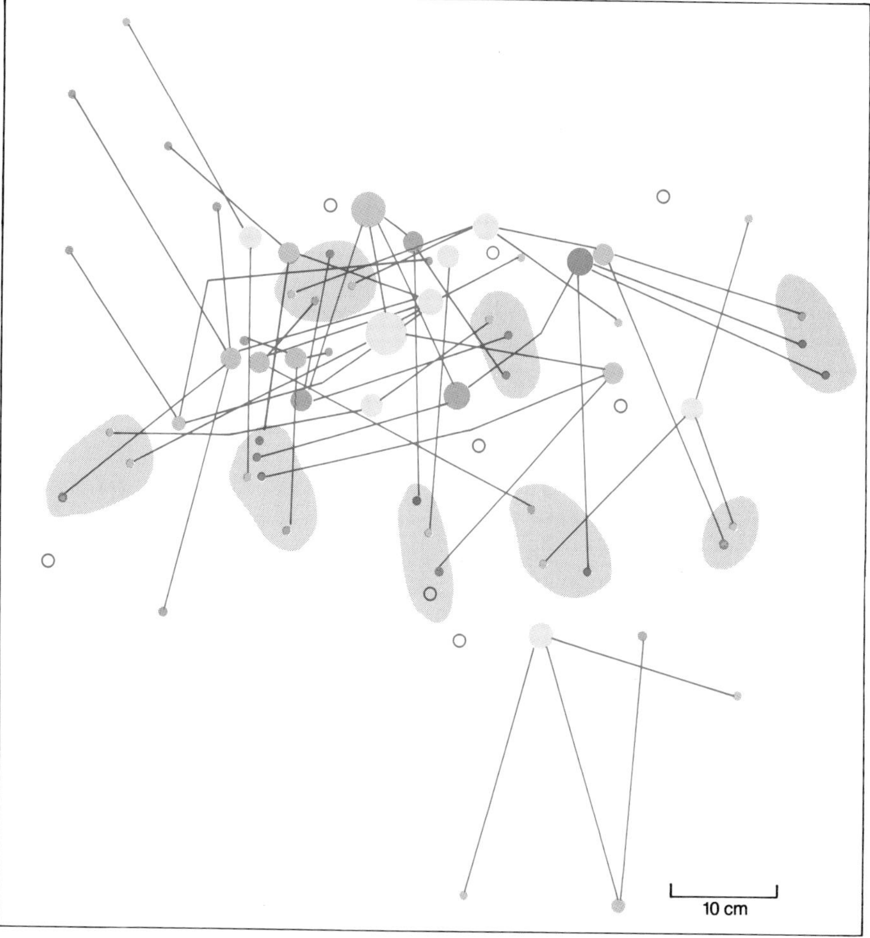

Figure 8. In this map of all ramets in a single population of *V. blanda*, the parent ramets are connected to daughter ramets by lines. The older the generation, the lighter the color of the dot; the size of the dot is correlated with the number of generations of daughter ramets. Shaded areas indicate clumps of ramets formed at the same time, and circles standing alone represent individuals which have yet to form any daughter plants. The fact that both daughter ramets from the same parent and ramets from different parents tend to converge on the same locations may be related to the presence of high concentrations of resources at these sites.

(1958, 1962) excavated populations of two grasses (*Agrostis tenuis* and *Calamagrostis neglecta*) and a sedge (*Carex biglowii*) and found similar widely spaced aggregations of ramets emerging from stolons that could be traced back to different parents. He suggested that stolons are capable of detecting gradations of nutrients in the soil and that they establish ramets in locations with higher concentrations.

In the forest understory where violet clones grow, the observer is greatly impressed with the heterogeneous matrix of half-buried rocks, tree roots, decomposing sticks, and litter, all strewn across the variable topography of the forest floor. Patches of resources for clonal herbs may be available only transiently, and may be rapidly depleted by competing species. In order to survive in such a habitat, a genet may have continually to discover and exploit new patches of resources. Violet clones appear to explore the surface opportunistically, maximizing the acquisition of such patches. Growing stolons may turn after leaving the parent, often rather abruptly, and the establishment of the daughter ramet is frequently associated with a piece of decomposing wood. In the deciduous forest the action of nitrogen-fixing bacteria may render such sites high in nitrogen (Roskowski 1980). Growing stolons are able to initiate adventitious roots over their entire length, and proliferations of roots are often seen in the vicinity of rotting organic matter and decaying twigs. Thus the paths followed by a violet clone may be as variable as the meanderings of a hungry amoeba.

Implications of clonal growth

Clearly, a knowledge of the genet structure of natural populations can greatly alter the interpretation of data which are based on the biology of ramets alone. Gene flow and natural selection, variation in life history, population regulation, competition, and predation are all current ecological and evolutionary problems affected by the precise clonal identity of ramets and the physiological relations among them. Future studies must attempt to identify the number of genets in plant populations, as well as their spatial configuration.

In particular, density mediated by competition has been logically considered responsible for the apparent regulation of birth and death among ramets of clonal species. Our work with *V. blanda* suggests that developmental processes may contribute significantly to this appearance of stability through the simultaneous senescence of older tissue and development of younger ramets. The clone may be an integrated, growing organism in which the flux of shoots at the surface is akin to the senescence of lower leaves on a plant and the growth of the apex as conserved nutrients are transported to it. External factors, including competition among ramets for available resources, can affect this flux, but a major determinant is a program of ramet development controlled within the clone.

The occupation of horizontal space, perhaps the most characteristic aspect of clonal growth, may be greatly affected by features of the habitat that cause the position of ramets to be highly variable. Many species, such as *V. blanda*, appear to grow opportunistically, influenced by a sphere of stimuli surrounding the ramet. In a habitat where resources are distributed in patches, the capacity to detect gradations of nutrients, initiate adventitious roots, and establish shoots may greatly increase the success of the clone.

As Dobzhansky has observed, nothing in biology makes sense except in the light of evolution (1973), and an understanding of the benefits of clonal growth and proliferation which underlie natural selection must await further studies of the biology of different clonal species. These benefits will include an enhanced capacity to harvest patchy resources (Harper 1977), the competitive ability to invade other clones and resist invasion by seedlings (Ovington 1953), and the lowering of the probability of genet extinction by spreading the risk of death among a number of independent ramets (Cook 1979b; Charlesworth 1980). Although one might construct stories supporting one or another of these explanations based on the available data, such accounts are likely to evaporate rapidly as we acquire a deeper knowledge of the natural history of genetic and physiological individuals living in clonal plant populations.

References

Abrahamson, W. G. 1980. Demography and vegetative reproduction. In *Demography and Evolution in Plant Populations*, ed. O. T. Solbrig, pp. 89–106. Oxford: Blackwell Scientific Publications.

Anderson, R. C., and O. L. Loucks. 1973. Aspects of the biology of *Trientalis borealis* Raf. *Ecology* 54:798–808.

Ashmun, J. W., R. J. Thomas, and L. F. Pitelka. 1982. Translocation of photoassimilates between sister ramets in two rhizomatous forest herbs. *Ann. Bot.* 49:403–15.

Barnes, B. V. 1966. The clonal growth habit of American aspens. *Ecology* 47:439–47.

Barrett, S. C. H., and K. Helenurm. 1981. Floral sex ratio and life history in *Aralia nudicaulis* (Araliaceae). *Evolution* 35:752–62.

Bawa, K. S., C. R. Keegan, and R. H. Voss. 1982. Sexual dimorphism in *Aralia nudicaulis* L (Araliaceae). *Evolution* 36:371–78.

Beattie, A. J., and D. C. Culver. 1979. Neighborhood size in *Viola*. *Evolution* 33:1226–29.

Bishop, G. F., A. J. Davy, and R. L. Jeffries. 1978. Demography of *Hieracium pilosella* in a Breck grassland. *J. Ecol.* 66:615–29.

Chapin, F. S. III. 1980. The mineral nutrition of wild plants. *Ann. Rev. Ecol. Syst.* 11:233–60.

Cahn, M. G., and J. L. Harper. 1976. The biology of leaf mark polymorphism in *Trifolium repens L.* 1. Distribution of phenotypes at a local scale. *Heredity* 37:309–25.

Charlesworth, B. 1980. The cost of meiosis with alternation of sexual and asexual generations. *J. Theor. Biol.* 87:517–28.

Cook, R. E. 1979a. Patterns of juvenile mortality and recruitment in plants. In *Topics in Plant Population Biology*, ed. O. T. Solbrig, S. Jain, G. T. Johnson, and P. Raven, pp. 207–31. Columbia Univ. Press.

______. 1979b. Asexual reproduction: A further consideration. *Am. Nat.* 113:769–72.

______. 1980a. The biology of seeds in the soil. In *Demography and the Dynamics of Plant Populations*. ed. O. T. Solbrig, pp. 107–27. Oxford: Blackwell Scientific Publications.

______. 1980b. Germination and size-dependent mortality in *Viola blanda*. *Oecologia* 47:115–17.

Dobzhansky, T. 1973. Nothing in biology makes sense except in the light of evolution. *Am. Biol. Teacher* (March):125–29.

Ducker, T. P. 1981. The population biology of two clonal violet species. Ph.D. diss., Harvard University.

Ginzo, H. D., and P. H. Lovell. 1973. Aspects of the comparative physiology of *Ranunculus bulbosus* and *R. repens*. II.Carbon dioxide assimilation and distribution of photosynthates. *Ann. Bot.* 37:765–76.

Goodall, D. W., and N. E. West. 1979. A comparison of techniques for assessing dispersion patterns. *Vegetatio* 40:15–27.

Harberd, D. J. 1961. Observations on population structure and longevity of *Festuca rubra L. New Phytol.* 60:184–206.

______. 1962. Some observations on natural clones in *Festuca ovina*. *New Phytol.* 61:85–100.

______. 1967. Observations on natural clones of *Holcus mollis*. *New Phytol.* 66:401–08.

Harper, J. L. 1977. *The Population Biology of Plants.* Academic Press.

———. 1981. The concept of population in modular organisms. In *Theoretical Ecology*, ed. R. M. May, 2nd ed., pp. 53–77. Princeton Univ. Press.

Harper, J. L., and A. D. Bell. 1979. The population dynamics of growth form in organisms with modular construction. In *Population Dynamics*, ed. R. M. Anderson, B. D. Turner, and L. R. Taylor, pp. 29–52. Oxford: Blackwell Scientific Publications.

Kemperman, J. A., and B. V. Barnes. 1976. Clone size in American aspens. *Can. J. Bot.* 54:2603–07.

Kershaw, R. A. 1958. An investigation of the structure of a grassland community. I. Pattern of *Agrostis tenuis. J. Ecol.* 46:571–92.

———. 1962. Quantitative ecological studies from the Landmannahellir, Iceland. II.The rhizome behavior of *Carex bigelowii* and *Calamogrostis neglecta. J. Ecol.* 50:171–79.

Kumar, D., and P. F. Wareing. 1972. Factors controlling stolon development in the potato plant. *New Phytol.* 71:639–48.

Leaky, R. R. 1981. Adaptive biology of vegetatively regenerating weeds. *Adv. Appl. Biol.* 6:57–90.

Legere, A., and S. Payette. 1981. Ecology of a black spruce (*Picea mariana*) clonal population in the hemiarctic zone, northern Quebec: Population dynamics and spatial development. *Arct. and Alp. Res.* 13:261–76.

Lovett-Doust, L. 1981. Population dynamics and local specialization in a clonal perennial (*Ranunculus repens*).I.The dynamics of ramets in contrasting habitats. *J. Ecol.* 69:743–55.

Newell, S. J. 1982. Translocation of ^{14}C-photoassimilate in two stoloniferous *Viola* species. *Bull. Torrey Bot. Club* 109:306–17.

Newell, S. J., O. T. Solbrig, and D. T. Kincaid. 1981. Studies on the population biology of the genus *Viola*. III.The demography of *Viola blanda* and *V. pallens. J. Ecol.* 69:997–1016.

Noble, J. C., A. D. Bell, and J. L. Harper. 1979. The structural demography of rhizomatous plants: *Carex arenaria* L. I.The morphology and flux of modular growth units. *J. Ecol.* 67:983–1008.

Oinonen, E. 1967a. The correlation between the size of Finnish bracken (*Pteridium aquilinum* [L.] Kuhn.) clones and certain periods of site history. *Acta For. Fenn.* 83:1–51.

———. 1967b. Summary: Sporal regeneration of ground pine (*Lycopodium complanatum* L.) in southern Finland in the light of the dimensions and age of its clones. *Acta For. Fenn.* 83:76–85.

———. 1969. The time table of vegetative spreading in the Lily-of-the-Valley (*Convallaria majalis* L) and the Wood Small-Reed (*Calamagrostis epigeios* [L.] Roth.) in southern Finland. *Acta For. Fenn.* 97:1–35.

Ovington, J. D. 1953. A study of invasion by *Holcus mollis* L. *J. Ecol.* 41:35–52.

Roskowski, J. P. 1980. Nitrogen fixation in a northern hardwood forest in the northeastern United States. *Plant and Soil* 54:33–44.

Sarukhan, J., and J. L. Harper. 1973. Studies on plant demography: *Ranuculus repens* L., *R. bulbosus* L. and *R. acris* L. I. Population flux and survivorship. *J. Ecol.* 61:675–716.

Schellner, R. H., S. J. Newell, and O. T. Solbrig. 1982. Studies of the population biology of the genus *Viola*. IV.Spatial pattern of ramets and seedlings in stoloniferous species. *J. Ecol.* 70:273–90.

Silander, J. 1979. Microevolution and clone structure in *Spartina patens. Science* 203: 658–60.

Tietema, T. 1980. Ecophysiology of the sand sedge, *Carex arenaria* L. II.The distribution of ^{14}C-assimilates. *Acta Bot. Neerl.* 29:165–78.

Vasek, F. C. 1980. Creosote bush: Long-lived clones in the Mohave Desert. *Am. J. Bot.* 67:246–55.

Waller, D. M. 1981. Neighborhood competition in several violet populations. *Oecologia* 51:116–22.

Wherry, E. T. 1972. Box-huckleberry as the oldest living protoplasm. *Castanea* 37:94–95.

Jeremy B. C. Jackson
Terence P. Hughes

Adaptive Strategies of Coral-Reef Invertebrates

Coral-reef environments that are regularly disturbed by storms and by predation often favor the very organisms most susceptible to damage by these processes

The life history of any organism is described by the sequence of developmental stages from birth to death, and by the schedule of associated vital processes, including growth, reproduction, and mortality. Since the resources required for these processes are never in infinite supply, and since environments are always variable, an organism's life history is, to some extent, a compromise. For example, if an animal concentrates on increasing its ability to survive by investing most of the energy at its disposal in building a sturdy, well-defended skeleton, then it cannot also invest maximally in fecundity by producing vast numbers of eggs. Resources must be budgeted to potentially competing processes. Thus, an organism's schedule of life-history events can be thought of as the consequence of a pattern of investment—an investment "strategy"—that has been maintained over evolutionary time by natural selection.

One of the central problems of evolutionary ecology is to explain the adaptive basis and origin of various investment patterns and to demonstrate their association with particular environmental conditions. This paper will describe how the characteristic patterns of life histories among different morphologies of sessile (attached, immobile) coral-reef invertebrates are related to the distributions of these animals on tropical reefs.

Most ideas about the ecology and life history of sessile organisms on the sea floor are based on studies of animals, such as barnacles, mussels, and oysters, that dominate the intertidal zone, estuaries, or other marginal marine environments. These animals have simple life cycles, termed aclonal; numbers of individuals in a population increase only through the sexual production of larvae, which typically are dispersed in the plankton before settlement (see Fig. 1). Individual animals are easily distinguished and counted, and there is ordinarily a genetically determined upper limit to body size, which may vary somewhat in different environments. In sufficiently favorable habitats, all surviving individuals grow to about this upper size limit, produce varying numbers of gametes or larvae, and eventually senesce and die. Such a life cycle is no different in its essential attributes from that of more familiar animals like humans or fruit flies.

Many marine environments, however, particularly coral reefs, are dominated by animals such as sponges, corals, and bryozoans whose life cycles are more complicated (Jackson 1977; Hughes 1984). These organisms are constructed of interconnected assemblages of genetically identical modules, such as coral polyps or bryozoan zooids, and are therefore termed clonal (Fig. 1). Single modules, like aclonal individuals, can often survive on their own, as is always the case immediately after larval settlement and before modular budding begins. Although the size of each module may be limited as in aclonal animals, the size of a collection of modules, such as a coral or bryozoan colony, is usually not intrinsically limited. The few exceptions are some mound-shaped and erect species that may suffer geometric constraints or accrue a burden of supportive tissues as they grow (e.g., Cheetham and Hayek 1983).

Sometimes, groups of modules from the same clone may become separated from one another by asexual division or through mortality of the modules between them (Hughes and Jackson 1980; Highsmith 1982). This tendency of clonal animals to live their lives in disconnected bits and pieces of their former selves greatly complicates the study of their demography. Unless clonal populations are observed repeatedly, it is frequently impossible to determine whether each physically separate colony is genetically distinct, or whether a single genotype is represented by more than just one colony.

We have studied the dynamics of populations of clonal corals, sponges, and bryozoans by revisiting the same colonies at regular intervals for almost 10 years. Our long-term measurements show that partial colony mortality and fission may increase numbers of colonies as much as or more than sexual reproduction does, and that clones can persist and increase in local abundance for long periods. In contrast to aclonal organisms, the clonal animals we have studied also lack any signs of physiological deterioration caused by senescence, except for turnover of individual modular components.

Jeremy B. C. Jackson, staff biologist and Marine Coordinator at the Smithsonian Tropical Research Institute in Panama, received his education in biology and geology at George Washington University and at Yale University (Ph.D. 1971). Terence P. Hughes attended Trinity College, Dublin, and Johns Hopkins University (Ph.D. 1984) and is currently a postdoctoral fellow at the University of California, Santa Barbara. *The authors wish to acknowledge the support of the NSF and of the Discovery Bay Marine Laboratory, which is operated by the University of the West Indies in Jamaica.* *Address for Dr. Jackson: Smithsonian Tropical Research Institute, Apartado 2072, Balboa, Republica de Panamá.*

Ecology of clonal organisms

Besides the greater complexity of clonal life cycles, two basic ecological differences distinguish clonal from aclonal benthos (bottom-dwelling organisms). First, it appears from the differences in the rates of positive and negative growth between colonies of clonal coral-reef invertebrates that their schedule of life-history events is closely tied to their size as well as to their age. The ability to overgrow or resist overgrowth by neighbors, survive predation, regenerate in response to injury, or reproduce sexually, all increase with colony size. A large colony or clone that has decreased in size is more likely to die and less likely to reproduce than when it was larger; thus, it may demographically resemble other small colonies, whatever their age, more than its contemporaries that have remained large. In contrast, life-history parameters of aclonal animals vary with size to a much more limited extent, because of generally smaller limits in body size, lower tolerance of injury or fission, and shorter life span. Since most demographic theory has been based solely on age-dependent models, the extreme size dependence of the life histories of clonal animals is an important problem requiring new demographic techniques (Hughes 1984; Caswell, in press).

That life-history parameters depend on colony size is clearly demonstrated by a population of the foliaceous (platelike) coral *Leptoseris cucullata* from our study site off Discovery Bay, Jamacia, at 35 m depth. Figure 2 shows the probabilities of transitions among 4 size classes by more than 100 colonies of this coral over 3 years. Consider first the fates of the corals in the smallest size class; 100% of these are accounted for by growth into the next two larger size classes (38%), by staying in the same size class (19%), and by death (43%). In contrast, the colonies in the largest size class increased to 317% of their original number, 67% of them remaining in the same size class while producing 250% of their original number in smaller colonies. Sex was not the primary means of producing new colonies in this population: over the 3 years of the study, 38 colonies formed by fission, compared to only 16 new colonies produced by the recruitment of larvae, which is the rare successful settlement of sexually produced larvae on the substratum. Similar relationships of size to survival and to fission have been shown for all corals, bryozoans, and other benthic clonal invertebrates for which adequate demographic data are available (e.g., Connell 1973; Bak et al. 1981; Sebens 1982; Winston and Jackson 1984).

The onset and continued capacity for sexual reproduction are also directly related to colony size in clonal animals. As an apparent response to initially high mortality, small colonies grow proportionately faster than larger colonies and do not begin to reproduce sexually until they reach a certain minimum colony size, presumably because of a limitation of resources (Connell 1973; Hughes and Jackson, in press; Jackson and Wertheimer, in press). Thus, following a decrease in size caused by partial mortality, small portions of formerly larger colonies may lose their capacity to reproduce sexually until they grow back again to a larger size (Wahle 1983).

Since few clones appear to die of old age, and the chances of survival and of successful reproduction both increase with colony size, undisturbed populations may become dominated by relatively few, long-lived, highly fecund individuals (Jackson, in press). Indeed, many reefs today are dominated by corals up to thousands of years old; and corals probably lived many times longer when sea levels were more stable (Potts 1984). There are numerous exceptions, but aclonal organisms on coral reefs are generally rather ephemeral compared to clonal species.

The second basic ecological difference between clonal and aclonal benthic animals is in their patterns of distribution. Just as one might expect, given their chances of long-term survival, clonal species are generally more abundant and diverse in more stable and predictable environments than are aclonal species (Jackson, in press). For example, the surfaces of sediments, which are physically less

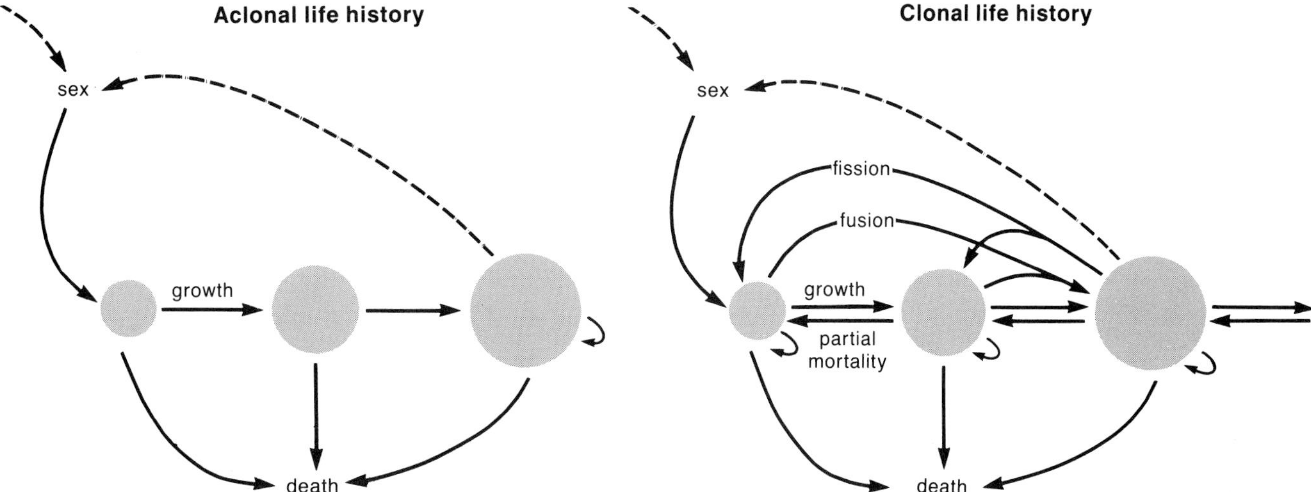

Figure 1. The two types of life histories of sessile organisms are represented schematically in terms of individual size, with the arrows indicating all possible transitions between size classes and mortality. Aclonal organisms, such as oysters or humans, have relatively simple life histories, increasing their numbers only by sexual reproduction and their size only by growth, and having an upper size limit. In contrast, colonies of clonal organisms, such as corals and bryozoans, can increase their numbers by fission and their size by fusion, can decrease their size, either by fission or by partial mortality, and can grow indefinitely. (After Hughes 1984.)

stable than most rocky substrata, are dominated by clams, worms, and other aclonal animals. In contrast, hard substrata are typically dominated by clonal groups such as sponges, corals, and bryozoans, except in environments that are highly variable physically, such as the intertidal zone and estuaries. Physical conditions in tropical seas are generally much less variable than in the temperate zone, and diversity and abundance of clonal taxa increase relative to aclonal taxa toward the equator. For example, stony corals and ascidians are groups with both clonal and aclonal species; the ratio of clonal to aclonal species in these groups was found to be only 0.75 in waters off Great Britain, compared to 2.97 in the Caribbean (Coates and Jackson, in press).

Similarly, on a more local scale, as environmental conditions generally become more stable and predictable with increasing depth, the ratio of numbers of clonal to aclonal species increases accordingly. Estuaries are especially variable environments, and clonal to aclonal ratios also increase from estuarine into more stable marine environments (Jackson, in press).

Beyond the striking differences that distinguish the life histories and distributions of clonal from aclonal invertebrates, within each group there are also persistent and ephemeral taxa. Among clonal animals, the characterization of these differences is more complicated than among aclonal animals, because there are more parameters to account for, such as the tendency of clonal organisms to live in bits and pieces and to grow in widely different shapes (Jackson 1979; Hughes 1984; Coates and Jackson, in press).

Nevertheless, there are several contrasting patterns of correlated life-history traits—including morphology, demography, and mobility—that appear repeatedly in different clonal taxa. Although quite variable, these different patterns can be observed both within a single growth form, such as sheetlike encrusting bryozoans, and among different growth forms, such as the branching, mound-shaped, and foliaceous forms of corals.

This paper will focus on one distinguishing trait of particular importance among sessile clonal animals, the extent to which they "move" about the substratum (Buss 1979; Jackson 1979). Although sessile organisms are incapable of active locomotion, sessile clones may still change position through growth and through passive dispersal of fragments, and different species vary enormously in the extent of this passive mobility. Rapidly growing clones are always on the move, exploring new areas of substratum, often far from where they settled as larvae. The youngest portions of these mobile clones tend to overwhelm most others they encounter, but they are rather weakly constructed and also tend to fall to pieces rapidly. In contrast, more stationary clones grow and take over space slowly, move about much less, and are built more sturdily. Despite these differences, however, individual clones of both mobile and stationary species may be extremely long lived and may cover quite large areas.

Mobile and stationary strategies occur commonly among all major taxa of clonal coral-reef invertebrates. We will first consider these strategies within single growth forms, using as examples foliaceous corals and encrusting bryozoans. Afterward, differences in life histories and mobility among different growth forms within the same taxa will be examined. Our examples are limited to Caribbean reefs, although we have observed equivalent patterns on reefs elsewhere.

Investment strategies of corals

One of the commonest patterns of growth among stony corals is in the form of foliaceous colonies, such as those shown in Figure 3, which extend out horizontally over the substratum from a restricted area of attachment at the base to form single or multiple flattened discs or lobes. On Caribbean reefs at least 10 species characteristically show this form. *Montastrea annularis* is one of the most abundant foliaceous corals, particularly in depths greater than about 20 m; in shallower water, it also grows abundantly as massive and encrusting colonies (Goreau 1959; Dustan 1975). *Leptoseris cucullata* is another foliaceous coral common on Caribbean reefs, where it always grows as thin plates.

These two species represent opposite extremes in growth rates among foliaceous corals. Foliaceous *Montastrea* colonies, although they are often very large, up to several meters across, grow very slowly, usually only 0.5 to 1.0 cm laterally per year. In deep water, they are highly persistent locally and long lived. In contrast, *Leptoseris* colonies may reach 1 m or more in diameter,

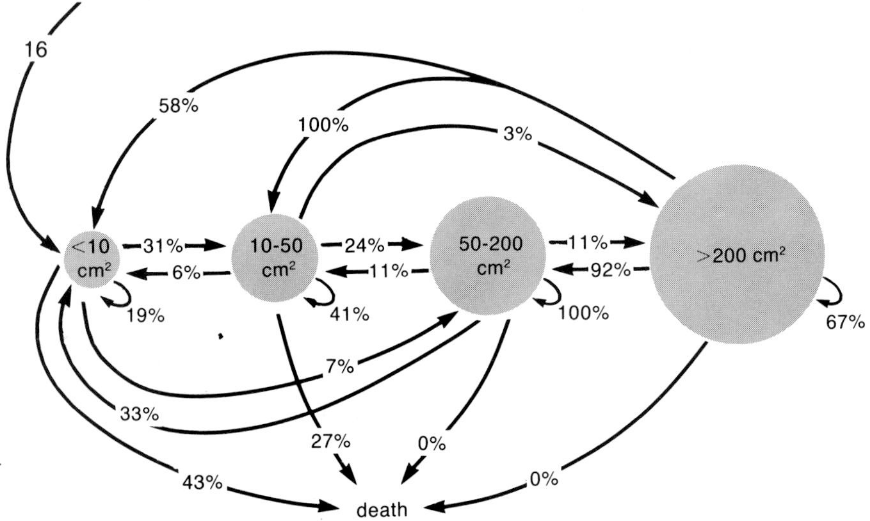

Figure 2. The correlation of the size of coral colonies with their ability to survive is indicated by these average probabilities of transition between four representative size classes of the foliaceous coral *Leptoseris cucullata* (shown in Fig. 3). Whereas only 57% of the smallest size class managed either to survive at the same size (19%) or grow into the next two larger size classes (31% and 7%), the transitions from the colonies in the largest size class amounted to 317% of their original number, with none of them dying. Over 100 colonies at 35 m depth off Discovery Bay, Jamaica, were observed over 3 years, during which time 16 new colonies were formed by sexual reproduction and 38 by fission. (After Hughes and Jackson, in press.)

but are usually much smaller; however, they grow laterally 5 to 10 times faster than *Montastrea*, at rates up to 5 cm per year.

Corals compete for space by direct and indirect means. Direct interactions are commonly aggressive: a dominant colony extends filaments armed with stinging cells onto the tissues of its less aggressive neighbors, which are subsequently digested away (Lang 1973). Indirect interactions involve growth of one colony above another, thereby inhibiting the loser's access to light or food (Connell 1973). *Montastrea* is the most aggressive Caribbean foliaceous coral, capable of digesting the tissues of any other foliaceous coral with which it comes into direct contact. In contrast, *Leptoseris* is among the least aggressive corals and loses most of its direct encounters with other clonal species. Instead, *Leptoseris* competes by rapidly growing up over its neighbors, as illustrated in Figure 4.

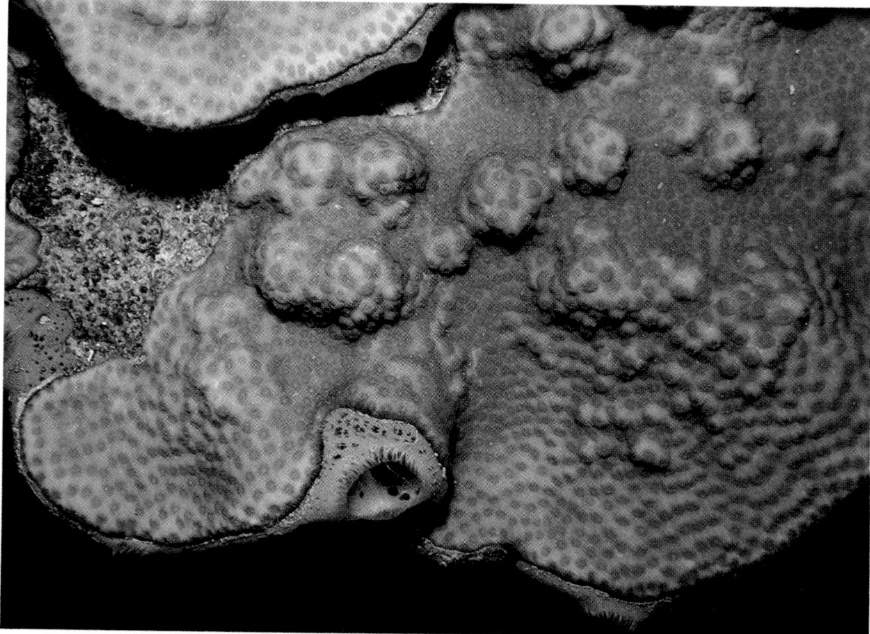

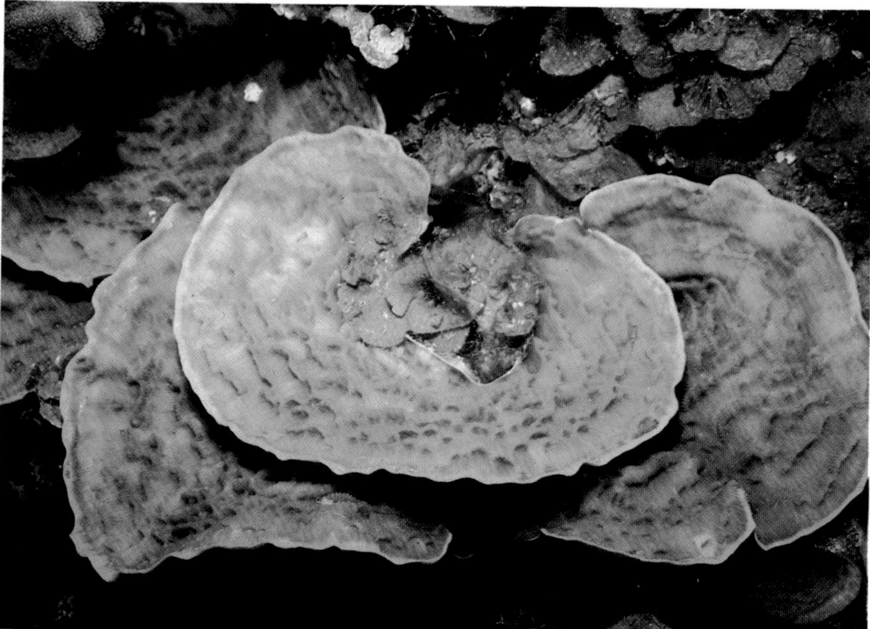

Figure 3. These two species of foliaceous corals represent opposite adaptive strategies among coral colonies that grow in this flattened, platelike form. *Montastrea annularis* (*top*) is relatively slow growing, but it is persistent in stable environments and successfully aggressive in direct encounters with other foliaceous colonies. (The orange creature near the bottom of the photograph is a sponge, *Mycale laevis*.) *Leptoseris cucullata* (*bottom*) is not aggressive and is much more ephemeral, but it nevertheless is more abundant than foliaceous *M. annularis* in highly disturbed environments, because of its greater mobility. (Photographs by G. Bruno.)

The two species of corals also have distinctive morphologies. Their skeletons, though similar in density, differ markedly in thickness; for average-size colonies at 30 m depth, foliaceous *Montastrea* skeletons have a mean thickness of 11 mm, versus 3 mm for *Leptoseris*. Consequently, *Leptoseris* colonies suffer far more in storms. For example, between 1977 and 1980 the mean percentage of tissue lost per year was about 5 times greater for monitored *Leptoseris* populations (112 colonies) than for *Montastrea* (75 colonies), with a rate of tissue turnover of 19% per year for the former compared to only 4% for the latter (Hughes and Jackson, in press). This suggests complete turnover of colony tissues after an average of every 5 years for *Leptoseris* versus every 25 years for *Montastrea*. Moreover, no *Montastrea* colony died completely in a single year, whereas whole *Leptoseris* colonies died every year. Another striking difference is in the larval recruitment of colonies, which was much higher for *Leptoseris* than for *Montastrea* (Rylaarsdam 1983); indeed, we have not detected a single larval recruit of *Montastrea* since 1977.

All these differences in life history are reflected in the contrasting distributions of foliaceous colonies of these species over the reef. Despite its more fragile construction, *Leptoseris* is relatively more abundant than foliaceous *Montastrea* at depths of 10 to 20 m; in these shallower waters, damage by storms and grazing predators, such as the sea urchin *Diadema antillarum* and damselfish, is greatest, and sponges, which are fierce competitors with corals, are relatively uncommon (Woodley et al. 1981; Jackson and Buss 1975). For foliaceous corals in shallow water, rapid growth is apparently more adaptive than strong skeletons and aggression. In contrast, at 35 m depth the foliaceous morphology of *Montastrea* is relatively more abundant than *Leptoseris*; in this deeper, less disturbed environment, the chance of persistence is increased by the reduced effects of storms and predators. Thus, just as they favor clonal over aclonal species, stabler environments seem to favor the more stationary foliaceous *Montastrea* over *Leptoseris.*

Investment strategies of bryozoans

A diverse community of sheetlike encrusting algae, bryozoans, sponges, corals, tunicates, and other sessile organisms lives on the undersurfaces of foliaceous corals such as *Montastrea* and *Leptoseris,* habitats where competition for space is so intense that the growth of any creature usually involves partial or complete death of another (Jackson and Buss 1975; Jackson and Winston 1982). Like their coral hosts, the encrusting organisms exhibit a wide variety of life histories that can be related to their patterns of distribution.

The two most abundant bryozoan species encrusting coral undersurfaces at our study sites in Jamaica are *Reptadeonella costulata* (Fig. 5) and a newly discovered species of *Steginoporella* (Fig. 6), both of which may occasionally grow to cover their entire substratum. *Reptadeonella,* a relatively stationary species, grows only 3 to 4 cm per year, forming fairly symmetrical colonies that are highly persistent. *Steginoporella,* in contrast, is extremely mobile, growing laterally at rates up to 11 cm per year, faster than any other encrusting bryozoan or foliaceous coral in Jamaica; the rapidly growing colonies form assymetrical lobes or fans that sweep across the coral undersurface, rarely remaining in any one place for long, as can be seen in the 10-month sequence in Figure 6. In the same 10-month period, a large *Reptadeonella* colony moved hardly at all, as shown in Figure 5.

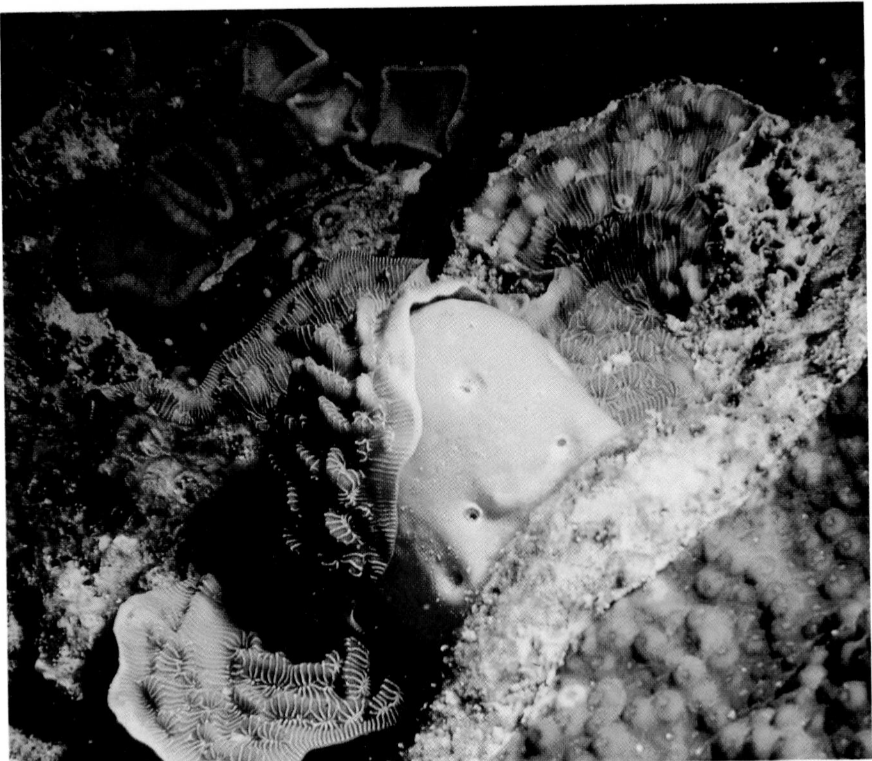

Figure 4. The coral *Leptoseris cucullata* is able to compete successfully by virtue of its rapid growth, a capacity that permits it to grow over and around competitors for space, such as this *Agelas* sponge. Note that the coral is not in contact with the sponge, which may be toxic to the coral. (Photograph by C. Arenson.)

Just as in the corals, these patterns of mobility are correlated with other demographic differences (Winston and Jackson 1984). Newly recruited *Steginoporella* colonies survive an average of only about 120 days, compared to an average of about 500 days for *Reptadeonella.* However, in addition to being more mobile, *Steginoporella* also recruits more than 1.5 times faster than *Reptadeonella*, and large colonies of both species survive for many years.

As with the corals, there are also striking differences in the architecture and related biomechanical properties of the individual zooids of the two species (Best and Winston 1984). Each zooid consists of a feeding organ and other organ systems enclosed within a boxlike skeletal

June 1983

May 1984

Figure 5. *Reptadeonella costulata*, one of the single-layered species of Bryozoa that encrust the undersurfaces of foliaceous corals, invests much of its resources in fortification and maintenance of its component zooids, but at the expense of rapid growth. Thus, this colony at 10 m depth near Rio Bueno, Jamaica, is relatively unchanged after 10 months, compared to the *Steginoporella* colony shown in Figure 6. The alligator clips are 5 cm long. (Photograph at left by J. Jackson; at right by G. Bruno.)

June 1983

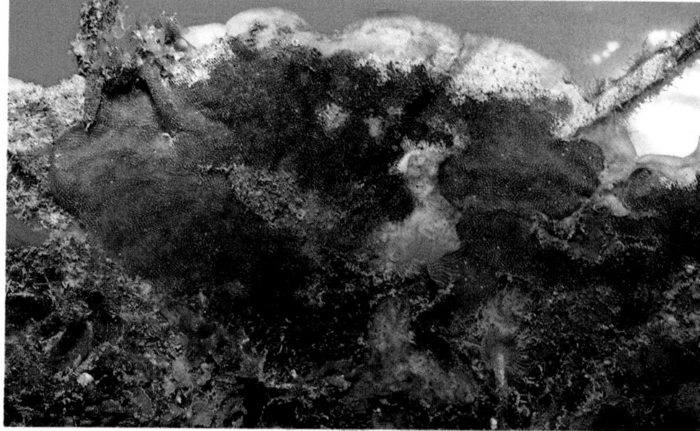

September 1983

Figure 6. *Steginoporella* is a highly mobile bryozoan, but at the expense of durability, as is indicated by these colonies—all probably members of the same clone—encrusting the undersurface of a foliaceous coral at 10 m depth near Rio Bueno, Jamacia. (First photograph by T. Hughes; other three by G. Bruno.)

chamber. As Figure 7 shows, *Steginoporella* zooids are rather simple in form, with an enormous orifice through which the feeding organ protrudes, and they have thin roofs and walls, without much extra calcification for protection. In contrast, zooids of the more stationary *Reptadeonella* are more complex, having obvious surface fortifications, a relatively small orifice, and thick roofs and walls; consequently, *Reptadeonella* zooids are nearly 15 times harder to puncture, and their colony surfaces nearly twice as hard to crush, as *Steginoporella*'s (Best and Winston 1984).

All of these differences in life history are reflected in the distributions of the two species under corals (Jackson 1984). *Steginoporella* is relatively more abundant than *Reptadeonella* nearer the coral edges, where new space is more or less regularly created by lateral coral growth, and by the grazing of the sea urchin *Diadema antillarum* and of fishes. This is also where sponges, generally the best competitors under corals, are least abundant. *Reptadeonella* shows the opposite pattern, becoming far more abundant than *Steginoporella* away from the edges. Thus, just as for foliaceous corals, the mobile species *Steginoporella* predominates in less stable, more disturbed environments, and the persistent species *Reptadeonella* predominates in more stable ones.

We can gain insight into the underlying basis of clonal life histories by comparing the functional biology of mobile and stationary species. Among the bryozoans, *Steginoporella* colonies display marked gradients between younger and older regions, correlated with the apparent senescence of older zooids, whereas *Reptadeonella* does not show regional senescence or differentiation. Younger regions of *Steginoporella* colonies are brightly colored, unfouled by other organisms, and unbroken; older regions are dark, fouled, and broken, and individual zooids are filled with waste.

This deterioration appears to affect *Steginoporella*'s ability to function in at least three important ways. First, the ability of *Steginoporella* colonies to overgrow their neighbors is greatly diminished in older regions; for example, in encounters between the two species, young regions of *Steginoporella* usually overgrow *Reptadeonella*, winning 90% of the time, whereas old regions always lose. Second, younger *Steginoporella* zooids feed more frequently than older zooids in the same colony. Third, younger regions of *Steginoporella* colonies regenerate injuries an average of more than 4 times faster than older regions. No such variation is apparent in *Reptadeonella* colonies, which regenerate injuries at rates almost equal to young *Steginoporella* (Palumbi and Jackson 1982, 1983).

Steginoporella colonies survive through a balance of spatially partitioned growth and decay. The extremely rapid growth of younger regions results in exceptional competitive and regenerative ability, but at the apparent expense of localized senescence of zooids, which prevents the colonies from holding on to any particular patch of substratum for very long. Furthermore, *Steginoporella* colonies first become sexually reproductive at 3 times the size necessary for *Reptadeonella* reproduction (a minimum of 8 versus 22 cm^2) and have less fecund zooids (Winston and Jackson 1984; Jackson and Wertheimer, in press).

These patterns of life history and morphology can be interpreted in terms of the allocation of limited nutritional resources within colonies (Palumbi and Jackson 1983). *Steginoporella* apparently invests so much in growth that it has few resources left for reproduction until colonies become large, whereas *Reptadeonella* invests less in growth and more in fortification, maintenance of zooids, and sexual reproduction. The mechanistic basis of these differences probably lies in the extent and direction of translocation between zooids of stored nutrients needed for growth maintenance, and reproduction; this is suggested by circumstantial evidence, particularly by the marked polarity of regeneration in *Steginoporella* colonies, and by analogy with other clonal animals and plants (Tardent 1963; Taylor 1977; Pitelka and Ashmun, in press).

Mobility and growth forms

Differences in mobility between different morphologies of corals and of bryozoans greatly exceed those within any single growth form. Among corals, for example, mound-shaped colonies are generally far more stationary, and branching colonies far more mobile, than any foliaceous corals. *Montastrea annularis*, which is an abundant foliaceous coral in deeper water, is also the most abundant mound-shaped coral species in the Caribbean, occurring generally in depths about 10 to 15 m

February 1984

May 1984

shallower than where its foliaceous growth form predominates (Goreau 1959). Colonies are often several meters high and across, but they grow slowly, as in the species' foliaceous morphology, thickening usually at rates less than 0.5 cm per year (Dustan 1975). Both morphologies of *Montastrea* are highly persistent locally and long lived.

Staghorn coral, *Acropora cervicornis*, and elkhorn coral, *A. palmata*, are two of the most abundant Caribbean branching corals (Adey 1978). Colonies commonly reach 1 to 3 m in height, and a single clone may extend laterally for 10 m or more (Neigel and Avise 1983). In contrast to either morphology of *Montastrea* or to *Leptoseris*, they grow extremely rapidly; branches of *A. palmata* may lengthen at rates up to 11 cm per year, and *A. cervicornis* at rates up to 20 cm per year. Also in contrast to *Montastrea*, they are relatively poor aggressors in direct interactions with other coral species and compete most successfully by rapidly growing over their neighbors, which they may then harm indirectly by shading (Porter et al. 1981). Clones formed by these two species are extremely mobile, both through growth and through dispersal of fragments, and they may also be very long lived (Tunnicliffe 1981).

Mound-shaped skeletons are inherently much more resistant to toppling or breakage during storms than are branching corals (Highsmith 1982). For example, after the passage of Hurricane Allen in 1980, extensive populations of the two *Acropora* species, observed at 6 m depth near Discovery Bay, were reduced to less than 1% of their original cover, whereas 91% of the massive *Montastrea* mounds at this depth endured (Woodley et al. 1981).

The functional biology of the different coral growth forms is intriguingly similar to what we have described for the two encrusting bryozoan species. Like *Steginoporella*, the branching *Acropora* species exhibit both very rapid growth at the edges of colonies—termed distal growth—and apparently higher rates of mortality (possibly due to local senescence) of older, basal regions (Tunnicliffe 1981). This correlates well with the extremely high rates of distal translocation of nutrients and calcium observed in *A. cervicornis;* foliaceous *Montastrea* colonies also exhibit predominantly distal translocation, although not so unidirectionally as in *A. cervicornis* (Taylor 1977). Moreover, despite their slow growth rate, *Montastrea* colonies quickly regenerate damaged regions over most of the colony surface, whereas only the distal portions of the branches in *Acropora* regenerate rapidly, if at all (Bak et al. 1977; Tunnicliffe 1981; Bak 1983).

Unlike corals, branching bryozoans are rare on coral reefs, and the most common forms other than sheetlike colonies are mound shaped. These grow by a series of processes, collectively termed frontal budding, that produce multiple layers of zooids, one on top of another, to form massive colonies superficially similar to those of mound-shaped corals (Cheetham and Cook 1983). Two abundant frontally budding bryozoans on Caribbean reefs are *Stylopoma spongites* and *Trematooecia aviculifera.* Both of these species grow more slowly and persist longer in any one place than do the single-layered colonies of *Steginoporella* or even of the more persistent *Reptadeonella.* Frontal budding is a strategy of persistence, providing greater resistance to both overgrowth and disturbance compared with single-layered growth (Jackson and Buss 1975).

The surfaces of *Stylopoma* and *Trematooecia* colonies also lack any obvious signs of regional senescence. Moreover, *Trematooecia* zooids have very thick frontal walls, as can be seen in Figure 7, which are $1\frac{1}{2}$ to 3 times harder to puncture than are those of *Reptadeonella* zooids, although colony surfaces are not harder to crush (Best and Winston 1984). Finally, *Stylopoma* and *Trematooecia* can regenerate injuries from live zooids below the surface layer, as well as laterally, which single-layered colonies cannot do (Lidgard, in press).

Investment strategies and distributions

The distributions of corals and bryozoans on reefs are strongly correlated with their life histories and growth forms, particularly as these relate to the animals' comparative mobilities over the substratum. In general, as we have seen, stationary growth forms are relatively more abundant in environments with low levels of disturbance, whereas mobile growth forms are relatively more abundant in environments with high levels of disturbance.

First consider two extremes in mobility, branching and mound-shaped corals. Branching *Acropora palmata* is by far the most abundant coral at and just below the reef crest, from 1 to 5 m depth, where the chances of damage caused by storms and by grazing predators is greatest (Adey 1978). Although mound-shaped corals are inherently much less susceptible to storm damage, they apparently cannot become established after hurricanes quickly

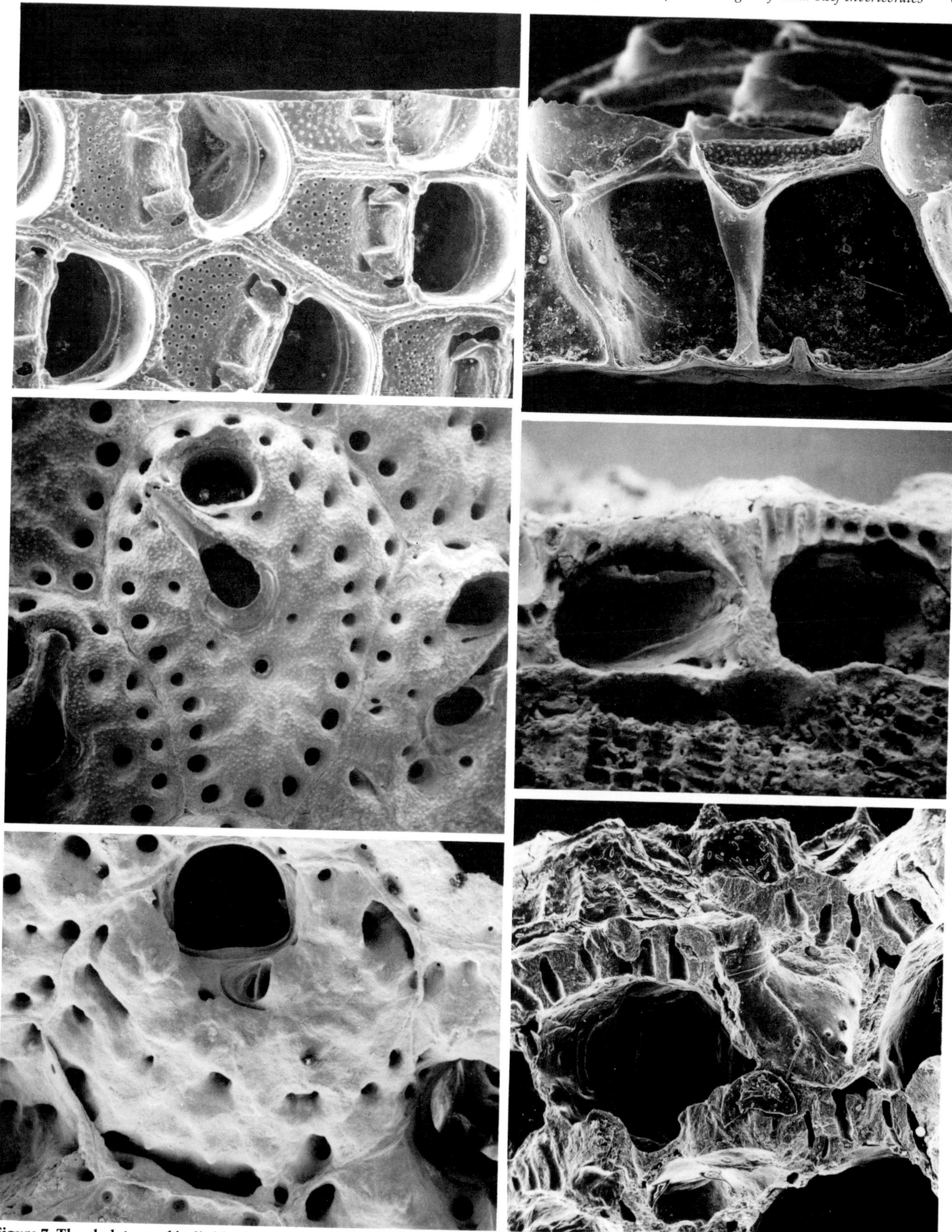

Figure 7. The skeletons of individual bryozoan zooids, shown in these scanning electron micrographs of their front surfaces (*left*) and transverse sections (*right*), clearly reflect the life histories of their respective species. The *Steginoporella* zooids shown in the top two photographs have a rather delicate skeletal structure, as would be expected in a highly mobile bryozoan that makes relatively little investment in the fortification and maintenance of its zooids (each approximately 1.1 mm long). In contrast, the more stationary and highly persistent *Reptadeonella costulata* (*middle*) has substantially fortified zooids (~0.7 mm). *Trematooecia aviculifera* (*bottom*) is even more stationary and persistent than *R. costulata*, and its zooids (~0.8 mm) are accordingly more fortified and can grow in several layers one upon another. (Upper two photographs by S. Lidgard; lower four by J. Winston.)

enough to prevent repopulation by more rapidly growing *A. palmata.* Thus, mound-shaped corals occur in slightly deeper water, typically from 3 to 15 m. The delicate branching species *A. cervicornis* overlaps considerably in distribution with massive corals; however, in general, the rapid, mobile growth of the two *Acropora* species seems more successful in shallow environments than the more stationary and massive skeletons of mound-shaped corals. During storms, both branching species are broken up, and live fragments are locally dispersed (Tunnicliffe 1981; Highsmith 1982). This appears to be their primary means of propagation, since larval recruitment by *A. palmata* and *A. cervicornis* is rare (Bak and Engel 1979; Rylaarsdam 1983).

However, although their strategy is highly successful, local *Acropora* populations can occasionally be so severely damaged by major hurricanes that recovery is extremely slow, or may not occur at all (Knowlton et al. 1981). Similarly, both *A. palmata* and *A. cervicornis* are rare in areas—such as most of the eastern Bahamas—that are swept each year by long-wave, open-ocean storms. Under these extreme conditions, all corals are excluded from shallow water, and massively crustose coralline algae predominate (Adey 1978, pers. com. 1981).

Foliaceous corals at shallow depths are typically less abundant than branching or massive corals and consist primarily of more mobile colonies, such as of *Leptoseris, Agaricia agaricites,* and *Porites astreoides.* Below about 20 m, the more stationary foliaceous colonies of *Montastrea, Agaricia lamarcki,* and *A. grahamae* become dominant, a pattern similar to the replacement of branching corals by more stationary massive species below about 5 m. However, at depths greater than about 50 m, where levels of light are very low, these foliaceous corals give way again to more fragile and more mobile foliaceous species such as *Agaricia undata* and *A. fragilis,* both of which grow 5 times faster than foliaceous *M. annularis.* Thus, on the deepest reefs, the generally positive correlation between levels of disturbance and mobility appears to break down, probably because of intrinsic physiological limitations on coral growth and morphology (e.g., Goreau 1963).

The distribution patterns of bryozoans on reefs change with depth in a manner very similar to those of corals. As we have seen, mobile *Steginoporella* is most common in the more disturbed habitats near coral edges, while the more stationary *Reptadeonella* is relatively more abundant farther back in the less disturbed sites. In deeper water, where predators are less prevalent and where foliaceous corals are also much longer lived, *Steginoporella* and *Reptadeonella* are replaced as the most abundant bryozoans under foliaceous corals by the potentially massive *Stylopoma,* which can form very large colonies 1 cm or more thick (Jackson and Wertheimer, in press).

If levels of predation are extremely high, however, single-layered encrusting species are excluded, and bryozoans are limited to a few massive, stationary species, such as *Trematooecia aviculifera,* which commonly forms colonies more than 10 cm thick. In contrast to other bryozoans in Jamaica, *Trematooecia* occurs most commonly on open reef surfaces, where most bryozoans are excluded by physical damage and by grazing sea urchins and fishes. Thus, *Trematooecia* seems similar to massive crustose coralline algae found on crests of exposed reefs, where branching corals are excluded by chronically high levels of physical and biological disturbance.

These patterns of distribution suggest that physical disturbance and predation, in addition to strongly affecting community structure and diversity of coral reefs, have profoundly influenced the evolution of different life-history strategies of bryozoans and corals (Connell 1979; Woodley et al. 1981). Patterns of growth form and life-history variation very similar to these have also been described for terrestrial plants (Lovett-Doust 1981; Cook 1983); this suggests that these kinds of investment trade-offs in space and time may be characteristic of all sessile clonal organisms. Although both clonal and aclonal animals and plants vary widely in their relative energetic investments in growth, maintenance, and sexual reproduction, the two additional components of life histories that are peculiar to clonal species are the spatial patterns of their growth and their local persistence or mobility over time.

References

Adey, W. H. 1978. Coral reef morphogenesis: A multidimensional model. *Science* 202: 831–37.

Bak, R. P. M. 1983. Neoplasia, regeneration and growth in the reef-building coral *Acropora palmata. Mar. Biol.* 77:221–27.

Bak, R. P. M., J. Brouns, and F. Heys. 1977. Regeneration and aspects of spatial competition in the scleractinian corals *Agaricia agaricites* and *Montastrea annularis.* In *Proceedings of the 3rd International Coral Reef Symposium* 1:143–48.

Bak, R. P. M., and M. S. Engel. 1979. Distribution, abundance, and survival of juvenile hermatypic corals (Scleractinia) and the importance of life history strategies in the parent coral community. *Mar. Biol.* 54: 341–52.

Bak, R. P. M., J. Sybesma, and F. C. van Duyl. 1981. The ecology of the tropical compound ascidian *Trididemnum solidum.* II. Abundance, growth and survival. *Mar. Ecol. Prog. Ser.* 6:43–52.

Best, B. A., and J. E. Winston. 1984. Skeletal strength of encrusting cheilostomes. *Biol. Bull.* 167:390–409.

Buss, L. W. 1979. Habitat selection, directional growth, and spatial refuges: Why colonial animals have more hiding places. In *Biology and Systematics of Colonial Organisms,* ed. G. Larwood and B. R. Rosen, pp. 459–97. London: Academic Press.

Caswell, H. In press. The evolutionary demography of vegetative reproduction. In *Population Biology and Evolution of Clonal Organisms,* ed. J. B. C. Jackson, L. W. Buss, and R. E. Cook. Yale Univ. Press.

Cheetham, A. H., and P. L. Cook. 1983. General features of the class Gymnolaemata. In *Treatise on Invertebrate Paleontology, Part G (Revised),* ed. R. A. Robison, pp. 138–207. Univ. of Kansas Press.

Cheetham, A. H., and L.-A. C. Hayek. 1983. Geometric consequences of branching growth in adeoniform Bryozoa. *Paleobiology* 9:240–60.

Coates, A. G., and J. B. C. Jackson. In press. Morphological themes in the evolution of clonal and aclonal marine invertebrates. In *Population Biology and Evolution of Clonal Organisms,* ed. J. B. C. Jackson, L. W. Buss, and R. E. Cook. Yale Univ. Press.

Connell, J. H. 1973. Population biology of reef-building corals. In *Biology and Geology of Coral Reefs,* ed. O. A. Jones and R. Endean, vol. 2, pp. 205–45. Academic Press.

______. 1979. Tropical rain forests and coral reefs as open non-equilibrium systems. In *Population Dynamics,* ed. R. M. Anderson, B. D. Turner, and L. R. Taylor, pp. 141–63. Oxford: Blackwells.

Cook, R. E. 1983. Clonal plant populations. *Am. Sci.* 71:244–53.

Dustan, P. 1975. Growth and form in the reef building coral *Montastrea annularis. Mar. Biol.* 33:101–07.

Goreau, T. F. 1959. The ecology of Jamaican coral reefs. I. Species composition and zonation. *Ecology* 40:67–90.

______. 1963. Calcium carbonate deposition by coralline algae and corals in relation to their roles as reef builders. *Annals N. Y. Acad. Sci.* 109:127–67.

Highsmith, R. C. 1982. Reproduction by fragmentation in corals. *Mar. Ecol. Prog. Ser.* 7: 207–26.

Hughes, T. P. 1984. Population dynamics based on individual size rather than age. *Am. Nat.* 123:778–95.

Hughes, T. P., and J. B. C. Jackson. 1980. Do corals lie about their age? Some demographic consequences of partial mortality, fission, and fusion. *Science* 209:713–15.

———. In press. Population dynamics and life histories of foliaceous corals. *Ecol. Monogr.*

Jackson, J. B. C. 1977. Competition on marine hard substrata: The adaptive significance of solitary and colonial strategies. *Am. Nat.* 111:743–67.

———. 1979. Morphological strategies of sessile animals. In *Biology and Systematics of Colonial Organisms*, ed. G. Larwood and B. R. Rosen, pp. 499–555. London: Academic Press.

———. 1984. Ecology of cryptic coral reef communities. III. Abundance and aggregation of encrusting organisms with particular reference to cheilostome Bryozoa. *J. Exp. Mar. Biol. Ecol.* 75:37–57.

———. In press. Distribution and ecology of clonal and aclonal benthic invertebrates. In *Population Biology and Evolution of Clonal Organisms*, ed. J. B. C. Jackson, L. W. Buss, and R. E. Cook. Yale Univ. Press.

Jackson, J. B. C., and L. W. Buss. 1975. Allelopathy and spatial competition among coral reef invertebrates. PNAS 72:5160–63.

Jackson, J. B. C., and S. P. Wertheimer. In press. Patterns of reproduction in five common species of Jamaican reef-associated bryozoans. In *Bryozoan Biology and Paleontology*, ed. C. Nielsen and G. P. Larwood. Fredensborg, Denmark: Olsen and Olsen.

Jackson, J. B. C., and J. E. Winston. 1981. Modular growth and longevity in bryozoans. In *Recent and Fossil Bryozoa*, ed. G. P. Larwood and C. Nielsen, pp. 121–26. Fredensborg, Denmark: Olsen and Olsen.

———. 1982. Ecology of cryptic coral reef communities. I. Distribution and abundance of major groups of encrusting organisms. *J. Exp. Mar. Biol. Ecol.* 57:135–47.

Knowlton, N., J. C. Lang, M. C. Rooney, and P. Clifford. 1981. When hurricanes kill corals; evidence for delayed mortality in Jamaican staghorns. *Nature* 294:251–52.

Lang, J. C. 1973. Interspecific aggression by scleractinian corals. II. Why the race is not always to the swift. *Bull. Mar. Sci.* 23:260–79.

Lidgard, S. In press. Budding process and geometry in encrusting cheilostome bryozoans. In *Bryozoan Biology and Paleontology*, ed. C. Nielsen and G. P. Larwood. Fredensborg, Denmark: Olsen and Olsen.

Lovett-Doust, L. 1981. Population dynamics and local specialization in a clonal perennial *Ranunculus repens* I. The dynamics of ramets in contrasting habitats. *J. Ecol.* 69:743–55.

Neigel, J. E., and J. C. Avise. 1983. Clonal diversity and population structure in a reef-building coral, *Acropora cervicornis:* Self-recognition analysis and demographic interpretation. *Evolution* 37:437–53.

Palumbi, S. R., and J. B. C. Jackson. 1982. Ecology of cryptic coral reef communities. II. Recovery from small disturbance events by encrusting Bryozoa: The influence of "host" species and lesion size. *J. Exp. Mar. Biol. Ecol.* 64:103–15.

———. 1983. Aging in modular organisms: Ecology of zooid senescence in *Steginoporella* sp. (Bryozoa: Cheilostomata). *Biol. Bull.* 164:267–78.

Pitelka, L. F., and J. W. Ashmun. In press. Physiology and integration of ramets in clonal plants. In *Population Biology and Evolution of Clonal Organisms*, ed. J. B. C. Jackson, L. W. Buss, and R. E. Cook. Yale Univ. Press.

Porter, J. W., et al. 1981. Population trends among Jamaican reef corals. *Nature* 294: 249–50.

Potts, D. C. 1984. Generation times and the Quarternary evolution of reef-building corals. *Paleobiology* 10:48–58.

Rylaarsdam, K. W. 1983. Life histories and abundance patterns of colonial corals on Jamaican reefs. *Mar. Ecol. Prog. Ser.* 13: 249–60.

Sebens, K. P. 1982. Competition for space: Growth rate, reproductive output, and escape in size. *Am. Nat.* 120:189–97.

Tardent, P. 1963. Regeneration in the Hydrozoa. *Biol. Rev.* 38:293–333.

Taylor, D. L. 1977. Intra-colonial transport of organic compounds and calcium in some Atlantic reef corals. In *Proceedings of the 3rd International Coral Reef Symposium* 1:431–36.

Tunnicliffe, V. J. 1981. Breakage and propagation of the stony coral *Acropora cervicornis.* PNAS 78:2427–31.

Wahle, C. M. 1983. The roles of age, size and injury in sexual reproduction among Jamaican gorgonians. *Am. Zool.* 23:961.

Winston, J. E., and J. B. C. Jackson. 1984. Ecology of cryptic coral reef communities. IV. Community development and life histories of encrusting cheilostome Bryozoa. *J. Exp. Mar. Biol. Ecol.* 76:1–21.

Woodley, J. D., et al. 1981. Hurricane Allen's impact on Jamaican coral reefs. *Science* 214:749–55.

PART II
The Dynamics and Consequences of How Species Interact

Ecology often grabs our attention when the population of some pest species explodes and ravages our crops, or when a disease sweeps through a population and causes massive death. But these dramatic examples are just special cases of a central theme in ecology: Species interact as competitors, as predators and prey, as pathogens and hosts, and as mutualists—and as a result of these interactions species can profoundly influence each other's abundance.

An ordinary backyard garden can be an arena for intense competition between species, and a source of inspiration for the development of ecological insights regarding species interaction. Bergelson probes the mechanisms of competition between species of weedy plants and shows how just a few days head start can make all the difference in the world with respect to who wins. Even more subtle is the importance of positioning for plants; for a seedling, finding oneself a few centimeters to the left or right can mean the difference between life and death. The difficulty lies in synthesizing all of this experimental detail into a predictive theory.

In general, the task of translating information on the natural history of species interactions into predictions about population dynamics has challenged theoretical ecologists for over fifty years. An incisive article by May makes it clear that the insights from building mathematical descriptions of host–pathogen interactions can be enormous. Using simple models, May shows how parasites and disease can both regulate animal populations and drive pronounced fluctuations in abundance. When these population models are applied to humans, they illuminate the role of disease in human history and help guide the design of vaccination programs. Ecological studies of interactions between humans and their pathogens also shed light on the emergence of new diseases such as hantavirus, HIV, and lyme disease. In a provocative paper, Levins and colleagues argue that the prominence of several of these new diseases can be traced to fundamental disruptions of ecological systems that alter the opportunities for contact between the pathogens and humans.

Two of the most exciting recent developments concerning studies of species interactions are: (1) increasing attention to mutualisms, and (2) a growing appreciation that, because most species are embedded in food chains, simplistic studies of pairwise interactions can be extremely misleading. Papers by Bertness and by Ball and colleagues discuss mutualistic interactions in ways that merge traditional population ecology with an ecosystem-level perspective. Bertness shows how the success of marsh cordgrass, and in turn marsh productivity, is largely due to fiddler crabs turning over subtidal mud and to mussels depositing nitrogen-rich feces into the soil. In return, the cordgrass provides shelter and anchorage to these animals. The elegant feature of Bertness' analysis is that it blends a modern recognition of the importance of cooperative interactions with a more traditional consideration of competitive interactions among marsh plants and shows how cooperation can be the key to tipping a competitive balance toward particular species.

Ball and colleagues examine what they believe is a much more tightly coevolved mutualism than the sorts of cooperative interactions discussed by Bertness. Specifically, Ball is concerned with the mutual interdependence of fescue grass and endophytic fungi, an association in which the grass provides nutrition for the fungus and the fungus produces toxins that protect the grass from grazing vertebrates. Ironically, agricultural scientists have only recently worked out this story and are thus just beginning to appreciate that there may be many undiscovered mutualistic fungal–grass interactions, and that this has major implications for the productivity of rangelands. The subtlety of the fungal–fescue association also carries with it a warning about studying pairwise species interactions, since the benefits provided by the fungi to grass are only evident if one steps beyond the study of pairs of species and considers a third "interactor": grazers. Similarly, Myers' examination of population outbreaks in forest lepidoptera reinforces the importance of interaction webs. She summons evidence suggesting that "boom-and-bust" cycles of caterpillar abundance, which in turn produce cycles of forest defoliation, are driven by viruses, with weather and plant defenses modulating the impact of viruses through effects on caterpillar sensitivity to infection.

Myers' synthetic treatment highlights how far ecology has come in the last decade. The science no longer relies on single-factor experiments, but seeks an understanding of ecological dynamics through a blend of mathematical models, chemical analyses, small-scale manipulative experiments, and large-scale geographic studies.

Competition between Two Weeds

Common groundsel and annual bluegrass battle each other in a war based on germination, litter and open patches in the environment

Joy Bergelson

Every gardener knows that plants compete with each other—especially if one of the plants is a weed. Although gardeners help their plants win competitive interactions by removing or poisoning weeds, plants in nature must fight on their own—finding ways to defeat rival plants. The question is: What causes competing plants to win or lose?

Competition between plants represents one of the best-documented interactions between biological species. Hundreds of examples show that plant species affect one another through the use of shared resources, including light, nutrients and water. A tomato seedling in your garden, for instance, may grow poorly in the shade cast by a mature oak tree. Another tomato plant may survive better in the shade, because some individual plants have different competitive abilities. The outcome of competition between plants may also depend on the composition of the local community and debris or chemicals left by plants from earlier generations.

Although botanists have assembled a virtual catalogue of examples of competition, we know very little about the mechanisms that govern differences in competitive ability. I shall describe a series of experiments that reveal an intricate—and somewhat surprising—web of competitive interactions between two particular plant species.

Joy Bergelson is an assistant professor of ecology and evolution at the University of Chicago. In 1986, she earned a master's degree in biology from the University of York in the United Kingdom, and in 1990 she earned a Ph.D. in zoology from the University of Washington. Her current research interests include the ecology and evolution of interactions between plants and their enemies. Address: Department of Ecology and Evolution, University of Chicago, 1101 East 57th Street, Chicago, IL 60637–1573. Internet: joy@pondside.uchicago.edu.

Garden Experiments

One of the best-studied examples of competitive dynamics involves two weedy annuals: common groundsel and annual bluegrass. These species thrive throughout Eurasia and the United States, where bluegrass grows more abundantly and persists admirably when faced with neighboring groundsel. About a decade ago, I started a series of experiments to unravel the factors that drive the interactions between these two species and to test the population-level repercussions of their competition.

I began by gathering 20 groundsel plants from the Seattle area. I collected seeds from each of those plants and established a "common garden" experiment to test the ability of the seeds to grow and reproduce with and without intense competition from bluegrass. In this experiment, I planted seeds from each group of siblings, or *sibship*, in a field plot. I planted some seeds alone, and I encircled others with 12 bluegrass seeds. The results from that experiment showed that some of the groundsel plants fared better than others when faced with competition from bluegrass. For instance, some of them produced twice as many flowers as others. Such differences in performance were not evident among the groundsel plants grown alone.

What caused some groundsel plants to perform better than others in the presence of competitors? A plant's genotype could affect its competitive ability. Nevertheless, my initial experiment did not prove a connection between genetics and competitiveness, because the seeds used in that experiment also came from plants that had experienced different environmental conditions. A number of studies indicate that a mother plant's environmental conditions can influence the characteristics of progeny—a phenomenon known as the *maternal effect.* Luckily, maternal effects can be controlled for rather easily, at least in this case. Groundsel produces seeds autogamously—meaning that a flower gets pollinated by its own pollen—and plants collected in the field can be propagated in a common environment for subsequent experiments. So I grew sibships from each plant for three generations in a greenhouse, and then repeated my original experiment. Again, different groundsel plants exhibited large differences in their ability to endure competition, but no differences appeared without competitors. In addition, if a field-collected mother plant performed well against competitors, so did her offspring. These results confirmed that there is a genetic basis for the varied tolerance to competition.

Early Winners

What features make a plant more competitive? That question can be addressed through a closer look at competitively superior and inferior groundsel genotypes. During my experiments, I tracked the date of germination and the height of each groundsel plant, which provided data that might explain differences in competitive ability.

Using a technique called path analysis, I assessed the relative importance of differences in early growth rates, late growth rates and emergence dates. Path analysis unravels the effects of several correlated characteristics, such as growth rates and date of emergence, on a plant's subsequent performance. That analysis suggested strongly that competitively superior genotypes gain their advantage by emerging early. In other

Nigel Cattlin (Photo Researchers, Inc.)

Dan Guravich (Photo Researchers, Inc.)

Figure 1. Common groundsel (*top*) and annual bluegrass (*bottom*) coexist throughout the United States and Eurasia. All coexisting plants compete for shared resources, including light and water. The outcome of competition may depend in part on the composition of the local community or remnants of earlier generations. The author examines how groundsel and bluegrass compete, as well as that competition's impact on the surrounding environment.

words, a rapidly emerging groundsel plant is a competitive one. Moreover, the emergence date of the plants in this experiments explained more than 70 percent of the variation in their success.

That result led to another experiment: manipulating relative emergence dates by planting seeds at different times. I planted slow-germinating genotypes before fast-germinating ones, so that the seedlings would emerge simultaneously. The resulting plants possessed comparable competitive ability—confirming my hypothesis that competitive superiority comes from getting a head start on germination. Other plant systems operate similarly; small differences in emergence dates translate into large differences in performance. The effects can be dramatic. For example, Thomas Miller of Florida State University reported that a three-day head start in emergence results in more than a 1,900-fold increase in adult performance in common lambsquarters.

If emergence date determines a seedling's success, we might expect to see the evolution of traits that enable seeds to modulate germination in response to competitors. In fact, seeds from some plant species reduce their probability of germination when adult plants are already growing in the area, evidently avoiding competition with established plants. Seeds also appear responsive to the presence of other seeds. In both groundsel and bluegrass, my colleagues and I have demonstrated a negative relation between seed densities and the probability that each seed will germinate, presumably a tactic for preventing high levels of competition among seedlings, which often leads to extensive mortality. Moreover, seed-seed communication appears even more complex than a simple density-dependent germination. A series of greenhouse experiments showed that groundsel and bluegrass seeds accelerate their rates of germination when they are grown in soil that had previously contained germinating seeds. Apparently, the previously germinating seeds excrete chemicals, which have not yet been isolated, that accelerate the emergence of later seeds.

Neighborhoods and Gaps

In the groundsel-bluegrass system, competition portrays a race in which small differences in emergence produce large differences in performance, and seeds respond vigorously to clues that indicate the presence of competitors.

Apparently, the threat of competition shapes the activity of these seeds. That raises broader questions: To what extent does competition influence the population dynamics of competitors, and to what extent does community context influence competitive interactions? Although biologists commonly assume that competition affects the dynamics of plant communities, we are just beginning to explore the interplay between individual performance, which responds to competition, and population- and community-level factors. Such research provides an important arena for testing our understanding of the forces that structure plant communities.

At this interface between individuals and communities of plants, spatial patterning receives a disproportionate share of attention. Two distinct theories have been developed that predict that spatial patterning can exert a strong influence on competitive interactions. Since plant populations exhibit very patchy arrangements, as opposed to random distributions, the question arises as to whether patterning drives the dynamics of plant populations in nature.

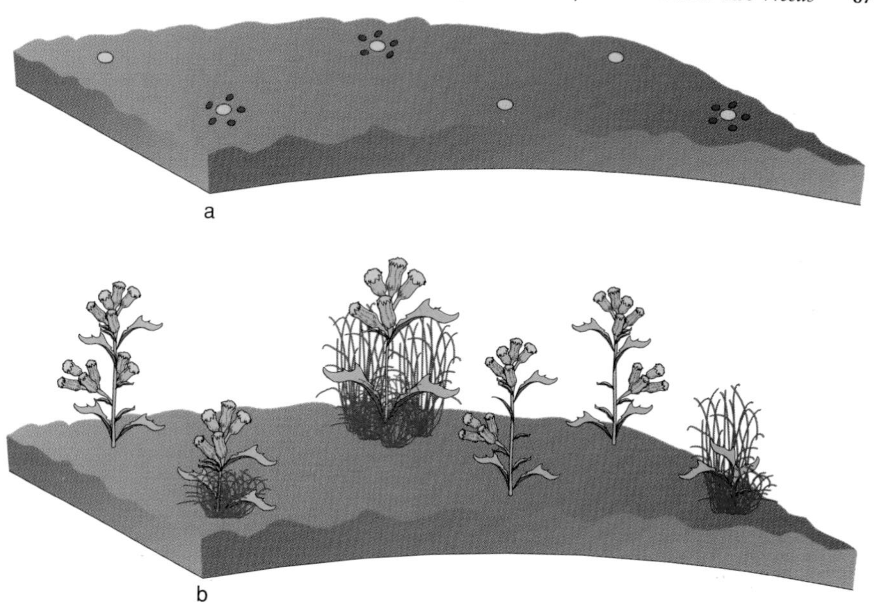

Figure 2. Early emerging groundsel seedlings compete more effectively against bluegrass. The author planted groundsel seeds (*yellow*) either alone or surrounded by a dozen bluegrass seeds (*green*) (*a*). The groundsel seeds planted alone all grew about the same in terms of size and number of flowers, but when faced with competition from bluegrass, some groundsel plants did better than others (*b*). Additional experiments showed that competitive success in groundsel depends on a genetic propensity to produce fast-emerging seedlings, which can get a head start on the competing bluegrass.

Stephen Pacala of Princeton University and John Silander of the University of Connecticut champion the *neighborhood-competition theory*. That theory assumes that plant growth and reproduction depend on the density of nearby competitors and that competitors beyond a designated distance exert no impact on a particular plant. Combining that assumption with an elaborate mathematical theory suggests that patchiness in a distribution of competitors can profoundly alter competition at the population level. Intuitively, this theory can be understood by recognizing that patchiness leads to considerable variation in the amount of competition experienced by any individual plant. Some plants will experience lots of competition and experience little reproduction; other plants will be virtually free from competition and contribute a disproportionately large share to the next generation. In this way, the neighborhood-competition theory suggests that the spatial pattern of plants influences competition through interactions between contemporaneous plants.

An alternative theory for patchiness, *gap colonization*, considers the competitive impact of one generation on another. Traditionally, this theory has been applied to the dynamics of forests, where the germination and growth of seedlings depends on gaps left after trees fall. Nevertheless, it might be equally relevant to the dynamics of grassland plants. In nearly all cases, established vegetation can overpower seedlings. In fact, seedling emergence and survival can be totally inhibited by the presence of adult plants, or even by the litter left from previous generations. This type of inhibition can arise from several factors: chemicals, shading or structural interference produced by the established plants. In some cases, these factors may restrict seedling survival to areas that lack established vegetation. Moreover, the gaps in a patchy spatial pattern promote the persistence of competitively inferior plants. Gap-colonization theory, then, suggests that an area's spatial pattern will influence competitive interactions because previous generations affect the survival of present seedlings.

Competition between common groundsel and annual bluegrass also depends on spatial patterns. Both of these species produce two generations each year—one in the spring and another in the fall. This rather unusual life cycle proves tremendously convenient for testing the neighborhood-competition and gap-colonization theories. In each fall generation, newly emerging seedlings compete with each other, which provides an opportunity for neighborhood competition, and they also interact with remnants from the spring generation, which could lead to gap colonization.

Litter Blockade

To explore the role of patchiness on competition between groundsel and bluegrass, I established artificial plots in the spring in which I created either a random or a patchy distribution of bluegrass. In the same plots, I also planted groundsel, but it was always distributed randomly at one-quarter the density of bluegrass. That relative abundance resembles natural communities, where groundsel is typically less abundant than bluegrass. I let the seedlings grow and compete, and then I counted the number of groundsel plants in the fall generation. The spatial distribution played a dramatic role in the growth rates of the groundsel: The fall generation contained nearly four times more groundsel plants when the bluegrass distribution was patchy rather than random. That experiment represented the first experimental manipulation of plant spatial patterns to assess competitive interactions, and it showed unequivocally that spatial patterning can have an enormous impact.

These results did not arise from neighborhood competition. Regardless of the distribution of the bluegrass, the spring generation of groundsel pro-

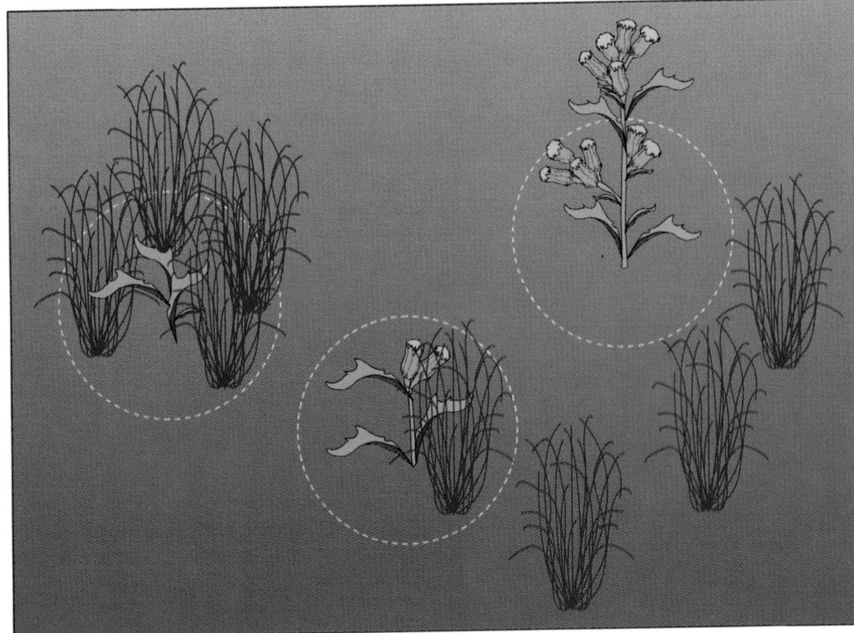

Figure 3. Community composition may explain some of the mechanics behind plant competition. The neighborhood-competition theory (*a*) suggests that a groundsel plant (*yellow-flowered*) would grow better if fewer competing bluegrass plants (*green*) grew nearby, or within a designated distance indicated by the color circle. Gap colonization (*b*) suggests that groundsel might grow best in bare patches—free from competition with bluegrass and the dead blades from previous generations.

duced essentially the same number of seeds. That is, competition within the spring generation did not generate the differences that appeared in the fall. Pacala and Silander obtained similar results when exploring how patchiness affected the growth rates in a competitive system of velvetleaf and pigweed.

On the other hand, gaps did affect the groundsel's growth. In the plots, a large fraction of all the surviving groundsel seedlings grew in areas of low bluegrass density. So patchy plots promoted the growth of groundsel populations by providing a greater number of gaps.

The difference in groundsel's success in the spring and fall generations suggested that the bluegrass litter—dead blades and roots from previous generations—could be a factor. To test that hypothesis, I repeated the above experiment with the patchy and randomly distributed bluegrass but removed the dead bluegrass from half of the plots and left the bluegrass litter intact in the other half. If litter drives the effect of spatial patterning on groundsel, then removing the litter should remove the effect, which is exactly what I found. With the litter intact, groundsel grew better where the bluegrass distribution was patchy; but with the litter removed, the groundsel grew about the same regardless of the distribution of bluegrass. This experiment indicated clearly that competition between generations, not within them, governs the

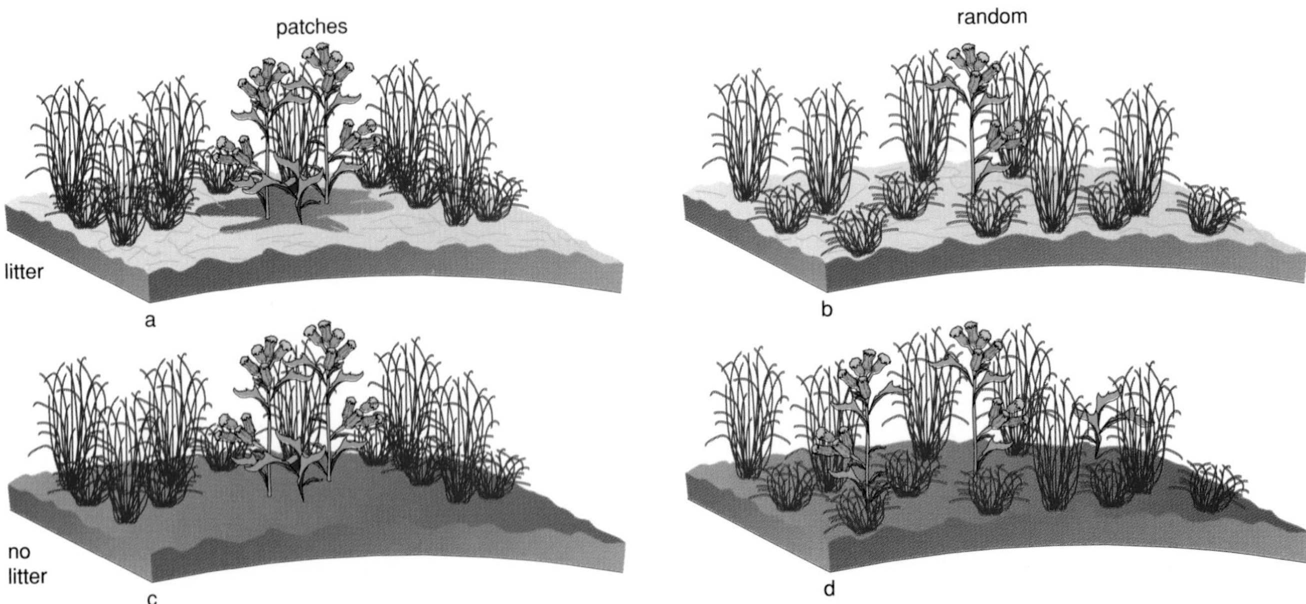

Figure 4. Between-generation competition controls the dynamics of a groundsel-bluegrass system. In the presence of live bluegrass blades and dead bluegrass litter groundsel grows much better in patchy environments (*a*), which include bare spots, than in environments that include random arrangements of bluegrass litter (*b*). After removing the litter from previous generations of bluegrass, groundsel grows equally well in patchy (*c*) and random (*d*) environments. So the previous generation of bluegrass, represented by the litter, determines groundsel's success.

dynamics of the groundsel-bluegrass system and explains the importance of the spatial pattern.

A series of greenhouse experiments revealed that the effect of bluegrass litter comes from the dead blades above ground. The presence of grass roots or chemicals that might have leached from the bluegrass did not affect the germination or survival of groundsel seedlings. Instead, litter inhibits groundsel seedlings, because emerging seedlings get trapped by the litter above them. The seedlings cannot penetrate the litter, which prevents them from capturing light or growing, and they die. This structural inhibition between bluegrass litter and groundsel seedlings provides the crucial competitive interaction. In addition, litter generates little trouble for the relatively slender morphology of a bluegrass seedling.

Figure 5. Bluegrass litter mechanically blocks groundsel seedlings. The seedlings cannot get through the litter to receive light, so they die. On the other hand, slender bluegrass seedlings can slip through their own litter as they emerge. (Photograph courtesy of the author.)

These investigations illustrate that the spatial pattern of bluegrass produces large effects on the success of the competitively inferior groundsel, and that the mechanism involves gap colonization, or interactions between generations. That conclusion has several additional implications. First, the interaction between groundsel and bluegrass includes a time lag—earlier generations affecting later ones. A variety of simple mathematical models illustrate that biological systems with time lags tend to have relatively more complex dynamics than systems without time lags. The second implication involves succession. As succession proceeds, a system's litter accumulates, which can shift the competitive balance from litter-intolerant species to litter-tolerant ones. In that way, litter can qualitatively alter the outcome of competition. Again, models support such a conclusion, showing that bluegrass should dominate whenever litter accumulates, and that groundsel dominates if the litter decomposes quickly.

Modeling Trade-offs

These small-scale experiments showed that groundsel grows more successfully in a patchy plot. In the simplest terms, one might say that greater amounts of bare ground favor groundsel, because the plant requires such gaps for establishment. Then one might ask: Given a particular amount of bare ground, how does its spatial distribution influence the success of invading weeds? I approached that question in collaboration with Jonathan Newman of Southern Illinois University and Ernesto Flores-roux, then of the University of Chicago. We performed experiments on a somewhat larger spatial scale—over a few meters—where we tried to determine how the dispersion of gaps influences how fast groundsel progresses through a field. These experiments reveal the community-level repercussions of between-generation competition.

We approached the effects of gap dispersion with a simple experiment. For each experimental plot in a field of ryegrass, we created six transects that were oriented like spokes on a wheel. On each transect, we created artificial gaps that covered one of three areas: 25, 225 or 900 square centimeters. To control the total amount of gap in a given transect, we created fewer large gaps than small ones. We distributed the gaps either uniformly or randomly, based on the distance between them. By analogy with the small-scale experiments described earlier, a large variance in the intergap distance corresponds to a patchy distribution, and equal intergap distances correspond to a uniform distribution. We introduced 12 invading groundsel plants in the center and then counted the number and position of all seedlings in two subsequent generations—hoping to determine whether the success of invasion depends on the spatial heterogeneity of the gaps.

One can assess a plant's success of invasion in two different ways. The number of individuals that get established provides one index, and the distance between a parent and its offspring—the rate of spread—provides another. In our experiments, larger gaps increased the number of established groundsel seedlings, even though we controlled for overall gap area. In addition, the invad-

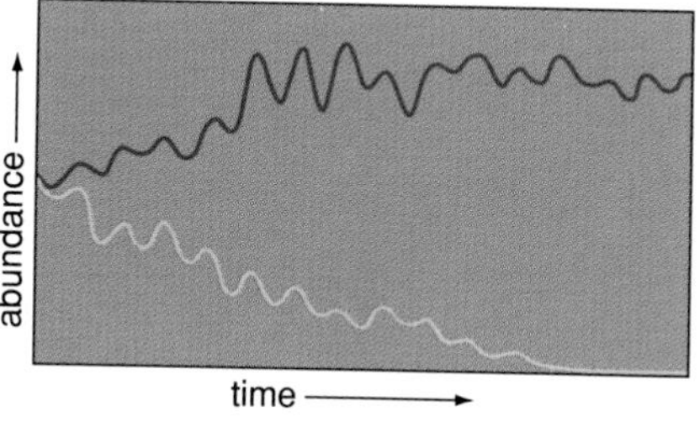

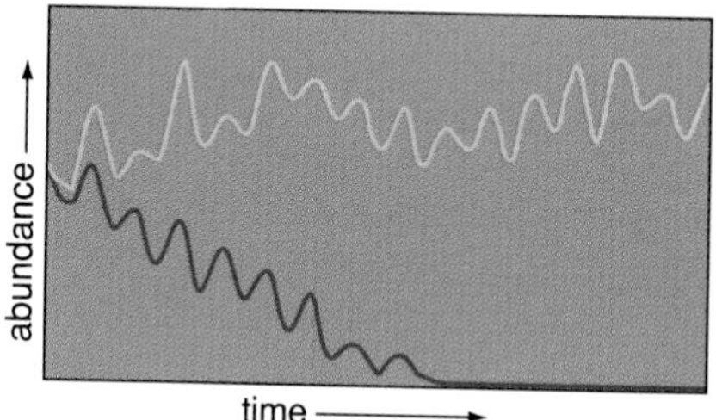

Figure 6. Mathematical model shows that the litter could control whether groundsel or bluegrass dominates a community. If the litter decomposes slowly (*left*), bluegrass (*green*) takes over and groundsel (*yellow*) nearly dies out, because it would only grow in bare patches generated by disturbances. If the litter decomposes quickly (*right*), the situation reverses: Groundsel thrives, because it would grow well regardless of the arrangement of bluegrass (Figure 4), and the bluegrass dies out.

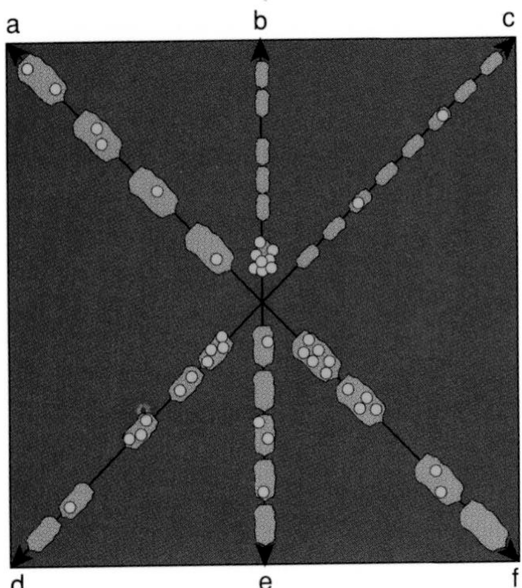

Figure 7. Groundsel's invasion of a field of bluegrass depends on the arrangement of gaps. The author and her colleagues created transects in a field of bluegrass, and they created artificial gaps, which were spaced either uniformly (*a*, *c* and *e*) or in patches (*b*, *d* and *f*). The investigators planted a dozen groundsel plants at the intersection of the transects and then counted the number and position of the groundsel seedlings over the next two generations. The invading groundsel (*yellow circles*) produced more plants along the transects with patchy gaps, but the groundsel traveled the farthest with uniformly arranged gaps. So depending on the competitor's spatial pattern, groundsel can either disperse widely or produce many offspring, but not both. Therefore, competition between plants depends on the characteristics of individual plants, the capabilities of competitors and the surrounding environment.

ing groundsel spread faster with large gaps. For example, large gaps produced nearly three times more distance between a parent and its offspring, as compared with small gaps. Moreover, the invading groundsel produced more established seedlings with patchy gaps than uniform ones, regardless of the size of the gaps. However, offspring traveled farther when the gaps were positioned uniformly, regardless of the size of the gaps.

We wondered if the way that a plant's "shower" of seeds would fall on such gaps would lead to similar results. One can imagine that a plant produces a seed shadow, which depicts the proportion of seeds that fall relative to the distance from the plant. In our transect experiment, some seeds would fall in gaps and germinate, and others would fall in vegetation and not germinate. By knowing a plant's seed shadow and a transect's arrangement of gaps, one can predict the expected distance between parents and seeds that land in gaps. A simple mathematical model of this scenario produced results that resembled what we found in our experiments. In other words, how the seeds disperse and the strong competitive dominance of established grass over seedlings explains what we observed.

These results point to an interesting trade-off: An invading plant can progress faster in a field that contains a uniform distribution of gaps, but fewer seeds land successfully in uniform gaps. This trade-off affects models of the persistence of competitively inferior species in patchy environments. In the past, such models suggested that a competitively inferior species can persist in a community by dispersing more effectively than its superior competitor, but models of that phenomenon ignore the spatial positioning of gaps. Our results, however, indicate that a competitively inferior species faces a more difficult challenge, because of the negative relationship between rates of dispersal and the probability that seeds land in gaps and establish successfully. Dispersing seeds that travel a far distance, on average, require uniformly positioned gaps; but patchy gaps lead to more established seedlings. In other words, a plant can either widely disperse its offspring or produce lots of them, but it probably cannot do both. Future research should address how competitively inferior species persist in realistic, spatially heterogeneous environments.

My work with two common weeds—groundsel and bluegrass—shows that competition between these plants depends on many factors. Competition between generations—later groundsel seeds battling established bluegrass—is the primary factor that governs the dynamics of this system. Nevertheless, groundsel's genotype determines largely when a seedling will emerge—a crucial factor in competitive success—and that suggests that contemporary plants must compete, as well. Moreover, the result of competition between groundsel and bluegrass also depends on the structure of the local environment, including the size and arrangement of gaps. In the future, ecologists hope to develop models and experimental systems that simultaneously examine how these factors contribute to plant competition.

Bibliography

Bergelson, J., and R. Perry. 1989. Interspecific competition between seeds: Complex effects of relative planting date and density on patterns of seedling emergence. *Ecology* 70:1639–1644.

Bergelson, J. 1990. Life after death: Site preemption by the remains of *Poa annua*. *Ecology* 71:2157–2165.

Bergelson, J. 1990. Spatial patterning in plants: Opposing effects of herbivory and competition. *Journal of Ecology* 78:937–948.

Bergelson, J. 1993. Details of local dispersion improve the fit of neighborhood competition models. *Oecologia* 95:299–302.

Bergelson, J., J. A. Newman and Ernesto Floresroux. 1993. Rates of weed spread in spatially heterogeneous environments. *Ecology* 74:999–1011.

Grace, J. B., and D. Tilman (ed.). 1990. *Perspectives in Plant Competition*. New York: Academic Press.

Hubbell, S. P., and R. Foster. 1986. Canopy gaps and the dynamics of a neotropical forest. In *Plant Ecology*, ed. M. J. Crawley. London: Blackwell Scientific Press.

Miller, Thomas. 1987. Effects of emergence time on survival and growth in an early oldfield community. *Oecologia* 72:272–278.

Pacala, S. W., and J. A. Silander, Jr. 1990. Field tests of neighborhood population dynamic models of two annual weed species. *Ecological Monographs* 60:113–134.

Parasitic Infections as Regulators of Animal Populations

The dynamic relationship between parasites and their host populations offers clues to the etiology and control of infectious disease

Robert M. May

Many field, laboratory, and theoretical studies have focused on the possibility that competition or prey-predator interactions may regulate animal numbers and influence the geographical distribution of species. On the other hand, the growing body of research on the transmission and maintenance of infectious diseases usually assumes that the host population is constant, not dynamically engaged with the disease. This article outlines recent work which weaves these two strands together, with the aim of understanding how parasitic infections may act as regulatory agents in natural populations of animals.

In what follows, the term *parasite* is used broadly to include viruses, bacteria, protozoans, and fungi along with the helminths and arthropods more conventionally defined as parasites. I begin with a number of anecdotes about the dynamical effects of parasites on particular host populations. Some useful concepts (including the intrinsic reproductive rate of the parasite and the effects of host density) are next introduced, and applied in an analytic discussion of a classic series of laboratory experiments in which mice populations were regulated by viral or bacterial infections. The discussion then broadens to consider general patterns in the regulation of natural populations by parasitic infections, and I go on to apply some of these ideas to that most fascinating of animal populations, *Homo sapiens*. The article concludes by touching very briefly on evolutionary aspects of host-parasite associations.

There is abundant evidence that parasites, in the broad sense defined above, cause many deaths in natural populations. Thus Delyamure's survey shows that helminths contribute significantly to the mortality rate in many populations of pinnipeds and cetaceans (*1*); one particularly careful study, for example, suggests that 11 to 14% of deaths among spotted dolphins (*Stenella* spp.) are caused by nematode infections in the brain (*2*). Lanciani has demonstrated that an ectoparasitic mite influences the population dynamics of the aquatic insect *Hydrometra myrae* (*3*). Lloyd and Dybas have suggested that the ultimate determinant of the population densities of the spectacular 13- and 17-year periodical cicadas is a fungal infection, *Massospora cicadina* (*4*). Many studies have shown that infectious diseases are an important, and possibly the predominant, mortality factor in bird populations (*5*).

Parasites can also affect the outcome of competition among species. This was shown in Park's (*6*) classic laboratory experiments on competition between two species of flour beetles: when the sporozoan parasite *Adelina* was present it dramatically reduced the population density of *Tribolium casteneum*, and in some situations reversed the outcome of its competition with *T. confusum*. The simian malarial parasite, *Plasmodium knowlesi*, is highly pathogenic for the rhesus monkey *Macaca mulatta*, but produces a chronic and much less lethal infection in *M. fascicularis; M. mulatta* is distributed widely throughout central, northern, and western India, where the malaria mosquito vector, *Anopheles leucosphyrus*, is absent, but is replaced by *M. fascicularis* in eastern India and parts of Bangladesh where *A. leucosphyrus* is present (*7*). On a grand scale, it seems that the geographical distribution of most artiodactyl species in East Africa today is determined largely by a pandemic of the rinderpest virus that occurred toward the end of the nineteenth century.

In many cases, different species of parasites combine to kill the host. Thus among bighorn sheep in North America the main cause of death is probably infection by the lungworms *Prostostronglylus stilesi* and *P. rushi*, which then predispose the hosts to pathogens causing pneumonia (*8*). More generally, it may be that the interplay between parasitic infections and the nutritional state of the host contributes importantly to the density-dependent regulation of natural populations.

Despite these illustrations of the devastating effects that diseases can have on natural populations, it is hard to assess the extent to which diseases are the primary regulators of such populations, as opposed to acting as occasional or incidental sources of mortality. For example, among certain species of wildfowl in North America some 80 to 90% of the individuals not shot by hunters die of diseases each year, yet it remains arguable that the essential factor regulating population density is the availability of breeding sites (*9*). In a study with implications for many insect populations, both temperate and tropical, Wolda and Foster (*10*) have documented outbreaks of the larvae of a tropical moth, *Zunacetha annulata*, which cause severe defoliation and are ended by a fungal infection; the key determinants of the overall population dynamics of this moth, however, remain

Robert M. May is Class of 1877 Professor of Zoology and Chairman of the University Research Board at Princeton University. After getting his doctorate in theoretical physics at Sydney University in 1960, he spent two years in the Applied Mathematics Department at Harvard University before returning to the Sydney University Physics Department. In the early 1970s, his interest turned to problems arising in community ecology; the work reported here is supported by the NSF. Address: Biology Department, Princeton University, Princeton, NJ 08544.

enigmatic. Delyamure (*1*) sums it up in his lament that "unfortunately, so far the influence of helminths on the population dynamics of pinnipeds and cetaceans has not been investigated at all" (p. 517), despite its likely importance.

In a recent review of the empirical evidence, Holmes (*9*) suggests that invertebrate populations and vertebrate populations in disturbed situations may more typically be regulated by parasitic infections than are natural populations of vertebrates. I now proceed to sketch an analytic framework within which these questions and suggestions may be pursued.

Microparasites and macroparasites

By classifying parasites according to their population biology rather than their conventional taxonomy, we can make a rough but useful distinction between microparasites and macroparasites (*11*).

Broadly speaking, microparasites are those having direct reproduction, usually at very high rates, within the host. As exemplified by most viral and bacterial, and many protozoan, infections, microparasites tend to be small in size and to have short generation times. Although there are many exceptions, the duration of infection is typically short relative to the average life-span of the host, and hosts that recover usually possess immunity against reinfection, often for life. Microparasitic infections are thus characteristically of a transient nature.

Macroparasites are broadly those having no direct reproduction within the host. This category embraces essentially all parasitic helminths and arthropods, which typically are larger and have much longer generation times than microparasites. When an immune response is elicited, it usually depends on the number of parasites present in the host, and tends to be of relatively short duration. Thus macroparasitic infections are typically of a chronic or persistent nature, with hosts being continually reinfected.

For microparasites, it generally makes sense to divide the host population into three distinct classes: susceptible; infected; and recovered and immune (*12*). Such a "compartmental" model for the dynamics of a host-microparasite system is shown schematically in Figure 1, and will be discussed further below. For macroparasites, on the other hand, the various factors characterizing the interaction—egg output per parasite, pathogenic effects upon the host, evocation of an immune response in the host, parasite death rates, and so on—all tend to depend on the number of parasites present in a given host (*13, 14*).

It follows that, for many macroparasites, a medically significant distinction can be made between infection (harboring one or more parasites) and disease (harboring a parasite burden large enough to cause illness). In other words, for an archetypal microparasitic infection such as smallpox it is reasonable to assume that a given individual either does or does not "have smallpox"; but for a macroparasite such as hookworm or schistosomiasis there is a real difference between being infected with one or two worms and carrying a worm load large enough to cause disease.

Reproductive rates and thresholds

For a discussion of the overall population biology of any organism, a central concept is its intrinsic reproductive rate, R_o. For parasites, R_o essentially measures the number of offspring that can be produced; R_o depends both on the basic biology of the parasite and on ecological, environmental, and behavioral factors that can influence transmission rates (*15–18*).

More precisely, for microparasites R_o is defined as the average number of secondary infections produced when one infected individual is introduced into a host population where everyone is susceptible. For macroparasites, R_o is the average number of female offspring produced by a mature female parasite over her lifetime that themselves achieve reproductive maturity in the absence of density-dependent constraints.

In either case, it is clear that R_o must exceed unity for a parasite to be capable of establishing itself within a host population. In simple situations, it may often be assumed that R_o is directly proportional to the total number of hosts that are candidates for infection, whence we can write

$$R_o = N/N_T \qquad (1)$$

Here N is the host population size, and the proportionality constant (which subsumes all manner of biological and environmental aspects of the transmission process) has been written as N_T^{-1}. The condition $R_o > 1$ for establishment of the infection thus translates into the requirement that the host population, N, exceed a given threshold magnitude, N_T (*12*).

More generally, R_o may be some nonlinear function of the host population, $R_o = f(N)$. The criterion $R_o > 1$,

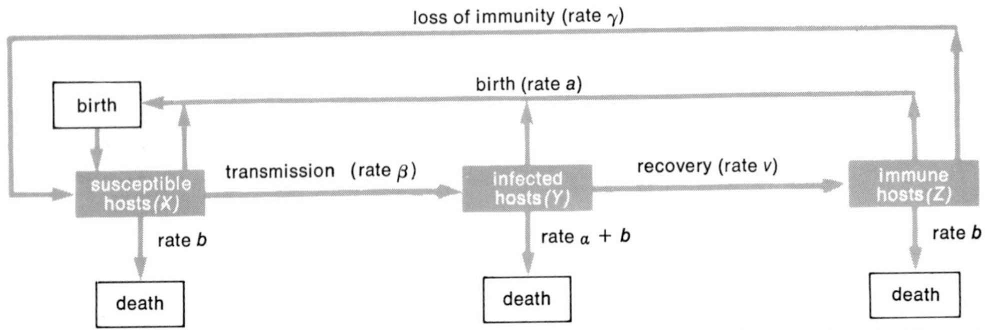

Figure 1. In this schematic representation of the dynamics of a host-parasite association, the host population is divided into susceptible (*X*), infected (*Y*), and immune (*Z*) classes. Susceptible individuals are gained through birth or through loss of immunity at rates *a* and γ, respectively, and are lost through natural mortality at rate *b* or by acquiring infection. Infected individuals are lost by disease-induced mortality, natural mortality, and recovery into the immune class at rates α, *b*, and *v*, respectively.

however, will still usually lead to some threshold condition, $N > N_T$. An important class of exceptions is sexually transmitted infections. Here R_o depends on the average rate at which new sexual partners are acquired, which usually has no direct dependence on N; doubling the population size does not affect sexual habits, except indirectly through possible social changes caused by greater crowding. Sexually transmitted parasites, which produce long-lasting infections and do not induce acquired immunity in recovered hosts, can be admirably adapted to persist in low-density populations of promiscuous hosts.

Direct assessment of the threshold density, N_T, is usually difficult. Some useful generalizations can, however, be made. Many microparasitic infections, such as smallpox and measles in man, are of very short duration and have relatively low transmission efficiencies; that is, the transmission stages may be short-lived in the external environment, and fairly direct contact may be needed to acquire infection. In this event, a large population of candidate hosts will be required before R_o can exceed unity, and thus the threshold density for the host population, N_T, will be large. Specifically, it has been estimated that the threshold population for maintaining measles in human communities is around 300,000 individuals or more (*19*, *20*). Conversely, the reproductive life-span of many macroparasites within a host is often an appreciable fraction of the host's life, and transmission pathways are often quite efficient, involving intermediate vector hosts or long-lived transmission stages. Threshold host densities for the maintenance of macroparasite populations can thus be small.

In short, directly transmitted microparasites typically require high host densities in order to persist, and should more commonly be associated with animals that exhibit herd or schooling behavior or that breed in large colonies. A certain amount of anecdotal support for these ideas comes from the observed abundance of such infections within modern human societies, large herds of ungulates, breeding colonies of seabirds, and communities of social insects (*11*). But there is a great need for comparative studies in which data are systematically compiled.

Laboratory populations

As a halfway house en route to applying analytic models to naturally occurring host-parasite associations, let us consider the artificial world of laboratory populations.

In the 1930s, Greenwood and his colleagues conducted a series of experiments on the way diseases can influence the dynamical behavior of populations of laboratory mice (*21*). Two infections were studied: a bacterial "mouse plague," *Pasteurella muris*, and a poxvirus, ectromelia. These experiments are remarkably detailed and well designed. Among other things, the cage space available to the mice was adjusted to keep the population density constant as absolute levels changed, thus avoiding complications arising from density dependences.

With one important exception, the mathematical model depicted schematically in Figure 1 may be used to analyze these host-parasite systems (*11*). The rates at which the numbers of susceptible, infected, and immune mice—$X(t)$, $Y(t)$, and $Z(t)$, respectively—change is described by a set of three differential equations, with the flows into and out of the three compartments being as indicated. The one important difference between Figure 1 and the actual experiments is that adult mice were introduced by the hand of the experimenter, at a constant rate of A adult mice per day, rather than by birth (as shown here). Susceptible mice are thus gained by "birth" or by recovered mice losing their immunity (at the per capita rate γ), and are lost by natural mortality (at the per capita rate b) or by acquiring infection. Both *P. muris* and ectromelia are transmitted directly, and we make the conventional epidemiological assumption that the net rate at which new infections appear is proportional to the number of encounters between susceptible and infected mice, βXY; the proportionality constant β represents the transmission rate. Thus infected mice appear at the net rate βXY and are lost by disease-induced mortality, by natural mortality, or by recovery into the immune class (at the per capita rates α, b, and v, respectively). The crucial difference between this system and conventional epidemiological ones is that the total host population, $N = X + Y + Z$, is not a predetermined constant but is itself a dynamic variable. In the absence of infection, A mice are added each day and bN are lost by natural deaths; the system thus settles to an equilibrium population, N^*, equal to A/b mice.

When N mice are in the cage, the intrinsic reproductive rate for the infection is

$$R_o = \beta N/(\alpha + b + v) \tag{2}$$

This follows from the assumption that if one infected mouse were introduced into a population of N susceptibles, new infections would appear at the rate βN for the duration of the infection, which on average is $1/(\alpha + b + v)$. Equation 2 reduces to equation 1 if we define the threshold mouse population density as $(\alpha + b + v)/\beta$. From the earlier discussion, we see that *P. muris* or ectromelia infections can establish themselves among these mice provided the population exceeds the threshold density N_T. But the disease-free mouse population is $N^* = A/b$. Thus if the introduction rate A exceeds bN_T the infection can be established, and not otherwise.

Once the infection is established, the mathematical model predicts that the mouse population will settle to a new equilibrium value, below the disease-free level. This disease-regulated population density will increase linearly with increasing introduction rate, A.

In their experiments with *P. muris*, Greenwood and his co-workers introduced new mice at rates ranging from 0.33 to 6 mice per day. As can be seen in Figure 2, the ensuing equilibrium populations of mice indeed depended linearly on A, as suggested by the theory. In fitting the theoretical line to the observed data points, the transmission parameter β has been treated as an adjustable parameter. The critical introduction rate, bN_T, below which *P. muris* cannot be maintained appears to be 0.11 mice per day, corresponding roughly to an equilibrium population of 19 mice. Unfortunately, the experiments were never conducted at so low a rate, so this conclusion remains untested.

The researchers also investigated the dynamical behavior of the infected mouse populations, $N(t)$, as a function of time, t, for two particular introduction rates,

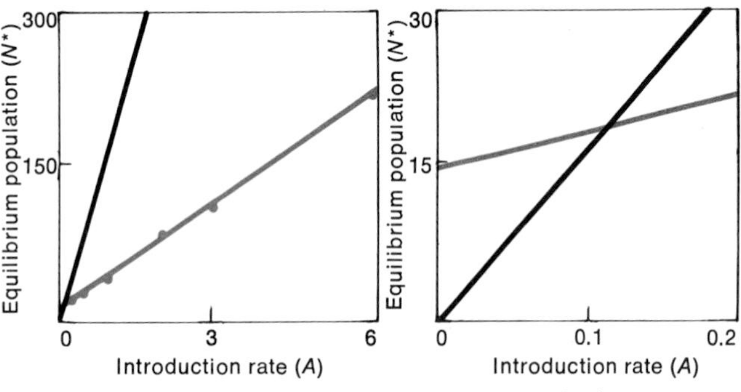

Figure 2. The equilibrium population of mice, N^*, is shown as a function of the introduction rate, A. The colored dots (*left*) represent the experimental results for populations infected with *P. muris*; the colored line indicates the relation given by a simple theoretical model, and the black line shows the relation in the absence of the infection. The intersection between the colored line and the black line – the disease-regulated and disease-free dependence of N^* on A, respectively – gives the critical rate of introduction, bN_T, below which *P. muris* cannot be maintained. An enlargement of the lower left-hand corner of the graph (*right*) suggests that the critical rate is 0.11 mice per day, corresponding roughly to an equilibrium population of 19 mice. (After ref. *11*.)

6 and 0.33 mice per day (Fig. 3). The theoretical curves here contain no adjustable parameters: the rates α, b, v, and γ can be estimated from independent studies of individual mice, while β (although it concatenates many biological and epidemiological factors in a way that defies direct estimation) is determined from Figure 2 as explained above. The agreement between the data and the theoretical curves, thus constructed, is encouraging.

The experiments with ectromelia were performed only for a single value of A, 3 mice per day. Consequently the analogue of Figure 3 now has one adjustable parameter (β). The fit between theory and data is again good. A comparison between the results for the bacterium *P. muris* and the virus ectromelia—which are biologically quite different organisms—is interesting, showing that the same basic mathematical model can apparently account for the essential dynamical features of both infections.

I have dwelt on this work for three reasons. First, it gives a hint of the kind of analysis that underlies the assertions in the next few sections. Second, it brings out some of the points about thresholds and regulation in an explicit fashion. Third, it shows that simple mathematical models can give a satisfactory account of observed facts, at least in the laboratory.

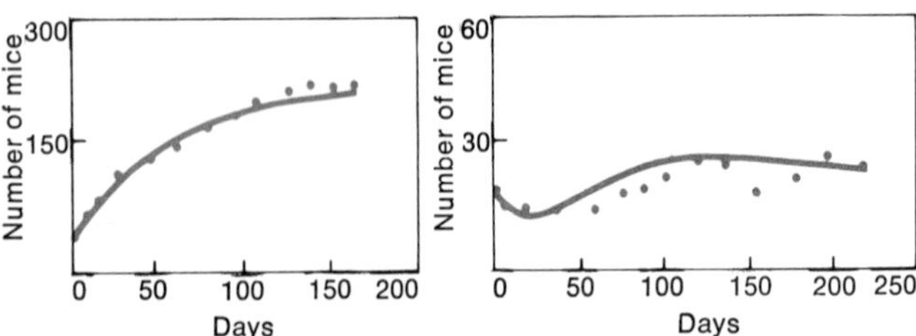

Figure 3. Changes in the number of mice in an experimental colony which was infected with *P. muris* are shown as a function of time for an introduction rate of 6.0 (*left*) and 0.33 (*right*) mice per day. The colored dots represent experimental observations, the colored lines the results of the simple theoretical model (in which all the parameters are estimated from other kinds of data). The agreement between theory and experiment suggests that the simple model does indeed capture the essentials of such a host-parasite association. (After ref. *11*.)

Natural populations

In the absence of infection, the host population depicted schematically in Figure 1 will grow exponentially at the per capita rate $r = a - b$, the difference between the per capita birth and death rates. Given this assumption of exponential growth, the host population will eventually exceed any threshold level for the establishment of a particular parasite. Under what circumstances may the additional mortality associated with the infection regulate the host population, holding it to some steady value? Under the simple circumstances represented in Figure 1 and assuming that the net transmission rate is βXY, the criterion for such regulation by a microparasite is (*11*)

$$\alpha > r[1 + v/(b + \gamma)] \quad (3)$$

where α is the virulence, or disease-induced mortality rate, as defined above.

The assumptions on which equation 3 are based are excessively simple in many ways. Vertical transmission (whereby infection is transmitted from a parent to an unborn offspring, as happens in the case of many microparasites of insects) does not affect the criterion represented by equation 3, but does lower threshold population levels. Latent periods of significant length, during which infected individuals are not yet infectious, both raise thresholds and alter the criterion in a way that requires α to be larger. Microparasites that depress the reproductive capacity of infected hosts are effectively increasing α, and thus obviously find it relatively easier to satisfy the criterion for regulation. Dependence of the virulence on the nutritional state of the host, and a multitude of other possible density-dependent effects, make for further complications (*11*). The possibility that a macroparasite may regulate a host population can be similarly assessed, using a formula that is related to equation 3 (*14*).

Using equation 3 as a basic guide, we see that it will be easier for a host population to be regulated by a microparasite if the hosts have a relatively small r value, and if there is no acquired immunity (corresponding to $\gamma \rightarrow \infty$, whence the criterion for regulation is simply $\alpha > r$). This latter observation implies that, other things being equal, invertebrate populations may be regulated by parasites more commonly than vertebrates, which typically do possess acquired immunity against many infections. This marches with Holmes's empirically based suggestion that regulation may be more prevalent in invertebrate populations (*9*).

Among univoltine insects (where $\gamma \rightarrow \infty$ and r is not too large), there appear to be many instances of associations with viral, fungal, or microsporidian protozoan parasites which are candidates for regulating their host population (*22*). In particular, several baculovirus and microsporidian parasites found among univoltine forest insects are highly virulent and possess transmission stages during which the parasites live a relatively long time in the external environment, free of their hosts, thus helping the host-parasite association to persist. The outcome of such regulation of an insect population by a parasite population with long-lived transmission stages

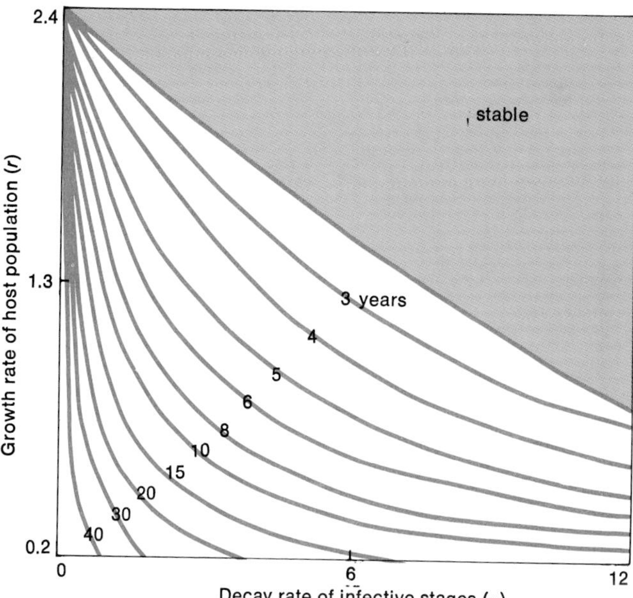

Figure 4. An association between a host population and a virulent parasite with free-living transmission stages can result either in a state of constant equilibrium or in stable cycles of varying length, depending on the intrinsic growth rate of the host (r) and the decay rate of the infective stages of the parasite (μ). In the upper right-hand corner of the figure, both r and μ are relatively high, leading to constant equilibrium values; at the lower left, r is relatively small and the transmission stages are long, resulting in cyclic periods ranging from a few years to a few decades, as indicated. (After ref. *22*.)

may be either a steady, constant value or a stable cycle in which the density of the host and the prevalence of disease rise and fall in a periodic fashion (*22*). Figure 4 illustrates how this range of outcomes depends on the intrinsic per capita growth rate, r, of the host insect population, and on the decay rate, μ, of the free-living infective stages of a parasite that is very virulent, so that $\alpha \gg r$.

Many univoltine forest insects have indeed been observed to exhibit such cyclic variations in abundance, with periods ranging from 5 to 12 years. In most cases, baculovirus or microsporidian pathogens have been found in these populations. But the possibility that the cycles are the result of host-parasite dynamics has only begun to be explored.

However, changes in the abundance of the larch budmoth, *Zeiraphera diniana*, in the Engadine Valley in Switzerland have been investigated in the light of the prevalence of infection with a granulosis virus in this population (*23*). As shown in Figure 5, all the main features of the data are in accord with the results of a theoretical model which is essentially that represented in Figure 1 (simplified by putting the recovery rate $v = 0$, and with the transmission process made complicated by accounting for the dynamics of the free-living infective stage of the parasite). Although this model explains both the 9- to 10-year period of the budmoth cycle and the pattern of prevalence of viral infection, it gives an amplitude of population oscillation that is an order of magnitude less than that observed. Other authors have explained the budmoth cycles as deriving from the interaction between this herbivore and the foliage it eats (*24*). I think the truth may lie in a combination of interactions, with the host-parasite dynamics setting the basic period and plant-herbivore effects enhancing the amplitude.

Such studies of the possible regulation of natural populations of invertebrates by viral, fungal, or protozoan parasites are of interest to those who seek a fundamental understanding of community dynamics. The studies may also eventually be useful in helping to specify the properties a pathogen should have, and in what quantities it should be distributed, to control an insect pest (*22*).

For an example involving vertebrate hosts, I turn to the current epidemic of rabies in Europe. This outbreak is thought to have originated in Poland in 1939, and is characterized by a high incidence in populations of red fox (*Vulpes vulpes*); of the roughly 17,000 cases of animal rabies reported in Europe in 1979, more than 70% were in this host species (*25*). Where rabies is now endemic in Europe, fox population densities appear to exhibit 3- to 5-year cycles around average levels that are significantly lower than the disease-free ones, which are set primarily by territoriality. An appropriately modified version of the host-parasite model described in Figure 1, allowing for a latent period in infected foxes and for density dependence in fox birth rates, indeed exhibits these dynamical features (*26*). Incorporating rate parameters estimated from field data on fox behavior and on the etiology of rabies in individual foxes, the model also helps explain the observed prevalence of infection in fox populations, and suggests a threshold density of around 1 fox per km^2 for maintenance of the infection (*26*). Such models may be applied to get a rough idea of

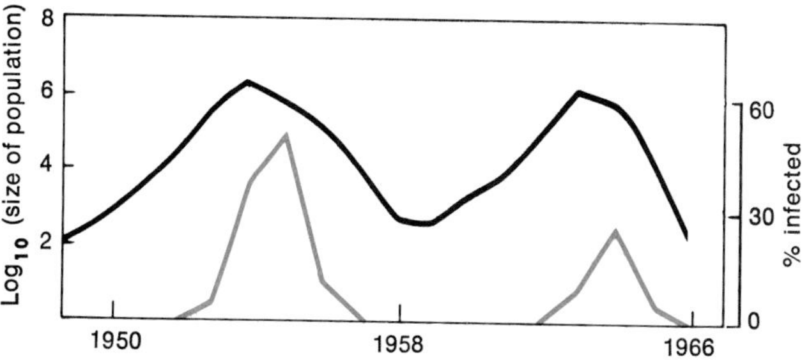

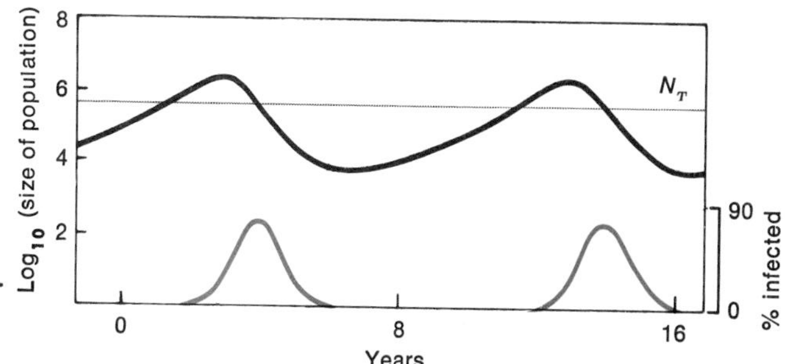

Figure 5. The observed patterns in the abundance of the larch budmoth, *Zeiraphera diniana* (*black line*), and in the prevalence of infection with a granulosis virus among this population (*colored line*), in the Engadine Valley in Switzerland from 1949 to 1966 are shown at the left. When the pertinent epidemiological and demographic parameters are used in a modified version of the model shown in Figure 1, the oscillations in the larch budmoth population in relation to the threshold level, N_T, and in the prevalence of infection are as indicated at the right. The theoretical model appears to give the correct period for the cyclic rise and fall of the budmoth population, although the actual oscillations have a larger amplitude than is predicted. (After ref. *22*.)

the possibilities of controlling rabies by culling foxes or by vaccinating them with live baits.

General patterns

The preceding discussion has dealt largely with microparasitic infections, where it is usually possible to classify hosts simply as infected or uninfected. For macroparasitic infections, on the other hand, the damage inflicted on a given host individual usually depends strongly on the magnitude of the parasite burden.

Figure 6 illustrates the typical relation between the average number of macroparasites per host and the host population density, N, that emerges from mathematical models: below the threshold host density (N_T, corresponding to $R_o = 1$) the parasite cannot persist; as N increases above N_T, the average parasite burden at first increases markedly; and for very large N the average burden tends to saturate to a constant level as density-dependent effects exert their influences. If relatively high parasite burdens produce a sufficiently high degree of host mortality, then the relation shown in Figure 6 can lead to the parasites regulating their host population: at low host densities, there is little parasite-induced mortality, and the population grows; at high host densities, the average parasite burdens rise and cause many deaths, thus halting host population growth (*14*).

Laboratory and field studies have shown that the general theoretical relation depicted in Figure 6 is indeed found in real host-macroparasite systems. In the field—if the deep sea may be so called—Campbell and his colleagues (*27*) have demonstrated that the mean parasite loads in deep-living benthic fishes are directly related to the fish population densities. In addition, they found that more species of metazoan parasites occur when fish population densities are higher, which arguably is because such high host densities are above the threshold level for an increasing number of parasite species. Similar examples of an increase in the average number of parasites per host with increasing host density have been found in several ungulate species (*9*).

A related study, with broader implications, has been carried out by Freeland (*28*), who examined the protozoan parasites in the intestines of primates. Freeland's study embraced four primate species, and defies brief summary. One set of results showed that in social groups of mangabeys the number of protozoan species in a given individual increases systematically with the size of the group of which it is a member (Fig. 7). (This is not a sampling effect, because all individuals in a given social group exhibit identical protozoan faunas.) The explanation may be, in part, that suggested above for similar phenomena among fish. Freeland advances the idea that this increase in parasite infestation with increasing group size may influence the evolution of group size and of intragroup behavior. Thus parasites may affect not only the dynamics of their host population (the theme of my article), but also the behavioral ecology.

To summarize, the general relation shown in Figure 6 is found in real host-macroparasite associations. Such a relation, in conjunction with sufficiently high host mortality produced by high parasite burdens, can regulate the host population. There are, however, very few empirical studies of this question.

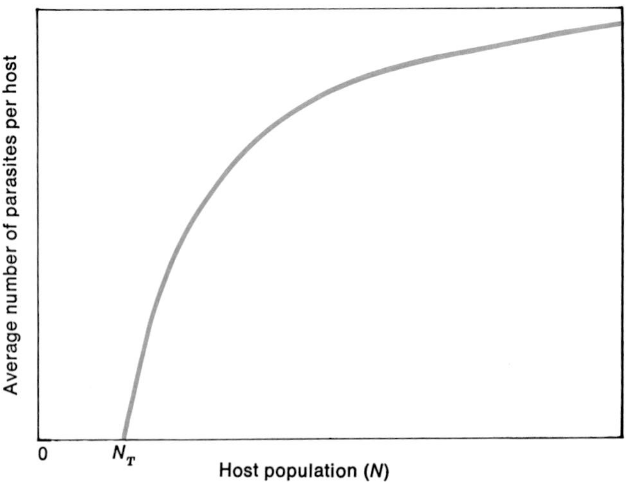

Figure 6. The theoretical relation between host population density and the number of macroparasites per host produces a characteristic curve which is supported by laboratory and field studies of real host-macroparasite systems. As the host density rises above the threshold level, N_T, the average parasite burden at first increases rapidly, but tends to rise fairly slowly or to settle to a steady value at very high values of *N*. At relatively high parasite burdens and correspondingly high levels of host mortality, this relation can lead to regulation of the host population by the parasite.

What fraction of all host deaths must be attributable to a particular microparasite or macroparasite, if the parasite is truly the regulator? A simple yet general answer can be given, provided only we assume that the per capita birth rate, a, and the per capita death rate from all other causes, b, are density-independent constants (which, of course, is not usually so):

$$\frac{\text{parasite-induced host deaths}}{\text{total host deaths}} = \frac{a-b}{a} \tag{4}$$

The result holds for both microparasites and macroparasites, independent of the mode of transmission and of the nature of immune processes. If a and b are not too disparate, so that $(a - b)/a$ is small, the parasite can be responsible for regulating the host population, even though few deaths will be laid at its door. Conversely,

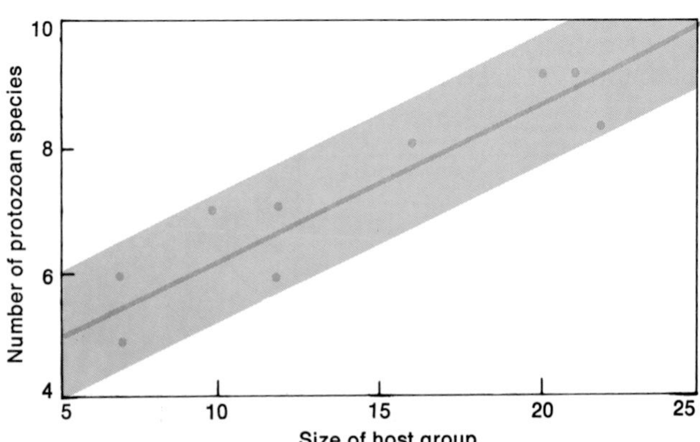

Figure 7. In a field study of mangabey social groups, the number of protozoan species exhibited by members of groups was found to rise systematically with increase in group size. Colored dots represent the observational data, and the colored line shows the best straight line that can be fit to the data; 95% of the results would be expected to fall within the shaded area if this linear fit is indeed significant. (After ref. *28*; © 1979 Ecol. Soc. of Am.)

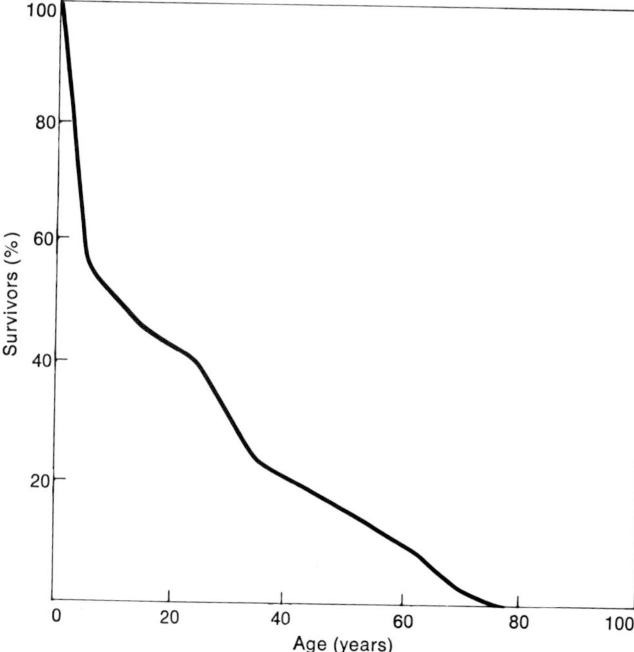

Figure 8. The survival curve for a group of hunter-gatherers who lived approximately 15,000 years ago on the Mediterranean coast shows high child mortality, with only about 50% of the population surviving past age 10. Although skeletal remains yield little definite evidence of parasitic infection, such a pattern suggests that this factor may have played a role. (After ref. *31*.)

if $a \gg b$ (as is the case for many invertebrates), the infection needs to be responsible for most of the observed mortality before it can be a candidate for consideration as the regulatory agent. My own equivocal view is that more often than not, no single factor can usefully be called *the* regulator.

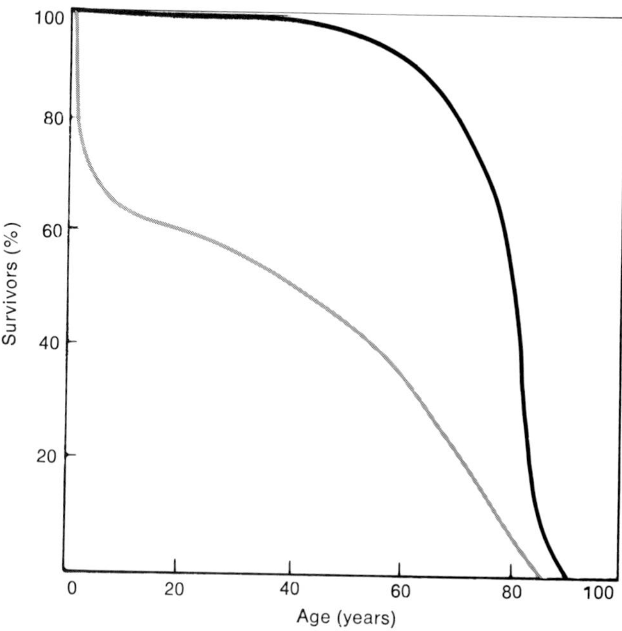

Figure 9. The survival curve for a typical developed country (*black line*) shows relatively low child mortality, with essentially all the population surviving to age 10, and more than 90% surviving past age 60. By contrast, the curve for a typical developing country (*gray line*) shows high child mortality, with only about 60% surviving past age 10, and less than 40% surviving past age 60; this higher mortality arises from a greater prevalence of parasitic infections combined, on average, with a lower nutritional state. (After ref. *38*.)

Human populations

During most of the million or so years that humans have existed, their numbers have been low and relatively constant. The processes of fertility and mortality that kept this rough balance among hunter-gatherer populations are still far from understood (*29, 30*).

Figure 8 shows a survival curve, deduced from skeletal evidence (*31*), for a group of Mediterranean hunter-gatherers who lived about 15,000 years ago; this curve is probably fairly typical for such pre-agricultural groups (*29*). There is little doubt that diseases contributed to the mortality patterns exemplified by Figure 8, but precise knowledge of the extent of the contribution is impossible. Although some parasitic infections leave traces on early human skeletons—usually the only remains available for examination—most, particularly those caused by viruses and bacteria do not.

Certain conclusions may, however, be drawn from the discussion of threshold host densities given above. Bands of hunter-gatherers probably ranged in size from around 20 to at most 100 individuals (*29*), and these populations may have been sufficient to maintain many macroparasites as well as microparasites with long periods of infectiousness or (as is the case with hepatitis, herpes simplex, gonorrhea, and other infections) with asymptomatic carriers (*32, 33*). Pathogenic organisms whose normal habitat is a host other than man or that are able to multiply and survive successfully in soil and other inanimate environments (such as tetanus, gas gangrene, and other *Clostridiae*) could also play a role among these sparse early populations. But the directly transmitted microparasites responsible for much mortality in historical times—smallpox, measles, cholera, and the like—have very high host threshold densities, and could not have been present in the pre-agricultural era (*32, 33*).

Starting about 10,000 years ago in the Old World, nomadic cultivation began to give way to true agriculture, leading to denser and denser aggregations of people. Many macroparasites now undoubtedly began to attain infection levels that could produce morbidity and mortality (see Fig. 6). Zoonoses associated with domestic animals (such as tuberculosis) probably became significant. More important, population levels now became high enough for the virulent microparasitic infections to establish themselves. Many authors (*32–35*) have discussed the various consequences of these changing patterns in the relation between "plagues and peoples" (*36*).

One large pattern deserves more attention than it has received. According to Deevey's necessarily rough estimates (*37*), the first 5,000 years after the beginning of the Agricultural Revolution saw human numbers increase about twentyfold, from about 5 million to about 100 million. A second, roughly equal period, from about 5,000 years ago to 300–400 years ago, saw only about a fivefold increase, to approximately 500 million. A possible explanation is that human conglomerations gradually rose to levels capable of maintaining directly transmitted microparasitic diseases, whose effect was then to slow population growth.

A more familiar application of these ideas about threshold population sizes lies in the history of the Western European conquest of the New World and Oceania, which was accomplished largely with biological weapons: smallpox, tuberculosis, and measles. It seems likely the peoples of the New World and Oceania had no similarly virulent microparasites of their own with which to counter because they were, or until recently had been, at densities too low to maintain such infections. A notable exception that may have been exported to the Old World (although even this is debatable) is syphilis, which is well adapted to persist at low host densities.

Figure 9 contrasts modern survival curves for a developed country and a developing one (*38*). The comparatively high death rate in the Third World comes not primarily from exotic tropical diseases, but rather—especially in the first few years of life—from diarrhea, measles, and the like, combined with poor nutrition. The figure makes it clear, however, that infectious diseases continue to take a substantial toll in developing countries today. Moreover, both curves would tell a gloomier story if they were plotted for earlier times: the increases in life expectancy in the developed world over the past two centuries, and in the developing world since World War II, are due almost wholly to reduced mortality from infectious diseases (*39*).

Although this broad conclusion is plain, the details are difficult to elucidate. In Western Europe, mortality from microparasitic infections fell throughout the late eighteenth and the nineteenth centuries, long before the advent of modern drugs or vaccines (excepting that for smallpox). There is an unresolved argument about the relative contributions made by better nutrition, improved hygiene due to innovations ranging from more use of soap to better sanitation, and other factors (*40, 41*). Similarly complex is the debate about the contribution made to overall population growth in European countries from about 1750 to 1850 by this decreased mortality, versus that made by increased fertility. Recent analyses suggest no single answer (*42*). In England, population grew mainly as a result of increased fertility, but with some help from declining mortality; in Sweden, population grew almost entirely as a result of decreased death rates, partly associated with smallpox vaccination; and in France, population remained roughly steady as mortality and fertility fell together.

In short, infectious diseases have been important sources of mortality throughout human history (*39*). In any epoch, however, their exact contributions to demographic trends are hard to pin down.

Control by vaccination

It is possible to be relatively precise about one technical aspect of the interplay between humans and directly transmitted microparasitic infections, namely the proportion of the host population that must be protected by an immunization program in order eventually to eradicate the infection.

We recall that a microparasite's intrinsic reproductive rate, R_0, is the number of secondary infections produced in a population where all are susceptible. Suppose now that a portion of the population, p, is protected by vaccination. The fraction remaining susceptible is at most $(1 - p)$, and, assuming that the host population mixes homogeneously, the reproductive rate of the parasite is diminished to $R_0(1 - p)$. For eradication, this rate must be driven below unity; that is, the fraction protected must exceed (*16, 17*)

$$p > 1 - 1/R_0 \tag{5}$$

For microparasitic infections in human populations, R_0 can be estimated accurately from serological studies, or roughly from knowledge of the average age at which infection is acquired (*16, 17, 43*). Table 1 displays the value of R_0, and the attendant value of p as derived from equation 5, for a variety of such infections (*43*).

Table 1. Estimates of the intrinsic reproductive rate, R_0, for infections of human populations

Infection	*Location and time*	R_0	*Approximate value of* p (%)[a]
smallpox	developing countries, before global campaign	3–5	70–80
measles	England and Wales, 1956–68	13	92
	US, various places, 1910–30	12–13	92
whooping cough	England and Wales, 1942–50	17	94
	Maryland, US, 1908–17	13	92
german measles	England and Wales, 1979	6	83
	West Germany, 1972	7	86
chicken pox	US, various places, 1913–21 and 1943	9–10	90
diphtheria	US, various places, 1910–47	4–6	~80
scarlet fever	US, various places, 1910–20	5–7	~80
mumps	US, various places, 1912–16 and 1943	4–7	~80
poliomyelitis	Holland, 1960; US, 1955	6	83
malaria (*P. falciparum*)	northern Nigeria, 1970s	~80	99
malaria (*P. malariae*)	northern Nigeria, 1970s	~16	94

SOURCES: See refs. *17* and *43*

[a] p = the proportion of the population that must be protected by immunization to achieve eradication

Smallpox appears to have one of the smallest values of R_0 of these tabulated infections, suggesting that vaccination of around 70 to 80% of the population in the neighborhood of known cases may be sufficient for eventual eradication. This fact, in conjunction with the obviousness of the disease and the availability of an effective vaccine, may help explain the success of the global eradication campaign. Although the etiology of measles is similar to that of smallpox, with the infection always being apparent and running a relatively short course, the tentative estimate that R_o is probably around 15 for measles in developing countries (corresponding to a p in excess of 93%) may make its global eradication much more difficult than was the case for smallpox.

The very high values of R_o for malaria suggest that its eradication is likely to be very difficult, whatever the control method. In particular, a campaign against *P. falciparum* based wholly on the use of an effective vaccine would appear to require that 99% of the population be protected for eradication to be achieved.

Coevolution of host and parasite

The received wisdom, set forth in most medical texts and elsewhere, is that "successful" or "well-adapted" parasites are relatively harmless to their hosts. This idea is reasonable, at first sight: all else being equal, it is to the advantage of both host and parasite for the parasite to inflict little damage. Thus in regions of Africa where trypanosomiasis is endemic, indigenous ruminants suffer mild infections with insignificant morbidity, while domestic ruminants that have been bred for a long time in the region suffer more severely, and recently imported exotic ruminants suffer virulent infections which are usually fatal if untreated (*7*). The fact that parasitic infections appear to be more effective as regulatory agents among newly introduced species of plants and animals, or when the parasites are introduced into new regions, further supports this conventional view (*9*).

On theoretical grounds, it would indeed appear that parasites evolve to be avirulent, provided that transmissibility and duration of infectiousness are entirely independent of virulence. This assumption, however, is not generally valid; the damage inflicted on their hosts by viral, bacterial, protozoan, and helminth parasites is often directly associated with the mechanism by which the organism produces its transmission stages. Once these complications are introduced into the theoretical models, it appears that many coevolutionary paths are possible, depending on the details of the interplay between the virulence and the transmissibility of the parasite (*14, 22, 44–46*).

There are circumstances where the evolutionary pressures on the parasite may promote virulence. The various baculoviruses, which kill their insect hosts and effectively turn them into masses of viral transmission stages, are likely examples.

The introduction of the myxoma virus into wild populations of rabbits in Australia and England in the early 1950s provides an unusually well-documented and interesting case study (*47*). At first the disease was highly virulent, but throughout the subsequent decade successively less virulent strains of the virus began to appear. Since the mid-1960s, the virus appears to have come to an equilibrium with its rabbit host in both Australia and England, with the predominant strain of the virus being one of intermediate virulence. The data can be analyzed to get a rough estimate of the relationship between the virulence, α, and the transmission rate, $\beta(\alpha)$, and host recovery rate, $v(\alpha)$, of the various strains of the myxoma virus (*44*). Substituting these empirical relations for $\beta(\alpha)$ and $v(\alpha)$ in equation 2 for the intrinsic reproductive rate R_o of myxoma virus, we obtain an estimate of the overall dependence of R_o on α. In this particular instance, it turns out that strains with intermediate virulence, α, have the largest R_o, and may thus be expected to predominate (*44*).

In brief, although parasite "harmlessness" may characterize many old and established associations, neither a priori theoretical arguments nor empricial evidence point to this being a general rule.

The reader may conclude from this survey that empirical evidence about the extent to which natural populations of animals are regulated by parasitic infections is scattered and equivocal, and that theoretical models are largely still in a formative stage. I think this is an accurate impression. The general subject of the regulation of animal populations by parasites is, as yet, one in which there are more questions than answers.

References

1. S. L. Delyamure. 1955. *The Helminth Fauna of Marine Animals in the Light of Their Ecology and Phylogeny.* Moscow: Izd. Akad. Nauk SSSR. (Translation TT67-51202. Springfield, VA: US Department of Commerce.)
2. W. F. Perrin and J. E. Powers. 1980. Role of a nematode in natural mortality of spotted dolphins. *J. Wildlife Man.* 44:960–63.
3. C. A. Lanciani. 1975. Parasite induced alterations in host reproduction and survival. *Ecology* 56:689–95.
4. M. Lloyd and H. S. Dybas. 1966. The periodical cicada problem: I. Population ecology. *Evolution* 20:133–49.
5. J. W. Davis, R. C. Anderson, L. Karstad, and D. O. Trainer, eds. 1971. *Infectious and Parasitic Diseases of Wild Birds.* Iowa State Univ. Press.
6. T. Park. 1948. Experimental studies of interspecies competition: I. The flour beetles *Tribolium confusum* and *T. castaneam. Ecol. Monogr.* 18:265–308.
7. A. C. Allison. 1982. Coevolution between hosts and infectious disease agents, and its effects on virulence. In *Population Biology of Infectious Diseases*, ed. R. M. Anderson and R. M. May, pp. 245–67. Springer Verlag.
8. D. J. Forrester. 1971. Bighorn sheep lungworm-pneumonia complex. In *Parasitic Diseases of Wild Mammals*, ed. J. W. Davis and R. C. Anderson, pp. 158–73. Iowa State Univ. Press.
9. J. C. Holmes. 1982. Impact of infectious disease agents on the population growth and geographical distribution of animals. In *Population Biology of Infectious Diseases*, ed. R. M. Anderson and R. M. May, pp. 37–51. Springer Verlag.
10. H. Wolda and R. Foster. 1978. *Zunacetha annulata* (Lepidoptera: Dioptidae), an outbreak insect in a neotropical forest. *Geo. Eco. Trop.* 2:443–54.
11. R. M. Anderson and R. M. May. 1979. Population biology of infectious diseases. *Nature* 280:361–67 and 455–61.
12. N. J. T. Bailey. 1975. *The Mathematical Theory of Infectious Diseases.* 2nd ed. Macmillan.
13. H. D. Crofton. 1971. A quantitative approach to parasitism. *Parasitology* 63:179–93.
14. R. M. Anderson and R. M. May. 1978. Regulation and stability of host-parasite population interactions. *J. Anim. Ecol.* 47:219–47 and 249–67.
15. G. Macdonald. 1952. The analysis of equilibrium in malaria. *Trop. Dis. Bull.* 49:813–29.
16. K. Dietz. 1975. Transmission and control of arbovirus diseases. In *Epidemiology*, ed. D. Ludwig and K. L. Cooke, pp. 104–21. Philadelphia: Society for Industrial and Applied Mathematics.
17. R. M. Anderson and R. M. May. 1982. Directly transmitted infectious diseases: Control by vaccination. *Science* 215:1053–60.
18. J. A. Yorke, N. Nathanson, G. Pianigiani, and J. Martin. 1979. Seasonality and the requirements for perpetuation and eradication of viruses in populations. *Am. J. Epidem.* 109:103–23.
19. M. S. Bartlett. 1957. Measles periodicity and community size. *J. Roy. Stat. Soc.*, ser. A, 120:48–70.

20. F. L. Black. 1966. Measles endemicity in insular populations: Critical community size and its evolutionary implication. *J. Theor. Biol.* 11:207–11.

21. M. Greenwood, A. Bradford Hill, W. W. C. Topley, and J. Wilson. 1936. *Experimental Epidemiology.* Special Report Series, no. 209, Medical Research Council. London: HMSO.

22. R. M. Anderson and R. M. May. 1981. The population dynamics of microparasites and their invertebrate hosts. *Phil. Trans. Roy. Soc.* B 291:451–524.

23. C. Auer. 1968. Ergenbnisse einfacher stochastischer Modelluntersuchungen uber die Ursachen der Populations beuegung desgrauen Larchenwicklers *Zeiraphera diniana* im oberengadin 1949/66. *Z. angew. Ent.* 62:202–35.

24. A. Fischlin and W. Baltensweiler. 1979. Systems analysis of the larch budmoth system: Part 1. The larch budmoth relationship. *Bull. Soc. Entomol. Suisse* 52:273–89.

25. World Health Organization. 1980. *Rabies Bulletin, Europe*, vol. 3, p. 5. Tubingen: WHO Collaboration Centre for Rabies Surveillance and Research.

26. R. M. Anderson, H. Jackson, R. M. May, and T. Smith. 1981. The population dynamics of fox rabies in Europe. *Nature* 289:765–71.

27. R. A. Campbell, R. L. Haedrich, and T. A. Munroe. 1980. Parasitism and ecological relationships among deep-sea benthic fishes. *Marine Biol.* 57:301–13.

28. W. T. Freeland. 1979. Primate social groups as biological islands. *Ecology* 60:719–28.

29. F. A. Hassan. 1981. *Demographic Archaeology.* Academic Press.

30. R. M. May. 1978. Human reproduction. *Nature* 272:491–95.

31. G. Y. Acsadi and J. Nemeskeri. 1970. *History of Human Life Span and Mortality.* Budapest: Ackademiai Kiado.

32. D. Brothwell and A. T. Sandison. 1967. *Disease in Antiquity: A Survey of the Disease, Injuries and Surgery of Early Populations.* Springfield, IL: C. C. Thomas.

33. T. A. Cockburn. 1971. Infectious diseases in ancient populations. *Curr. Anthrop.* 12:45–62.

34. G. T. Armelagos and A. McArdle. 1975. Population, disease and evolution. In *Population Studies in Archaeology and Biological Anthropology*, ed. A. C. Swedlund, pp. 1–10. *Am. Antiquity*, Memoire 30.

35. J. B. S. Haldane. 1949. Disease and evolution. *La Ricerca Sci. Suppl.* 19:68–76.

36. W. H. McNeill. *Plagues and Peoples.* Doubleday.

37. E. S. Deevey. 1960. The human population. *Sci. Am.* 203(g):195–204.

38. D. J. Bradley. 1974. Stability in host-parasite systems. In *Ecological Stability*, ed. M. B. Usher and M. H. Williamson, pp. 71–87. London: Chapman and Hall.

39. M. P. Hassell. 1982. Impact of infectious diseases on host populations (group report). In *Population Biology of Infectious Diseases*, ed. R. M. Anderson and R. M. May, pp. 15–35. Springer Verlag.

40. T. McKeown. 1976. *The Modern Rise of Population.* London: Edward Arnold.

41. P. E. Razzell. 1974. An interpretation of the modern rise of population in Europe: A critique. *Pop. Studies* 28:5–17.

42. E. A. Wrigley and R. S. Schofield. 1981. *The Population History of England, 1541–1871.* Harvard Univ. Press.

43. R. M. May. 1983. Ecology and population biology of parasites. In *Tropical and Geographical Medicine*, ed. K. S. Warren and A. F. Mahmoud, chap. 24. McGraw-Hill.

44. R. M. Anderson and R. M. May. 1982. Coevolution of hosts and parasites. *Parasitology.* 85:411–26.

45. S. A. Levin and D. Pimentel. 1981. Selection of intermediate rates of increase in parasite-host systems. *Am. Nat.* 117:308–15.

46. B. R. Levin. 1982. Evolution of parasites and hosts (group report). In *Population Biology of Infectious Diseases*, ed. R. M. Anderson and R. M. May, pp. 213–43. Springer Verlag.

47. F. Fenner and K. Myers. 1978. Myxoma virus and myxomatosis in retrospect: The first quarter century of a new disease. In *Viruses and Environment*, ed. E. Kurstak and K. Maramorosch, pp. 539–70. Academic Press.

The Emergence of New Diseases

Lessons learned from the emergence of new diseases and the resurgence of old ones may help us prepare for future epidemics

Richard Levins, Tamara Awerbuch, Uwe Brinkmann, Irina Eckardt, Paul Epstein, Najwa Makhoul, Cristina Albuquerque de Possas, Charles Puccia, Andrew Spielman and Mary E. Wilson

As recently as 25 years ago, the threat of plague seemed old-fashioned, even medieval. Death from infectious disease was thought to be the result of poor hygiene and a lack of good antibiotics and vaccines, problems that by the mid-1970s had been largely overcome in the United States and most other industrialized nations. Medical practitioners were confident that infectious disease would represent a vanishingly small percentage of their concern. Wrote one prominent biologist in 1975, "During the last 150 years the Western world has virtually eliminated death due to infectious disease."

At the time, his optimism seemed justified. Smallpox had nearly been eradicated; tuberculosis and polio were on the decline and, with the exception of malaria, so were all of the other major infectious health threats of the 20th century. Scientists believed that, thanks to improved hygiene and sanitation, immunizations and antibiotics, all remaining infections of human beings and domestic animals would soon be eradicated.

Of course, skepticism was expressed even then. Agents of disease ranging from bacteria to insects had started to show resistance to the drugs and chemicals that had once so successfully killed them. And the optimistic projections were not consistent with what scientists knew to be true about the remarkable malleability of pathogens. For example, scrapie, a relatively mild disease of sheep, could somehow be transmitted to cattle, where it is devastating. Plants were also known to become afflicted with new diseases as old ones were eliminated. But the enthusiasm of the medical profession in general was not dampened by these examples, which after all, came from other disciplines and seemed too remote from the urgencies of medical practice.

Then came Lyme disease (1975), Legionnaire's disease (1978), toxic-shock syndrome (1978) and, more recently, AIDS (1981), chronic-fatigue syndrome (1985), and hantavirus (1993). The seventh cholera pandemic began in Indonesia in 1961, spread to Africa in the 1970s and reached South America in 1991, and now a new variant has emerged. Malaria reemerged in regions where it had been eliminated. Dengue and yellow fever are spreading. The incidence of tuberculosis started to climb in countries that had previously reported declines. Diphtheria reemerged in adults in the former Soviet Union. Suddenly, the proclamation of freedom from infection seemed, at best, premature.

These days, scientists no longer predict that the history of human infection will progress steadily toward the total elimination of infectious disease. More likely, the pattern will be one of disease turnover. With a new acceptance that infectious diseases will always be part of the human experience comes the realization that scientists will have to adopt a new approach to understanding the patterns of disease evolution. Rather than place sole confidence in measures we would use to fight infectious diseases after they arise, we, the members of the Harvard Working Group on New and Resurgent Diseases, are trying to identify the factors that encourage the emergence and spread of new diseases. To do that, we integrate complex social, epidemiological, ecological and evolutionary processes to understand how events in these various dimensions interact under changing circumstances to produce radically new health problems. In exploring potential threats to human health, we examine recent trends as part of epidemiological history and explore the progression of human diseases, as well as those of plants and animals. In order to anticipate new disease problems, including diseases that have not yet emerged, we have to examine the patterns of existing diseases and vectors and also look at the gaps in epidemiology. We apply current concepts and reexamine the conceptual framework that guides our present strategy of disease control. It is one of our principles that the emergence of new diseases can not be fully understood without understanding the social context in which they emerge.

What Is a New Disease?

At the start of our work we had to make two major decisions. First we had to define when a disease would be considered "new." Toward that end we identified ways in which a new disease may be recognized. A disease is recognized as new when its symptoms are distinct from any disease that has come before, or when a previously tolerated condition becomes unacceptable, as was the case with chronic exhaustion. A disease also becomes recognized when a previously marginal

The authors are members of the Harvard Working Group on New and Resurgent Diseases, scientists and practitioners who have been working together since 1991. The members are: ecologist Richard Levins, biomathematician Tamara Awerbuch, international health epidemiologist Uwe Brinkmann, philosopher of science Irina Eckardt, medical practitioner Paul Epstein, sociologist of science Najwa Makhoul, social scientist Cristina A. de Possas, systems ecologist Charles Puccia, public-health entomologist Andrew Spielman and clinical infectious-disease specialist Mary E. Wilson. They have recently compiled Disease in Evolution, *a book of conference proceedings to be published in 1994 by the New York Academy of Sciences. After this article was completed, Uwe Brinkmann died suddenly in June 1993 while on a site visit in Brazil. Address for Levins: Department of Population and International Health, Harvard School of Public Health, Boston, MA 02115.*

Figure 1. -Macabre images of plague-as-killer, such as this one from an 1845 wood carving by European artist Alfred Rethel, seemed, by the mid 1970s, old-fashioned. The near-elimination of smallpox and many other infections made many scientists confident that death from infectious disease was a thing of the past. Then came the emergence of new diseases and the resurgence of some old ones, and many scientists began to reassess the importance of infection in the future of human medical history.

population gains a public voice, which is what happened with black-lung disease. Diseases that are slow to develop may be newly identified in a population whose life span is increasing. In addition, conditions are identified as new when a local disease becomes widespread, a rare disease becomes common or a mild disease becomes severe. Diseases also become recognized when cases cluster in a locality or social group, which was the situation with Legionnaire's disease. Sometimes a disease is identified as new when a new human population is medically examined or when improved diagnostic techniques allow a unique infectious agent to be seen.

We also had to settle on a definition of infectious disease. We agreed that it would be defined as a disease in which infection is brought about by one or more kinds of parasite invading a susceptible animal. These parasites, commonly called pathogens (literally, the origin of suffering) can be microorganisms such as bacteria and viruses, or they can be multicellular organisms, such as worms. In spite of their diverse classifications, they all contribute to disease by a similar mechanism. They all carry out part of their life cycle inside another ani-

Recent New Diseases		
	viruses: newly identified agents	
1957 India	Kyasanur Forest	severe systemic infection and fever; transmitted by tick
1959 Uganda	O'nyong–nyong	explosive outbreaks of acute illness with fever and severe joint pain; transmitted by mosquito
1983	HIV-1	Acquired Immunodeficiency Syndrome
1989	Hepatitis C virus	transfusion-related and sporadic hepatitis
1990	Hepatitis E virus	acute hepatitis water-borne epidemics and sporadic hepatitis
1991	Venezuelan hemorrhagic fever Guanarito virus	outbreaks of severe hemorrhagic illness
	viruses: old agent, new location	
1992–1993 Kenya	Yellow fever	severe hepatitis and hemorrhagic fever; mosquito-borne virus
1993 southwestern U.S.	Hantavirus	new syndrome with pulmonary distress
	rickettsial diseases: newly identified agent	
1986	*Ehrlichia chaffeenis*	moderate to severe systemic infection with fever, headache, low white-blood-cell count
	bacterial diseases: newly identified agents	
1975	*Borrelia burgdorferi*	Lyme disease manifestations include arthritis and skin rashes
1976	*Legionella pneumophila*	Legionnaire's disease typically severe pneumonia. water-associated bacterium
1978	*Staphylococcus aureus*	toxic-shock syndrome; profound shock, kidney failure
1983	*Afipia felis*	cat-scratch disease; mild infection with enlarged, tender lymph nodes usually; acquired from cat
1992	*Vibrio cholerae* 0139	new variant of cholera

Figure 2. In spite of antibiotics and vaccines, a number of new diseases have emerged in the past 40 years. Here, a partial list demonstrates that the future of human health history will be one of disease turnover. New diseases come about through a complex interaction of biological, social, economic, evolutionary and ecological factors.

Rapid urbanization is one of the socioeconomic factors that puts people in contact with an unknown vector or pathogen

Creating new habitats—for example, by bulldozing forests—permits rare or remote microorganisms to become abundant and gain access to people

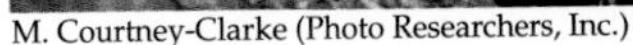

M. Courtney-Clarke (Photo Researchers, Inc.)

Alan D. Carey (Photo Researchers, Inc.)

mal—the host—where, for a time, they live, eat and reproduce. Disease is often a by-product of these activities. The parasite may produce a chemical that is toxic to the host, or it may damage the cells it infects. In addition, the host's own response may damage tissues.

But the parasite is actually only one of many factors that contribute to disease. Many parasites require an accomplice to facilitate their spread from host to host. Often this accomplice, called a vector, is an insect. Frequently the vector bites an infected animal and ingests some of its blood. It then bites a second animal and deposits into that animal's tissues some of the parasite-ridden blood from the first. Sometimes insect vectors themselves require carriers. Rodents, in addition to harboring pathogens, can also serve as hosts to fleas, ticks or other insects in which the pathogens reside. An animal's contact with parasites and vectors will obviously be an important factor in determining whether it will become sick. A potential host's general state of health and nutrition as well as its genetic predisposition to control infection will also determine the outcome of an encounter with a pathogen. Furthermore, social processes shape pathways of infection and disease. In the end, it is likely that the classification of diseases into infectious, environmental, psychosomatic, autoimmune, genetic and degenerative will prove to be applicable only to a sample of cases where one factor overwhelms all others. The more accurate viewpoint will encompass full complexity of this network of factors that leads to recognizable disease.

Recognizing New Diseases
• a previously tolerated condition becomes unacceptable (chronic exhaustion)
• a marginal population acquires public voice (black lung)
• increase of life span allows a slow disease to develop
• a local disease becomes widespread
• a rare disease becomes common
• a mild disease becomes severe
• symptoms are clearly distinctive
• contagion is high and latency short, so that cases cluster in a locality or social group
• health service examines a new population
• diagnostic techniques permit visualization of a new pathogen

Figure 3. Criteria for recognizing a disease as "new" include changing attitudes as well as changes in the organisms responsible for disease.

Identifying the Pathogen

Most bacteria are not pathogens, most arthropods are not disease vectors and most mammals are not a source of human disease. What, then, are the characteristics of a successful pathogen or vector, and where should we look for them?

The potential for a nonpathogenic species to become a pathogen can be assessed by examining the present distribution of diseases, symptoms and virulence across groups of pathogens. We call this field systematic epidemiology and focus on ecological and social processes that influence disease emergence.

Systematic epidemiology asks, for example, questions about the range of hosts a particular pathogen can infect as well as the types of pathogen a particular host can support. It also explores unique and shared characteristics among related species of pathogen, symptom variability for the same pathogen in different hosts, modes of transmission and the epidemiological potential for different species groups.

For a sample of 247 infections, we note that, relative to other pathogens, fungi are less common in serious primary human infections, but are prominent pathogens of fish and plants. We also note that viruses depend much more on arthropod vectors than other groups, while fungi typically require no vectors at all. Viruses are smaller and more fragile than fungal spores, which probably accounts for their reliance on vectors.

Will Kastens, a student at the Harvard School of Public Health, prepared a preliminary survey of 412 human infections. Of these, 180 were exclusively human diseases, 118 were primarily hu-

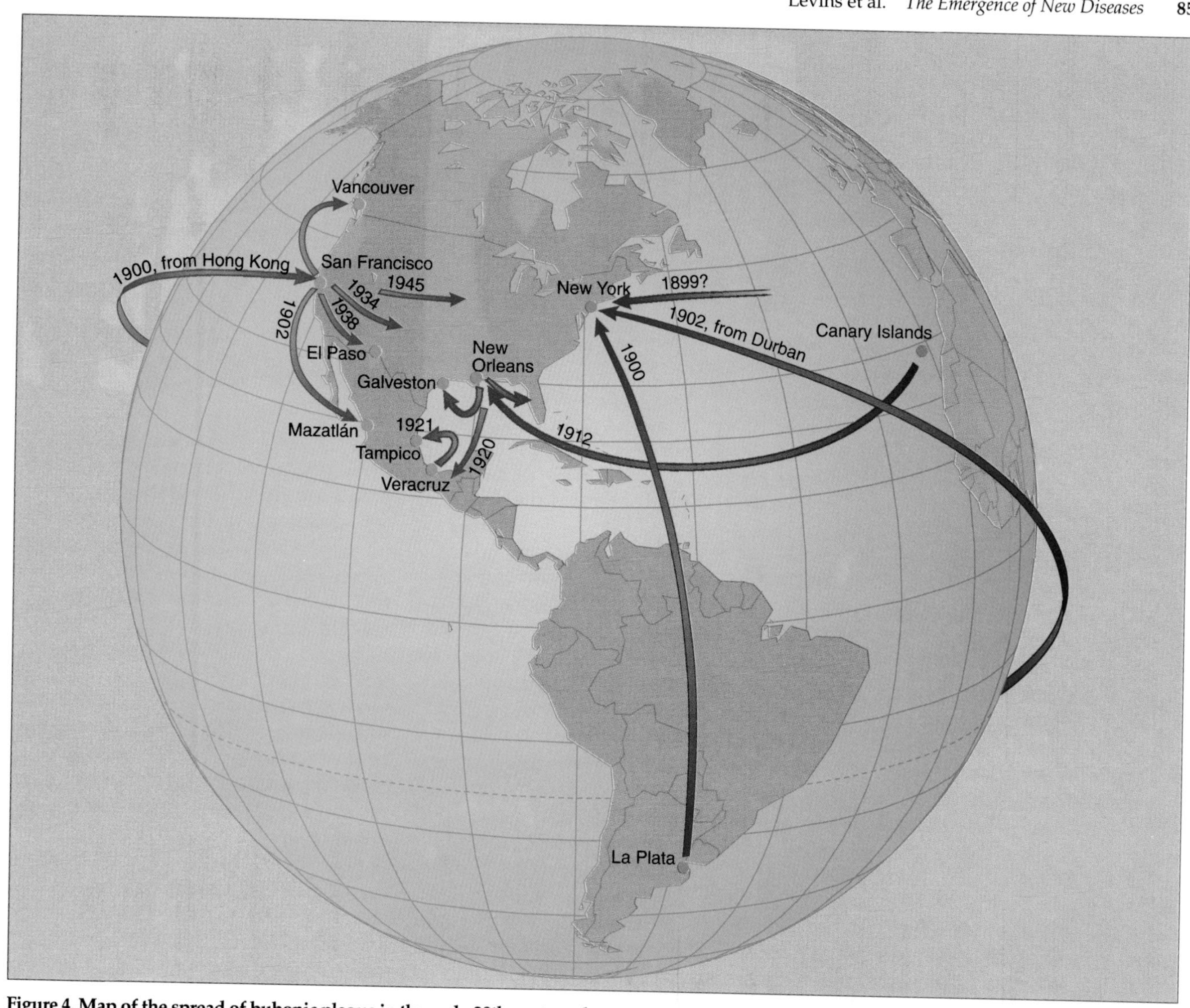

Figure 4. Map of the spread of bubonic plague in the early 20th century demonstrates how modern, rapid transportation can turn a local disease into a worldwide pandemic. When travel time is shorter than the incubation time of the disease, pathogens are more likely to move between countries and even continents. The spread of disease shown on this map would have been improbable during Christopher Columbus's time.

man diseases but could be found in other animals and 62 are principally found in other animals but can occasionally be found in people. The remaining pathogens are not normally pathogens of people or animals. Usually they are free-living microbes that cause disease when a chance encounter places them in contact with people. Of the diseases shared between people and other animals, 35 are widespread among mammals, about a dozen are shared with livestock and domestic animals, and approximately seven are shared with nonhuman primates. A few are shared between people and birds, and a smattering of pathogens can infect fish, people, shellfish and insects. Finally, at least two species of *Vibrio*, the microorganism that causes cholera, can be associated with plankton.

Kastens's data also show some interesting trends with regard to virulence. Those pathogens that mostly affect animals but sometimes invade human populations include a greater proportion with high virulence. We speculate that this is because their evolution is dominated by selection in their nonhuman hosts. On the other hand, diseases that are commonly or exclusively human have evolved in part in the human context, and some of them evolve to be gentler to people. Paul Ewald of Amherst College has suggested that vector-borne diseases tend to be more virulent than those requiring direct contact between people. His argument is that where the mobility of the patient is a requirement for transmission, it is to the pathogen's advantage that the infected person remain mobile, even if this limits the pathogen's rate of reproduction. But these are only tendencies. More detailed examinations of natural selection in pathogens reveal many exceptions to these examples.

Adaptive Potential

Although we give names to different species of pathogens, vectors, reservoirs and their associated diseases, these are not static entities. Pathogens as organisms undergo natural selection both within the host and in the course of transmission between hosts. And a pathogen's success in adapting to conditions within and between hosts will determine, in part, its success in spreading throughout a population.

A pathogen is confronted with three sometimes conflicting demands. It must obtain its nourishment to develop and reproduce, avoid being killed by the body's defenses and find a satisfactory exit to another host. Meeting these de-

A pathogen is introduced to a human population, as when carrier rodent populations increased and exposed people to hantavirus

John Gerlach (Animals Animals)

People moving into a new country may encounter pathogens against which they have no resistance

Wide World Photos

mands may require that a pathogen localize to a particular site in the body. For example, the blood is an optimal site for feeding, but it is a site of high immune activity. A pathogen in the central nervous system is relatively secure from destruction by the immune system but has no easy exit. The skin is also relatively safe from the immune system and can be exited fairly easily, but it is not a good site for reproduction.

Some pathogens adopt strategies for dealing with the immune system, so they are freer to choose sites in the body where immune activity is high. The human immunodeficiency virus, which causes AIDS, can remain in the blood because it destroys part of the immune system. Trypanosomes, which cause sleeping sickness, can also remain in the blood because they are adept at changing their protein coat, and in this way dodge detection by the immune system.

From the point of view of the pathogen, the symptoms suffered by the host are merely by-products of the pathogen's life-style. For example, in diarrheal diseases, the most obvious symptom arises when the pathogen exits one host in search of another. The pathogens remaining in the original host invade the gastrointestinal mucosa so they are not whisked away during the diarrheic episode.

Pathogens face other strategic decisions as well. Should they reproduce rapidly and exit quickly, or should they prolong the infection in the face of uncertain success in infecting someone else? The strategy adopted will depend on the relative rates of pathogenic reproduction, contagion possibilities and the danger of strong and effective treatment of the infection.

The role of drugs—antibiotics and antivirals in particular—in directing natural selection in the pathogen makes the intervenor a part of the system being intervened in. The host's behavior in effect becomes part of the selection pressure and affects the characteristics of the pathogen in the next outbreak. For example, if the host uses antibiotics, some of the pathogens may develop resistance to the drugs. During future outbreaks, these drug-resistant pathogens may predominate, and other antibiotics will have to be used to eliminate them.

Vectors, like pathogens, also undergo evolutionary change. Currently, a new biotype (or possibly, sibling species) of the whitefly *Bemisia tabaci*, carrier of bean golden mosaic virus, is spreading at the expense of the previous biotype. The new biotype has a wider range of host plants to feed on and is therefore spreading viruses to new plant species. In this case, a change in host range of the vector makes new species of plants serve as reservoirs for infections of crop plants. Reservoirs can maintain pathogens at low levels in small wild populations without being noticed, until a change in the environment or vector opens up new opportunities.

Pathogens on the Move

To cause a disease, a pathogen must first find a potentially receptive population of hosts. Sometimes the pathogen is required to travel. Various measures have been proposed to indicate the likelihood of a disease invading a population, among them the number of new cases derived from a given case, which epidemiologists refer to as the reproductive rate.

We have identified several key factors affecting the introduction of diseases into new populations. When diseases are carried from one area to another, it is important to establish the travel time needed to reach the new population relative to the rate of progression of the disease. For example, in Columbus's time, crossing the Atlantic was slow compared to the progression of smallpox. Since all carriers of the virus manifest symptoms of the disease, most of the infected travelers would have either become sick and died or recovered before reaching the New World. As a result, smallpox probably did not reach the Americas until several decades after Columbus's voyage. If Columbus were to begin his journey today, the situation might be different. Modern transportation has cut travel time to almost anywhere in the world to a few days at most, less than the average incubation time of many pathogens. Travel time, therefore, presents a less significant barrier to the spread of disease than it once did.

So does travel itself. Populations are much more easily moved than before. Left to their own devices, species spread into new areas very slowly. It has taken just over a century for rodent-borne plague to reach Mississippi from the West Coast. Fire ants spread at a rate of only a couple of miles a year. But man-made transport can speed the process so that pathogens can travel many thousands of miles in a few days. Political and economic oppression and opportunity are prime motivators for large-scale movements of people across countries and continents. The net effect of so much human migration is that diseases once confined to small regions of the globe can potentially spread to many regions.

Large-scale global commerce increases the probability of introducing vectors (often insects) and nonhuman carriers of disease to naive populations—a situation that may have touched off and accelerated the seventh cholera pandemic. For example, a freighter is thought to have transported the cholera vibrio in its ballast water from China to Peruvian coastal waters. Vibrio flourished in al-

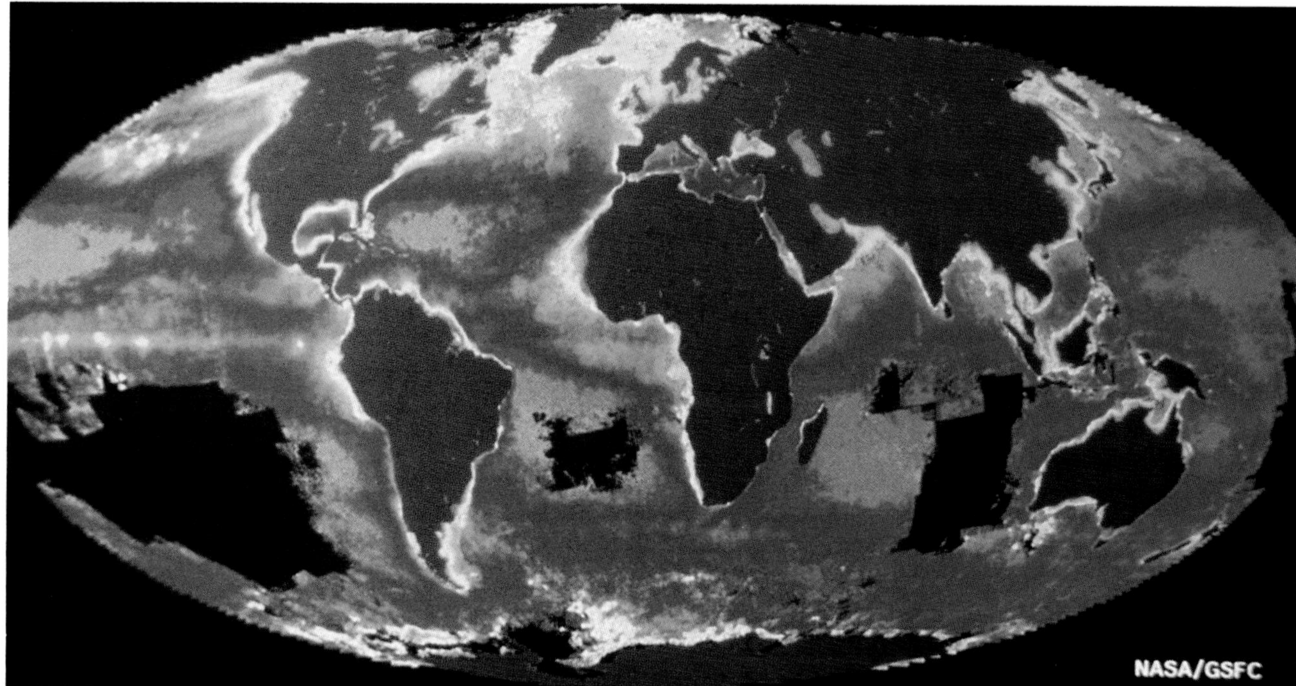

Figure 5. Satellite image shows the global distribution of coastal algal blooms, which are breeding grounds for pathogens, in particular the microorganism that causes cholera. Sewage and fertilizer pouring into marine ecosystems, overharvesting of fish and shellfish and the loss of wetlands, combined with climatic changes, have conspired to cause a worldwide explosion in the number of such blooms. (Photograph courtesy of NASA.)

gal blooms enriched with nitrogen and phosphorus from sewage and fertilizers. Algae are filtered and eaten by molluscs, crustaceans and fish that are, in turn, eaten by people. Once it entered, the infection in Latin America spread rapidly, as social and economic conditions provided a fertile environment for infection. Rapid urbanization, foreign debt and political changes strained the economy made sanitation and public health low priorities and paved the way for the epidemic spread of cholera. As of August 1992, more than 500,000 Latin Americans had become ill, and 5,000 of those people had died.

Changing Ecosystems

Just arriving in a new location does not ensure that a pathogen will take hold there. In fact, most introductions do not result in colonization because the species does not land in a hospitable niche. To successfully colonize new terrain, the invader must find a suitable environment and, if it is a pathogen, a receptive host population.

In general, invasion is easiest in regions of low biological diversity, where the invader faces less competition from native species. Oceanic islands are notoriously vulnerable to invasion. They have been known to be devastated by invasions of rats, goats or weeds, because their few native species could not compete.

Also vulnerable to invasion are habitats that have been disturbed by natural events or human activity. These events eliminate predators and competitors and create opportunities for new species to take up residence. For example, the spread of Lyme disease in New England is related to a number of human activities that have dramatically altered the region's ecology. During the past century, the forests were cleared to make way for agriculture. This eliminated from the area both the deer and their predators. The forests returned eventually, as did the deer. But the deer's predators did not. The deer tick, carrier of the infection, could spread, unimpeded, throughout the deer population and into the human populations that came into contact with them.

Vectors of human disease generally thrive in newly established habitats. Piles of used tires around the edges of rapidly growing cities collect water in which the mosquito *Aedes aegypti*, a vector for the organism that causes dengue and yellow fever, reproduces. Irrigation ditches, borrow pits, construction sites, poorly drained water pumps and puddled river bottoms each may serve as breeding sites for the mosquitoes that carry malaria. The pathogens carried by the mosquitoes can feed in these man-made habitats without being diverted to other animals, who are less successful in shuttling the pathogen to human hosts. In this manner, whole new niches have been created beyond the original geographic and ecological range of the vectors.

Of course, the successful spread of a human pathogen requires a vulnerable human population. The vulnerability of a group of people to a pathogen depends, on one hand, how contagious the pathogen is and how quickly it is transmitted from person to person, versus the population's immunity on the other. In this equation, all environmental changes are potentially reflected epidemiologically since conditions can affect the opposing processes of contagion and recovery, acquisition and loss of immunity.

The contagion rate depends on the number of pathogens that leave an infected individual and enter the environment. It also depends on the number that survive in that environment and gain contact to and ultimately infect other people. Each of these steps is complex and combines biological and social factors that are not constant. For example, no two people are equally susceptible to infection. A person's general state of health is as much determined by social, nutritional, age and gender factors as by genetics. Personal habits, such as smoking, sexual practices, alcohol consumption and food avialability and preferences can also contribute to a person's susceptibility to a particular disease.

Malnutrition, as well as immunosuppresive drugs and environmental stressors, makes people vulnerable to infection

Bruce Brander (Photo Researchers, Inc.)

Animal pathogens mutate and acquire the ability to infect people; the human immunodeficiency virus may have evolved from a monkey virus

Renee Lynn (Photo Researchers, Inc.)

In addition, there is now widespread concern about the potential effects of climatic change on disease. Changes in global temperature would carry with them changes in wind and precipitation patterns, humidity, soil composition and vegetation. All of these affect human activity and movement of populations.

Finally, the environment of a pathogen includes other parasites. In developing countries, it is not uncommon for people to harbor two to four simultaneous infections. Within their shared host, these pathogens may interact in familiar ecological patterns. They may compete for nutrients, or they may alter immune function in such a way as to benefit one while deterring another. They may alter their shared environment by causing fever or by damaging cells. The symptoms of one infection, say, sneezing, may facilitate the spread or mask the symptoms of the other. What this suggests is that the most effective way to deal with disease in the clinic is to consider the entire epidemiological profile, rather than consider one disease at a time.

Hantavirus

The emergence of a disease within a changing ecosystem was dramatically illustrated when a mysterious illness emerged in the Four Corners region of the southwestern United States earlier this year. In August 1993, a 37-year-old farmer who worked in the Four Corners area sought medical help when an illness he had had for six days took a turn for the worst. At first, the farmer experienced flu-like symptoms, including fever, nausea and vomiting, which progressed to coughing and shortness of breath. An xray showed fluid in both of the farmer's lungs. After 12 hours, he developed acute respiratory distress and died. Several weeks and several cases later, scientists at the Centers for Disease Control in Atlanta linked the mysterious disease to a new strain of hantavirus, viruses that have been associated with hemorrhagic fevers and renal disease in Europe and Asia, but that had not previously been known to cause disease in North America. Where had the virus come from, and why did it suddenly emerge?

That answer came serendipitously from studies conducted by Robert Parmenter and colleagues at the University of New Mexico. Parmenter and his team had been interested in the sudden increase in deer mice, which, as it turns out, are carriers of the hantavirus. Six years of drought ended in the spring of 1992, when heavy rains deluged the area. The abrupt change disturbed the ecological balance in the region, producing an abundance of piñon nuts and grasshoppers, food for the mice. The deer-mouse population flourished, but the drought had virtually eliminated all of the mouse's predators. In the year between May 1992 and May 1993, the deer-mouse population increased ten fold. By October 1993, the deer-mouse population had declined sharply, and the epidemic apparently came to an end. It had taken its toll. Forty-two cases of hantavirus pulmonary syndrome were reported in 15 states, mostly clustered in the Southwest. Twenty-six of those cases were fatal.

One of the lessons to be learned from such case studies is how disruption of stable ecosystems can alter an existing disease and facilitate its spread. For that reason, we are particularly concerned with recent disruptions of marine ecosystems, which are undergoing dramatic changes. Sewage and fertilizer pouring into marine ecosystems, overharvesting of fish and shellfish and the loss of wetlands, combined with climatic changes, have conspired to cause a worldwide ex-

Pollutants or radiation increase the mutation rate of pathogens

Earl Roberge (Photo Researchers, Inc.)

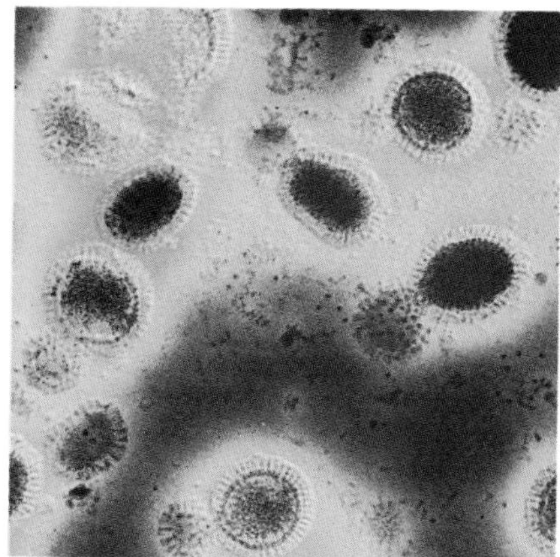
CNRI (Science Photo Library)

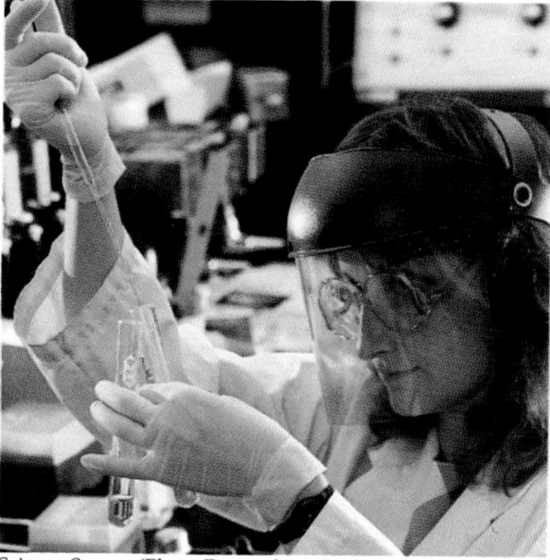
Science Source (Photo Researchers, Inc.)

Influenza virus and other pathogens may evolve toward greater virulence

Although it has not happened yet, bioengineered organisms may contribute to infection in the future

plosion in coastal algal blooms around coastal regions, providing a rich environment for diverse communities of microorganisms. The sea-surface temperatures in these environments are frequently elevated, which shifts organisms towards more toxic species, possibly by increasing their mutation and reproduction rates.

Among the new species that have been identified in these algal blooms is a new variant of the cholera vibrio. Now present in 10 Asian nations, this new variant seems to be distinct from previous forms of cholera, based on immunological tests. Antibodies that recognize other known variants do not recognize this one. This newly emergent, environmentally hardy form of cholera threatens to become the agent of the eighth pandemic. Monitoring algae and other microscopic marine organisms for vibrios offers the opportunity for establishing an early warning system for this new pathogen. Images from remote-sensing satellites can help guide this operation.

Confronting Uncertainty

The ultimate goal of these types of analyses is to anticipate the onset of new diseases and to eliminate the situations that facilitate their spread. In meeting these challenges, however, we confront a high degree of uncertainty. We can make some short-term predictions about new disease outbreaks by recognizing conditions that favor outbreaks of known disease and by anticipating ecological changes associated with human activity. We are developing models based on evolutionary ecology that would allow us to make some longer-range predictions about new disease outbreaks.

Just as organisms adapt rapidly to a new condition, we too can respond rapidly to new disease threats. But to do that, we have to recognize them quickly. An analysis of the factors facilitating or delaying recognition along with improved diagnostic techniques, might allow us to recognize problems sooner than is now possible.

We also must be very general in our response to disease. Our current therapeutic tools—immunization and antibiotics— are highly specific for the diseases they fight and can be developed only after the pathogen has been studied. Other organisms develop very nonspecific resistances that confer protection against a range of threats. For human beings, measures could be taken to boost the body's defenses in general. These would include good nutrition, pollution control, biodiversity for vector control and social arrangements that ensure these measures reach the entire population.

We advocate a mixed strategy, which offers back-up protection in case the first plan turns out not to be the best. A mixed strategy combines the tried-and-true approaches with newer ones. It may rely on some short-range predictions and monitoring, but it encourages longer-range ecological and evolutionary surveillance. It may concentrate on public health measures and seek to create healthful social systems that renegotiate our relations with the rest of nature.

Ultimately, we may have to reevaluate our notion of disease. We must see disease as the outcome of multiple conditions arising from changes not only within cell nuclei, but also around the globe, including changes in climate, economic patterns and communities of species. Any effective analysis of emerging diseases must recognize the study of complexity as perhaps the central general scientific problem of our time.

Bibliography

Anderson, P. K. 1991. Epidemiology of insect-transmitted plant pathogens. Dissertation in the Department of Population Sciences, Harvard School of Public Health.

Cairns, J. 1975. *Cancer: Science and Society.* San Francisco: W. H. Freeman.

Centers for Disease Control. 1993. Hantavirus pulmonary syndrome. *Morbidity and Mortality Weekly Report* 42:816–820.

Ewald, Paul. 1988. Cultural vectors, virulence, and the emergence of evolutionary epidemiology. *Oxford Surveys in Evolutionary Biology* 5: 215–245.

Epstein, P. R. 1992. Commentary—pestilence and poverty—historical transitions and the great pandemics. *American Journal of Preventive Medicine* 8:263–265.

Epstein, P. R., T. E. Ford, and R. R. Colwell 1993. Marine ecosystems. *Lancet* 342:1216–1219.

Garret–Jones, C. 1964. Prognosis for intervention of malaria transmission through assessment of the mosquito's vectorial capacity. *Nature* 204:531–545.

Hughs, J. M., C. J. Peters, M. L. Cohen and B. W. J. Mahay. 1993. Hantavirus pulmonary syndrome: an emerging infectious disease. *Science* 262:850–851.

LeDuc, J. W. , J. E. Childs, G. E. Glass and A. J. Watson. 1993. Hantaan (Korean Hemorrhagic Fever) and Related Rodent Zoonoses. In *Emerging Viruses*, S. S. Morse, ed. New York: Oxford University Press, p. 149.

MacDonald, G. 1952. The analysis of equilibrium in malaria. *Tropical Disease Bulletin* 49:813–828.

Marshall, E., and R. Stone. 1993. Hantavirus outbreak yields for PCR. *Science* 261:832–836.

Nichol, S. T., F. C. Spiropoulou, S. Morzunov, P. E. Rollin, T. G. Ksiazek, H. Feldmann, A. Sanchez, J. Childs, S. Zaki and C. J. Peters. 1993. Genetic identification of a hantavirus assoicated with an outbreak of acute respiratory illness. *Science* 262:914–917.

Possas, C. A. 1992. A sociological approach to epidemiological analysis: a tool for future health scenarios in developing countries. Boston: Takemi Program in International Health, Harvard School of Public Health, Research paper 71.

Settergren, B., P. Juto, A. Tollfors, G. Wadell and S. R. Norrby. 1989. *Reviews in Infectious Disease* 11:121.

Wilson, M. E. 1991. *A World Guide to Infection: Disease, distribution, diagnosis.* New York; Oxford University Press.

Figure 1. Lush vegetation at Rumstick Cove on the coast of Rhode Island illustrates how productive a salt marsh can be. Yet salt-marsh plants and animals live in a harsh environment

The Ecology of a New England Salt Marsh

In the harsh environment of a salt marsh, plants and animals not only compete, but also cooperate to survive

Mark D. Bertness

The salt marshes of North America's coastal regions are home to some of the most productive biological communities in the world. In these broad expanses of mud and sand, covered by a lush carpet of grasses and rushes, life seems to thrive on the soothing rhythm of the tides. The thick growth hints that here is found fertile soil, washed by the sea—an environment where plants should thrive.

Yet a salt marsh is in fact a harsh environment, where survival is difficult for plants and animals alike. The receding tides leave the soil soaked with

in which the extent of tidal inundation defines distinct ecological zones. Close to the sea is the low marsh, the area covered each day by the high tide. Cordgrass *(tall grass bordering the water)* dominates this area. Beyond the mean high-tide line is the high marsh, which begins at a distinct line between cordgrass and salt-meadow hay *(lower left)*. The arrangement of plants and animals in the marsh is affected not only by environmental constraints such as the tides but also by competition and cooperation among species. (Except where noted, photographs courtesy of the author.)

salt; waves break violently over the plants during storms. At latitudes where winter wraps the marsh in ice, the plant carpet is easily uprooted by the movement of chunks of the ice sheet, pried loose by winter storms. The animals of the littoral zone—the shore between the high-tide and the low-tide lines—likewise endure environmental extremes, spending part of the day underwater and the rest of the day exposed to the air.

These physical constraints have a visible impact on the marsh. Plant life in a salt marsh is organized into zones, the grasses forming distinct strips between the tide marks. The most obvious explanation for the patterns that form across the intertidal landscape is a physical one: Organisms vary in their ability to tolerate conditions at different tide levels, and the competition for space in the marsh is won by species that do best at a given tide level. The clear patterning of life in salt marshes makes them a good field laboratory for the study of the forces shaping vascular-plant communities, but the simplicity of these patterns is deceptive. To survive under extreme conditions, plants and animals often cooperate. The ecology of a salt marsh is shaped not simply by adaptation and competition, but by a combination of physical forces and

Mark D. Bertness is professor of biology at Brown University, where he teaches in the program of ecology and evolutionary biology. He received a Ph.D. in 1973 from the University of Maryland at College Park. His interest in coastal communities began during his childhood as he wandered the beaches surrounding Puget Sound. Address: Box G-W201, Brown University, Providence, RI 02912.

Figure 2. High marsh is partitioned by the extent of tidal flooding. Closer to the sea, salt-meadow hay *(right)* dominates the high marsh. The terrestrial border of the high marsh is covered by black rush *(left)*.

biological interactions, including cooperation. I have spent more than a decade studying the thriving salt marsh at Rumstick Cove, near my home in Barrington, Rhode Island. Along with others who have taken a close look at salt-marsh communities in New England and elsewhere, I have come to appreciate their underlying complexity.

Walking the Marsh

A walk from land to sea at Rumstick Cove begins at the edge of the high marsh, an area flooded only by monthly extreme tides. Black rush (*Juncus gerardi*) dominates this zone. This deep-green grass is a dense turf, and is tall enough to scrape your calves. Partway through the high marsh, however, there is a line: The black rush ends, and salt-meadow hay (*Spartina patens*) begins. This tidal line marks the mean highest tides of each month. Salt-meadow hay is a light-green grass, about as tall as your lawn would be without mowing. A few other grasses, such as spikegrass (*Distichlis spicata*), also grow here. And there are distinct grass-free patches, often filled with a succulent annual—slender glasswort (*Salicornia europaea*)—that adds color to the high marsh. Slender glasswort is green until fall; then it turns reddish. But in any season, glasswort is a very salty plant that can be pickled or taken as is and placed in a salad.

Just beyond the salt-meadow hay there is another line. This is the boundary between the high marsh and the low marsh—the mean high-water line of the tide. Below this line, there is no salt-meadow hay. It is replaced by the cordgrass (*Spartina alterniflora*) that dominates the low marsh, the area covered by each day's high tides.

Cordgrass, the tall, wispy grass that rolls back and forth in waves under a gentle breeze, is the plant that dominates the common image of a marsh.

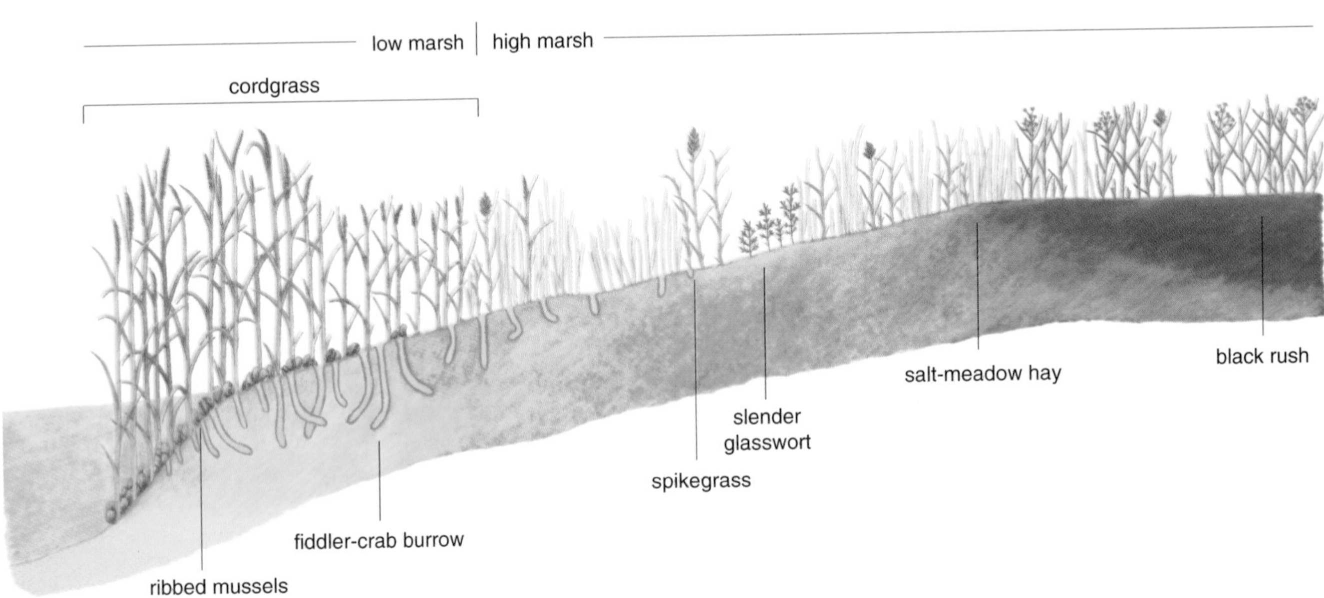

Figure 3. Distinct zones characterize a New England salt marsh, for animals as well as plants. On the coastal side of the low marsh, shown in cross section, thick beds of ribbed mussels are attached to the roots of cordgrass. The mussels decrease in abundance as one moves inland, and the marsh soil becomes dotted with small holes, the burrows of fiddler crabs. These too, however, are largely limited to the low marsh. Even the cordgrass of the low marsh is divided into two zones: The cordgrass closer to the sea is tall and the cordgrass farther from the sea is short because soil there is composed of compacted peat produced by the decay of the grass. The high-marsh zones of salt-meadow hay and black rush are primarily monocultures, but disturbed areas support small populations of spikegrass and slender glasswort.

At the edge of the border marked by the high-tide line, the cordgrass is short, less than 30 centimeters tall. But as you move closer to the sea the cordgrass becomes taller, sometimes reaching nearly two meters in height. As you bend down to look at the muddy soil around the roots of the tall cordgrass, small holes are evident in the ground, perhaps as many as 175 in a square meter of soil. These are openings to the burrows of fiddler crabs. The fiddler crab is small, less than three centimeters across, and the male has one very large claw that he holds the way a violinist carries his instrument—hence the name.

If you push through the last stems of cordgrass during low tide, you come to a muddy beach. Densely packed beds of ribbed mussels cover the transition from the cordgrass to the sea. Sometimes as many as 1,500 mussels can be found in a single square meter, attached near the roots of the cordgrass.

By the end of your journey, it is clear that the zonation of a salt marsh is precise, a distribution dictated by the limits of the tides. There is no cordgrass on the terrestrial side of the high marsh. There is no black rush near the boundary between the high marsh and the low marsh. Specific constraints must impose this arrangement.

Border Disputes

Early investigators believed that environmental variation caused the distri-

Figure 4. Plant morphology affects competitive ability. Spikegrass stems are separated by lengths of belowground runners *(left)*. This makes for a less densely growing plant, but one well adapted for colonizing new areas. Black rush *(right)* has turf morphology—dense groups of stems arising from a belowground mat of roots and rhizomes. Turf grasses do not rapidly invade new areas, but turf morphology is competitively superior to runner morphology over time.

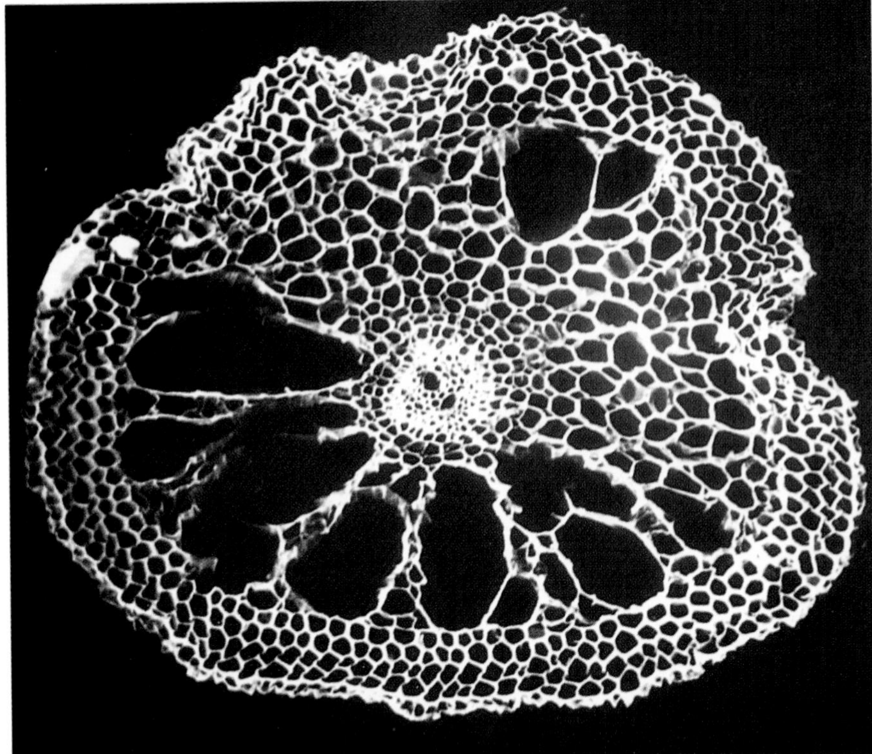

Figure 5. Aerenchymal tissue provides oxygen to the roots of cordgrass by creating a passageway through the plant. In this photomicrograph showing a cross section of a stem, the tissue is visible as a series of circular structures along the perimeter. Repeated flooding by the tide keeps the low-marsh soil waterlogged and thereby deficient in oxygen. Without oxygen in the soil, plants have difficulty using the available nutrients. Cordgrass partially solves this problem by transporting oxygen from its leaves, through the aerenchymal tissue to the roots to oxygenate the soil. (Photograph courtesy of Irv Mendelssohn, Louisiana State University.)

Figure 6. Fiddler crabs enhance the productivity of cordgrass by digging burrows in the low marsh for protection from predators during high tide. The crabs' burrowing aerates the soil, increases the soil's drainage and helps decompose belowground debris.

bution of plants across marshes. The distinct borders between patches of salt-marsh vegetation were seen as clear examples of the varying ability of species to adapt to tidal conditions. Experiments on rocky beaches, however, had suggested that other biological factors—specifically predation and competition among species—significantly affect the zonation of plants and animals in littoral habitats. Could the same forces create zones of plants in a salt marsh? Because herbivores had been shown to have little influence on the survival of marsh plants, I suggested that marsh-plant zonation might arise from a combination of environmental variation and competitive interactions.

I tested my ideas at Rumstick Cove. After describing the plant-zonation pattern in detail, my students and I experimentally examined the distribution of plants across the marsh; we transplanted salt-marsh plants from one zone to another, both with and without neighboring competitors. Through this work, we began to understand what controls the salt-marsh borders.

In one respect, cordgrass controls itself. Beneath tall cordgrass, near the water, the soil has little peat. But the growing cordgrass continuously produces more belowground debris, and the burrowing crabs hasten its decomposition. As this material compacts, it becomes peat. Peat decreases substrate drainage, and the consequently waterlogged soil is low in oxygen. These factors combine to limit the productivity of cordgrass. As a stand of cordgrass matures, it effectively destroys its own habitat. This process creates the variation in the productivity of cordgrass that is seen as you move down the shore. Higher on shore, cordgrass is short; its soil is rich in peat, and thus its growth is stunted. Cordgrass grows better near the water—uninhibited by deposits of peat—and pushes ever farther into the sea.

Other interactions in a salt marsh involve competition between species. In fact, cordgrass is limited to the low marsh more by competition than by its production of peat. If cordgrass is transplanted to an area free of neighboring plants, it grows rather well in the high marsh. But if cordgrass is transplanted near salt-meadow hay or black rush, these plants quickly eliminate the cordgrass.

Likewise, salt-meadow hay and spikegrass both grow best in the terrestrial portion of the high marsh, the black-rush zone; but black rush eliminates its competitors.

Aaron Ellison of Mount Hollyoke College and I showed that competitive relationships among salt-marsh plants are primarily determined by two factors: morphology and the timing of spring emergence. Both black rush and salt-meadow hay have dense mats of roots, rhizomes and aboveground tillers (shoots). Cordgrass and spikegrass tillers, however, are separated by relatively long lengths of belowground runners. In most competitive interactions, turf morphology—the mats of roots, rhizomes and tillers—defeats runner morphology. As a consequence, salt-meadow hay and black rush exclude cordgrass from the high marsh, and spikegrass is limited to disturbed areas within the high marsh. On the other hand, runner morphologies are more mobile than turf morphologies; therefore, cordgrass and spikegrass can rapidly colonize an area. That increases the success of cordgrass in the ever-changing low marsh. These bits of information explain most zones, but leave one question: How does black rush exclude salt-meadow hay from the terrestrial portion of the high marsh? Black rush wins this battle through timing. It emerges in March, nearly two months ahead of any other perennial grass in the marsh, and thereby defeats salt-meadow hay, even though the two plants have similar morphologies.

Environmental variation alone does explain some patterns in the marsh. In some cases, a plant simply cannot tolerate a specific environment. For example, black rush, salt-meadow hay and spikegrass all die within a single season if transplanted to the low marsh, even if they are planted in an area with no competition with cordgrass.

Cooperation and Cordgrass

The low marsh is the most difficult marsh environment for plants. The soil is extremely salty and lacks oxygen. Irv Mendelssohn of Louisiana State University showed that anoxic soil can prevent plants from using the nutrients in the soil. Moreover, the low marsh is continually battered by physical disturbances, especially erosion and ice damage. The lapping of waves throughout the year eats away the shoreline. In the

Figure 7. Sheets of ice cover the low marsh during winter. During high tide, the sheets of ice can uproot large patches of cordgrass that are attached to them, disturbing the edge of the low marsh. For the marsh to survive, each year's production of cordgrass must exceed the amount of cordgrass lost to winter ice damage.

Figure 8. Ribbed mussels protect cordgrass from erosion. The mussels attach themselves to the roots of the cordgrass by strong filaments called byssal threads. In this way, mussels join a series of plants into a common and thereby stronger structure. Moreover, the attachment of the mussels stimulates the cordgrass to produce more belowground roots, further strengthening the plants that protect the marsh from the constant battering of the tides.

Figure 9. Mats of dead cordgrass disturb growth in the high marsh. During the winter, the action of the tides and floating sheets of ice mow down thousands of stems of cordgrass. By spring, the intertidal basin is covered by a layer of floating cordgrass stems. Then, high tides can carry the mats—sometimes larger than a football field—onto the high marsh, leaving them stranded there. Sometimes a mat of cordgrass is stranded in the high marsh for months before another tide washes the mat away or it decays. The mat of cordgrass can kill all underlying vegetation, leaving a bare substrate.

winter, sheets of ice cover the low marsh in wave-protected areas. In Rhode Island I have found these sheets of ice to be 10 to 30 centimeters thick and frozen to the underlying cordgrass. During severe high tides, large chunks of ice, up to 10 meters across, and the incorporated substrate of the marsh can be torn away and rafted offshore. For the marsh to survive, the growth of cordgrass must exceed the destructive effects of erosion and ice damage.

Cordgrass survives in the low-oxygen soil of the low marsh largely because of a morphological specialization. Cordgrass contains aerenchymal tissue—an internal pathway that allows air to move from the tips of the leaves to the ends of the roots, oxygenating the soil. And dense stands of cordgrass move disproportionately more oxygen into the soil. In other words, cordgrass plants thrive by cooperation; dense stands do better than sparse plantings because the increased oxygen makes the soil more hospitable for the plants.

Nevertheless, cordgrass has other help in dealing with the low-oxygen soil. As I said earlier, fiddler crabs are prolific in the low marsh, and dig many burrows. A burrow is usually 10 to 30

Figure 10. Slender glasswort *(left)*, the most salt-tolerant of the high-marsh plants, invades bare spots *(right)* that are created when stranded mats of cordgrass kill vegetation. In areas with no vegetative cover, the sun evaporates water from the soil, leaving a layer of salt on the surface. Slender glasswort, which can germinate and develop in hypersaline conditions, is the first plant to colonize bare spots.

centimeters deep and provides shelter from predators during high tides. These burrows, however, are in a constant state of flux. Some are abandoned; others collapse. Old burrows are often modified or enlarged. Through this process, the crabs work over much of the top 10 centimeters of soil during each season. This increases the drainage of the soil, the decomposition of belowground debris and the oxygen content of the soil. The fiddler crab is the earthworm of the low marsh, and can be largely responsible for the high productivity of the tall cordgrass.

Likewise, ribbed mussels contribute to the survival of cordgrass. Tom Jordan of the Smithsonian Environmental Research Center and Ivan Valiela of Boston University calculated that a mussel can filter as much as five liters of seawater per hour in search of plankton. This results in the deposition of nitrogen-rich feces that can increase the growth of cordgrass by 50 percent in a single season. Moreover, mussels buffer cordgrass against physical disturbance. A special gland in mussels secretes strong, proteinaceous filaments called byssal threads. Mussels use a series of these threads to attach to the roots of cordgrass, binding marsh soil from erosion. The cordgrass, in response, grows more belowground roots. This, indeed, is a team effort.

Figure 11. Spikegrass colonizes bare soil by sending out runners belowground. In the high marsh, spikegrass is largely limited to bare spots because of the superior competitive abilities of salt-meadow hay and black rush. Like slender glasswort, spikegrass can survive in hypersaline conditions; but spikegrass does so by receiving water, through its runners, from plants in less saline areas. Once established, spikegrass shades the soil, reducing evaporation and soil salinity.

Dealing with Disturbance

When you are standing on the seaward edge, the low marsh seems a more difficult environment than the high marsh because its soil is waterlogged and constantly eroded by the tide. Nevertheless, walking through the high marsh you find bare patches in the midst of the stout green grasses. And particularly in the spring, large mats of entangled, dead cordgrass stems are scattered throughout the high marsh, where they can kill the underlying vegetation. Some of these mats are bigger than a football field and more than 10 centimeters thick. The high marsh, too, is a demanding habitat.

Floating plant debris creates chronic physical disturbances in the high marsh. During the winter, sheets of ice driven back and forth by the tides clip cordgrass stems. By early spring, floating mats of tangled cordgrass skeletons cover the intertidal low marsh. Extreme high tides in the spring raft these mats of dead cordgrass onto the high marsh and leave them stranded when the tide ebbs. The mats may clutter the high marsh for months before decaying or being washed away. Plants trapped under the stranded mat often die, leaving bare soil.

Although at first a bare patch is a competition-free environment, it rapidly becomes a challenging habitat. The sun heats the exposed surface. As the water in the soil evaporates, salinity increases, particularly near the surface. How high the salinity goes depends on the size and location of the bare patch. Larger bare patches have more surface exposed; more water evaporates, and the soil becomes more saline. The highest salinity is found in large bare patches near the border between the salt-meadow hay and the black rush. Here the salinity of the soil can be 30 times the soil salinity under the dense perennial vegetation in the high marsh. Closer to the sea, frequent flooding limits the accumulation of surface salt. And farther into the black rush, rainwater dilutes any salty soil.

The hypersaline soil in bare spots creates water-balance problems for most plants. Scott Shumway of Wheaton College demonstrated that a hypersaline environment prevents germination of the seeds of most marsh plants. Even a healthy seedling often dies if transplanted to hypersaline soil, because the salty soil pulls water from the plant.

Glasswort, however, thrives in hypersaline bare patches. This plant readily germinates in extremely salty soil. Ellison and I found that seeds or seed-bearing skeletons of slender glasswort are often conveniently carried to the bare spots by the cordgrass mats. Because of these two factors, slender glasswort typically dominates bare spots during the first year of regrowth.

Glasswort's reign, however, is tem-

first year

second year

third year

fourth year

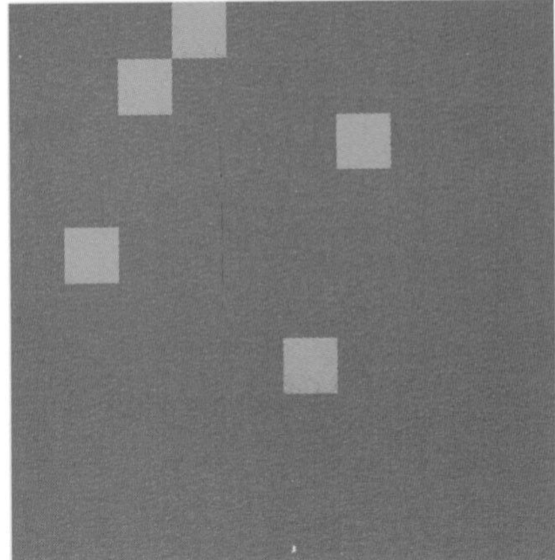

porary. Spikegrass moves in shortly, invading bare spots through asexual means. This salt-tolerant New England marsh perennial produces long rhizomes, just a few centimeters below the surface, that invade the bare spots. Shumway showed that these invading rhizomes survive the saline soil and produce young shoots because, through the rhizomal connection, they receive water from surrounding plants growing in less saline soil. The spikegrass prospers and shades the soil, reducing evaporation and thereby reducing soil salinity. As the salinity of the soil decreases, the habitat becomes more hospitable to other high-marsh perennials, particularly those with turf morphologies. Within two to four years, either salt-meadow hay or black rush displaces both slender glasswort and spikegrass.

Much as the crabs, mussels and cordgrass cooperate to occupy the seaward edge of the low marsh, high-marsh perennials work together to reclaim hypersaline bare spots. Salt tolerance and competitive ability are inversely related in plants of the high marsh in New England. The less competitive but more salt-tolerant plants first invade bare spots. This facilitates succession, making the habitat livable for the more competitive but less salt-tolerant perennials. Here again, a cooperative effort among the plants emerges only if the bare spot is hypersaline, and thereby stressful. If a bare spot has a low level of salt, the spot is filled through strictly competitive interactions among the high-marsh plants. So local conditions determine whether a patch is invaded cooperatively or competitively.

Figure 12. Succession of species in a bare spot in the black-rush zone of the high marsh results from cooperation and competition among plants. New bare spots in the black-rush zone become extremely hypersaline because of direct exposure to sunlight. By the end of the first year, more than one-third of a disturbed spot remains bare *(brown)*. Slender glasswort *(red)* and spikegrass *(light green)*, however, are relatively salt-tolerant and cover more than half of the ground, providing shade that makes the soil more habitable for black rush *(dark green)*, which cannot tolerate the hypersaline soil. In the second and third years, after the physical conditions are improved by the initial invades, black rush invades the area. The results of competition become clear by the third year. Black rush covers approximately three-fourths of the area and competitively displaces the initial invaders. After four years, no bare soil remains, slender glasswort is completely excluded and spikegrass covers only a few percent of the zone. The once-bare spot is dominated by black rush.

Cooperation Is Important

Ecologists have overestimated the universal role of competition in nature. Phrases such as "survival of the fittest" certainly evoke and perpetuate this misconception. Over the past 20 years, many ecologists have examined natural communities in search of competition, and have found it to be particularly pervasive in physically mild habitats. Early in this century, ecologists accepted, uncritically, that species can cooperate to reclaim disturbed habitats. Because of this *untested* acceptance, contemporary ecologists largely reject this idea. Nonetheless, many recent experiments show that positive interactions among organisms often play a large role in natural communities.

Physically harsh environments generate, as a matter of course, cooperation among organisms. David Wood of California State University at Chico and Roger Del Moral of the University of Washington found facilitation among plants during early succession in subalpine habitats on Mount St. Helens. Other reports show that interactions among the same organisms can be competitive in benign environments and cooperative in harsh environments. For example, Mark Hay of the University of North Carolina's Marine Science Center showed this in turf-forming seaweed. Examples such as these reveal that a New England salt marsh is one habitat among many in which biotic interactions are competitive under mild physical conditions and cooperative under harsh physical conditions. Both competitive and cooperative forces likely play major roles in the organization of most natural communities.

Bibliography

Bertness, Mark D. 1984. Ribbed mussels and the productivity of *Spartina alterniflora* in a New England salt marsh. *Ecology* 65:1794–1807.

Bertness, Mark D., and T. Miller. 1984. The distribution and dynamics of *Uca pugnax* burrows in a New England salt marsh. *Journal of Experimental and Marine Biology and Ecology* 83:211–237.

Bertness, Mark D. 1985. Fiddler crab regulation of *Spartina alterniflora* production on a New England salt marsh. *Ecology* 66:1042–1055.

Bertness, Mark D., and Aaron M. Ellison. 1987. Determinants of pattern in a New England salt marsh community. *Ecological Monographs* 57(2):129–147.

Bertness, Mark D. 1988. Peat accumulation and the success of marsh plants. *Ecology* 69:703–713.

Bertness, Mark D., L. Gaugh and Scott Shumway. In press. Salt tolerances and the distribution patterns of New England marsh plants. *Ecology*.

Ellison, Aaron M. 1987. Effects of competition, disturbance, and herbivory in *Salicornia europaea*. *Ecology* 68:576–586.

Harper, J. L. 1977. *Population Biology of Plants*. San Diego: Academic Press.

Hay, Mark. 1981. The functional morphology of turf-forming seaweeds: persistence in stressful marine habitat. *Ecology* 62:739–750.

Jordan, Tom E., and Ivan Valiela. 1982. A nitrogen budget of the ribbed mussel, *Geukensia demissa,* and its significance in nitrogen flow in a New England salt marsh. *Limnology and Oceanography* 27:75–90.

Mendelssohn, Irv A. 1979. Nitrogen metabolism of the height forms of *Spartina alterniflora* in North Carolina. *Ecology* 60:574–584.

Mendelssohn, Irv A., K. L. McKee and W. H. Patrick. 1981. Oxygen deficiency in *Spartina alterniflora* roots: metabolic adaptation to anoxia. *Science* 214:439–441.

Shumway, Scott. 1991. *Secondary Succession Patterns in a New England Salt Marsh.* Dissertation, Brown University, Providence, Rhode Island.

Wood, David M., and Roger Del Moral. 1987. Mechanisms of early primary succession in subalpine habitats on Mount St. Helens. *Ecology* 68:780–790.

The Tall-Fescue Endophyte

Evolution meets economics in the tale of the nation's most popular planted grass, which owes many of its qualities to a fungus toxic to livestock

Donald M. Ball, Jeffrey F. Pedersen and Garry D. Lacefield

Figure 1. Rural landscapes of the upper South get much of their lush green color from tall fescue, the most widely grown pasture and turf grass in the humid central and eastern United States. But tall fescue's reputation is mixed. The qualities that have made the cool-season perennial popular—its adaptability to various soils and difficult climate conditions and its resistance to insect and nematode attacks—derive largely from an internal fungus, or endophyte, that has

evolved with its host. Unfortunately the endophyte also has toxic and costly effects on grazing animals. (Painting by Judy Kitzman.)

A motorist traveling the rural roads of the upper South and lower Midwest is likely to get an eyeful of green—much of it a certain hue that a farmer might call "fescue green." Although Kentucky is known for its bluegrass pastures, in fact another grass dominates the abundant turf and pasture land throughout most of Kentucky and nearby states, especially those to its south, east and west. The European cool-season perennial known as tall fescue now occupies more acreage in the United States than any other introduced grass, giving the plant an enormous role in the nation's agricultural economy. But it has become clear that tall fescue, planted for its vigor, lush beauty, pest resistance and tolerance of drought and poor soils, is associated with a problem—perhaps as costly as a billion dollars a year—in the livestock industry. The effort to identify and eradicate the source of the problem is a story of importance both to science and to the nation's agricultural productivity.

Not long after tall fescue became popular in the late 1940s, farmers began noticing declining health and productivity among cattle, horses, sheep and dairy cows that grazed on the new forage. Despite the grass's high nutritional quality, beef-cattle producers talked of "fescue toxicity," which showed up as rough hair coats, intolerance to heat and poor weight gains among their animals. A gangrenous condition that sometimes develops on the extremities, especially the rear feet, during cold weather became known as "fescue foot." Bovine fat necrosis, the deposition of masses of fat in the abdomens of cattle, was observed among cattle grazing nearly pure stands of tall fescue that were highly fertilized with nitrogen.

Horse producers found that mares grazing tall fescue often abort, have foaling problems, give birth to weak foals and produce little or no milk. Weight gains among sheep showed the declines observed in beef cattle, and dairy animals often showed sharp depressions in milk production.

Although farmers had eagerly adopted tall fescue as a multipurpose grass, many livestock and dairy producers turned away from it once these problems were recognized. In many areas there were no other productive perennial pasture grasses, and so the need for a solution for the animal-toxicity problems associated with tall fescue was obvious.

In 1973, the crucial clue to the animal disorders was found at a farm near Mansfield, Georgia. The cattle in one tall-fescue pasture were exhibiting fescue-toxicity symptoms, but animals in a nearby pasture were not. J. R. Robbins, Charles W. Bacon and J. K. Porter, scientists at the U.S. Department of Agriculture's research station at Athens, began an investigation of this puzzling situation. Eventually they took samples from both pastures to test their hypothesis that a fungus might be at fault. In the laboratory they found strong support for their idea: Virtually all the grass in the pasture where the cattle were sick was infected with an internal fungus, or endophyte, whereas only about 10 percent of the grass in the other pasture was infected.

The identification of a culprit in the tall-fescue endophyte, whose role in livestock disorders was to be confirmed by many experiments that followed at many locations, was a breakthrough of remarkable importance. But like many research findings, it pointed the way to new questions. Could, and should, the fungus be eradicated, or at least controlled, or its toxic effects ameliorated, and how? Could pastures be kept endophyte-free? What is the mechanism by which a grass endophyte affects livestock health? How is endophyte infection maintained in the grass? What effect does the endophyte have on the grass itself?

The answers to these questions cross the boundaries that separate science, economics and agricultural practice. The endophyte turns out to be a splendidly adapted fungus that has coevolved with its host, to their mutual benefit. Its animal-toxicity effects are among several protective advantages the fungus provides to the grass, and are part of a complex plant-animal-fungus relationship that may actually be common among grasses and weeds.

Donald M. Ball is a professor of agronomy at Auburn University. He is a graduate of Auburn University, where he received a Ph.D. in 1976. He has specialized in forage crops since 1976 and has done agronomic consulting work in several states. Jeffrey F. Pedersen is a research geneticist with the U.S. Department of Agriculture's Agricultural Research Service and a professor of agronomy at the University of Nebraska, where he received his Ph. D. in 1981. Garry D. Lacefield is a professor of agronomy at the University of Kentucky and holds a Ph.D. from the University of Missouri (1974). Address for Ball: Department of Agronomy and Soils, Auburn University, AL 26849.

Figure 2. Kentucky 31, the vigorous tall-fescue variety responsible for the rapid adoption of tall fescue in the United States in the late 1940s and 1950s, was collected from this hillside in Menifee County, Kentucky, in 1931 by the late E. N. Fergus, a professor at the University of Kentucky. Fergus and his colleagues planted and evaluated the ecotype before the 1943 release of the new variety, now used widely for forage, turf and conservation. The seed was infected with the endophyte *Acremonium coenophialum*, which was implicated in animal toxicity in the mid-1970s. Because most tall fescue in the U.S. is of this variety, most is endophyte-infected.

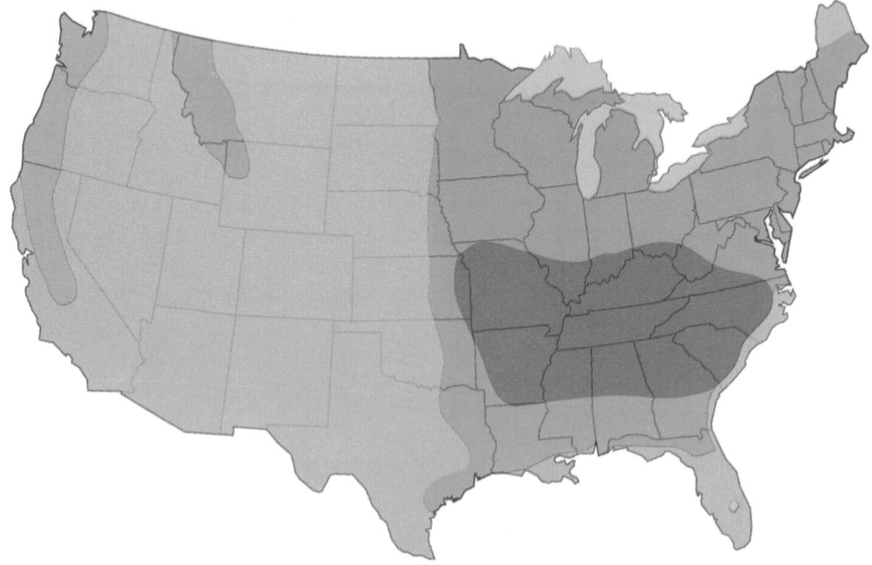

Figure 3. Range of tall fescue's adaptation *(light green)* includes most of the eastern U.S. and parts of the Pacific Northwest, but its area of primary use *(dark green)* is in the upper South and lower Midwest, where it is the dominant grass. In much of this area, climate and soil conditions are not suited to other grasses; thus tall fescue has been the overwhelmingly preferred grass for hay and livestock production. (Adapted from Buckner and Bush 1979.)

The forage producer's fight against it is, in a way, a small battle against the forces of evolution.

Tall Fescue and Its Endophyte

Now a popular introduced grass in many countries, tall fescue (*Festuca arundinacea* Schreber) was planted in the United States in the late 1800s, but it did not come into widespread use until well into the 20th century. In 1931, E. N. Fergus, a professor of agronomy at the University of Kentucky, visited the farm of William O. Suiter in Menifee County, Kentucky, where he found a particularly vigorous tall-fescue ecotype. The seed he collected and evaluated was released in 1943 as the cultivar Kentucky 31. It was widely adapted, had a long growing season, resisted pests and persisted under a wide range of conditions, including drought, poor soils and variable soil *p*H.

The new cultivar grew swiftly in popularity, becoming the grass of choice for forage, turf and soil conservation in the humid south-central states. Tall fescue, overwhelmingly dominated by the Kentucky 31 cultivar, is now grown on more than 35 million acres in the United States, on parks, playgrounds, athletic fields and lawns and along highways and waterways in addition to fields used for pasture and hay production.

But the reports of animal disorders sent agricultural scientists into the laboratories in the 1960s. Early chemical analysis showed that alkaloid compounds found in the grass might have effects on livestock, in the same way that the alkaloids produced by the ergot fungi of the *Claviceps* genus cause the similar, often severe disorder called ergotism when ergot-infected grain is ingested. Causal relationships were not clear, however, and the source of the alkaloids was not known.

When the endophyte was identified in the mid-1970s by microscopic analysis of the pith tissue of the infected Georgia pasture grass, the patterns of animal disorders began to be explained. At first the endophyte was identified as *Sphacelia typhina*, the asexual state of *Epichloë typhina*, a fungus found on many grasses and known as choke disease. But taxonomists later decided that the endophyte found in tall fescue is sufficiently distinct to deserve a new species name—*Acremonium coenophialum* (Morgan-Jones and Gams). Grazing studies by Carl Hoveland of Auburn University soon confirmed the association of the fungus with fescue toxicity. More recent tests of samples from tall-fescue pastures throughout the U.S. have shown that over 90 percent have high levels of fungus infection. It is estimated that three-fourths of the tall fescue in the country is infected.

An endophyte grows within a plant, and the tall-fescue endophyte is a particularly elusive fungus. Because it is not visible externally, diagnosis in the field is not possible, and laboratory analysis is required to detect its presence. The fungus's mode of reproduction makes it more elusive. *A. coenophialum* does not disseminate spores that spread infection through the field; rather, it is transmitted by seed from one generation to the next. This mode of transmission might not have allowed the fungus to become so prevalent, had it not been for a mutualistic relationship that gives infected plants a competitive advantage. The fact that Kentucky 31 has been such a successful grass may stem in large part from the fact that the original seed almost certainly contained the endophyte.

The simple life of *A. coenophialum* begins in the seed, where its mycelium—the fungus's vegetative stage, consisting of a mass of convoluted filaments, or hyphae—can be seen under the microscope (with the help of a stain) between the embryo and the starchy endosperm that is the embryo's food. The fungus can survive in stored seed for a year or so. Shortly after the seed germinates, mycelial filaments begin growing with the plant—moving through the intercellular space between both stem and leaf cells in the emerging shoot. After seven days, mycelia can be detected in practically all infected seedlings. In the maturing plant the endophyte becomes concentrated in the leaf sheaths (rather than the blades), but as reproduction approaches the hyphae begin moving upward between the rapidly elongating cells of the flower stem. Finally the fungus becomes concentrated in apical regions where flower heads are to form, and then in the flower heads themselves, where it penetrates the tissues of ovaries and ovules. The fungus has been detected only in the mycelial state, and its spores have not yet been detected in the field.

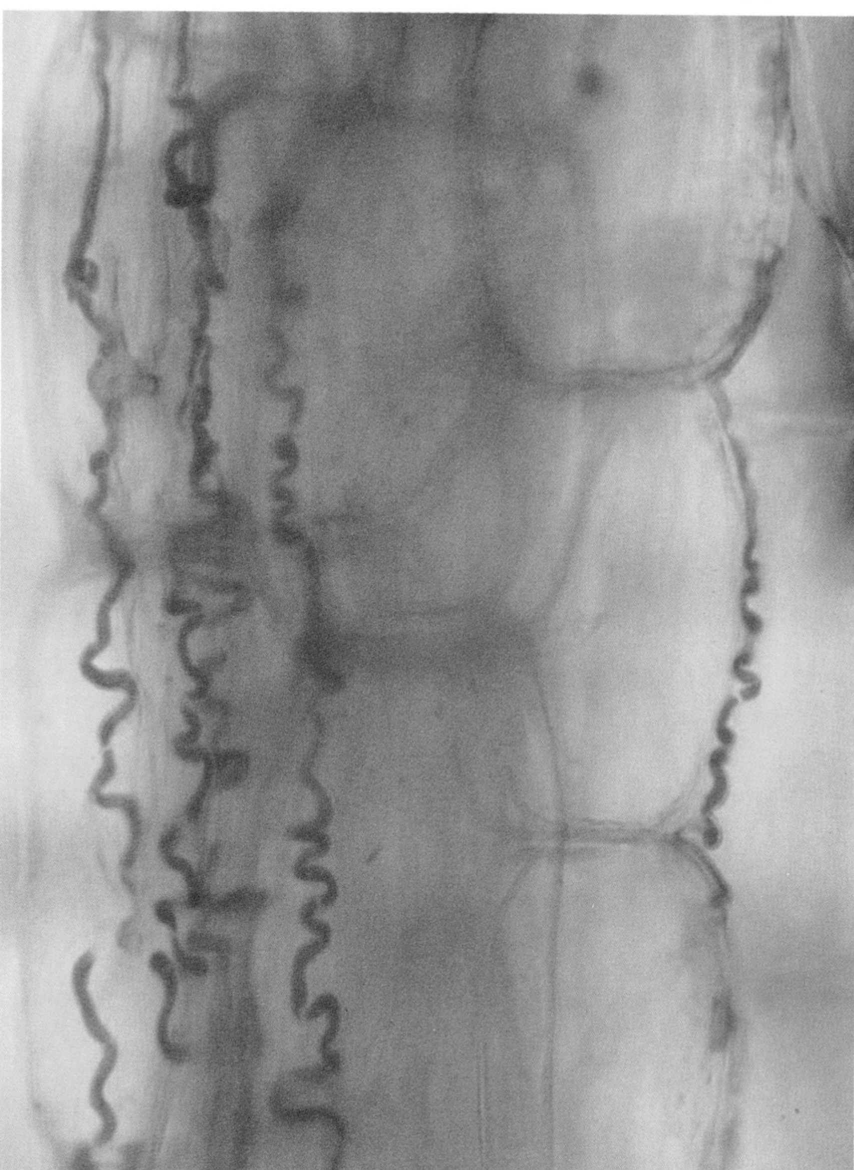

Figure 4. Mycelium of the tall-fescue endophyte grows as the plant grows, extending through the intercellular space between plant cells. In this microscopic view of a tall-fescue tissue sample, the filaments of the fungus can be seen with the help of a blue stain. Because the fungus is internal and causes no visible symptoms in the grass, laboratory analysis is required to diagnose endophyte infection in tall fescue. (Except where noted, photographs courtesy of Donald M. Ball.)

A Symbiotic Relationship

The simple and efficient life cycle of the endophyte seems to put very little stress on its host, from which it apparently derives nutrition by way of the intercellular fluid. The fungus does not penetrate or alter the cells of the host plant, and tests comparing the growth of infected and uninfected tall fescue indicate that infection confers a variety of favorable attributes to tall-fescue plants.

Various studies have shown that infected tall fescue, at least in some environments, has more-rapid germination rates. Infected seed produces more tillers and seedlings that are heavier and more likely to survive. Infected plants have been shown to be capable of higher seed production. In greenhouse studies, infected tall fescue showed greater overall resistance to drought, and drought-stressed infected plants were more likely than their endophyte-free counterparts to to exhibit leaf roll—a response to drought that reduces the amount of exposed surface area, and thus the amount of water lost by the leaves. In field studies, infected plants lost less water to evaporation. In addition, plant-parasitic nematodes of several species have been shown to be present in greater numbers in the vicinity of the roots of uninfected tall fescue.

Several studies conducted in stressful environments have shown that established stands of uninfected tall fescue are more prone to decline under stress. Some producers who have planted commercially available tall fescue that is endophyte-free have often had difficulty establishing or maintaining stands of the grass.

In nature, unpalatability is a common protective mechanism for plants. In greenhouse studies several species of insects appear to prefer uninfected tall fescue; not coincidentally, they survive and reproduce better when consuming the uninfected grass. Grazing livestock and laboratory animals alike prefer uninfected tall fescue, and eat more of it.

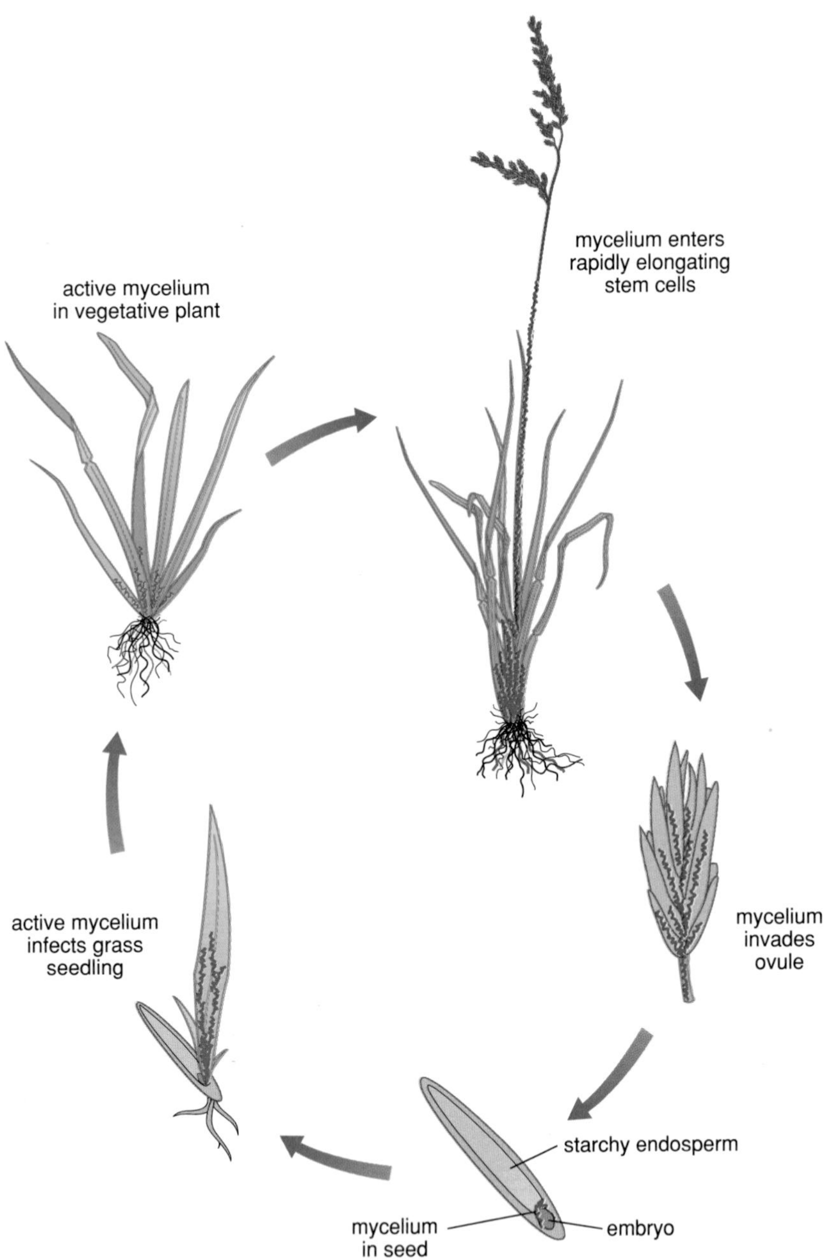

Figure 5. Life cycle of the endophyte *Acremonium coenophialum* is simple and follows the growth and reproduction of the host plant, tall fescue or *Festuca arundinacea* Schreber. In tall-fescue seed, the endophyte can remain viable for about a year. When the seed germinates, the fungal mycelium begins to grow, extending into the young shoot by growing through the space between plant cells. In the vegetative plant, the mycelium is concentrated in the leaf sheaths, rather than the blades, but as reproduction approaches, its filaments grow into the flower stem and ultimately invade the plant's ovaries and ovules, where they become encapsulated in the seed. Spores have not been detected in tall-fescue fields, but some *Acremonium* fungi are known to reproduce asexually by spores. The tall-fescue endophyte is transmitted solely through the seed.

Thus the toxicity that accompanies endophyte infection can be seen as another protective advantage acquired during the coevolution of the grass and its endophyte. By selecting seed from vigorous plants, plant breeders and seed producers have taken advantage of the benefits conferred by the endophyte; they have also unwittingly selected for animal toxicity.

The evidence that the animal symptoms now associated with endophyte infection in tall fescue are caused by ergot alkaloids produced by the fungus is not yet conclusive, but the circumstantial evidence is strong. When J. K. Porter, one of the USDA scientists who associated the tall-fescue endophyte with livestock-toxicity problems, analyzed his laboratory cultures, he found that the endophyte does produce several ergot alkaloids. More recent analysis has shown that the tall-fescue endophyte produces members of a chemical subgroup, the peptide ergot alkaloids, that are more active biologically than most ergot alkaloids. Although the fungus does not grow into the leaf blades, the alkaloids are regularly detected there, indicating that they are translocated throughout the plant after they are synthesized by the fungus. Experiments have shown that the parent alkaloid of the peptide subgroup, ergotamine, produces symptoms in cattle and sheep similar to fescue toxicity. In rats the daughter compounds found in tall fescue have been shown to be two to four times more toxic than ergotamine. And in recent tests one of the peptide alkaloids, ergovaline, was shown to cause fescue-toxicity symptoms when fed to steers. (No alkaloids have been found, however, in meat or milk from fescue-fed animals; the compounds appear to be fully degraded by the animals.)

Problems in the Pasture

In an infected tall-fescue pasture, a rise in temperature in the summer can have a strong and visible effect on cattle. Instead of grazing, the cattle spend an inordinate amount of their time seeking shade and water. Their body temperatures and respiration rates are high, they salivate excessively and they gain little weight. These striking effects are sometimes called "summer slump." The condition is costly for livestock producers.

Experiments in several states have shown that for each 10 percent increase in the endophyte infection rate, the dai-

Figure 6. Leaf roll is a mechanism by which grass can reduce its surface area during dry conditions and thus minimize the loss of water by evapotranspiration. In experimental plots, tall-fescue plants that are not endophyte-infected exhibit little leaf roll *(left)*; because precious water is lost through the thinner, wider leaf blades, uninfected plants have trouble surviving the stress of drought. During drought, infected tall fescue *(right)* is more likely to exhibit leaf roll and to survive.

ly weight gain among young beef cattle feeding on tall fescue drops by about one-tenth of a pound. The effects are greatest when the ambient temperature exceeds 31 degrees Celsius, but studies have shown that the endophyte has an effect on livestock productivity at any temperature. Because the summer effects are far more visible, cool-weather effects sometimes go unnoticed, and productivity losses are often greater than producers realize, especially in cooler climates.

In a number of recent studies, pregnancy rates, weight gains and milk production of cows and sheep, as well as the weaning weights of calves and lambs, have been lower on infected tall fescue. The reproduction of animals under nutritional stress is especially likely to be adversely affected.

Finally, the tall-fescue endophyte is now known to be a common cause of reproductive problems among mares. In a 1986 experiment at Auburn University, 22 pregnant thoroughbred, quarter-horse, Arabian and Morgan mares were fed either infected or non-infected tall-fescue pasture grass and hay during gestation, foaling and the immediate postpartum period. All but one of the mares eating the infected fescue had foaling problems, only three of 11 foals were born alive and only seven of the mares survived. Furthermore, only one foal from this group survived beyond the first week. The 11 mares eating endophyte-free fescue had normal pregnancies; all mares and all foals survived. The normal rate of reproductive problems among mares grazing non-fescue forages has been estimated at 11.5 percent.

The economic impact of these effects is hard to estimate, but it is significant. The largest cost has been incurred by beef producers in the form of slower weight gains and reduced numbers of calves. Losses associated with horse reproduction, with milk production among dairy animals and with the growth and reproduction of sheep and other grazing animals are somewhat less. Fescue foot and bovine fat necrosis have had an impact that has not been quantified. A conservative estimate places the total livestock-related losses nationwide related to tall fescue at between $500 million and $1 billion a year.

Figure 7. Maintaining stands of tall fescue is difficult in many areas without the advantages conferred by the endophyte. In an experiment in Americus, Georgia, plots of uninfected *(left)* and infected *(right)* plants from the then-experimental variety Georgia 5 were established side by side. Although the plants were genetically identical, the far greater stress tolerance of the endophyte-infected grass resulted in dramatic differences in the stands. (Photograph courtesy of Carl Hoveland, University of Georgia.)

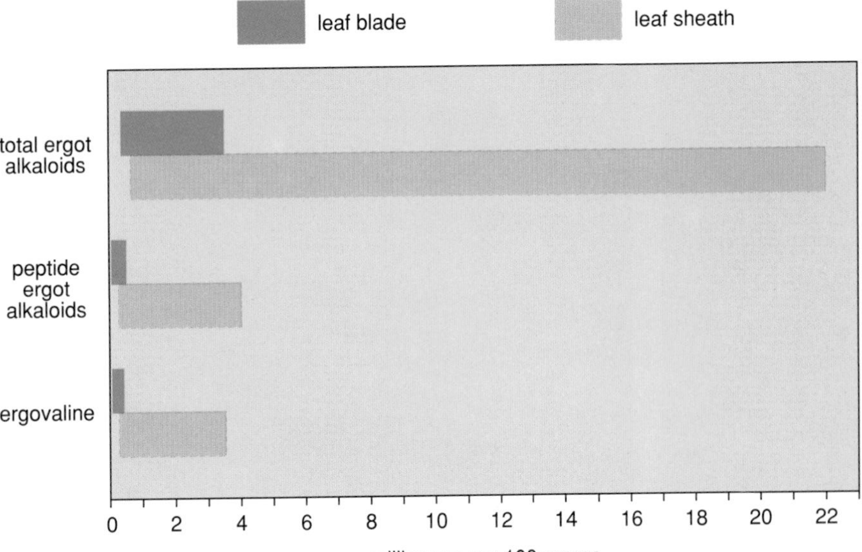

Figure 8. Ergot alkaloids are produced by the tall-fescue endophyte and can be detected in the leaf sheaths and leaf blades of infected grass, although the endophyte itself is found only in the sheath. Above are the ranges of some alkaloid concentrations detected in 15 samples of infected tall fescue collected from field and greenhouse. Ergot alkaloids are so called because they are produced by the ergot fungi *(Claviceps)* and are toxic to animals ingesting *Claviceps*-contaminated grain. They are believed responsible for the slow growth, reproductive problems and gangrenous disorders seen among animals grazing on endophyte-infected tall fescue. The most biologically active subgroup is the peptide ergot alkaloids, which are found in tall fescue; one of these compounds, ergovaline, has been shown to cause symptoms of fescue toxicity when fed to steers. (Data from Bacon et al. 1986.)

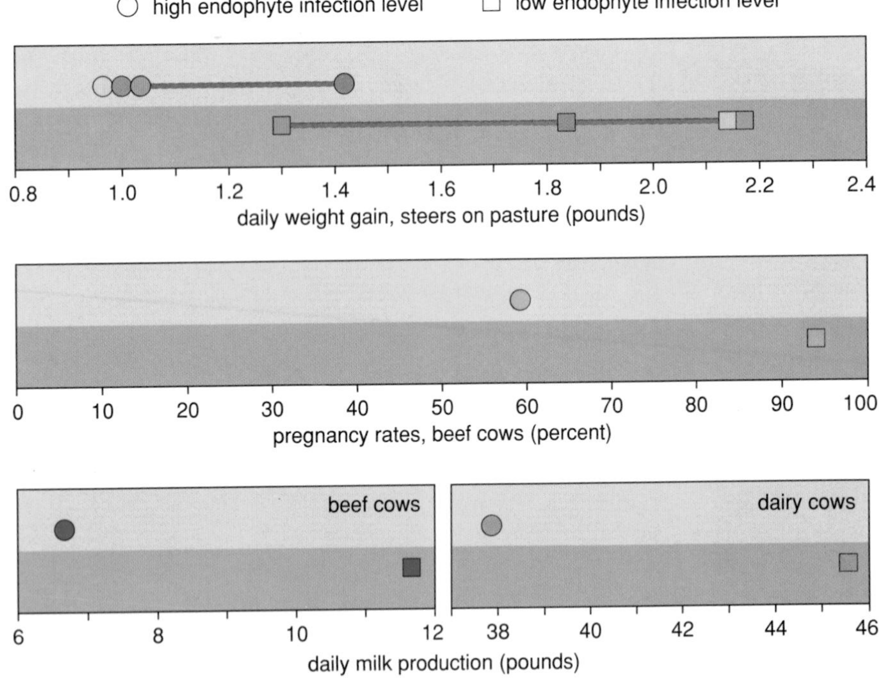

Figure 9. Animal performance is affected by the level of endophyte infection in tall-fescue pasture grass and hay. Experiments in numerous states have measured the average daily weight gain of beef steers in pastures that had low or high levels of endophyte infection, and consistently found that weight gains were significantly depressed by endophyte infections. Results from four grazing experiments, distinguished by color, are shown in the top panel. Beef cows grazing highly infected pastures are likely to have lower pregnancy rates *(middle)*, and the daily milk production of both beef and dairy cows fed on tall fescue *(bottom)* is greatly reduced in the presence of endophyte infection. (Data from various studies, as reported in Stuedemann and Hoveland 1988; also Gay et al. 1988.)

Planting Decisions

The information emerging from research on the tall-fescue endophyte has presented livestock producers with a dilemma. If they graze animals on infected grass, performance will be less than optimum. If they attempt to establish and maintain a stand of endophyte-free tall fescue, they know it will be more difficult and expensive, and that the stand may be prone to fail under a stress such as drought. Other grasses are not well adapted in many areas, especially the warmer parts of the region, which is why tall fescue became so popular so quickly. Along the edges of the area of primary adaptation, higher levels of environmental stress are most likely to result in poor stands or stand losses if producers try to establish endophyte-free pastures.

Short-term considerations complicate the decision for an individual forage or livestock producer. If a pasture of infected grass is to be destroyed and replaced with a new grass, there are issues of cash flow, erosion hazards and sustaining the livestock during the period of re-establishment. Thus, a producer must balance a complicated set of risks and uncertainties when deciding whether to replant.

Producers who plant endophyte-free tall-fescue seed must expect to put far more effort into managing their stands. During the year that the new grass is established, for instance, grazing should be restricted so that the plants are grazed no closer than 3 inches from the ground. Prudence dictates that overgrazing of established stands of endophyte-free tall fescue also be avoided, especially when subsequent severe climatic stress is likely.

In states where tall fescue is grown, livestock producers now have ways of monitoring endophyte infection in their fields to help in making management decisions. Producers typically collect plant samples and take them to university or private laboratories, where an estimate is made of the prevalence of fungal infection of tall-fescue plants in the field. The level of infection tends to increase in a field over time, presumably because the infected plants tend to outcompete uninfected plants.

As endophyte infection increases, producers must consider re-establishing the stand to maintain animal productivity. But there are several ways to manage the problem in the interim, and awareness of the endophyte has

Figure 10. Visual comparisons of animals grazing uninfected tall fescue *(left)* and endophyte-infected grass *(right)* are striking. Symptoms of fescue toxicity in beef cattle, especially evident during hot weather, include rough hair coats, intolerance to heat and poor weight gains. During cold weather a gangrenous condition sometimes develops on the extremities of animals fed with infected tall fescue. Because symptoms are less visible during fall and spring and because most producers do not have the benefit of side-by-side comparisons, many underestimate their losses from endophyte infection. Annual livestock-related losses related to the endophyte have been conservatively estimated at $500 million to $1 billion.

changed the management of infected tall fescue in forage fields. Traditional agronomic-management practices tend to thicken stands and increase the proportion of infected tall fescue in the diet of animals, and thus exacerbate animal disorders. Some producers now adjust the timing and levels of fertilization to favor other volunteer or planted pasture species, encouraging a mixture of grasses to "dilute" the effect of the toxins on animals. And since it has been shown that the toxic effects of the endophyte may be less in a pasture that is heavily grazed, some producers now use higher stocking rates—a greater density of grazing animals—in pastures of infected tall fescue.

Since the endophyte is concentrated in the seedheads of the grass, grazing on seedheads is a particular concern. There is an additional reason to discourage such grazing: It has been shown that between 1 and 2 percent of the seed ingested can pass through an animal's body with both seed and endophyte remaining viable. Animals thus can spread infected seed from one area to another in the 48 hours or so after it is consumed; by this method the endophyte can easily be transferred into a uninfected field by a grazing animal moving from one pasture to the next. Producers are therefore encouraged to use techniques that reduce the amount of seed eaten by livestock.

In conventional pasture management, plant species are often mixed to extend the grazing season or to reduce the need for nitrogen-containing fertilizer through the use of legumes that add nitrogen to the soil. Clovers and some other legumes provide nitrogen fixation and good nutritional value and have long been grown with tall fescue. It now appears that some of the good results achieved by these mixtures come about because the other crops reduce the concentrations of endophyte toxins in a grazing animal's body. Therefore, mixed plantings are another technique available to producers trying to minimize animal problems.

The question of mixed plantings is not, however, a simple one when the endophyte is involved. Legumes have vigorous seedlings and are easy to establish, and when planted with tall fescue—especially uninfected grass—they may initially compete so strongly that they overwhelm the grass. But in an established stand, infected tall fescue tends to dominate. In addition to its sheer competitiveness, one interesting explanation could be allelopathy, the production by one plant of chemical compounds that suppress the growth of a competing plant. Recent studies have indicated that another *Acremonium* endophyte, found in perennial ryegrass, may have allelopathic effects on the establishment of legumes. Should the tall-fescue endophyte turn out to have similar effects, it might help explain the difficulty livestock producers have had in maintaining forage legumes with infected tall fescue.

A final ecological issue in planting decisions is the possible effect of the tall-fescue endophyte on other organisms. The list of herbivores affected by endophyte-produced toxins may extend beyond insects and large mammals. The reproduction of rats and mice (in addition to several insect species) has been shown to be adversely affected by feeding on infected tall fescue.

The advantages of the endophyte to the grass continue to make endophyte-infected seed a popular choice for turf

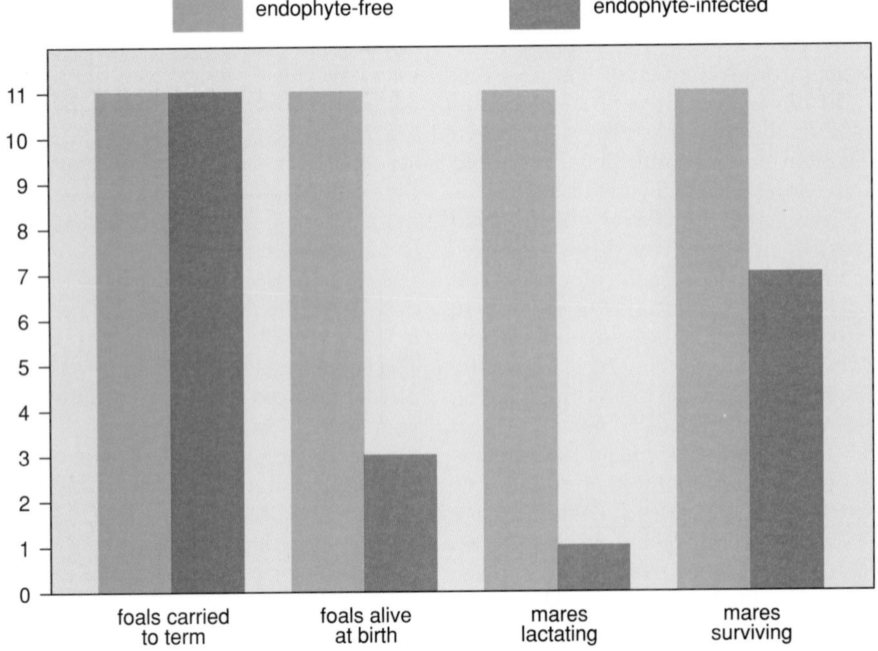

uses. Seed producers now find themselves serving one or both segments of a split market, producing uninfected seed for livestock producers and infected seed for turf and conservation uses.

For most markets, seed marketers now must test tall-fescue seed for the endophyte, a matter of some inconvenience and expense. Many are opposed to requirements that seed tags state the proportion of infected seed in a quantity offered for sale, but a strong argument can be made that this is in the best interest of consumers. Such requirements are in effect in a few states, and there have been proposals to implement a federal seed-labeling requirement.

Additional issues now surround the production and marketing of tall-fescue seed. Since the endophyte loses its viability during prolonged storage, companies marketing seed for turf purposes have begun to test for and monitor endophyte levels and to consider special handling to ensure endophyte viability.

The discovery of the endophyte's role also has called into question much of what was believed to be established knowledge about tall fescue. Forage and livestock data from tall-fescue experiments prior to the early 1980s must now be considered suspect. It can be assumed that most early experimental tall-fescue pastures were infected, but the level of infection and the impact of the endophyte on experimental results cannot be known. Studies that had been thought to have established important facts about tall fescue may need to be repeated using seed whose infection status is known.

Implications

The implications of the tall-fescue endophyte story for agricultural science, and biological science in general, are broad and fascinating. It is now

Figure 11. Reproductive problems are common among mares that feed on infected tall fescue. In a study by scientists at Auburn University, the results of pregnancies among mares grazing infected and uninfected fescue were compared. The 11 mares fed endophyte-free grass had successful pregnancies, but 11 mares in an infected pasture failed to lactate, and most produced stillborn foals during difficult deliveries. The foaling problems proved fatal to four of the mares. Shown at top are a healthy mare and foal in an Alabama tall-fescue pasture. Among livestock, horses are most affected by the reproductive problems associated with the tall-fescue endophyte. (Data from Putnam et al. 1990.)

Figure 12. Planting clover and other legumes with infected tall fescue, the approach used in this Alabama pasture, can greatly reduce the toxic effects of the tall-fescue endophyte. Establishing a mixed pasture can be difficult. Infected tall fescue competes strongly with legumes and is more persistent. In addition, an *Acremonium* endophyte in perennial ryegrass has been found to produce allelopathic compounds (compounds that suppress the growth of companion plants) that discourage the establishment of clover seedlings. It is possible that the endophyte of tall fescue may also have an allelopathic effect on clover. Prospects for reducing the impact of the endophyte on livestock production focus on techniques for using companion species with tall fescue, the development of more vigorous varieties of infection-free tall fescue, treatments to offset the effects of the endophyte and the genetic engineering of the fungus to reduce the production of toxic compounds while keeping the other advantages it confers on its host.

known that many species of grass, both wild and cultivated, contain endophytes. Undoubtedly many of these also have coevolved with their hosts; they may have played a large role in determining the ecological significance of the host grass and its evolutionary relationships with animals and many other organisms. Because tall fescue is a species of great economic importance, it has been the subject of extensive experimentation. But the effects of endophytes in other grasses on livestock performance, plant persistence and ecological interactions are largely unknown.

Research in this area has the potential for a significant impact on animal performance and the profitability of the livestock industry. Genetic studies of *Acremonium* might, for instance, lead to the identification or the development, possibly by genetic engineering, of strains that confer great benefits on tall fescue but that are not toxic to economically important animals. Plant breeders may be able to develop uninfected tall-fescue varieties that are more stress-tolerant, and veterinary scientists might be able to find treatments to offset the effects of grazing on infected tall fescue. Agricultural scientists are challenged to find ways to kill the fungus in existing plants without having to destroy and then re-establish stands, and to develop management techniques that minimize the toxic effects of the endophyte.

Meanwhile many producers face substantial losses, because their only option in the face of endophyte-induced livestock problems may be to turn to forage grasses without the stress tolerance and low management demands they have come to expect. The discovery of the endophyte has been one of the most interesting and important developments in the history of agricultural research, but much work remains to be done.

Bibliography

Bacon, C. W., J. K. Porter, J. D. Robbins and E. S. Luttrell. 1977. *Epichloe typhina* from toxic tall fescue grasses. *Applied and Environmental Microbiology* 34:576.

Bacon, C. W., P. C. Lyons, J. K. Porter and J. D. Robbins. 1986. Ergot toxicity from endophyte-infected grasses: A review. *Agronomy Journal* 78:106–116.

Ball, D. M., C. S. Hoveland and G. D. Lacefield. 1991. *Southern Forages.* Atlanta, GA: The Potash and Phosphate Institute and the Foundation for Agronomic Research.

Buckner, R. C., and L. P. Bush, eds. 1979. *Tall Fescue.* Madison, WI: American Society of Agronomy.

Fribourg, Henry A., C.S. Hoveland and Pascal Codron. 1991. Tall fescue and *Acremonium coenophialum*—review of current situation in the United States. *Fourrages* 126:209–223.

Gay, N., J. A. Boling, R. Dew and D. E. Miksch. 1988. Effects of endophyte-infected tall fescue on beef cow-calf performance. *Applied Agricultural Research* 3:182.

Hoveland, C. S., R. L. Haaland, C. C. King, Jr., W. B. Anthony, E. M. Clark, J. A. McGuire, L. A. Smith, H. W. Grimes and J. L. Holliman. 1980. Association of *Epichloe typhina* fungus and steer performance on tall fescue pasture. *Agronomy Journal* 72:1064–1065.

Joost, R., and S. Quisenberry, Eds. 1993. *Acremonium/Grass Interactions.* Vol. 44 in Agriculture, Ecosystems and Environment. Elsevier: Amsterdam.

Morgan-Jones, G., and W. Gams. 1982. An endophyte of *Festuca arundinacea* and the anamorph of *Epichloe typhina*, new taxa in one of two sections of *Acremonium*. *Mycotaxon* 15:311.

Pedersen, J. F., G. D. Lacefield and D. M. Ball. 1990. A review of the agronomic characteristics of endophyte-free and endophyte-infected tall fescue. *Applied Agricultural Research* 5:188–194.

Pedersen, J. F., and D. A. Sleper. 1988. Considerations in breeding endophyte-free tall fescue forage cultivars. *Journal of Production Agriculture* 1:127–132.

Putnam, M. R., D. I. Bransby, J. Schumacher, T. R. Boosinger, L. Bush, R. A. Shelby, J. T. Vaughan, D. M. Ball and J. P. Brendemuehl. 1990. The effects of the fungal endophyte *Acremonium coenophialum* in fescue on pregnant mares and foal viability. *American Journal of Veterinary Research* 52:2071.

Read, J. C., and B. J. Camp. 1986. The effect of the fungal endophyte *Acremonium coenophialum* in tall fescue on animal performance, toxicity, and stand maintenance. *Agronomy Journal* 80:811–814.

Rice, J. S. , B. W. Pinkerton, W. C. Stringer and D. J. Undersander. 1990. Seed production in tall fescue as affected by fungal endophyte. *Crop Science* 30:1303–1305.

Siegal, M. R., G. C. Latch and M. C. Johnson. 1985. *Acremonium* fungal endophytes of tall fescue and perennial ryegrass: significance and control. *Plant Disease* 69:179–193.

Stuedemann, J. A., and C. S. Hoveland. 1988. Fescue endophyte: History and impact on animal agriculture. *Journal of Production Agriculture* 1:39–44.

Population Outbreaks in Forest Lepidoptera

Viruses may help to explain the remarkable population cycles of tent caterpillars and other forest insects

Judith H. Myers

Almost everyone has a vivid childhood memory of an insect outbreak, a year when hordes of June bugs ran amok on suburban lawns or the entire countryside seemed to be festooned with the silken tents of caterpillars. Population ecologists are as awed and delighted by these episodes as are children. What kind of regulating mechanism, they wonder, would allow such fantastic excursions from normal population levels?

Among insects that undergo population outbreaks, there are a few that have even more improbable population dynamics: their outbreaks seem to occur on a regular schedule. Of the roughly 80 outbreak species among forest species of the Lepidoptera—the moths and butterflies—in North America and Europe, at least 18 are cyclic. Some better-known examples of cycling caterpillars are the tussock moth that feeds on Douglas fir in western North America, the spruce budworm that feeds on balsam fir and spruce in eastern North America, and the forest tent caterpillar that inhabits deciduous forests across North America.

Outbreak and cycling species constitute only 1 or 2 percent of forest Lepidoptera, but they have received a disproportionate share of the attention given to these species. Although population levels are influenced by many different variables, a cyclic pattern seems to imply a dominant force that should be easy to identify and quantify. That driving force, however, has proved surprisingly elusive.

Historically, population cycles have often been studied by measuring the mortality caused by different agents and by examining the progression of mortality through the life stages of the insect. The studies of mortality agents in Lepidoptera, however, have made disappointing progress. Moreover, experience has shown that population cycles are exceedingly robust and difficult to perturb from outside. In short, the evidence implies that these insect populations, if not self-regulating, may at least be regulated by an agent more intimately connected with the insect than are predatory birds or parasitoids (parasites with only one generation in their host).

Recent work suggests that this more intimate agent may be a virus. Viral disease has been reported in declining populations of caterpillars for many years, but it has usually been considered to have contributed to the decline once it was under way rather than to have initiated it. The development of laboratory techniques that allow viruses to be identified at low concentrations and in individuals that are asymptomatic is altering our understanding of their role in population cycles. Ecologists currently suspect that nuclear polyhedrosis viruses, which account for over 40 percent of the described insect viruses, may be the long-sought driving force of population cycles in forest Lepidoptera.

One of the attractions of this hypothesis is its generality. Remarkably, many forest caterpillars have population cycles with periods of similar length, between 8 and 11 years. The similarity suggests the cycles may have a common explanation. Although nuclear polyhedrosis viruses are host-specific, many species of lepidopterous larvae are infected by them. The similarity of the insects' population cycles might be explained by the similarity of their experiences with viruses.

Finding, and Modeling, Insect Cycles

Most forest Lepidoptera remain at very low densities and never significantly damage their host plants. For example, of 74 insect species collected from *Eucalyptus* stands in Australia, only four species had ever been recorded in outbreak numbers. Why are some species capable of outbreaks, whereas others are not? And why are some outbreak species erratic and others cyclic? Grasshoppers break out erratically, but outbreaks of larch budmoth occur at regular intervals.

Neither question has a clear answer. Attempts to find common ground among species in either category have met with little success. For example, Alison Hunter, formerly of McGill University and now at the University of British Columbia, recently reviewed the traits that distinguish outbreak species. The majority lay their eggs in clusters. Roughly half are colonial, whereas most non-outbreak species are solitary. Outbreak species tend to have high fecundity, to feed early in the spring and to feed on a variety of tree species. About half of the outbreak species surveyed are poor fliers or do not fly at all, whereas almost all non-outbreak species are good fliers. Finally, outbreak species tend to be unpalatable to predators either because they are hairy or be-

Judith H. Myers is a professor in the Departments of Plant Science and Zoology as well as an associate dean of science for the promotion of women in science at the University of British Columbia. Population fluctuations have been her addiction since she began studying the cycles of small mammals while completing a Ph.D. at Indiana University. In 1972 she joined the UBC faculty and turned her attention to the population cycles of forest Lepidoptera. Her work has taken her into local schools to discuss the beauty and usefulness of insects, and she has worked in several programs employing insects in the biological control of weeds. Address: Department of Zoology and Plant Science. 6270 University Boulevard, University of British Columbia, Vancouver, BC V6AT 2A9 Canada.

Figure 1. Tent caterpillars are particularly good subjects for population studies because of their colonial habits. Obtaining an accurate count of insect species with solitary habits is difficult, particularly when population densities are low. The conspicuous webs made by tent caterpillars, such as the western tent caterpillars shown here, are much easier to count. Moreover, a tent's ultimate size is correlated with the number of caterpillars in a family group that survive to late-instar stages. As soon as they hatch, the young caterpillars from one egg mass stay together and begin to build a web. The caterpillars do not feed inside their tents, but instead crawl to neighboring branches, spinning a fine strand of silk behind them wherever they go, so that the tents grow with time. Excrement and molted skins are usually suspended within a tent's dense sheets of silk. (Photograph courtesy of the author.)

cause they taste bad. Although there are suggestive tendencies, no simple pattern emerges.

Are the fluctuations in populations of forest Lepidoptera truly cyclic? The data on the periodicity of outbreaks of forest Lepidoptera were recently summarized by P. J. McNamee, then a graduate student at the University of British Columbia. McNamee's analysis detected patterns of population change, but they were not invariable. An example can be found in the outbreaks of western tent caterpillars on the Saanich Peninsula of British Columbia. High densities were noted in 1936, 1945, 1956 and 1964, intervals of from 9 to 11 years. The next two peaks occurred after intervals of only 5 and 7 years, but the population cycle then returned to a 10-year period with peaks in 1976 and 1986. To the population ecologist, of course, both the pattern and the deviation from the pattern are of interest, since a regulating mechanism must be able to account for both.

A rough idea of the type of force needed to drive population cycles can be gained from experiments with mathematical models. In a simple system, population density would increase to an environment's carrying capacity and then remain at that level. It would be prevented from increasing above the carrying capacity by some form of negative feedback. For example, an increase in population density might reduce the availability of food, shelter and hiding places, slowing the density increase. As a general rule, such negative density-dependent feedback has a stabilizing effect on population levels. Although the regulated population does not necessarily remain constant, it oscillates around an equilibrium level and tends to return toward this level when it is disturbed.

The feedback process may, however, overshoot or undershoot the equilibrium population density, or there may be a delay in the loop. Mathematical models show that time lags, high reproductive rates and over-compensating density dependence are all capable of causing population fluctuations of the kind observed in forest insects.

Of course, a realistic model of insect population dynamics would include a variety of negative feedback processes that influenced reproduction and survival at different rates and different densities. It would probably also include positive feedback loops and stochastic (density-independent) processes that would tend to destabilize population dynamics. Indeed, the central difficulty with the mathematical approach to population dynamics is that there are altogether too many ways of tweaking or tuning a control system to achieve a desired population curve.

The population cycles of the black-headed budworm provide an example

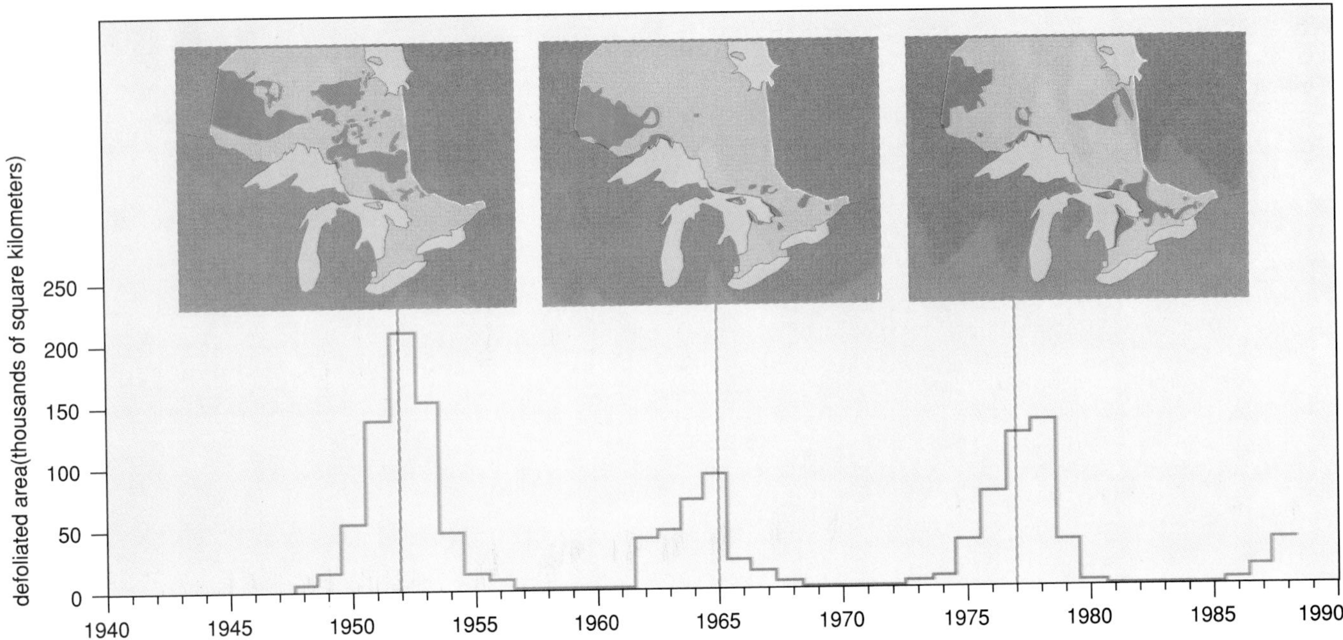

Figure 2. Population cycles in forest insects are illustrated by the outbreaks of forest tent caterpillars in Ontario. The insect populations were tracked by means of aerial surveys of defoliation carried out by the Ontario Forest and Insect Disease Survey. The forest tent caterpillar is the major defoliator of deciduous forests in this region. The data show that the caterpillar population has peaked roughly every 13 years for the past 40 years. The three maps show the extent of defoliation during the peak years of 1952, 1965 and 1977. Strong periodicities in population data such as this have elicited much creative interpretation, but definitive explanations have proved elusive. The data were analyzed by Colin Daniel.

of the difficulties this creates. R. F. Morris, formerly of the Canadian Department of Forestry in Fredericton, New Brunswick, conducted a long-term study of this species and concluded that its population cycles are due to parasitoids that attacked caterpillars. Parasitoidism and larval density were found to be highly correlated, and populations declined before food became limiting. McNamee used Morris's data to model budworm dynamics and included the effects of parasitoids, food depletion, weather and bird predation. The addition of bird predation resulted in two stable equilibria: an equilibrium at low density determined by bird predation and one at high density determined by food limitation. Alan A. Berryman of Washington State University also re-analyzed Morris's data. He concluded there was no evidence for two equilibria and that a delayed feedback process, such as a numerical response by a para-

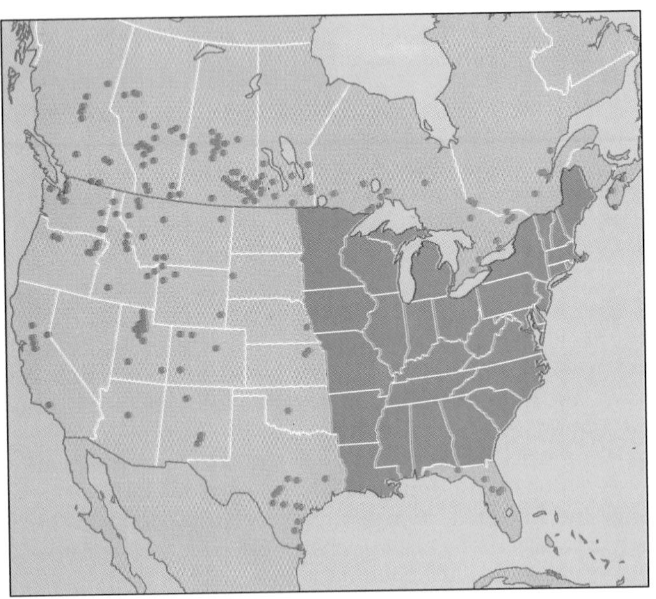

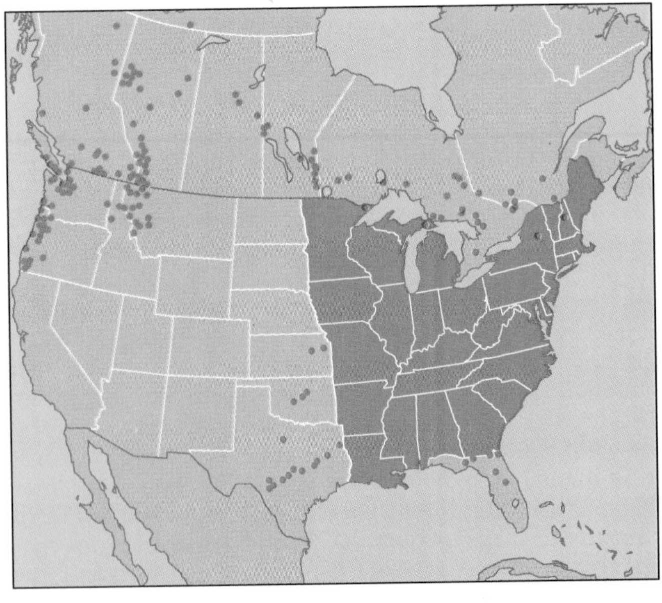

Figure 3. The six species of tent caterpillar found in North America all belong to the genus *Malacosoma*. The distributions of the three most commonly encountered species are mapped here (solid colors indicate states with widespread infestations). The forest tent caterpillar is found throughout the United States and adjacent Canadian provinces *(purple)*. The eastern tent caterpillar is found throughout the eastern United States and the southern part of eastern Canada *(blue)*. The western tent caterpillar consists of six subspecies that occupy well-defined geographic areas *(orange)*. The ranges of the insects correspond roughly with the ranges of their host trees. All three species feed on a variety of hardwoods, but the eastern tent caterpillar prefers trees in the family *Rosaceae*, such as black cherry, crabapple and apple trees; in Canada, the forest tent caterpillar's preferred host is the trembling aspen.

sitoid, adequately explained the relation between density and rate of population increase. His solution, however, led to damped population cycles. To maintain cyclic behavior, he added a random weather variable.

Which is more realistic: a model that produces cyclic dynamics with the addition of a function for bird predation, or one that produces cyclic behavior only if a random weather variable is included? Both models might produce cyclic behavior for the wrong reason. In general, although models can show what might happen to insect populations, they cannot show what *does* happen. And in practice mathematical models have rarely produced testable predictions that have allowed hypotheses to be validated or refuted by fieldwork.

The naive onlooker might assume that the population dynamics of forest Lepidoptera must be Malthusian in nature. As the density of insects increases in a forest, less food is available to each individual. Eventually many insects die of starvation or, if they survive, produce fewer offspring. The difficulty with this hypothesis is that population cycles do not track the food supply in any straightforward way. Cycling insects sometimes decline well before forests are defoliated. And they typically remain at low densities for several years after host trees have recovered from an infestation.

There has been surprisingly little study of the impact of cyclic forest defoliators on forests, but the insects cause much less damage than might be expected on the basis of the amount of defoliation that is observed. Most of the species of forest Lepidoptera that are considered to be pests feed on new foliage early in the spring. Deciduous trees, such as aspens and alders, respond to defoliation by refoliating later in the summer, and this allows them to compensate for insect damage. Evergreen trees are less able to compensate for the loss of their leaves and are more vulnerable to attack.

Although spruce-budworm attacks kill coniferous trees, widespread mortality of deciduous trees rarely occurs, even when defoliation is extensive and complete and recurs for several years. The main consequence of repeated annual infestation is the loss of wood increment during years of severe defoliation. Species of forest Lepidoptera with outbreak population dynamics appear to have evolved to have little long-term detrimental effect on forests.

Figure 4. Markings distinguish the three most common tent caterpillars in North America, whose distributions are shown in Figure 3. From top to bottom are shown the larval and adult forms of the forest tent caterpillar (*Malacosoma disstria*), the eastern tent caterpillar (*Malacosoma americanum*) and the western tent caterpillar (*Malacosoma californicum*). The actual length of the last-instar caterpillars shown here is about three inches.

Figure 5. Tent-caterpillar defoliation can denude large portions of forests, but the insects cause much less damage than the visual evidence might suggest. These trembling aspens in Prince George, British Columbia, have lost their spring foliage to the caterpillars, but they are likely to refoliate in the summer. Widespread mortality of deciduous trees is rare. (Photograph courtesy of the author.)

Mortality, Weather and Climate

Traditionally, most of the emphasis in the study of insect population fluctuations has been on mortality. The mortality caused by different agents, and the progression of mortality through the life stages of forest Lepidoptera, were typically summarized in life tables. A classic example of this type of research is a 1963 monograph by R. F. Morris on the spruce budworm.

Initially it was thought that analysis of life tables could reveal which mortality agents were most important in determining population trends. But although life-table analysis was originally seen as the epitome of a quantitative approach to population ecology, the results have been disappointing.

The methodological problems with life-table studies have been summarized by Peter Price of Northern Arizona State University in Flagstaff. One problem is that measurement errors are likely to be large when population densities are low. And because stage-specific mortalities are not additive, small changes in mortality can have large effects on population density. Field measurements may not be precise enough to measure relevant shifts in mortality.

Moreover, as Price showed, a single life-table data set can lead to very different conclusions, depending on the method of analysis. For example, J. P. Dempster, formerly with the Institute of Terrestrial Ecology in Britian, reviewed life tables of 21 species of Lepidoptera and concluded that parasitoids rarely act in a density-dependent manner. Other authors have shown, however, that temporal and spatial variation can make the demonstration of density dependence very difficult.

A final drawback is that mortality studies are unlikely to yield a general hypothesis because different insect species typically have different mortality agents. Although birds prey on the spruce budworm, for example, they avoid the hairy tent caterpillars. Life-table studies have therefore led to an intellectually dissatisfying proliferation of regulating mechanisms.

Another variable that received much attention in early studies of population cycles of forest Lepidoptera was the weather. How could a random factor such as the weather be responsible for regular population cycles? A typical explanation was that the survival of early-instar caterpillars is affected by the relation of the timing of egg hatch to the emergence of leaves, since the quality of leaves changes as they mature. (Instars are the developmental stages of caterpillars, separated by molts, early being the stages soon after hatching and later being the approach to the pupal stage that precedes winged adulthood.) Several years of favorable weather, it was suggested, might therefore precipitate an insect outbreak.

Alternatively, the weather was invoked as the means by which dispersed insect populations were periodically resynchronized. Insect outbreaks sometimes cover enormous areas. Outbreaks of eastern spruce budworm can extend from Ontario to Nova Scotia, and dense populations of forest tent caterpillars can extend from Manitoba through Ontario. Given the amount of jiggle in population cycles, it is surprising that dispersed populations are not completely asynchronous. The suggestion was that episodes of severe weather cause populations that might have drifted out of phase to collapse simultaneously, re-synchronizing them.

Peter J. Martinat, formerly of the University of Maryland and now at Abbott Laboratories in Chicago, has summarized studies that attempt to relate forest-insect outbreaks to climatic conditions and has examined the methodological problems associated with testing these hypotheses. In the literature, the following conditions are reported to favor outbreaks of forest tent caterpillars: a warm, sunny spring; a continuously warm, humid and cloudy spring; weather that is not unusually warm; a warm, early spring followed by cooler weather; a cool fall, cold winter and warm spring. With this varied a list of favorable conditions, the probability of poor conditions is very low. Martinat concludes that most studies that claim support for the climatic-release hypothesis "would not hold up under rigorous examination."

Studies that associate the collapse of insect populations with inclement weather have similar defects. Although it is tempting to attribute significance to specific conditions, these explanations lack generality among sites, among outbreaks and certainly among species.

Recent studies at the University of British Columbia also cast doubt on the climate and weather hypotheses. For the past 40 years forest entomologists with Environment Canada, a government agency that includes the forest service, have collected extensive data on the defoliation of deciduous forests in Ontario. In 1990 Colin Daniel, then a graduate student at the university, examined these data for associations between defoliation and weather variables at a number of sites. He found no correspondence between increases in defoliation and warm spring conditions. Nor were declines in defoliation associated with freezing conditions.

My fieldwork also suggests that weather is an unlikely explanation for widespread synchrony among insect populations. Tent caterpillars in coastal British Columbia inhabit sites that range in elevation from sea level to over 600 meters above sea level. There are large variations in temperature and rainfall with elevation. High-elevation populations hatch a month to six weeks after those at sea level and may be exposed to quite different weather conditions in any given spring. And caterpillar eggs at high elevations are exposed to prolonged freezing, whereas those at low elevations only rarely experience freezing. Despite these differences, populations at different elevations cycle in synchrony.

Insects and Plants: Active Controllers?

A few authors have suggested that the susceptibility of insects to mortality agents may be more important than the mortality agents themselves in explaining population dynamics. In other words, population cycles may be driven by variation in the "quality of the insect" rather than variation in an external agent. Others have explored the bio-

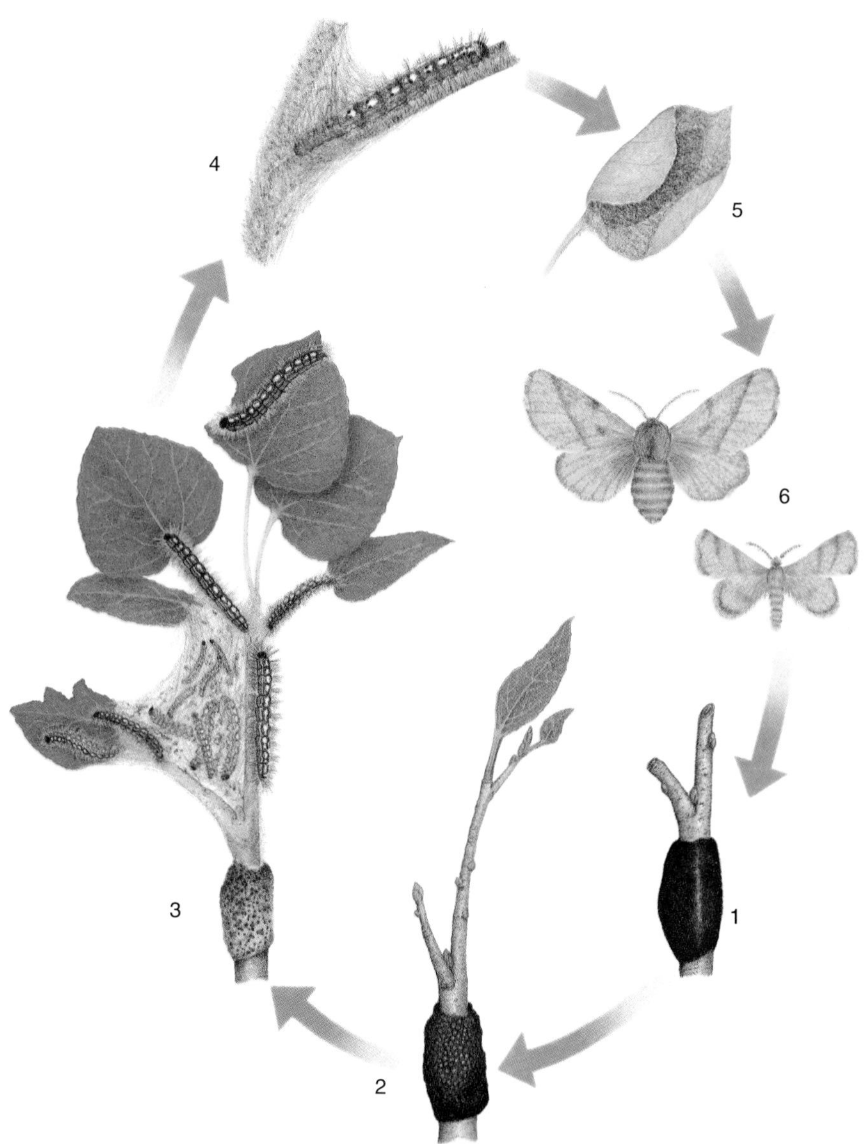

Figure 6. Life cycles of all species of *Malacosoma* are very similar, the only significant exception being that forest tent caterpillars do not construct tents. The eggs (typically 150 to 250) are laid as a mass that encircles or partially encircles a twig of the tree, or as a relatively flat mass on a larger branch. The eggs *(1)* are held in place by a dark, frothy substance called spumaline that erodes over time. A few weeks after the eggs are laid, the young caterpillars are fully formed within them. They then enter a period of diapause for a few months, followed by dormancy, which lasts until the next spring *(2)*. Hatching takes place about the time first new leaves appear in spring. The hatched caterpillars together form a tent in and on which they live, except while they are feeding. Eastern tent caterpillars spend most of their time in the tent, whereas western tent caterpillars spend much time on their tents, basking in the sun on sunny days *(3)*. The developmental stages delineated by the molts are called instars; most tent caterpillars pass through five instars. Most of the damage to trees is caused by the last two instars. By the time the last instar has been reached, the caterpillars have become less gregarious and travel more widely. After the final molt they leave *(4)* and select a protected site, such as a folded leaf, for constructing a cocoon *(5)*. The pupal stage takes up to two weeks. The adults have no functioning mouthparts, and they mate and die within a few days after emerging *(6)*.

chemical responses of trees to caterpillar attacks to develop an alternative theory.

William G. Wellington, formerly of the University of British Columbia, attributed population fluctuations in the western tent caterpillar to behavioral and physiological variation he observed in the populations. He found that a behavioral spectrum—from active to sluggish larvae—was reflected in the shape of the tent constructed by the later instars. A group of larvae with a high proportion of active individuals constructed elongate tents. The tents of groups with a high frequency of sluggish individuals were compact and usually small. This classification of tents allowed Wellington to characterize the condition of field populations of tent caterpillars.

According to Wellington, an infestation begins with a predominance of active colonies. As the population increases, active moths disperse from the area of increasing density, and sluggish colonies accumulate there. The sluggish colonies are susceptible to poor weather, starvation, viral disease and parasitoid attacks. Any or all of these destroy the populations and cause the cyclic decline.

Wellington's hypothesis had the valuable effect of directing attention to the possibility that the insects were playing a role in their own population cycles instead of acting strictly as the dependent variable. Wellington contended that the activity phenotype of a larva was determined by the distribution of maternal food reserves; an egg receiving a generous amount of yolk developed into an active larva, and one with less yolk produced a sluggish larva.

My work on tent caterpillars suggests that viral disease could be related to the "maternal influence" described by Wellington if sublethal infection reduces the amount of nutrient available for partitioning among eggs. Transmission of virus on egg masses could also reduce the survival of caterpillars such that fewer survive to late instars, causing the tents to be smaller and more compact.

In the past 10 years we have learned much more about the trees caterpillars attack, and another possibility has arisen: that the trees themselves regulate caterpillar population levels. This hypothesis, known as the induced-defense hypothesis, was originally formulated by David F. Rhoades of the University of Washington and Erki Haukioja of the University of Turku in Finland. They suggested that trees respond to herbivore damage by producing high levels of phenolic compounds, such as tannins, lignins and resins, that are deterimental to insect growth and survival and that therefore cause a decline in insect populations. The induced-defense hypothesis has great appeal because it seems to provide the delayed density-dependent feedback needed to explain the observed population dynamics.

The hypothesis is also attractive because it seems eminently testable. The quality of foliage from plants attacked by insects could be compared to that from undamaged plants by means of either chemical assays or bioassays that examine the growth, survival and ultimate size of insects. Many of the experiments that have looked for foliage-quality changes, however, can be faulted for design problems. Some of the problems have been outlined in articles by Haukioja and Seppo Neuvonen of the Finnish Forest Research Institute.

The response of leaves to damage is frequently evaluated by measuring levels of total phenols. Higher levels of these so-called "defense chemicals" are interpreted to indicate increased plant defense. Without an evaluation of the impact of total or particular phenols on insects, however, speculation about their role in regulating insect populations is unwarranted. Many studies have shown that insects do not always respond to different levels of phenols in their host plants. The growth and survival of the tree locust *Anacridium melanorhodon* actually improved when tannic acid was added to its diet.

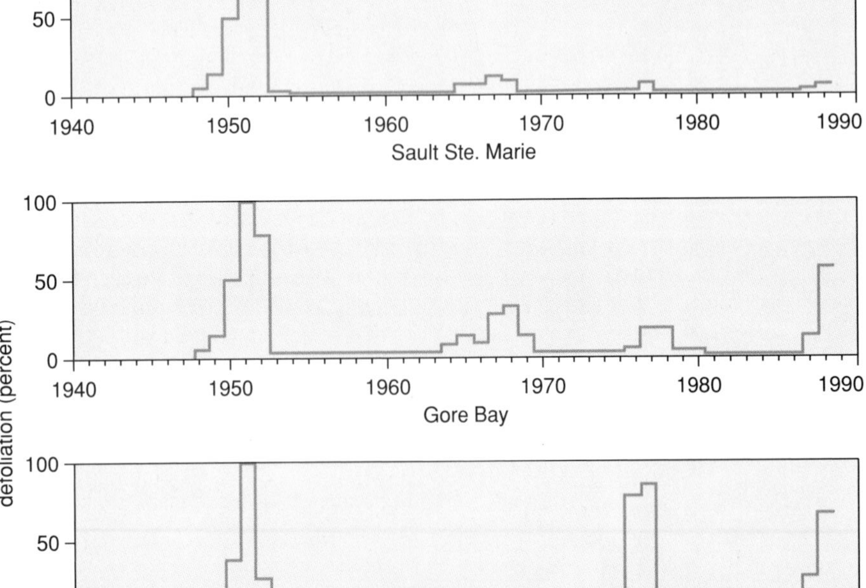

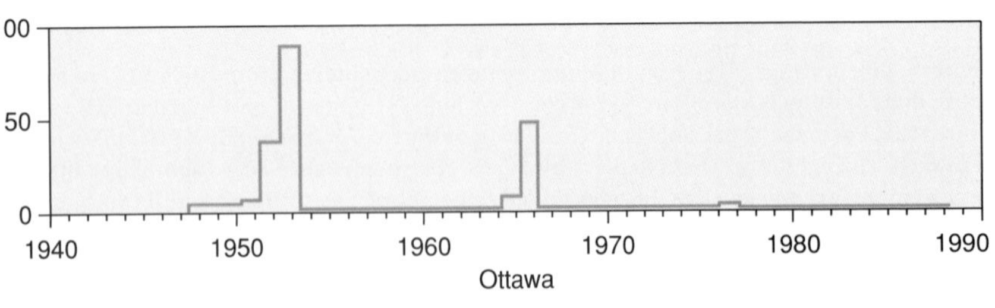

Figure 7. Food-supply exhaustion is one obvious explanation for population cycles: Populations might be expected to explode when food is available and collapse once a region is defoliated. Observational evidence, however, contradicts this hypothesis. The record of defoliation in four areas of Ontario shown here demonstrates that the degree of defoliation varies among sites and among outbreaks, and collapses often occur after only partial defoliation, before food is depleted. Notice that the populations at all four sites are synchronized. The mechanism by which synchrony is maintained over such long distances is still a mystery. The data were collected by the Ontario Forest and Insect Disease Survey and analyzed by Colin Daniel.

There is also some question whether changes in leaf biochemistry are adaptations to insect attack. When leaves are damaged, concentrations of the enzyme phenylalanine ammonia lyase increase, causing a shift toward the production of phenolic acids and away from the production of phenylalanine. These biochemical changes can be interpreted as a reallocation of resources from the production of compounds containing nitrogen to non-nitrogenous ones. If phenolic compounds are defensive, then stressed trees should be better defended against insects. But many authors propose just the opposite, pointing out that environmental stress appears to trigger insect outbreaks.

It is also very difficult to evaluate the impact of changes in phenolic compounds on insects in a natural situation, because leaf damage is associated with a suite of chemical modifications. Leaves from damaged trees lose water and nitrogen and become more fibrous and tougher. Any and all of these changes could be detrimental to insect growth, but they are more likely to be passive responses with an accidental relation to herbivore damage than active defenses that have evolved to control herbivore populations.

The upshot of these and other subtleties is that studies of the induced-defense response have been difficult to replicate. A recent collection of essays on induced defense summarizes three studies of caterpillars feeding on birches that epitomize the problem. Haukioja found a rapid induced response in birch foliage that reduced the growth and survival of caterpillars of the autumnal moth. The caterpillars' response, however, seemed to vary with the degree of crowding during the laboratory trials and with the geographic location of the trees. S. E. Hartley of the Institute of Terrestrial Ecology and J. H. Lawton of the Imperial College at Silwood, both in the United Kingdom, found that the response of birches to damage was minor and variable. Insect damage caused a stronger chemical response than cutting. P. J. Edwards, S. D. Wrattan and R. M. Gibberd of the University of Southampton found that the induced response varied with the time of year. The lack of consistency among the studies makes it difficult to assess their bearing on the induced-defense hypothesis.

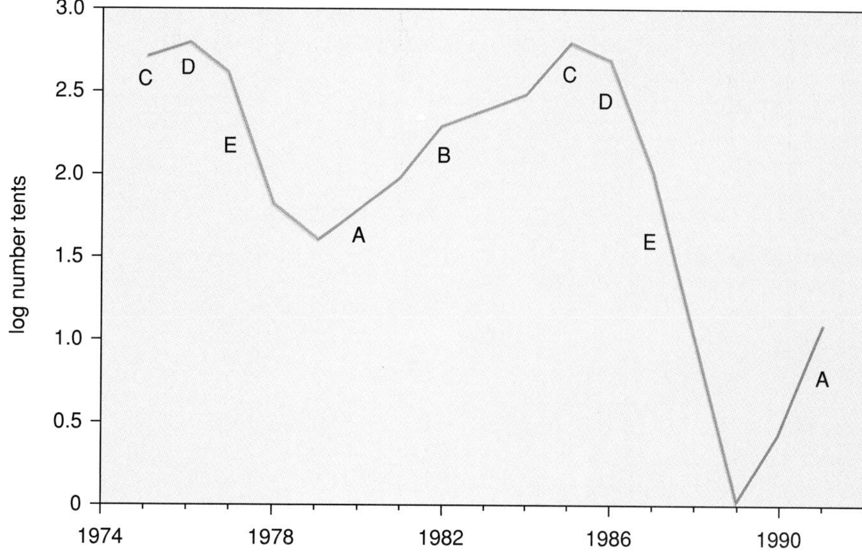

Figure 8. Population cycle of western tent caterpillars on Mandarte Island off the coast of British Columbia is typical of the tent-caterpillar populations studied for 15 years by the author. At the beginning of the increase phase *(A)*, the fecundity of the moths increases and the survival rate of the caterpillars is high. Later in the increase phase *(B)*, the survival rate begins to decline. At the population peak *(C)*, fecundity is still high, but attacks by parasitoids are also high and the incidence of disease increases. New populations are often initiated at this point in the cycle. In the year when the decline begins *(D)*, the survival of early-instar caterpillars is typically good, but that of later caterpillars and pupae is comparatively poor. There are fewer moths at the end of the summer, and those moths are less fecund than before. In mid-decline *(E)*, survival is poor and fecundity is low.

Experiments with Population Cycles

Much of the original speculation about the responses of trees to attacks arose from work David Rhoades did on the western tent caterpillar. In a study published in 1983, Rhoades found that western tent caterpillars that were fed detached leaves of red alder trees that were undergoing attack by tent caterpillars grew more slowly, died at a faster rate, and produced fewer egg masses than insects fed leaves from control trees.

At about the same time K. S. Williams, now at San Diego State University, and I undertook a similar experiment. From 1982 (early peak density) to 1986 (early decline) we reared in the laboratory tent caterpillars collected each year as early instars from a field population. Some caterpillars were fed red alder foliage collected from an area that had seen prolonged and intense insect herbivory, and others were fed foliage from another area (on the campus of the University of British Columbia) where trees had not been attacked. This experiment had two remarkable results.

The history of attack on the trees had little influence on the growth or survival of the caterpillars. Instead the growth and survival of the caterpillars was strongly influenced by the history of the field population. As the field population reached peak density and began to decline, it became increasingly difficult to rear caterpillars from eggs collected in the field, regardless of what they were fed. In other words, the experiment did not bear out the induced-defense hypothesis but suggested instead that "insect quality" was driving the population cycle. In 1986 many of the laboratory-reared caterpillars died of virus.

At this point it might be worth pausing to note how population cycles are studied closely in the field. Tent caterpillars are particularly good subjects for the study of population dynamics because the peculiarities of their life cycle allow scientists to monitor not just the mortality of different life stages but also the caterpillars' survival rates and the fecundity of the moths. The female moth lays all of her eggs in one batch, and the eggs stay on the trees from late summer of one year through the spring of the next year, even after hatching. Fecundity can be monitored by collecting these egg masses and counting the eggs.

Studies of insect population cycles tend to be plagued by large measurement errors when densities are low. Here also tent caterpillars offer an advantage. The colonial habits of the tent caterpillar make it much easier to monitor the survival of larvae. In all species except the forest tent caterpillar, the caterpillars of family groups form silken tents that can be readily seen and counted. Moreover, the tent's size is correlated with the number of fourth-instar caterpillars that constructed it. The survival of caterpillars can therefore be estimated from the number that hatched from the egg mass and an index of tent size.

I have scored the changes in the density, survival and fecundity of western tent caterpillars in British Columbia for the past 15 years. The fieldwork has made me aware of several unexpected and intriguing features of the population cycles.

One of the most interesting findings is that the fecundity of the moths increases as the population density increases. Crowding or reductions in the quantity or quality of food should reduce fecundity, yet I found that the fecundity of the moths is at its highest in the year before the cyclic peak.

Another interesting finding is that in the year of peak density, early larval survival is relatively good, but the flight of moths is low. Inclement early-spring weather or poor food quality might be expected to have the greatest influence on the survival of young larvae. It is intriguing that population declines seem to begin in years where there is good early larval survival.

Much as one might expect, caterpillar survival decreases as population density increases. But when the population density declines, survival does not rebound as expected. For several years after a population peak, the fecundity of moths and the survival of caterpillars continue to be low. It is this lag in recovery that leads to cyclic population dynamics.

A final intriguing feature is that new populations of tent caterpillars initiated by dispersers crash at the same time as the settlements from which they arise. The timing of the decline, in other words, is more influenced by the history of the insects than by the duration of insect attack at a specific site.

Some of the most intriguing experiments I have done were ones in which I deliberately attempted to manipulate the population cycles of tent caterpillars, to erase in one or another way the history of the population and to restart the cycle at a different point. It turns out to be difficult to do; the cyclic pattern is impressively resilient. In one such experiment, I attempted to initiate outbreaks out of phase by introducing tent caterpillars to sites with no or few caterpillars. The source populations were in the increase phase of the population cycle from 1981 until 1986. Between 1984 and 1986 I introduced egg masses to seven sites in southwestern British Columbia. Remarkably, the introduced populations declined at the same time as their source populations; in no case was the decline more than a year out of phase.

In a second experiment, I attempted to keep an island population of western tent caterpillars in the outbreak phase by cropping it. Even though density was kept low, this population declined at the same time as surrounding control populations. When I reviewed the literature on attempts to disrupt the cycles of forest Lepidoptera, I found that others had encountered similar resilience.

A satisfying mechanism for tent-caterpillar population cycles would explain all of these characteristics of the population cycle: increasing fecundity with increasing density; relatively good survival of early-instar caterpillars in the first year of the decline, but poor survival of late instars and pupae; prolonged poor survival and low fecundity after population decline; synchronous decline of populations with different histories; and resistance to experimental manipulation of insect densities.

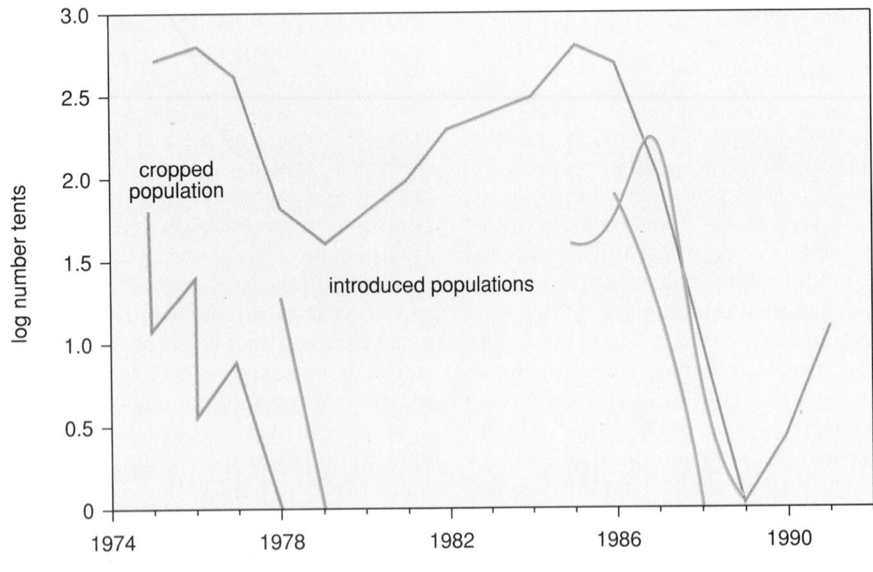

Figure 9. One of the most remarkable features of tent-caterpillar population cycles is their resistance to manipulation. Shown here are the results of a few of the experiments in which the author attempted to prolong an outbreak by cropping (*blue-green line*) or to initiate an outbreak by introducing caterpillars to a new area (*green lines*). Despite the continuing availability of food, the cropped and introduced populations followed the trajectory of surrounding populations, represented here by the Mandarte Island population (*purple line*).

The Nuclear Polyhedrosis Viruses

Previous hypotheses are not consistent with the observed characteristics of population cycles, but many characteristics could be explained by viral disease. Disease has frequently been reported in declining populations of forest Lepidoptera. As early as 1940, Harvey L. Sweetman of Massachusetts State College suggested that "a wilt disease produced by a virus" combined with unfavorable spring weather to produce "abundance cycles" in the eastern and forest tent caterpillars. He decided, however, that "it is probable that the weather is the dominant factor in production of the cyclic abundance of these pests."

A number of difficulties prevented scientists from concluding that viruses are the driving force behind population cycles. One was that disease is typically observed in only a small proportion of the population, perhaps because the cadavers of caterpillars that die of viral disease tend to disintegrate rapidly. The method by which the disease was transmitted between generations was not known, nor was it clear that it could persist in low-density populations. Viral diseases have received renewed attention now that the techniques of molecular biology allow viral DNA to be detected at low concentrations both in the environment and in asymptomatic insects.

Most insect viruses belong to the genus *Baculovirus*, a diverse group of viruses that are primarily pathogens of Lepidoptera. The most common baculoviruses belong to a subgroup that are called nuclear polyhedrosis viruses because they replicate in the nucleus of host cells and because the viral DNA is surrounded by a polyhedral-shaped protein crystal during much of the virus's life cycle.

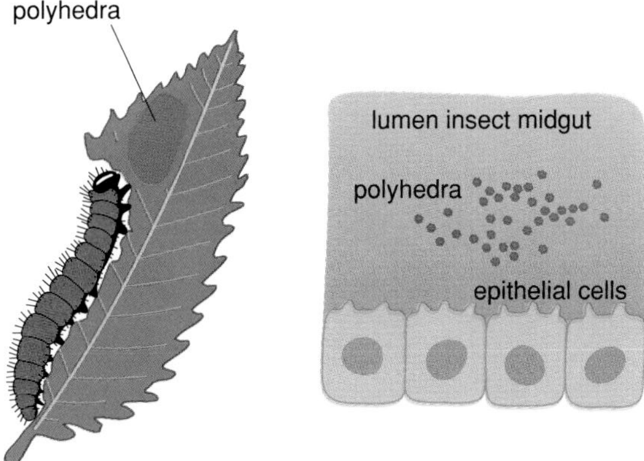

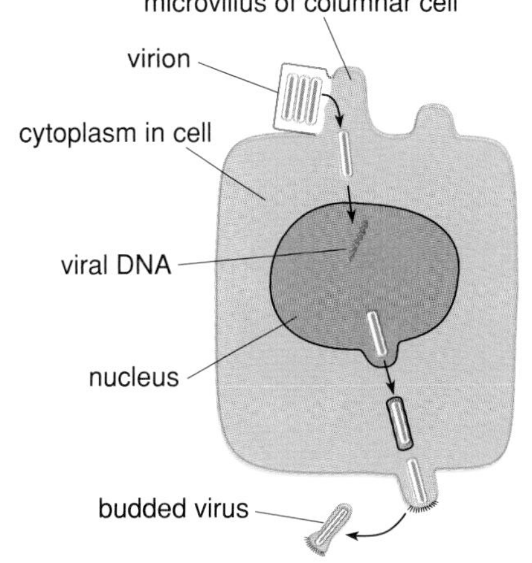

Figure 10. Nuclear polyhedrosis viruses may be the long-sought agent that drives population cycles in forest Lepidoptera. These viruses, which are primarily pathogens of lepidopterous insects, consist of many virions occluded within protein crystals called polyhedra. The polyhedra are commonly found on plants and in the soil following the death of caterpillars at high density. When an insect ingests contaminated material, the polyhedra dissolve in the high *p*H of the midgut and release virions (which contain the viral DNA). The virions enter the midgut epithelial cells when the viral envelope fuses with the cell's microvilli. After reaching the nucleus of an epithelial cell, the viral DNA is released from the virion and undergoes an initial round of replication. Some progeny virus moves out of the nucleus into cytoplasm and buds through the plasma membrane, enveloping itself in a piece of this membrane as it goes. This "budded virus" may infect many types of cells, including fat bodies, muscle and hemocytes. Late in the course of the infection of a cell, viral genes start to be expressed that direct the synthesis of the protein, called polyhedrin, that makes up the crystal. The polyhedrin protein accumulates in the cell's nucleus and crystallizes around newly formed virions. As a result, the virus can no longer exit the cell, but instead becomes embedded within the crystals of polyhedrin protein. When the insect dies and disintegrates, the very durable polyhedra may remain for many years, where they can be ingested by another insect and reinitiate the cycle.

When the polyhedral virus is ingested by a caterpillar, the protein crystal is dissolved in the insect's gut and virions, which contain particles of viral DNA are released. The virus penetrates the lining of the gut and replicates in the nuclei of cells. The time from ingestion of the polyhedra to death of the caterpillar varies from four days to three weeks. In the last stages of disease, millions of polyhedra are produced and the caterpillar ruptures, releasing the polyhedra onto the leaves or bark of the tree. Nuclear polyhedrosis disease is commonly called wilt disease because the insect's cadaver hangs limply from the tree.

The environmental reservoir of virus may well be the modulating filter that produces cyclic population dynamics. The virus is killed by exposure to ultraviolet light, but polyhedra protected from direct sunlight can remain virulent for a number of years. They could build up in the environment and be spread to neighboring areas by rain, wind and other insects. The gradual buildup might explain the good survival of early instars in the first year of the decline but reduced survival in following years.

Populations initiated by dispersers (or by scientists) might decline in synchrony with their source populations because they carry virus with them. There is some evidence that virus is carried on eggs; caterpillars reared in the laboratory from eggs collected from declining populations often develop disease. One intriguing but speculative possibility is that the disease has a latent form, and that insects without overt disease may still be carriers.

The increase in fecundity as the population approaches a cyclic peak is puzzling and may be unrelated to viral infection. The lag in recovery, however, might be explained by sublethal effects of disease that reduce the size and fecundity of moths. It may be that caterpillars that ingest virus shortly before they pupate do not die but instead become smaller, less vigorous adults. So far, the evidence for this is circumstantial. For example, pupae in populations of forest tent caterpillars with high mortality from virus are significantly smaller than those from populations with less virus. Some experiments have shown that caterpillars exposed to virus produce smaller, less fecund moths than those that are not exposed, but the relation varies with the dose and time of infection.

Mathematical simulations of insect disease systems produce population cycles similar to those observed in the field, although they do not reproduce the cycles perfectly *(see Figure 11)*. The best evidence that viral disease is the elusive driving force comes not from simulations but from experience: It is the only truly effective means of controlling population cycles. Experience with insecticide spray programs has shown that foliage must be sprayed yearly if it is to be protected during the upswing of a population cycle. For example, seven years of spraying with DDT during the 1958 outbreak of spruce budworm in northern New Brunswick prevented extensive tree mortality, but budworm populations declined at the same time in sprayed and unsprayed areas.

To my knowledge, an increasing population of a cyclic species has been permanently suppressed by a spray program only once. In 1981, early in an outbreak, forest entomologists with Environment Canada treated tussock moths with aerial and ground spraying of nuclear polyhedrosis virus. Measured densities declined to zero by the next year in six experimental plots and in two of three control plots to which virus may have spread. Mean densities at 10 other sites peaked in 1982 and declined through 1983, in association with an increased incidence of naturally occurring disease. In short, spraying with virus early in the outbreak terminated the outbreak.

As this instance suggests, nuclear polyhedrosis virus may have potential as a biological insecticide for forest caterpillars. Because the virus particles are protected by the protein crystal, they are robust and can be applied to foliage in much the same way as a chemical spray. Unlike *Bacillus thuringiensis (B.t.)*, a bacteria highly pathogenic to many

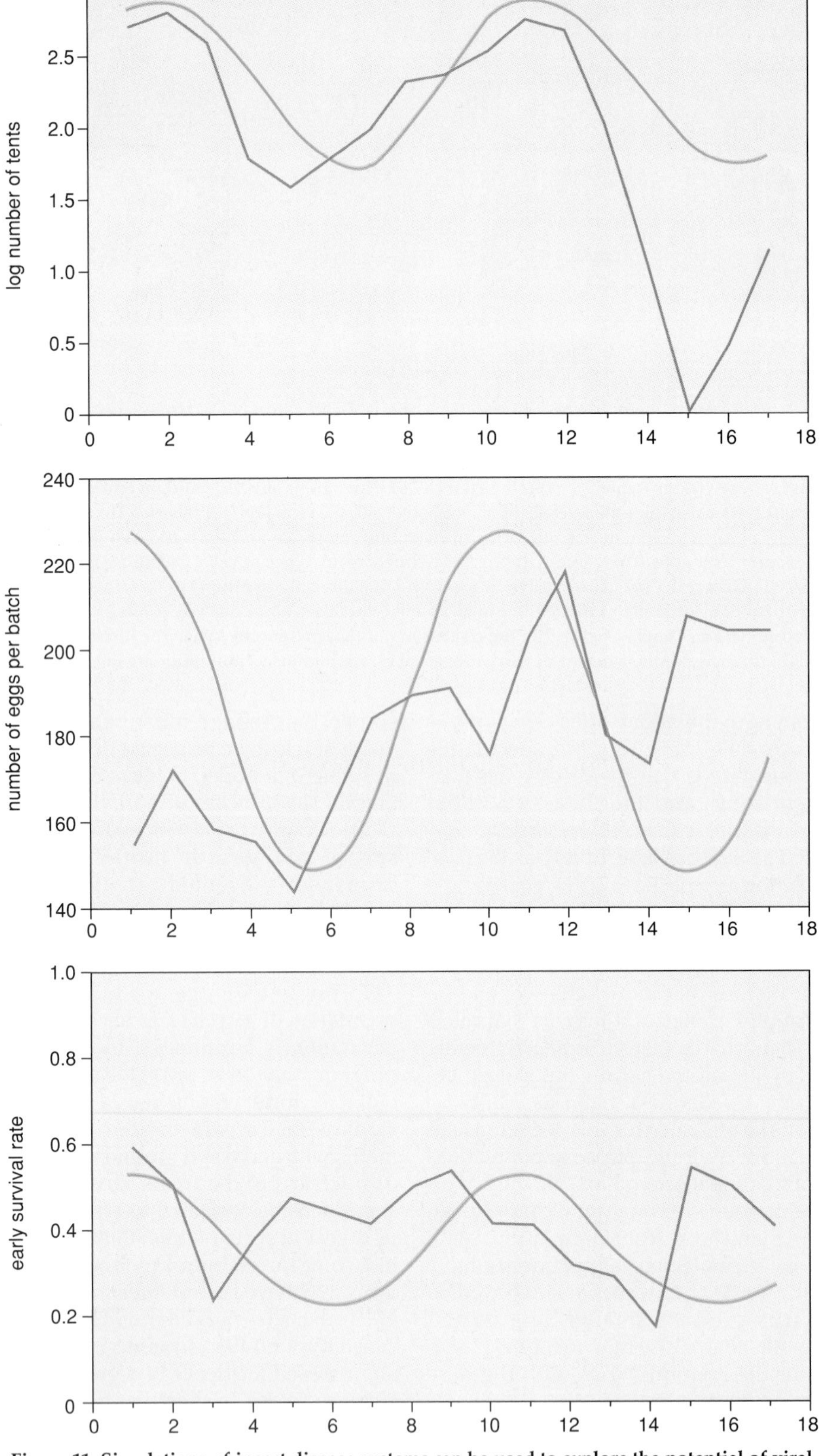

Figure 11. Simulations of insect disease systems can be used to explore the potential of viral disease to generate cyclic population dynamics. Shown here are the results of a model that examined whether vertical transmission of the virus (from mother to offspring) and the reduction of fecundity caused by sublethal effects of the virus would result in the patterns of caterpillar survival and moth fecundity observed in field populations *(purple lines)*. The simulated populations developed cyclic population dynamics and the associated changes in fecundity and early caterpillar survival *(green lines)*. However, in the simulated populations, the egg masses were largest early in the cyclic peak, and in natural populations they are largest late in the peak. The survival of caterpillars began to decline earlier in the natural populations than in the simulated populations. The simulation was done by Sarah Bukema at the University of British Columbia.

lepidopterous caterpillars that has been widely used for control, nuclear polyhedrosis viruses are very specific. They usually infect only a single species or a few closely related species and do no harm to other useful insects, much less humans, fish or wildlife.

Two minor problems are that the virus is deactivated by exposure to sun and that caterpillars usually feed for about a week after exposure, so that they continue to cause damage. The major problem, however, is that the virus is difficult to produce in commercial quantities. *Bacillus thuringiensis* can be grown on artificial media, but the viruses have to be grown on their insect hosts, and so production is much more expensive.

Even with a complete understanding of population cycles of forest Lepidoptera, we will not necessarily be able to keep populations at "acceptable" levels. As this article has emphasized, the population cycles are extremely resistant to manipulation. Given the comparatively minor consequences of the outbreaks and the difficulty of controlling them, a sensible policy would combine control efforts with modifications in forest-industry practices and reduced expectations. More than one author has remarked that efforts to control forest Lepidoptera sometimes seem disproportionate to the damage the insects do. It is worth remembering that these insects seem to have been adapted by years of evolution to ensure that outbreaks do not devastate host trees.

Acknowledgments

I am grateful to Colin Daniel and Sarah Bukema for providing unpublished results from their Master of Science theses and to the North Carolina State University Department of Entomology for the loan of their specimens for aid in illustration.

Bibliography

Berryman, A. A. 1986. On the dynamics of blackheaded budworm populations. *The Canadian Entomologist* 118:775–779.

Daniel, Colin J. 1990. Climate and outbreaks of the forest tent caterpillar in Ontario. M.Sc. thesis. Department of Zoology, University of British Columbia, Vancouver, Canada.

Dempster, J. P. 1983. The natural control of populations of butterflies and moths. *Biology Review* 58:461–481.

Edwards, P. J., S. D. Wratten and R. M. Gibberd. 1991. The impact of inducible phytochemicals on food selection by insect herbivores and its consequences for the distribution of grazing damage. In *Phytochemical Induction by Herbivores*, ed. D. W. Tallamy and Michael J. Raupp, pp. 205–222. New York: John Wiley & Sons.

Hartley, S. E., and J. H. Lawton. 1991. Biochemical aspects and significance of the rapidly induced accumulation of phenolics in birch foliage. In *Phytochemical Induction by Herbivores*, ed. D. W. Tallamy and Michael J. Raupp, pp. 105–132. New York: John Wiley & Sons.

Haukioja, E., and S. Neuvonen. 1987. Insect population dynamics and induction of plant resistance: The testing of hypotheses. In *Insect Outbreaks*, ed. P. Barbosa and J. Schultz, pp. 411–432. New York: Academic Press.

Hunter, Alison. 1991. Traits that discinguish outbreaking and nonoutbreaking Macrolepidoptera feeding on northern hardwood trees. *Oikos* 60:275–282.

Jones, R. E., V. G. Nealis, P. M. Ives and E. Scheermeyer. 1987. Seasonal and spatial variations in juvenile survival of the cabbage butterfly *Pieris rapae*: evidence for patchy density-dependence. *Journal of Animal Ecology* 56:723–737.

Kaupp, W. J., and S. S. Sohi. 1985. The role of viruses in the ecosystem. In *Viral Insecticides for Biological Control*, ed. K. Maramorosh and K. E. Sherman, pp. 441–465. New York: Academic Press.

McNamee, P. J. 1979. A process model for eastern black-headed budworm. *The Canadian Entomologist* 111:55–66.

McNamee, P. J. 1987. The equilibrium structure and behavior of defoliating insect systems. Ph.D. diss., Department of Zoology, University of British Columbia, Vancouver, Canada.

Martinat, P. J. 1987. The role of climatic variation and weather in forest insect outbreaks. In *Insect Outbreaks*, ed. P. Barbosa and J. Schultz, pp. 241–268. New York: Academic Press.

Morris, R. F. 1959. Single-factor analysis in population dynamics. *Ecology* 40:580–587.

Morris, R. F. 1963. The dynamics of epidemic spruce budworm populations. *Memoirs of the Entomological Society of Canada* 31:1–332.

Myers, J. H. 1988. Can a general hypothesis explain population cycles of forest Lepidoptera? *Advances in Ecological Research* 18:179–284.

Myers, J. H. 1988. The induced defense hypothesis, does it apply to the population dynamics of insects? In *Chemical Mediation of Coevolution*, ed. K. Spencer, pp. 530–557. New York: Academic Press.

Myers, J. H. 1990. Population cycles of western tent caterpillars: experimental introductions and synchrony of fluctuations. *Ecology* 71(3):986–995.

Myers, J. H., and K. S. Williams. 1987. Lack of short or long term inducible defenses in the red alder-western tent caterpillar system. *Oikos* 48:73–78.

Neuvonen, S., and E. Haukioja. 1985. How to study induced plant resistance. *Oecologia* 66:456–457.

Neuvonen, S., and E. Haukioja. 1991. The effects of inducible resistance in host foliage on birch-feeding herbivores. In *Phytochemical Induction by Herbivores*, ed. D. W. Tallamy and Michael J. Raupp, pp. 277–292. New York: John Wiley & Sons.

Price, P. 1987. The role of natural enemies in insect population dynamics. In *Insect Outbreaks*, ed. P. Barbosa and J. C. Schultz, pp. 287–313. New York: Academic Press.

Rhoades, D. F. 1979. Evolution of plant chemical defense against herbivores. In *Herbivores: Their Interaction with Secondary Plant Metabolites*, ed. G. A. Rosenthal and D. H. Janzen, pp. 3–54. New York: Academic Press.

Rhoades, D. F. 1983. Herbivore population dynamics and plant chemistry. In *Variable Plants and Herbivores in Natural and Managed Systems*, ed. R. F. Denno and M. S. McClure. New York: Academic Press.

Rhoades, D. F. 1983. Responses of alder and willow to attack by tent caterpillars and webworms: Evidence of pheromonal sensitivity to willows. *American Chemical Society Symposium Series* 208:55–68.

Shepherd, R. F., I. S. Otvos, R. J. Chorney and J. C. Cunningham. 1984. Pest management of Douglas-fir tussock moth (Lepidoptera: Lymantriidae): prevention of an outbreak through early treatment with a nuclear polyhedrosis virus by ground and aerial applications. *The Canadian Entomologist* 116:1533–1542.

Sweetman, Harvey L. 1940. The value of hand control for the tent caterpillars *Malacosoma americana* Fabr. and *Malacosoma disstria* Hbn. (Lasiocampidae, Lepidoptera). *The Canadian Entomologist* 72(12):245–250.

Wellington, W. G. 1957. Individual differences as a factor in population dynamics; the development of a problem. *Canadian Journal of Zoology* 35:293–323.

Wellington, W. G. 1960. Qualitative changes in natural populations during changes in abundance. *Canadian Journal of Zoology* 38:290–314.

Wellington, W. G. 1962. Population quality and the maintenance of nuclear polyhedrosis between outbreaks of *Malacosoma pluviale* (Dyar). *Journal of Insect Pathology* 4:285–305.

PART III
Process and Pattern in Communities and Ecosystems

Ecologists seek to understand entire communities and ecosystems, not just individual adaptations and population dynamics. Unfortunately, as one moves to questions about ecosystem dynamics, experiments need to be supplemented by a wide variety of less direct techniques, especially when examining large-scale ecological processes or patterns. One widely heralded success story for "large-scale ecology" is island biogeography theory—a suite of models that explains how the number of species in a region varies with the size and isolation of a habitat. Case and Cody use more than 30 islands in the Gulf of California to test predictions from island biogeographic models, using distributional records for everything from plants to reptiles to mammals. They find pronounced reductions in extinction rates with increasing island size, and some effect of isolation on extinction (although the isolation effect varies markedly with the history of the island and the taxon under consideration).

But analyses of observational data sets such as records of species' occurrences are fraught with the danger of spurious correlation and misplaced causality, a theme addressed brilliantly by Connor and Simberloff. These authors emphasize the importance of statistical models that can distinguish between "patterns" arising by chance alone versus patterns representing fundamental biological organization. The "null model" approach advocated by Connor and Simberloff is now ingrained in community ecology; as a result, the use of nonexperimental data is much more rigorous than it was only a decade ago.

While null models and biogeographic observations certainly play a major role in community ecology, many of the field's greatest advances are still due to manipulative experiments. The power of such manipulations is especially evident in a paper by Peterson regarding marine intertidal communities. Using transplant experiments, and experimental exclusions of competitors or predators, ecologists have found that the strength of biotic interactions is generally mediated by physical stress and disturbance—a theme elaborated by Peterson's discussion of benthic marine communities. The importance of disturbance, in particular, becomes the focus of a provocative paper by Reice. It is apparent that as the role of disturbance in communities is increasingly appreciated (in contrast to the more traditional view of a placid "balance of nature"), environmental managers will need to re-think their policies regarding fires, floods, and other natural catastrophes. Indeed, Reice suggests that for both aquatic and terrestrial communities, disturbance can be so crucial in maintaining biodiversity that land management aimed at reducing disturbance may ultimately cause local extinctions.

Community ecology is perhaps most satisfying when its patterns find their explanation in terms of simple life history trade-offs and behavioral ecology. Thus Skelly shows that the distribution of frogs makes sense in light of a fundamental trade-off between growth and vulnerability to predation. For example, chorus frogs thrive in small, temporary ponds because they are very active and eat enough to have a speedy development; however, the same activity that is good for feeding draws the attention of predators and makes these frogs especially vulnerable to the sorts of predators often found in large, permanent ponds. In general, trade-offs such as that between activity and exposure to predation may explain both the diversity of organisms we find on our planet and their patterns of distribution.

Large-scale community processes cannot be understood without attention to biogeochemical and physical events as well as species interactions. Larson traces the recovery of Spirit Lake, which was effectively sterilized by superheated volcanic rock and mudflows during the Mount St. Helens eruption. Phytoplankton, zooplankton, and eventually aquatic plants and snails were able to recover in Spirit Lake only after hospitable levels of nutrients and oxygen were regained.

Nowhere are large-scale community processes more clearly seen than in the Antarctic ecosystem, with its dramatic population fluctuations and concentrations of krill, whales, seals, and seabirds. Laws, who is the Director of the British Antarctic Survey, shows how the upwelling of nutrient-rich water at the Antarctic ice shelf drives the entire vertebrate food web in this extraordinary community.

Attention to species interactions and distribution patterns, and the study of biogeochemistry and the physical environment: these provide the foundation of community and ecosystem ecology. However, rarely are these two approaches combined—an unfortunate schism that is evident in this collection of papers as well as in the field of ecology at large.

Testing Theories of Island Biogeography

Ted J. Case
Martin L. Cody

Islands hold a special fascination for biogeographers, ecologists, and evolutionists, highlighting as they do questions of distribution and evolution. Historical biogeographers have been concerned mainly with problems related to dispersal and with the geographic and phylogenetic origins of island species. Ecological biogeographers are more concerned with the island setting and the resulting communities. The two groups share an interest, however, in a number of central questions. What factors determine the number of species on a given island and the particular set of species that occurs there? Why are certain species frequent and successful colonists, whereas other species rarely or never reach islands? To what extent does a given species vary ecologically and morphologically among islands, and how is such variation best interpreted?

Islands in the Sea of Cortez suggest that not one but many models may be required to explain patterns of colonization and extinction

The answers to these questions might depend on historical factors such as whether an island was ever connected to the mainland by a landbridge. But if the turnover of species on an island is high—that is, if colonization and extinction events are rapid there—then historical legacies might be quickly erased, making interpretation of the island biota dependent on current physical attributes of the island and on the biological attributes controlling the dispersal and persistence of its potential colonists.

Before much was known of continental drift and plate tectonics, former landbridges were often invoked to account for discontinuous distributions of species on widely separated land masses. It seemed implausible that many island-dwelling taxa, such as lizards or mammals, could colonize distant islands by oversea dispersal. "I do not deny that there are many and grave difficulties in understanding how several of the inhabitants of the more remote islands, whether still retaining the same specific form or modified since their arrival, should have reached their present home," Darwin commented in *Origin of Species* (1859, p. 396). It was commonly believed that past geological events had provided temporary terrestrial avenues for colonization, and that once these avenues disappeared all colonization ceased. While a few isolated forms might subsequently become extinct, islands were generally thought to serve as a kind of "museum" of archaic taxa, preserving such species as the giant tortoises of the Galápagos Islands, the tuatara of New Zealand, the Cuban *Solenodon*, the Tasmanian wolf, the laurel-dominated woodlands of the Canary Islands, and (until recently) the dodo of Mauritius. Moreover, it was believed that island extinctions were more than compensated for by the internal production of new species—and indeed, a large part of the floras and faunas of many archipelagos is comprised of genera and even families that occur nowhere else in the world. The Galápagos finches, the Hawaiian honeycreepers and silverswords, and the Malagassy vangashrikes and tenrecs are just a few examples.

This view sees island biotas as made up largely of evolutionarily modified species that are relics of past historical events. It contrasts with views that emphasize the role of ongoing dispersal to islands from continental areas, a phenomenon for which there is a considerable body of evidence. The colonization of new or recently active (and sterilized) volcanic islands such as Surtsey near Iceland or Krakatoa in the Java Straits has been well documented. Long Island, off New Guinea, was devastated about two centuries ago by a volcanic explosion that deposited a layer of ash 30 m thick, but today the island has a diverse flora with nearly as many resident species of land birds as islands without any such recent cataclysm (Diamond 1974). Other examples of unaided dispersal within and beyond continents are the recent spread of the collared dove, Cetti's warbler, and the firecrest northward through Europe to Britain, the 1975 colonization of Minorca by Dartford warblers, the colonization of Norfolk Island by blackbirds from New Zealand, some 250 km away (Turner et al. 1968), and the recent spread of African cattle egrets throughout much of the world (see Cody 1985 for details of several of these examples).

The debate over the relative importance of dispersal over water versus dispersal over land across temporary landbridges was heated at a 1961 symposium (Gressitt

Figure 1. The islands of the Sea of Cortez, located between the peninsula of Baja California and mainland Mexico, offer a varied and relatively undisturbed setting in which to observe biogeographical patterns. The some two dozen larger islands and numerous smaller islets differ not only in size and isolation but in age and geological formation. Shown here is Partida Norte, a small island of about 1.2 km^2 off the east coast of Baja, which is seen in the distance. In spite of its barrenness, Partida Norte supports four species of reptiles and seven species of land birds. Larger islands in the group are characterized by richer vegetation and many more species of vertebrates. (Photo by Martin L. Cody.)

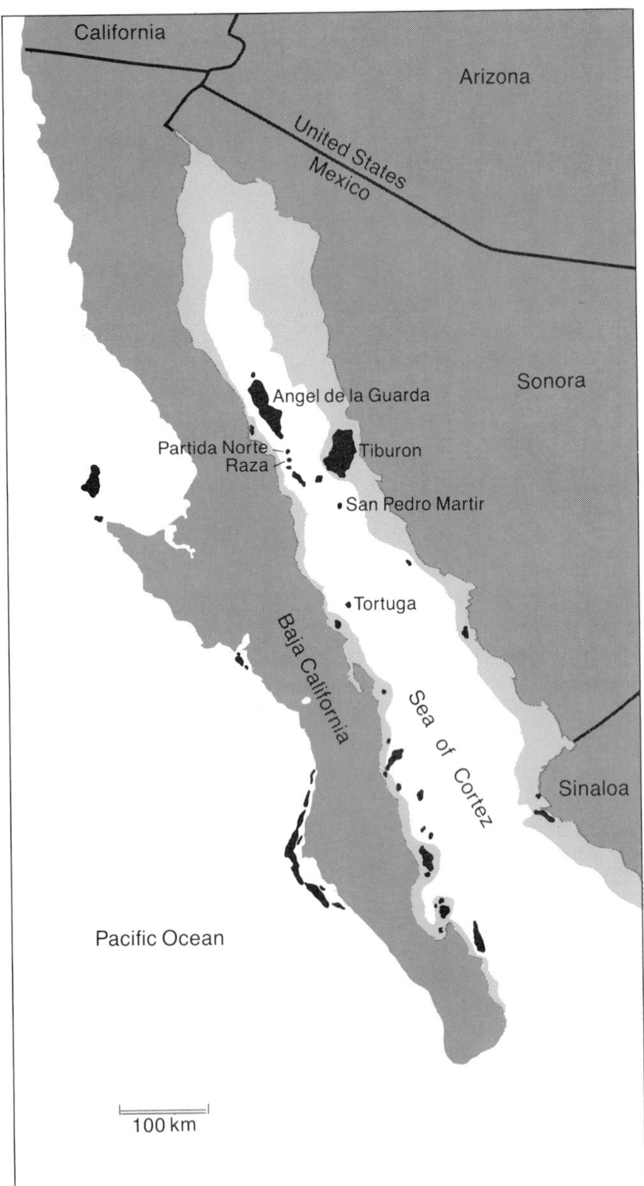

Figure 2. The varied geological history of the islands in the Sea of Cortez, also known as the Gulf of California, aids biogeographical studies. The islands within the 110 m depth contour (*light gray*) were connected to the mainland within the last 13,000 years. By contrast, the islands of Raza and Tortuga, although about the same age, arose as oceanic volcanoes without any landbridge connection. The remaining islands are at least a million years old.

1963), but a 1982 symposium on the same subject proceeded without polemics (see Diamond 1982). Three scientific developments in the intervening period had dramatically altered the nature of the debate. The first was the analysis of new geological data on continental drift and seafloor spreading, which revealed that some isolated and deepwater islands such as New Zealand, the Seychelles, and New Caledonia were formerly parts of much larger landmasses. At the same time, it became clear that many other volcanic islands had never been connected to any other landmass (Kennett 1982; Brown and Gibson 1983). The ancestors of present-day plants and animals on such islands as Hawaii, Tahiti, the Marquesas, the Galápagos, the Canary Islands, and the Azores all must have arrived by oversea dispersal.

Second, major advances have been made in applying molecular techniques to systematics and in calibrating the ages and relationships of species when the fossil record is insufficient. The use of these techniques has solved some longstanding problems. For example, molecular dating reveals that the endemic Hawaiian honeycreepers, Drepanididae, are descended from fringillid finches, and that they diverged from their mainland relatives 15 to 42 million years ago—long before the present-day Hawaiian Islands even existed (Sibley and Ahlquist 1982). We now know with reasonable certainty that the two species (and genera) of endemic Galápagos iguanas diverged around 20 million years ago on the South American mainland, each probably colonizing the Galápagos much later (Wyles and Sarich 1983).

The third major development, more conceptual in nature, was MacArthur and Wilson's equilibrium theory of island bigeography (1963, 1967), which realigned in a single sweep much of the history-oriented thinking on the ecology and biogeography of island systems. In the following sections we describe this theory, relating it to other current theories and especially to the view that historical factors are paramount. The remainder of the paper will compare the theories using a recent synthesis of data primarily from one source, the islands in the Sea of Cortez (Fig. 1; Case and Cody 1983).

The equilibrium theory

MacArthur and Wilson viewed island biotas not simply as relics of past historical events but as ever-changing entities. Species numbers, they reasoned, represented a dynamic equilibrium determined by the extinction rate of earlier, established species and the immigration rate of new species to the island. For a given taxonomic group such as land birds, a given island will support a certain number of species. The number of species will be continuously augmented by the arrival of new species from mainland sources and continuously depleted by extinctions. All else being equal, immigration and extinction rates will be specific to the island (as well as to the taxon), with higher immigration rates for closer, more accessible islands and lower extinction rates for larger islands that can support larger populations. On any given island, the balance between the immigration rate and the extinction rate will be a function of the island's size and degree of isolation, and will set the equilibrium number of species in each taxon to be found there.

Very few assumptions are needed to show that the existence of such an equilibrium per se is virtually guaranteed. Whether or not such equilibriums are reached or even approached with any frequency in nature is, however, quite another matter. Various arguments shape the immigration and extinction curves, but the most parsimonious of them will produce curves continuously decreasing and increasing respectively over much of their range, and will thus produce an intersection and a stable equilibrium number of species S.

For ecologists, the question is not whether an equilibrium species number S exists, but rather whether the circumstances that determine it are constant over a sufficient time to render the equilibrium attainable or even approachable. For this to be so, the sizes and other characteristics of source biotas, the magnitude and shapes of immigration and extinction curves, and the sizes and positions of the islands must remain relatively constant over the "ecological" time required for population establishment, growth, and decline, even though we know that over geological time all of these factors will vary.

Some of this variation can be examined within the premises of the equilibrium model. The equilibrium is characterized by both the number of species and the rate of turnover; islands similar in the former may differ in the latter, since a small island near a colonization source might support the same number of species as a large, isolated island, but with a much higher rate of turnover. Further, more recent landbridge islands created by rising post-Pleistocene sea levels may still be losing species as a result of reduced immigration rates and increased extinction rates, and may be "supersaturated" relative to their size and isolation; such islands may require more time for the process of equilibration than the few thousands of years since their island status was achieved.

Figure 3. Studies of the abundant reptiles of the Gulf islands have shown that landbridge islands are especially rich in species, as might be expected given the low mobility of this group. On oceanic islands reptiles are distinguished by high endemism, unlike the more mobile taxa of plants, shorefish, and land birds. Here *Sauromalus hispidus*, an endemic species, basks in a jatropha shrub on the island of Angel de la Guarda. In the absence of mammalian competitors and predators, these lizards reach population levels and body sizes much greater than those experienced by their mainland relatives. (Photo by Ted J. Case.)

On the other hand, new islands that arise from beneath the sea following dramatic geological events will approach equilibrium conditions from below (*S* increasing) rather than from above (*S* decreasing), and again the time necessary for the number of species to increase through colonization might not yet have elapsed. More important, islands move on the backs of oceanic plates, with changing areas and degrees of isolation, through changing climates and habitats over time. Can an island's biota constantly remain "on track," in continuous equilibrium with its changing circumstances, or will there be a serious time lag between the physical changes and biotic responses? It is not difficult to envision a variety of circumstances in which the historical legacy might prevail.

Other biogeographical models

Since the appearance of the equilibrium theory, researchers have been busy offering alternatives to remedy some of its omissions and oversimplifications. Among these alternatives are the historical legacy models, which explain the occurrence of species in terms of unique geological or climatic events in a region. Historical legacy models emphasize a slow modification of the original island biota, and regard short-term immigration and extinction events as insignificant relative to the island's historically determined attributes. Such models might well apply to islands that have undergone dramatic, relatively recent geological changes to which the biota has not yet adjusted, and to taxonomic groups in which species persist well on islands (for example, lizards or plants) or are poor colonizers (for example, mammals; see Brown 1971; Wilcox 1978; Lawler 1983).

For mobile species with rapid colonization rates—most birds, bats, and perhaps winged insects—an equilibrium should be reached relatively quickly, and it is unlikely to be coincidental that these taxa have provided some of the best evidence for the equilibrium theory. For terrestrial mammals, amphibians, freshwater fish, and many woody plants with limited dispersal powers, such an equilibrium may require time periods equalling those of major geological or evolutionary events. Accordingly, examples of historically determined island biotas seem to predominate in the literature of these taxa (see, for example, Raven and Axelrod 1974; Rosen 1975, 1978; MacFadden 1980).

The discovery of extinctions on islands is consistent with any number of island biogeographical theories. What is critical is whether the extinctions appear relatively constant over time (as predicted by equilibrium theory) or are episodic and associated with major geological or climatic changes. Monitoring the turnover rates of species might decide the issue, but many human lifetimes would be needed to distinguish regularity from spurts related, for instance, to climate. The fossil record could be brought to bear, but often the geological circumstances on oceanic islands are poorly suited to preserving fossils. For most small islands, only the relatively narrow time window provided by Holocene subfossils is available, and even this evidence is sporadic.

Subfossil documentation of extinctions has, however, been found on a number of islands, including Puerto Rico, the Bahamas, Barbuda, Antigua, Henderson, New Zealand, and the Canary and Hawaiian islands (Arnold 1976, 1980; Pregill 1981; Olson and James 1982; Steadman 1982). Although vertebrate extinctions are common, in many cases the temporal pattern and the interpretation of their causes are ambiguous. Man and his entourage of introduced mammals are strongly implicated in many instances (Steadman et al. 1984; Steadman and Martin 1984; Steadman and Olson 1985), as are climatic changes in others (Pregill 1982). Does any "equilbrium turnover" remain after the effects of such agencies are removed? As we explain below, the islands in the Sea of Cortez provide an important clue to this puzzle, since significant extinction has occurred there over the last 10,000 years in the absence of much human or climatically induced environmental change.

Besides the possibility that immigration and extinction rates are dominated by long-term historical effects,

Does any "equilibrium turnover" remain after the effects of agencies such as man and climatic changes are removed?

they may be poor predictors for other reasons. If such curves cannot be drawn as quasi-constant functions of the number of species on islands but are wholly stochastic, sporadic, and episodic, or if they are in general unrelated to island size and isolation, then an equilibrium species number either will not exist or will not be predictable from a simple description of the island. In this case patterns of diversity will appear vague and will be essentially uninterpretable by the equilibrium model.

A second group of models augments the equilibrium theory by focusing on species interaction and coevolution. One way in which immigration and extinction curves might be functionally dependent is by dint of interspecific interactions such as competition. Thus established species might reduce the chances that ecologically similar species can invade successfully, and may allow access only to ecologically compatible species in diminishing numbers over time. This is precisely what happens with laboratory "islands" in experimental aquariums, where a controlled sequence of colonization from a species pool of various bacteria, algae, and small metazoans results in a particular community which is stable and resistant to colonization by any of the missing species (Drake 1984).

Over longer periods of time, island communities might be altered by coevolutionary adjustments such that eventually the island becomes resistant to invasion, immigrations and extinctions approach zero, and the turnover of species declines steadily with time as island assemblages become more stable and finely tuned. This view was espoused by Lack (1976), who found evidence for it in the birds of the Canary and Caribbean islands, and who argued that short-term competitive exclusion and long-term coevolutionary adjustment of niches would produce balanced and stable sets of species. According to Lack, small islands have fewer and more generalized species because of the absence of richness of ecological opportunities, not because of higher extinction rates.

Coevolutionary models seem to find their best application to old islands that are intermediate in size and isolation. Random extinction must not be so great as to muddle the deterministic extinction predicted from interspecific competition, a requirement that excludes smaller islands. On the other hand, colonization rates must be both low enough that gene flow will not impede local evolutionary adjustments yet high enough that colonists are available, a requirement that points to combinations of taxa and islands characterized by intermediate isolation, such as lizards on the Canary Islands, the Lesser Antilles, and the islands in the Sea of Cortez (Case 1979; Roughgarden et al. 1983; Case and Bolger, unpubl.).

A third model hypothesizes that, as in Lack's theory, island communities are closely related to the diversity and abundance of the island's resources and have low to zero rates of species turnover, but (in contrast to Lack's hypothesis) are composed of species that do not interact. If resources are distinct and not shared among species there will be no interspecific competition for them, no increased densities where some species are absent, and no negative effect on immigration curves from existing island residents. There will be, however, a tight coupling of the island's resources with its consumer species that is free of species interaction. In such a situation, similar islands with similar resources would support not just similar numbers of species but the same kinds of species or even precisely the same species. Further, the densities of individual species would be predictable from a knowledge of the island's resources, which is not possible when species interactions permit complementary relationships and density compensation. With such a nonoverlapping allocation of resources, chance extinctions would allow neither community reorganization nor easier access to new species but only a chance for recolonization by the same (recently extinct) species from mainland sources.

This noninteractive, resource-coupled model might be most appropriate where the diversity of the mainland species pool is low and its members ecologically distinct. Islands drawing their colonists from this pool should support various subsets of the mainland communities, subsets dependent on the degree to which the resources of the mainland habitats are represented on the islands. Islands of increasing size would presumably support increasingly large subsets of the mainland community, while similarly sized islands, given their similar resource base, would support nearly identical communities. If isolation is a factor in colonization, then more distant but similarly sized islands would be reached by fewer colonists. This model is likely to be supported where three conditions exist: a dearth of species that are ecologically similar, a single source of colonizing species (rather than several sources with ecologically equivalent species), and source habitats in which ecological segregation arises from interspecific differences in resource use and associated morphological or physiological differences (rather than from subtle behavioral differences, such as foraging in different subhabitats).

From this discussion it is clear that to expect a single biogeographical model to account for distribution and density patterns on all islands is hardly reasonable. Before backing a specific model, one would like to know the relative time frames both of biological events such as colonization, population establishment, and extinction and of geological events governing the island's origin, age, history, and climate. One would need to know, also, the way in which climate, substrate, and topography combine to produce the island's particular resources, as well how biological interactions affect the consumers of these resources.

The biogeography of the Gulf islands

The Sea of Cortez, also known as the Gulf of California, separates the peninsula of Baja California from mainland Mexico (Fig. 2). There are about two dozen larger islands and even more numerous smaller islets in the Gulf, and these vary not only in size and isolation but in age and geological formation. Thirty-two of the islands have an area of more than 0.50 km^2; the two largest, Tiburon and Angel de la Guarda, are about 1,000 km^2. A majority of the islands lie closer to the eastern coast of Baja than to mainland Mexico, with more than 77% within 20 km of the peninsula. San Pedro Martir, at a distance of 50 km, is the most isolated. Some of the islands were connected to the mainland or to larger islands 6,000 to 13,000 years ago, during the last glacial maximum. Other islands are much older, dating from 1 to 4 million years ago, and arose either as volcanoes from the seabed or by tectonic separation from the mainland, with many beginning as blocks faulted off the trailing edge of the northward-drifting peninsula. There are also two recent volcanic islands, Raza and Tortuga, both estimated to date from the Holocene, about 10,000 years ago.

The islands in the Sea of Cortez are exceedingly dry, hot, and rocky. Like the Baja peninsula to the west and the mainland Mexican states of Sonora and Sinaloa to the east, they support a predominantly Sonoran Desert vegetation, sparse, low, and open except for the few trees and taller cacti in and around the deeper arroyos. Plants represent the largest of the reasonably well-known taxonomic groups on the islands, with over 570 species recorded. About 100 of these are limited to the large island of Tiburon, only 2 km from the Sonoran mainland across a 3-fathom channel (Cody et al. 1983; Moran 1983). Approximately 300 plant species occur on Tiburon and nearly 200 on Angel de la Guarda, but totals are much more modest on the smaller islands. By contrast, the Sonoran Desert as a whole contains about 3,500 plant species.

Over 100 species of fish are primary residents of the shallow reefs around the islands, with up to two dozen occurring at a given collection site (Thomson and Gilligan 1983). Reptiles are diverse and conspicuous in deserts, and 130 taxa (including subspecies) have been recorded on the islands (Fig. 3). Twenty or more species are found on four of the larger islands, some of which were once connected to the mainland by landbridges (Case 1983a; Murphy 1983). More conspicuous, perhaps, but less diverse are the islands' land birds (Cody 1983). Thirty species breed on the islands, with 18 to 21 species found on six larger islands, as opposed to 21 to 31 species recorded at mainland census sites, comparable to Tiburon's total of 26 species. Finally, mammals are the least conspicuous and the least diverse, with scattered records of some larger species—deer, coyote, fox, ring-tail cat, rabbits—for a few landbridge islands only (Lawlor 1983). Most island records show only pocket mice (*Perognathus*), white-footed mice (*Peromyscus*), or wood-rats (*Neotoma*). Twenty-four mammal species in all are found in the islands, but most islands support only one to three species, with four to seven species recorded on landbridge islands and thirteen counted on Tiburon.

In both their geological variety and their relative freedom from the impact of man's activities, the islands of the Sea of Cortez offer an ideal opportunity to compare island biogeographical theories. Recent studies of the biogeography of these islands, summarized in Table 1, have given us an index of the overall diversity of six taxonomic groups (the fifth group, lizards, is a subset of the fourth, land reptiles). According to equilibrium theory, islands formerly connected to the mainland by way of landbridges approach equilibrium from above, with extinction exceeding immigration until the two rates eventually balance. As seen in the first column of Table l, however, after roughly 10,000 years this histori-

Table 1. Biogeographical comparison of taxonomic groups on islands in the Sea of Cortez

Taxa	*Average no. of species on 50 km² oceanic islands*	*Are landbridge islands species-rich relative to oceanic islands?*	*Are islands species-poor relative to the mainland?*	*Are Holocene oceanic islands species-poor compared to old oceanic islands?*	*Does number of species decrease with distance?*		*Endemism (%)*	
					Landbridge islands	*All oceanic islands*	*Landbridge islands*	*All oceanic islands*
Land plants	105	no*	no, for islands >3 km²; yes, for all others	no	no	no	0	~2
Shorefish	–	no	no	no	no	no	0	0
Land birds	13	no	slightly	no	no	no	0	0
Land reptiles	9.3	yes	yes	no	no	no	5	35
Lizards	3.5	yes	yes	no	no	yes	0	47
Land mammals	1.3	extremely	extremely	slightly	slightly	yes	16	69

* The answers "yes" and "no" imply statistical significance or its absence at $p < 0.05$. "Slightly" means that an effect exists but its magnitude is small, even though it may be statistically significant; "extremely" implies an effect of large magnitude.

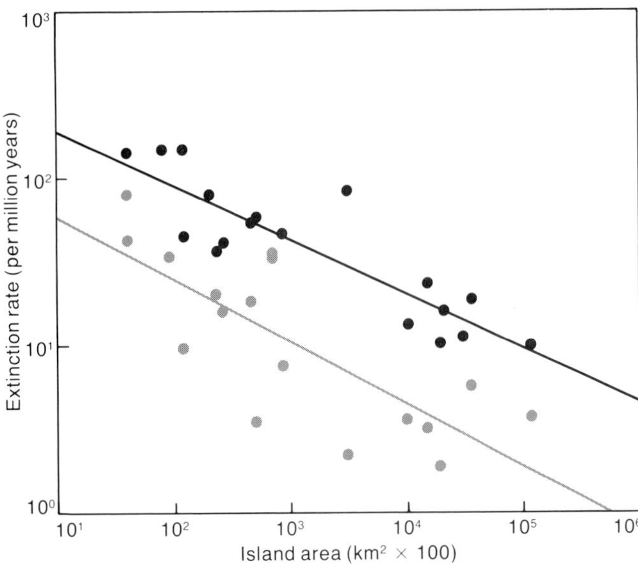

Figure 4. Estimated extinction rates for reptiles and mammals on various landbridge islands off both sides of the Baja peninsula suggest that small islands lose reptile and mammal species (*gray and black dots, respectively*) at rates substantially greater than large islands, and that mammals become extinct faster than reptiles. The presumed initial number of species on each island was calculated from the observed relationship between area and number of species on the mainland today. The lines show the best fit through the points, which represent taxa on individual islands. The extinction rate is determined as K from the model $ds/dt = -ks^2$ (see Richman et al., in press).

cal difference between recent landbridge and old oceanic islands has been erased for the highly mobile taxa of plants, shorefish, and land birds. For the less mobile groups, on the other hand—reptiles and especially mammals—landbridge islands are currently richer in species, dramatically so in the case of mammals, and now appear supersaturated compared to oceanic islands of similar size.

Is there an "island effect" at all—that is, does the isolation that characterizes islands in general affect the diversity of different groups? Column three of the table shows that in plants this effect is seen only on very small islands, which with their small catchment areas and lack of drainage channels are qualitatively different from mainland areas, supporting less diversity; in larger islands the effect disappears. It is absent also in shorefish and is minor in birds, which are nearly as diverse on larger islands with vegetation comparable to mainland sites as on the mainland itself, regardless of the degree of isolation of the island. Only on the smaller islands does the diversity of birds fall, most likely because of the impoverishment of vegetation. In the case of reptiles, however, the paucity of species on islands is immediately apparent, and with mammals the island effect is even more pronounced. Here the fieldworker immediately recognizes differences between island and mainland, whereas with plants and birds a lower diversity is not obvious in a superficial survey. In these more mobile taxa, recolonizations from the mainland must be sufficiently frequent that local extinctions are compensated for either before or shortly after they occur. Column four of the table indicates that the two newest oceanic islands, Raza and Tortuga, have received ample immigration; only in the case of mammals are there fewer species than would be expected for islands of their size.

The phenomenon of frequent recolonization can produce lower extinction rates on islands that are closer to the mainland as opposed to those that are more isolated (cf. Schoener 1983), resulting in a "distance effect"—that is, on more isolated islands there will be fewer species in taxa with more limited mobility. Column five of the table confirms that on landbridge islands this effect is detectable only in mammals; on oceanic islands it is more pronounced, but only in the least mobile taxa, lizards and mammals.

Landbridge islands that have retained their historical legacy and still remain supersaturated thousands of years after gaining (or regaining) island status can be used to calculate the relationship between extinction rates and island size (see, for example, Diamond 1972; Wilcox 1980). The necessary condition of low to zero immigration is most closely satisfied in reptiles and mammals on Gulf landbridge islands; the relationship of extinction rates to island size, calculated in Figure 4, is similar in the two taxa, although reptiles persist much better over time than do mammals.

These results are important because of the controversy surrounding the cause of island extinctions. While many biogeographical theories predict extinctions with changes in the physical and climatic setting, equilibrium theory predicts the turnover of species even in the absence of such physical changes. In theory this essential difference provides a way of testing the various models, but in practice it is next to impossible to find a set of islands without environmental change or evidence of the severe impact of man's activities. The islands in the Sea of Cortez probably come closer than any to fulfilling the requisite conditions. There was never any permanent aboriginal population except on Tiburon, and the presence of modern man is restricted to a few small settlements on three of the larger islands. Relatively few islands have any plants or animals introduced from elsewhere, and the effect of such imports has been severely limited by the inhospitable climate and the absence of standing water.

One warning seems clear: future island biogeographers must be very cautious in generalizing beyond the bounds of their taxon and the island system in which it has been studied

Richman and his colleagues (in press) have shown that extinction rates for reptiles of the Gulf islands are nearly identical to those calculated for relatively undisturbed landbridge islands off South Australia during the same period. The extinction rates for Gulf reptiles are, moreover, not very different on average from rates for other islands based on the direct evidence of subfossil deposits. The important difference is that small oceanic islands with substantial human disturbance exhibit significantly elevated extinction rates compared to undisturbed islands. Calculating probable separation times ranging from 6,000 to 14,000 years for the landbridge islands in the Sea of Cortez, Wilcox (1978) showed that the number of lizard species "relaxed" over this interval: older landbridge island have lost relatively more lizard species than have islands isolated for shorter periods of time. While there were some procedural and statistical problems with Wilcox's original methods (Faeth and Connor 1979), a "time effect" still emerges after correcting for these. This effect is not seen in mammals (Lawlor 1983), however, perhaps because mammalian extinctions occur more rapidly, so that the number of species on even the youngest landbridge islands has already declined to levels appropriate for their present areas.

The question of the emergence of new species on islands via evolution rather than colonization is examined in the last column of Table 1. For endemic populations to form on islands, colonization rates must be sufficiently low that island populations are not genetically swamped by immigrants with mainland genes, and in addition island populations must persist long enough for evolution to act upon them. If these two conditions are satisfied, a third is necessarily met: since islands for which colonization rates are low for a given taxon will be species-poor, selective pressures for genetic change will be enhanced for the successful colonists. Table 1 shows that endemism at the level of species or above is zero for shorefish and birds, and zero on landbridge islands and extremely low (≈2%) on oceanic islands for plants. It reaches appreciable levels only in the least mobile taxa, reptiles and mammals. On landbridge islands, where swamping by mainland genomes is most recent, levels of endemism are modest, but on oceanic islands levels are high, increasing from 35% for reptiles to 46% for lizards to 69% for mammals.

The effect of island size on the level of endemism is equivocal. Although larger islands will support the longer persistence times necessary for endemics to form, they will also have higher colonization rates and a community structure more similar to the mainland, and thus more swamping by incoming genes and less impetus for evolutionary change. In fact, in lizards, the only group large enough to test for effects of island size on endemism, there is no such effect (Case 1983a). The effect of island age on endemism can be studied by comparing the two Holocene islands, Raza and Tortuga, with other, much older islands. Again, the expected degree of endemism is equivocal. Endemism might be expected to be low because Raza and Tortuga are young oceanic islands but high because they are well isolated, with low colonization rates. In fact, levels of endemism in lizards on these two islands—50% and 25%, respectively—are typical of those on much older oceanic islands and significantly greater than those of equally young landbridge islands, a result which suggests that isolation is more important than time in producing speciation.

Finally, let us examine evidence for species interactions and ask whether such interactions have contributed to the observed patterns of distribution and density on the Gulf islands. When records on land birds on the Gulf islands are compared, going from the smallest to the largest islands, the sets of species are simply subsets of the species occurring on the next larger island. This is

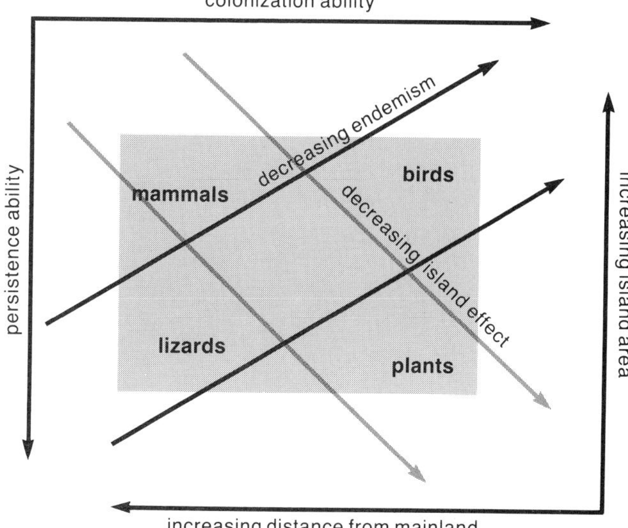

Figure 5. In this chart of elements governing diversity, the axes of colonization ability and persistence ability are paralleled by axes of increasing distance and increasing island area, respectively. Levels of endemism are shown as a joint function of these axes, as is the "island effect"—the effect of insularity on diversity. The biological attributes of colonization ability and persistence tell most of the story. High colonization ability and low persistence result in minimal endemism (birds), whereas the reverse pattern produces maximal endemism (reptiles and mammals). Where both colonization ability and persistence are high, minimal island effect is found (plants); where both elements are low, the island effect is at its maximum (mammals). As the chart suggests, islands of similar size should be relatively poor in mammals but rich in plant species; lizards will have high levels of endemism and birds relatively low levels. For all taxa, increasing island area will increase both the richness of species and the level of endemism, whereas increasing distance from the mainland will decrease the richness of species but increase endemism.

just the sort of pattern that would be expected from the noninteractive species model and from the Sonoran Desert vegetation, with its collection of relatively specialized bird species showing dissimilar morphologies and little ecological overlap in foraging habits. One of the few possibilities for interspecific competition occurs between two kinds of small foliage insectivores, verdins (*Auriparus flaviceps*) and gnatcatchers (*Polioptilus* spp.), but in fact these insectivores show either identical (on the southern islands) or nearly identical (on the northern islands) distributions as well as similar variations in density over island size, with verdins about twice as dense as gnatcatchers. Another possible instance of interaction is found in three species of the thrasher *Toxostoma*. Out of ten islands apparently large enough to support the taxon, six have only a single species—a result that does not differ significantly from what would be expected from chance if no interaction were present.

The same high degree of predictability of particular species from island size that occurs among the land birds (Murphy 1983) is found also in lizards, indicating that many of the lizard species are likewise noninteractive and resource-coupled. But this does not preclude the possibility that some lizard species are competitively interactive, with overlapping resource requirements. For instance, body sizes and population densities of *Uta stansburiana*, the most ubiquitous species, decline with increasing numbers of other related species with which it coexists (Soulé 1966; Case 1983a). Moreover, the overall co-occurrence of lizard species on the islands cannot be interpreted simply as random subsets from the mainland species pool; certain lizard species with low ecological overlap occur together on islands statistically more often than chance would dictate (Case 1983a, 1983b). Particular pairs of lizard species reveal strong patterns of negative distribution, most obviously in the case of *Uta* and *Sator*, which occur together on no islands—a result that would be expected by chance less often than once in a million cases. In laboratory cages, the larger *Sator* dominates and frequently kills and eats the smaller *Uta*.

Another example of interspecific interaction, of a milder nature, is seen in the whiptail lizards (genus *Cnemidophorus*), which are diurnal, actively foraging insectivores. On the peninsula two species fill this niche, both occupying the same habitats, but *C. hyperythrus* is small and *C. tigris* is substantially larger; unlike *Sator* and *Uta*, these two species rarely fight. Their ecological segregation depends on the larger lizards eating larger prey (see Case 1979). Knowing the size distribution of available prey, it is possible to predict the optimal sizes of lizards in one- or two-species sets at a given locality. Some islands support just one *Cnemidophorus* species; on the six oceanic islands where the single species is *C. tigris*, it is smaller in size than on the mainland, and on two islands where *C. hyperythrus* or its derivatives occur alone, it is larger. On some landbridge islands *C. tigris* occurs alone, but no reduction in body size is apparent. Perhaps *C. hyperythrus* has only recently become extinct there, or perhaps the evolution of smaller body size requires more time than has been available on these recent islands; alternatively, gene flow from the mainland might have precluded a shift to smaller body size in these lone *C. tigris* populations.

A diversity of models

With a diversity of physical circumstances and a diversity of biological attributes in potential colonizing groups, evidence exists to support a diversity of models. We find it helpful to organize this diversity in the fashion depicted in Figure 5. Historical effects seem pronounced in mammals and some reptiles, since low colonization and good persistence yield low propensities for erasing historical legacies. The role of species interaction in shaping community structure remains one of the most controversial areas. The consistent patterns of species sets and subsets across islands suggest that birds, and to a lesser extent lizards, are by and large noninteractive, resource-coupled species. But although the number of species and their identities and densities are largely predictable in these taxa, and although the turnover of species must be low, some species interactions are strong, at least in lizards, and tend to obscure the patterns.

As work continues on these islands, it will be interesting to see whether the generalities of Figure 5 survive, and how other taxa might fit into the scheme. Ongoing work by Gary Polis on the scorpions of the islands and the mainland suggests that their colonization abilities are generally better than those of reptiles but not as good as birds, and that their extinction rates are similar to those of reptiles. The average Gulf island of 50 km^2 has about six scorpion species, but (unlike any other taxa studied so far) there is a sharp geographic increase in the richness of species from north to south on the islands, a latitudinal trend not paralleled on the peninsula. Endemism is not quite as high for scorpions as for reptiles. Similar work is needed on other invertebrates; ants and land snails would make exciting candidates, because the peninsular fauna is rich and highly endemic. The time is also ripe for the use of molecular techniques to clarify the ages and relationships of the many endemic species of the region. From what we have learned so far, one warning seems clear: future island biogeographers must be very cautious in generalizing beyond the bounds of their taxon and the island system in which it has been studied.

Ted J. Case is professor of biology at the University of California at San Diego, where he has been since 1978. He received his Ph.D. from the University of California at Irvine in 1973. He is associate editor for the journals Evolution *and* Oecologia, *and has been investigating reptiles in the Sea of Cortez since 1970. Martin L. Cody is professor of biology at the University of California at Los Angeles. He received his Ph.D. from the University of Pennsylvania in 1966. His research in the Sea of Cortez has focused on the ecology of land birds and plants. The work reported here was sponsored by the NSF and the National Geographic Society. Address for Professor Case: Department of Biology, C-016, University of California at San Diego, La Jolla, CA 92093.*

References

Arnold, E. N. 1976. Fossil reptiles from Aldabra Atoll, Indian Ocean. *Bull. Brit. Mus. Nat. Hist.* 29:83–116.

———. 1980. Recently extinct populations from Mauritius and Reunion, Indian Ocean. *J. Zool.* 191:33–47.

Brown, J. H. 1971. Mammals on mountaintops: Non-equilibrium insular biogeography. *Am. Nat.* 105:467–78.

Brown, J. H., and A. C. Gibson. 1983. *Biogeography*. St. Louis: Mosby.

Case, T. J. 1979. Character displacement and coevolution in some *Cnemidophorus* lizards. *Fortschr. Zool.* 25:235–82.

———. 1983a. The reptiles: Ecology. In *Island Biogeography in the Sea of Cortez*, ed. T. J. Case and M. L. Cody, pp. 159–209. Univ. of California Press.

———. 1983b. Assembly of lizard communities on islands in the Sea of Cortez. *Oikos* 41:427–33.

Case, T. J., and D. Bolger. Unpubl. The role of species interactions in the island biogeography of reptiles.

Case, T. J., and M. L. Cody, eds. 1983. *Island Biogeography in the Sea of Cortez*. Univ. of California Press.

Cody, M. L. 1983. The land birds. In *Island Biogeography in the Sea of Cortez*, ed. T. J. Case and M. L. Cody, pp. 210–45. Univ. of California Press.

Cody, M. L., ed. 1985. *Habitat Selection in Birds*. Academic Press.

Cody, M. L., R. Moran, and H. Thompson. 1983. The plants. In *Island Biogeography in the Sea of Cortez*, ed. T. J. Case and M. L. Cody, pp. 49–97. Univ. of California Press.

Darwin, C. R. 1859. *On the Origin of Species by Means of Natural Selection*. London: Murray.

Diamond, J. M. 1972. Biogeographic kinetics: Estimation of relaxation times for avifaunas of Southwest Pacific Islands. *PNAS* 69:3199–203.

———. 1974. Colonization of exploded volcanic islands by birds: The supertramp strategy. *Science* 184:803–05.

———. 1982. The biogeography of the Pacific Basin. *Nature* 298:604–05.

Drake, S. A. 1984. Ecological communities as assembled structures. Ph.D. diss., Purdue University.

Faeth, S. H., and E. F. Connor. 1979. Supersaturated and relaxing island faunas: A critique of the species-age relationship. *J. Biogeography* 6:311–16.

Gressitt, J. L., ed. 1963. *Pacific Basin Biogeography*. Honolulu: Bernice P. Bishop Museum Press.

Kennett, J. P. 1982. *Marine Geology*. Prentice-Hall.

Lack, D. 1976. *Island Biology Illustrated by the Land Birds of Jamaica*. Univ. of California Press.

Lawlor, T. 1983. The mammals. In *Island Biogeography in the Sea of Cortez*, ed. T. J. Case and M. L. Cody, pp. 265–89. Univ. of California Press.

MacArthur, R. H., and E. O. Wilson. 1963. An equilibrium theory of insular zoogeography. *Evolution* 17:373–87.

———. 1967. *The Theory of Island Biogeography*. Princeton Univ. Press.

MacFadden, B. J. 1980. Rafting mammals or drifting islands?: Biogeography of the Greater Antilles insectivores *Nesophontes* and *Solenodon*. *J. Biogeography* 7:11–22.

Moran, R. 1983. Vascular plants of the Gulf islands. In *Island Biogeography in the Sea of Cortez*, ed. T. J. Case and M. L. Cody, pp. 348–81. Univ. of California Press.

Murphy, R. W. 1983. The reptiles: Origins and evolution. In *Island Biogeography in the Sea of Cortez*, ed. T. J. Case and M. L. Cody, pp. 130–58. Univ. of California Press.

Olson, S. L., and H. F. James. 1982. Prodromus of the Fossil Avifauna of the Hawaiian Islands. *Smithsonian Contributions in Zoology*, no. 365: 1–79.

Pregill, G. K. 1981. An appraisal of the vicariance hypothesis of Caribbean biogeography and its application to West Indian terrestrial vertebrates. *Syst. Zool.* 30:147–55.

———. 1982. Fossil amphibians and reptiles from New Providence Island, Bahamas. *Smithsonian Contributions in Paleobiology*, no. 48:8–21.

Raven, P. H., and D. I. Axelrod. 1974. Angiosperm biogeography and past continental movements. *Annals of the Missouri Botanical Garden* 61:539–673.

Richman, A., T. J. Case, and T. Schwaner. In press. Natural and unnatural rates of extinction. *Am. Nat.*

Rosen, E. D. 1975. A vicariance model of Caribbean biogeography. *Syst. Zool.* 24:431–64.

———. 1978. Vicariant patterns and historical explanations in biogeography. *Syst. Zool.* 27:159–88.

Roughgarden, J., D. Heckel, and E. Fuentes. 1983. Coevolutionary theory and the biogeography and community structure of *Anolis*. In *Lizard Ecology: Studies of a Model Organism*, ed. R. B. Huey, E. R. Pianka, and T. W. Schoener, pp. 371–410. Harvard Univ. Press.

Schoener, T. W. 1983. Rate of species turnover decreases from lower to higher organisms: A review of the data. *Oikos* 41:372–77.

Sibley, L. G., and J. E. Ahlquist. 1982. The relationships of the Hawaiian honeycreepers (Drepaninae) as indicated by DNA-DNA hybridization. *Auk* 99:130–40.

Soulé, M. 1966. Trends in the insular radiation of a lizard. *Am. Nat.* 100:47–64.

Steadman, D. W. 1982. Fossil birds, reptiles, and mammals from Isla Floreana, Galápagos Archipelago. Ph.D. diss., Univ. of Arizona.

Steadman, D. W., and P. S. Martin. 1984. Extinction of birds in the late Pleistocene of North America. In *Quaternary Extinctions: A Prehistoric Revolution*, ed. P. S. Martin and R. G. Klein, pp. 466–77. Univ. of Arizona Press.

Steadman, D. W., and S. L. Olson. 1985. Bird remains from an archaeological site on Henderson Island, South Pacific: Man-caused extinctions on an "uninhabited" island. *PNAS* 82:6191–95.

Steadman, D. W., G. K. Pregill, and S. L. Olson. 1984. Fossil vertebrates from Antigua Lesser Antilles: Evidence for late Holocene human-caused extinctions in the West Indies. *PNAS* 81:4448–51.

Thomson, D., and M. R. Gilligan. 1983. The rocky-shore fishes. In *Island Biogeography in the Sea of Cortez,* ed. T. J. Case and M. L. Cody, pp. 98–129. Univ. of California Press.

Turner, J. S., C. N. Smithers, and R. D. Hoogland. 1968. *The Conservation of Norfolk Island.* Australian Conservation Foundation Special Publ. 1.

Wilcox, B. A. 1978. Supersaturated island faunas: A species-age relationship for lizards on post-Pleistocene land-bridge islands. *Science* 199:996–98.

———. 1980. Insular ecology and conservation. In *Conservation Biology,* ed. M. E. Soulé and B. A. Wilcox, pp. 95–118. Sinauer.

Wyles, J. S., and V. M. Sarich. 1983. Are the Galápagos iguanas older than the Galápagos? Molecular evolution and colonization models for the archipelago. In *Patterns of Evolution in Galápagos Organisms*, ed. R. I. Bowman and A. E. Leviton, pp. 177–86. San Francisco: AAAS.

Edward F. Connor
Daniel Simberloff

Competition, Scientific Method, and Null Models in Ecology

Because field experiments are difficult to perform, ecologists often rely on evidence that is nonexperimental and that therefore needs to be rigorously evaluated

The controversy over the importance of interspecific competition in natural communities of organisms has sharply divided ecologists. In recent years the debate has increasingly involved such issues as the appropriateness of scientific methods in ecology (Salt 1984). Specifically, since ecological field experiments are very difficult to perform, much of ecological theory and much of the evidence for the importance of interspecific competition is nonexperimental. Therefore, the essence of the debate lies in the problem of evaluating nonexperimental evidence (Lewin 1983).

The ecology of Darwin's finches in the Galapagos Islands, particularly the ground finches (*Geospiza*), exemplifies these interrelated issues. There are 6 species, and various combinations of from 1 to 5 species are found on the 32 islands that have been studied. Furthermore, each species varies intraspecifically, so that its population on each island is morphologically distinct from its populations on other islands. In 1945 David Lack argued that this intraspecific variation is primarily nonadaptive and arose from such random processes as genetic drift. Later, however, he concluded that the variation is adaptive and is produced primarily by interspecific competition; his 1947 monograph argued that, for any species, the population on any particular island is subject to competition for food from a different subset of species than on other islands, and that the selection pressure engendered by this competition is responsible for the morphological variation (Lack 1983).

For example, Lack determined that three ground finches, *G. fuliginosa, G. fortis,* and *G. magnirostris,* are all seedeaters in coastal zones, but with beaks of very different size (Fig. 1). All three species are present, with other species, on many central islands, but on remote islands *G. fortis* is always absent, while the other two species are sometimes absent, and the morphologies of other *Geospiza* appear modified in a way that Lack interprets as a response to the absence of competitors. Thus on Culpepper Island all three species are absent, and *G. conirostris* has evolved a stronger and deeper beak than on other islands, superficially similar to that of *G. magnirostris* (Fig. 2).

Lack also considered—and rejected—the hypothesis that beak morphology varies intraspecifically because beaks serve for recognition. However, another hypothesis that this variation is adaptive is not as easily dismissed: Bowman (1961) noted that the islands differ greatly both physically and vegetationally, with humid forest, for example, found only on larger and higher islands. Would not these differences select for different morphologies, even in the absence of interspecific competition?

If such a dispute arose in an experimental science like medicine, one would devise a controlled experiment to test each causal hypothesis—competition and vegetational differences. One might also combine both putative forces and test them together against each separately and against a control with neither force. This procedure is impossible for the Galapagos finches. Even if one performed an experiment and found no effect, the result would not be conclusive; a skeptic could always claim that the competitive forces or vegetational differences exerted their selection pressure over eons, and that their effect cannot be duplicated in a matter of a few years. In any case, one could not perform the experiments in the first place. It would be illegal and immoral to perform the requisite manipulations in setting up the control and test islands; one cannot remove species from islands, or add others, or substantially modify existing vegetation. Furthermore, even if one were allowed to do such things, there are too few islands for appropriate replication. As an example, there are 20 possible 3-species combinations alone, and at least 2 levels of vegetation (if we view vegetation as an experimental treatment), but only 32 islands.

Since experiment is impossible, various analyses of observational data have been deployed to argue for one or another of these hypotheses—usually Lack's competition hypothesis. Grant (1981) summarizes many of these analyses, in which measurements of the birds as they exist now are examined for consistency with the notion that a particular force has been responsible for morphology. Grant and his co-workers have focused, as did Lack, on beak morphology. For example, they found that all pairs of *Geospiza* species on the same island differ by 15% or more in beak depth, length, or width. All pairs that differ by less than this amount in one dimension differ by a greater amount in another dimension. This difference can conceivably be attributed to interspecific competition, and thus it could be argued that two species more similar in beak morphology could not continue to coexist, because one species would be

Edward F. Connor is Assistant Professor of Environmental Sciences and Director of the Blandy Experimental Farm and Orland E. White Arboretum at the University of Virginia. He received his B.A. from New College and his M.S. and Ph.D. in biology from Florida State University. Daniel Simberloff, Professor of Biology at Florida State University, received his A.B. and Ph.D. in biology from Harvard University. Address for Dr. Connor: Department of Environmental Sciences, University of Virginia, Charlottesville, VA 22903.

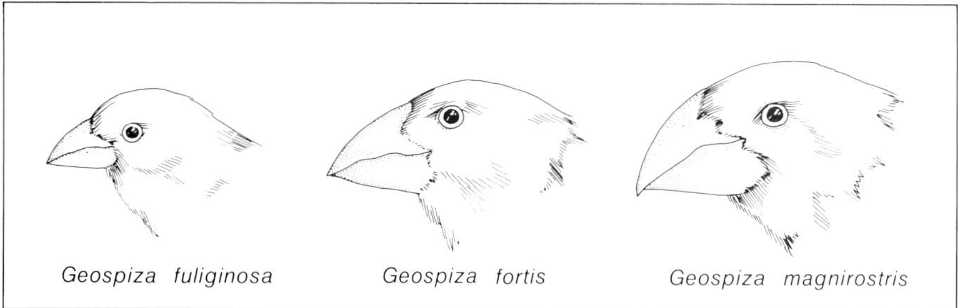

Figure 1. The considerable difference in beak morphology between these three species of Darwin's finches, *Geospiza,* which coexist on many Galapagos islands, has been the subject of much debate concerning its cause. Advocates of the central importance of interspecific competition argue that each particular variation in beak morphology, which enables that species to specialize on a certain source of food, evolved as a result of selection pressure from other species that would otherwise, in the absence of variation, compete for the same food. The opposing position maintains that interspecific competition is only one among a number of causal factors—such as genetic drift, predation, environmental differences—which are equally or more significant. (After Grant 1981.)

more efficient at monopolizing available food. Similarly, the principle of competitive exclusion can easily be used to rationalize the fact that of 20 possible 3-species combinations of *Geospiza,* only 2 combinations are observed on the 4 islands that have exactly 3 species. Other combinations of 3 species contain at least one species that would be outcompeted to the point of extinction.

With our colleagues, we have pointed to a deficiency in this sort of approach, arguing that one might have predicted the same sort of patterns if forces other than interspecific competition were causal—for example, if the birds had occupied islands independently of which other species were present (Connor and Simberloff 1979; Strong et al. 1979). In short, we suggested that one should ask if the observed distribution of a trait would have been expected even if interspecific competition were not operative. That is, the force under consideration should be tested in the form of a "null" hypothesis, which is simply a conjecture that a particular causal process is not evidenced in the observations under study.

Our null hypotheses (or null models) posited the independence of species from one another, while our alternative hypothesis envisioned the result that would obtain if interspecific competition were the main force. These attempts have been criticized on two different grounds: that our particular null hypotheses are unrealistic (Grant and Abbott 1980; Grant 1981), and that the entire approach of positing null hypotheses is unlikely to be useful in ecology (Gilpin and Diamond 1984).

Different forms of these issues—the importance of competition, scientific method, and the use of null models—have recurred frequently in ecology. The arguments of Clements and his colleagues (1929) and of Gleason (1939), with their contrasting views of a natural community, respectively, as a system dominated by complex interaction between species, or as a fortuitous mélange of independent species' autecologies, are an intellectual antecedent to our present concern over the roles of competition, predation, parasitism, mutualism, and the physical environment as processes shaping the natural world. The same is true of the debates over the importance of density-dependent factors (interactions between species and among individuals of the same species) and density-independent factors (interactions between organisms and the environment) in determining fluctuations in the abundance of populations. Just as with the present debate, methodological arguments and even null models were intertwined in these earlier dialogues on processes that shape ecological relations.

Evidence for competition

As Jackson (1981) points out, ecologists have long recognized that competition occurs, that there are a variety of competitive mechanisms, and that detecting and measuring competition under field conditions is problematic. However, the kind of evidence adduced as support for the presence of competition has been continuously debated (Elton 1946; Williams 1951; Simberloff 1970; Grant 1972; Diamond 1975; Connor and Simberloff 1979; Grant and Abbott 1980; Lawton and Strong 1981). The problem is that most evidence interpreted as indicating the presence of competition in nature is nonexperimental. Observations of pairs or larger groups of species having divergent morphologies, behaviors, and habitat use, or displaying patterns of geographical distribution that do not overlap, are adduced as prima facie evidence of either current or past interspecific competition. Patterns such as these, particularly when observed in groups of closely related species, are referred to as instances of resource partitioning.

The logic implicit in these conjectures is that competition between species, acting through natural selection, leads species to diverge in ecology because individuals of different species that overlap in their use of the environment and resources are at a selective disadvantage. The notion that these patterns logically entail a particular causal mechanism, interspecific competition, has been criticized; the alternative has been proposed that these patterns may be caused by a variety of forces—such as predation, the physical environment, and chance dispersal—in addition to interspecific competition (Connell 1975; Haila 1982; Connor and Simberloff 1983).

On the other hand, many field experiments have been carried out in an effort to test for interspecific competition. These experiments are based on the assumption that competition constrains a species' breadth of resource use, reduces individual growth rates, reduces population sizes, and adversely affects fecundity, survival, and other demographic parameters. Most experiments designed to test for competition involve either removing a putative competitor and observing the response of the remaining species, or introducing species into field enclosures, singly and in groups, and comparing individual growth rates and the demographic characteristics of a species' population under both sets of conditions.

Many of these experiments show that no competition occurs or that only very weak competitive interactions are present. Schoener (1983) reviewed the literature on field experimental tests of competition and found that 90% of the experimental studies and 76% of the species he examined "demonstrate some competition," and that 57% of these species show competition in every experiment. However, Connell (1983), conducting a similar review, arrived at a much different estimate of the frequency of competition, concluding that the percentage of experiments averaged across species showing competition was 43% at most. Connell's method for estimating the frequency with which field experiments detect competition accounts for spatial and temporal variability in the intensity of interspecific competition. For this reason, and because of rigid procedures he established in examining this literature, Connell's estimate is more likely to be free from sampling biases and to reflect the frequency of competition in nature than are Schoener's estimates.

Regardless of the exact value, however, these are poor estimates of the frequency of competition in nature. Many investigators perform experiments because they believe that competition is important for their subject organisms. Such preselection will tend to inflate estimates of the frequency of the competition in nature. To estimate accurately the frequency of competition in nature, species must be tested without reference to the likelihood that competition is operating. The best way to ensure this is to perform community-wide experiments designed to detect competition between each pair of species in each trophic level (each level on the food chain) in a community.

Very few field studies have attempted community-wide assessments of the frequency of interspecific competitive interactions, and these studies have been restricted both taxonomically and trophically. Seifert and Seifert's (1979) studies of herbivorous insects in *Heliconia* bracts, while focusing on only one trophic level, are an experimental attempt to test all pairs of species that share a resource for the presence of competitive interactions. Among the 4 species they studied, they estimated that between 17% and 25% of the pairs exhibited competitive interactions. Fowler (1981) examined a grassy-field community with more than 50 plant species; she removed singly and in groups approximately 20 of these species and found that of the 72 pairings examined, 14 showed significant interactions.

Other attempts at assessing the frequency of competitive interactions throughout a community or among species in a subset of a community have been nonexperimental. Studies such as these are predicated on the belief that taxonomically or trophically related organisms are more likely to compete. Species that are more distantly related, ecologically dissimilar, or both may be expected to compete still less frequently and strongly, on average, and thus to lower the frequency of competition in nature. Yet it is no trivial matter to determine a reasonable subset of species within which to seek competition. Even species that seem a priori very likely to compete may well fail to do so—or else why perform the empirical scan at all? Conversely, such surprising competitors as ants and rodents (Brown et al. 1979) and hummingbirds and insects (Brown et al. 1981) may emerge from careful analysis. It is unlikely that one can easily perform many community-wide experiments that would directly address how frequently competition occurs in nature, and the expedient of limiting the study to likely competitors yields even less reliable estimates.

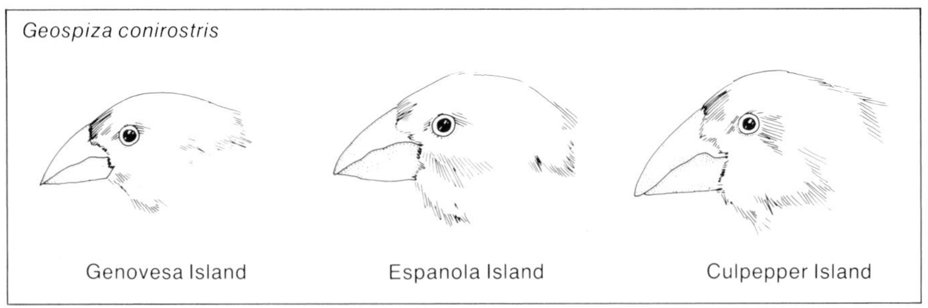

Figure 2. The beak morphology of *Geospiza conirostris* shows significant variation on different islands in the Galapagos. One theory suggests that this variation can be interpreted to be the result of interspecific competition from different species on different islands. For example, the shape of *G. conirostris*'s beak on Culpepper Island is similar to that of *G. magnirostris* (see Fig. 1), which is absent from that island. (After Grant 1981.)

The importance of competition

Beyond the problem of the evidence adduced for competition, there has also been considerable debate over the importance of competition in structuring plant and animal communities and in influencing the dynamics and adaptations of organisms (Gleason 1939; Wiens 1977; Diamond 1978; Connell 1980; Brown 1981; Schoener 1982). The magnitude, consistency, and nature of the ecological and evolutionary impact of competition relative to other processes are integral facets of an assessment of the importance of competition itself. However, it is difficult to conceive of a system in which to rate the relative importance of competition, predation, and other processes. This is particularly true if we wish to incorporate information on the magnitude and kinds of effects each process exerts, as well as to estimate the relative frequencies of each process.

For example, we might wish to use reproductive survivorship as the basic currency for a comparison of the relative importance of interspecific competition and the physical environment to a particular finch species. If, for the sake of illustration, we consider the physical environment to have the greater impact on reproductive survivorship, it is conceivable that it would lead to the evolution of certain feeding specializations. For example, a dry climate, through its effects on seed production by different plant species, could select for finches that specialize on certain seeds. At the same time, the lesser but still substantial effect of interspecific competition might lead to altered patterns of habitat use through changes in behavior, morphology, or both. In this instance the physical environment has a greater impact on reproductive survivorship, but competition has an apparently greater effect on the ecological relations of species.

Which process should then be considered more important? Although reproductive survivorship may seem to provide a common basis for comparing the importance of different processes in nature, such a comparison is ill-founded. It is the outcome of the common mechanism through which these processes operate to shape nature, evolution by natural selection, that one would wish to use to assess the relative importance of these processes. However, we *assume* that certain behaviors or morphological changes evolved in response to the physical environment and that others evolved in response to interspecific competition, but such assumptions are rarely buttressed with direct study of the selective pressures. In many instances such direct study is impossible because the necessary range of phenotypic variance does not exist in nature. For example, we may believe that a particular finch species has a thick bill because other species of the same genus on the same island have thinner bills and so would outcompete the thick-billed species for certain kinds of food if it had a thinner bill. To study this proposition directly we would need to have available thin-billed individuals of the thick-billed species to see if they are indeed outcompeted, but nature may not provide us with such individuals to study.

The prognosis for a comprehensive assessment of the relative importance of interspecific competition, predation, and other processes in nature is bleak, if we wish to estimate their past effects. However, there is a basis for optimism in that it is possible to determine when certain processes are important now in nature and to compare the relative magnitudes of particular processes in specific systems. Connell (1980, 1983) outlines experimental designs capable of assessing elusive competitive effects and of contrasting the magnitudes of inter- and intraspecific competition. Experiments have been conducted that illustrate the tractability of questions concerning when and where particular processes are important. For example, Paine (1969) and Dayton (1971) have shown that the effects of competitors, predators, and herbivores in rocky intertidal communities are contingent on the frequency of physical disturbances caused by high wave energies. Their work also demonstrates that it may be possible to achieve some degree of generalization, such as about the forces that are likely to be important in most local variants of the community of a particular habitat type or under particular environmental conditions. But such generalizations will be guides at best, and the idiosyncrasies of specific communities will necessitate detailed, probably experimental, field study (Pielou 1982).

Scientific method in ecology

The subject of interspecific competition and the questions about scientific method in ecology have a close association that stems from what some have viewed as excesses and failures of competition theory, and from the use of observational and nonexperimental evidence as a basis for inferring competition. These issues are only symptoms of the deeper problem besetting any science that has a strong historical component and, because the processes of interest cannot be isolated in a laboratory, must address important questions with limited experimentation. In response to this dilemma, a number of ecologists have offered suggestions on how to evaluate evidence.

The process of evaluating a scientific theory or hypothesis has four possible outcomes, two distinct correct conclusions and two kinds of errors. We can correctly reject a hypothesis when it is false or fail to reject it when it is true, and we can incorrectly reject a hypothesis when it is true or fail to reject it when it is false. In the testing of statistical hypotheses these two kinds of errors are termed Type I and Type II errors, respectively, and procedures to control the chances of making these errors are well developed.

Scientific theories are about processes, and scientific hypotheses, in the sense defined by Popper (1959), are just verbal forms of these theories (James and McCulloch 1985). Statistical hypotheses are simply the precise formulations of the consequences of alternative scientific (or other) theories and require data that are samples from clearly defined populations (Dolby 1982). The result of a test of a statistical hypothesis may be used, alone or with other evidence, to decide whether to reject a scientific hypothesis. Testing a statistical hypothesis, then, is not identical to testing a scientific hypothesis, although both sorts of tests allow the kinds of correct conclusions and errors described above. This limited analogy suggests that we would like to adopt a scientific method that minimizes the chances of making both kinds of errors, or that at least is more effective at minimizing the chance of making the more costly of these kinds of errors, just as statisticians have sought to minimize errors in their hypothesis tests. Which kind of error is more costly depends on the hypothesis being examined and on the motivation for the inquiry.

A test of a statistical hypothesis is often in the form of a null hypothesis. While the tested hypothesis could also posit a natural causal process, most often some process is proposed as an alternative to this null hypothesis, which posits the absence of the process. The widespread use of null hypotheses in statistical tests arises because of an asymmetry in the costs of the two types of errors, because of the ease with which the probability of making these errors can be controlled, as well as because the set of null hypotheses can be considered to consist of a single member. The uniqueness of the null case makes it a natural yardstick against which to evaluate any alternative hypothesis. For this reason, statistical theory and convention have developed around such a comparison.

In tests of statistical hypotheses, the probability of a Type I error, rejection of a true hypothesis, is set by the investigator and is equal to the significance level (α) of the stated test. The probability of a Type II error, failure to reject a false hypothesis, is indirectly controlled by the experimenter by specifying α, the sample size, and the alternative hypothesis, but it also depends on the sampling distribution of the test statistic. The chance of making a Type I error is therefore known prior to performing a test, whereas the probability of making a Type II error cannot be specified in advance, unless some prior information on the distribution of the phenomenon of interest is available. Of course, the probability of making a Type II error is also a function of the degree to which the null hypothesis is false. If the null hypothesis is false to a high degree, the probability of making a Type II error will be lower than that of making a Type I error. However, one is likely to recognize null hypotheses that are so extremely false and not employ them in testing alternative hypotheses. Making a Type II error is therefore usually more probable than making a Type I error.

If the tested hypothesis is a null hypothesis, then by setting α we can closely control the likelihood of concluding that some process is manifested by the evidence when actually it is not. However, if we fail to reject the tested null hypothesis, thereby concluding that some process is not operating, it is more difficult to know the chances that we do so incorrectly. While an effort can be made to minimize this risk, it will generally be greater than that of finding a process to be important when it is not, assuming that we are testing plausible null hypotheses in the first place. Thus, given a tested hypothesis that is a null hypothesis, the fact that one tends to make Type II errors more often than Type I errors means that one tends to err on the side of ignorance by finding no experimental effect. In contrast, if the tested hypothesis posits the operation of some process, then the implications of Type I and Type II errors are quite different, and we will then tend to err more often by finding an unimportant process or relationship to be important.

But which error is more costly? As we said above, this decision depends on the motivation for a particular inquiry. In the evaluation of scientific theories, there are two reasons, we believe, one would usually rather err on the side of ignorance. First, if we set more stringent standards for concluding that a theory is true than not true, we are less apt to champion a pet theory that may actually be false. Second, this stringency will tend to prevent us from proceeding to propose more elaborate, comprehensive, or ancillary theories based on the incorrect acceptance of a prior theory. The basic tenet of American jurisprudence, that the accused is innocent until proven guilty, parallels our contention that the rules for testing a scientific hypothesis should be designed to err on the side of ignorance. On the other hand, had our inquiry involved an applied problem such as screening drugs for toxicity to humans, the costs of a Type II error, failing to conclude that a drug is toxic when it truly is, will generally be greater than the costs of concluding that a drug is toxic when it is not. In this instance, using a non-null tested hypothesis would allow tighter control of the chances of making the more costly error.

Another important component of the elimination of errors in testing hypotheses is the specification of the alternative hypotheses. Seldom can we construct an experiment in ecology where the negation of a null tested hypothesis logically entails the acceptance of a specific alternative hypothesis, a "strong inference" (Platt 1964). Our inability to use such inferences in ecology should be viewed as both a mandate for more carefully planned experiments and as a warning that stating the full class of reasonable alternative hypotheses is as important as stating the tested hypothesis (Chamberlin 1897; Simberloff 1983a). This class of alternative hypotheses should contain hypotheses of multiple causation and of complex causation—that is, factor interaction—in addition to single-factor theories (Hilborn and Stearns 1982; Quinn and Dunham 1983).

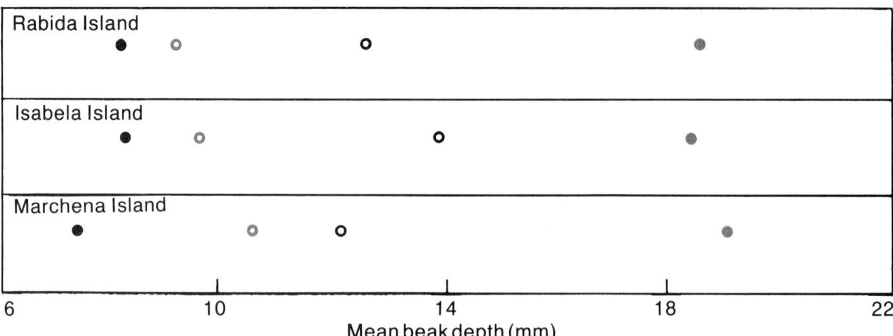

Figure 3. The average beak depths of four species of Darwin's finches on three islands where they coexist show considerable variation from island to island, even though the same set of possible competitors occurs on each island. Thus, this variation cannot be viewed as the result of interspecific competition. The four species represented here are *Geospiza fuliginosa* (*black dots*), *G. scandens* (*color circles*), *G. fortis* (*black circles*), and *G. magnirostris* (*color dots*). (After Simberloff 1983b.)

In the case of the Galapagos finches, for example, one can guess that several factors exert selection pressure on beak morphology. Lack (1983) originally suggested both competition and recognition, and Bowman (1961) suggested differences between islands, but one can easily imagine that sexual selection, allometric relationships between beak dimensions and other traits, and other forces might all affect beak morphology. The argument really is about the relative extent to which different forces are important. If a proposed hypothesis stated that a single force is operative, it would certainly be easier to falsify it, but the falsification would not be much of an advance. The main problem in the example of the Galapagos finches would probably be to achieve adequate replication for performing a statistical test; but even with just 32 islands certain single-factor hypotheses can be rejected, if the islands are viewed as independent replicates. For instance, as Figure 3 illustrates, one can reject the hypothesis that a species' beak morphology on any particular island is determined only by which competing species are present, because among a few combinations of species that happen to occur on more than one island, beak morphology varies even when the same combination of species is present (Simberloff 1983b). However, this observation does not force us to conclude that competition is less important than physical or vegetational differences between islands.

Null models in ecology

Null models were introduced into ecology by Cole (1951), Palmgren (1949), and Williams (1951) because many observational studies lead to very complex results beyond the capabilities of traditional statistical theory to evaluate. Nevertheless, their recent use has generated a vigorous reaction and considerable misunderstanding. Gilpin and Diamond (1984) and Harvey and his colleagues (1983) catalogue what they believe to be a plethora of sins committed by advocates of null models. Yet in spite of these criticisms, the use of null models has increased in ecology.

As outlined above, a null hypothesis is one that posits no effect, a conjecture that nothing in the data at hand or the data to be obtained would lead one to propose an explanation other than chance for the observed result. In the sense used here, *chance* is not meant to imply an absence of deterministic forces acting to produce the observed result; it means simply that, with respect to some specific hypothetical process, the range of variation in the observed results is random. When this is so, we have no reason to believe that the particular process is operative. For example, if a null hypothesis posits that species do not affect one another's geographical distributions, and the alternative hypothesis is that certain species exclude one another, a finding that some appropriate test statistic is consistent with the null hypothesis does not mean that the distribution of each species is simply a matter of chance. It does mean that, whatever the forces are that determine each species' distribution, there are no reasons from the geographical data themselves to conclude that among those forces is the presence or absence of other species.

Only at the level of Heisenberg's principle is there true indeterminacy in nature. Otherwise, in principle, there are always nonrandom causal forces, even in such a quintessentially "random" process as raindrops falling on a marked grid. In theory, one could explain why each raindrop falls in a particular square—by measuring wind, friction, size of the drop, and so on—even though we are satisfied in saying that the distribution of drops in the squares is random (Simberloff 1980).

In testing a statistical hypothesis in experimental research, results are derived from specific treatments or manipulations and are compared with results obtained in the absence of the treatment, the control situation. A statistic that combines the information obtained from the treatment and control situations is elaborated so that its probability distribution is known when the null hypothesis is true. The tested null hypothesis can be evaluated, since the probability of rejecting the null hypothesis when it is true can be determined from this distribution.

However, in nonexperimental research, data are collected and evaluated for their consistency with specific theories or with the operation of particular causal processes, but without knowledge of what values similar data might take in the absence of these causal processes. This is the problem referred to above regarding the observed pattern of beak morphology in Darwin's finches. In this instance, the consistency of data with theory is evaluated either by means of a narrative explanation that makes no attempt to assess the risk of erring when concluding that the data are consistent or inconsistent with theory, or by comparison to values that are expected for the observations and that are derived from a dynamic model of the proposed causal process. In the latter instance, the correspondence between the data and the expectation derived from the model is evaluated either graphically or by a strict assessment of the statistical fit of data and expectation. Thus, in nonexperimental research, a non-null hypothesis often serves as the tested hypothesis. The inconsistency of observation with the expectations of the putative causal process leads to rejection of that process, and consistency leads to a failure to reject the causal process.

A null model is an attempt to generate the distribution of values for the variable of interest in the absence of a putative causal process. As such, a null model is contrived to generate the expected distribution of the variable of interest under the null hypothesis positing no experimental effect. It can, therefore, approximate the role of the control in order to test a hypothesis involving nonexperimental evidence. Null models can then be seen as attempts to place data generated by nonexperimental procedures in an experimental context so that the rules used in testing a statistical hypothesis, as outlined above, can be applied.

From such an analysis, failure to reject the null hypothesis, given a low probability of making a Type II error, leads to the conclusion that nothing in the data at hand requires a causal explanation, and rejection of the null hypothesis requires that some additional causal explanation be invoked. As noted above, failure to reject a null hypothesis and its concomitant null model does not indicate that nature is random or totally without structure. It merely indicates that evidence of structure or nonrandomness is not apparent in the data at hand. It is conceivable that other evidence would lead to the rejection of the tested null hypothesis. On the other hand, repeated failure to reject the tested null hypothesis, given a

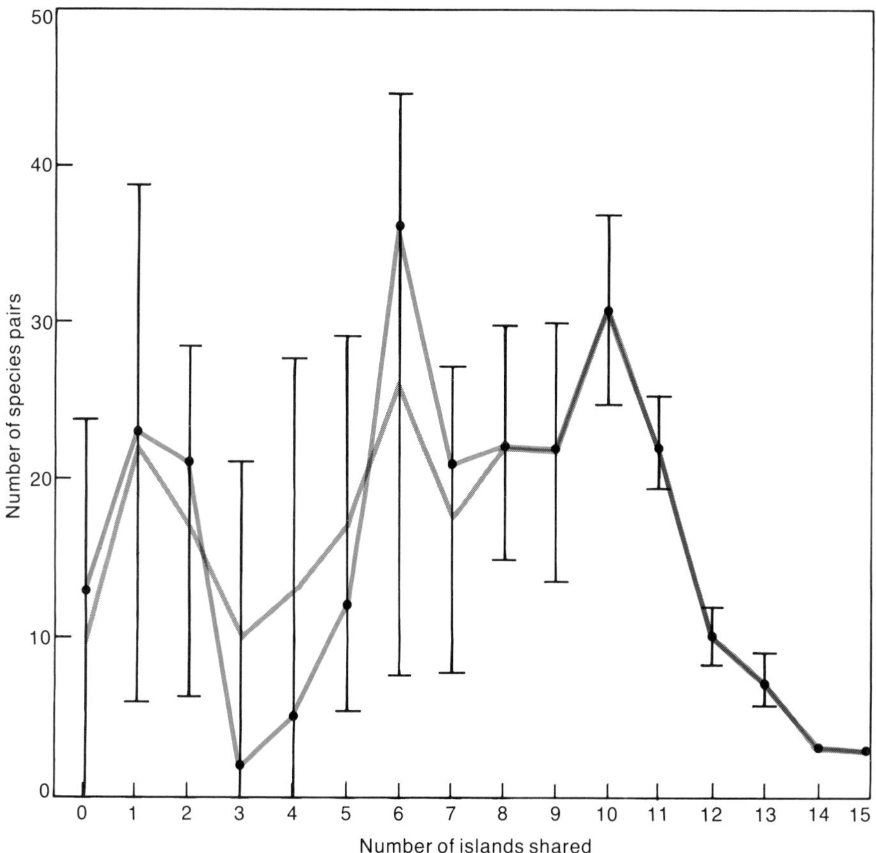

Figure 4. The observed numbers of different possible pairs of land birds that coexist on islands in the Galapagos (*color*) are compared to the numbers that would be expected if different species were distributed independently of each other (*gray*). The similarity of the expected null distribution and the observed one indicates that interspecific competition is not needed to account for them. The vertical lines represent two standard deviations about the expected values. (After Simberloff and Connor 1979.)

low probability of making a Type II error, may lead one to abandon future attempts to seek structure in similar kinds of data.

Another example of a null model that has been designed to allow nonexperimental evidence to be evaluated with the precepts used to test statistical hypotheses—and that has generated much criticism about the form and utility of null models in general—is one we proposed for the geographical distribution of animals among islands (Connor and Simberloff 1979). This model posits that the presence or absence of a species on individual islands does not depend on which other species are present. In other words, the geographical distribution of species among islands is largely unaffected by interspecific competition or by other interactions between species. We proposed this model as an alternative to the theory of Diamond (1975) that interspecific competition acting through competitive exclusion is responsible for nonoverlapping distributions of species.

We conjectured that if competition were unimportant, then the observed pattern of co-occurrence of species would not differ from what would be expected for a group of islands with numbers of species per island and numbers of islands per species identical to what is observed, but with species distributions set independent of each other. We examined the co-occurrence patterns of species of birds and bats on several archipelagos for consistency with this hypothesis and found that we could not reject this null hypothesis for birds in the New Hebrides and the Galapagos, but that we could do so for birds and bats in the West Indies. Figure 4 illustrates the observed and expected co-occurrence pattern for the Galapagos avifauna. Because of the similarity of the observed and expected co-occurrence patterns, and because the deviation between observation and expectation tended to indicate that species co-occurred more often than expected, we concluded that in these instances competition had little effect on the co-occurrence pattern of species.

Gilpin and Diamond (1984) criticize our model, and null models in general, on a number of grounds, arguing that they are difficult to interpret biologically, are not truly null, and are weakened by the inclusion of irrelevant data. We have responded to these criticisms in detail (Connor and Simberloff 1983); however, the contention that our model is not "truly null" refers implicitly to the dependence of all models on assumptions and so deserves further comment. Gilpin and Diamond note that the co-occurrence patterns of species one would expect in the absence of competition are generated in our null model by assuming that each island have as many species as observed and that each species occur on as many islands as observed; they contend that these assumed conditions of the null model could themselves be caused by competition, the process for which the null model purports to serve as a control, thereby making the model difficult to reject. However, although our null model does assume such conditions and does examine species' co-occurrence patterns subject to these assumptions, we believe that these are reasonable assumptions because not all islands have equal numbers of species, because not all species are equally widely distributed, and because these patterns can be caused by processes other than competition.

In any event, the hypothesis we tested was not that competition was occurring, or that competition determined how many islands a species occupies, but rather that the co-occurrence patterns do not differ from those expected if species are distributed among islands independently. For this null model, and for null models in general, if one is unwilling to make some assumptions to account for structure in the data that can reasonably be attributed to causal processes not under investigation, then posing and rejecting null hypotheses will be trivially easy and uninteresting. Yet, without null models we see no means of evaluating nonexperimental evidence for consistency with theory.

This example indicates that the role of null models is closely akin to that of any model in science. Both null and non-null models are conjectures about how nature operates and both may be explicitly tested by experiments. They both can serve as yardsticks with which to gauge the consistency of nonexperimental observations with the operation of some putative causal process. Both are weakened by their dependence on unverified assumptions; they are equally prone to this criticism.

Models and null models

When should one use a null model or a non-null model? Are there advantages in using one approach over the other? The comments above regarding the testing of null and non-null hypotheses apply equally to null and non-null models, at least when the models serve, as does a statistical hypothesis, as devices to transform conjecture into numerical expectation. Again, because of our ability to control the probability of rejecting the tested hypothesis when true (Type I error), and because of the cost of erring by finding that a causal process is important when it is not, we believe that the use of a null tested hypothesis, and therefore of a null model, should usually be the first step in evaluating nonexperimental evidence. Failure to reject the tested null hypothesis, given a low probability of making a Type II error, mandates more critical tests, examination of the assumptions of the null model, and possibly the use of some alternative variable as a criterion. Rejection of the tested null hypothesis again leads to scrutiny of the assumptions of the tested null model for clues as to the causes of its failure, and ultimately to the elaboration of more complicated models, both null and non-null.

On research questions where a tradition of experimentation at the scale of inference exists—that is, where field rather than laboratory experiments have been employed to study field-scale processes—null models are unnecessary since the operations of particular causal processes have usually been documented or discounted already. One may then proceed to examine non-null models directly. However, even in this instance care must be taken in elaborating the full class or reasonable models and in stringently evaluating the consistency of these models with observation (Caswell 1976).

Models used in ecology, which have always elicited skepticism, are increasingly being criticized as self-consistent, irrelevant intellectual endeavors. However, other kinds of observational and nonexperimental evidence are also being viewed with increasing skepticism. We agree that controlled experiments can provide a sounder basis for inferring causal relationships. However, many interesting and important ecological questions are simply beyond the scope of experimentation that is ethical and logistically feasible. Most of these questions fall within the subdisciplines of community and ecosystem ecology. If there is to be a community ecology, then our arsenal of methods must include approaches other than controlled experiments. The problem we face is integrating these alternative approaches in a manner able to expose their results to scrutiny and skepticism at least as intense as that with which we view the results of controlled experiments. If we fail to do so, the inferential basis these approaches provide will be compromised. This has been our contention regarding nonexperimental evidence adduced as support for competition.

The methodological procedures described above—posing null tested hypotheses, attempting to refute them, delineating the full class of reasonable alternative hypotheses, and using null models—are suggested in order to show how nonexperimental evidence may be effectively combined with the results of controlled experiments in an effort to understand nature. It is by no means an exhaustive list of the procedures and precautions that might be used when considering nonexperimental evidence, nor is this methodological prescription without pitfalls. However, coupled with the realization that ecological nature is more complex and idiosyncratic than we have been willing to concede, we may at least begin to learn how some communities work, and to build a firm foundation for more general ecological theory.

References

Bowman, R.I. 1961. Morphological differentiation and adaptation in the Galapagos finches. *Univ. of Calif. Publs. Zool.* 58:1–302.

Brown, J. H. 1981. Two decades of homage of Santa Rosalia: Toward a general theory of diversity. *Am. Zool.* 21:877–88.

Brown, J. H., D. W. Davidson, and O. J. Reichman. 1979. An experimental study of competition between seed-eating desert rodents and ants. *Am. Zool.* 19:1129–43.

Brown, J. H., A. Kodric-Brown, T. G. Whitham, and H. W. Bond. 1981. Competition between hummingbirds and insects for the nectar of two species of shrubs. *Southwestern Nat.* 26:133–45.

Caswell, H. 1976. The validation problem. In *Systems Analysis and Simulation in Ecology,* ed. B. C. Patten, vol. 4, pp. 313–25. Academic Press.

Chamberlin, T. C. 1897. The method of multiple working hypotheses. *J. Geol.* 5:837–48.

Clements, F. E., J. F. Weaver, and H. C. Hanson. 1929. *Plant Competition.* Carnegie Inst. Washington Publ. 398.

Cole, L. C. 1951. Population cycles and random oscillations. *J. Wildlife Mgmt.* 15:233–52.

Connell, J. H. 1975. Some mechanisms producing structure in natural communities: A model and evidence from field experiments. In *Ecology and Evolution of Communities,* ed. M. L. Cody and J. M. Diamond, pp. 460–90. Harvard Univ. Press.

———. 1980. Diversity and coevolution of competitors, or the ghost of competition past. *Oikos* 35:131–38.

———. 1983. On the prevalence and relative importance of interspecific competition: Evidence from field experiments. *Am. Nat.* 122:661—96.

Connor, E. F., and D. Simberloff. 1979. The assembly of species communities: Chance or competition? *Ecology* 60:1132–40.

———. 1983. Interspecific competition and species co-occurence patterns on islands: Null models and the evaluation of evidence. *Oikos* 41:455–65.

Dayton, P. K. 1971. Competition, disturbance, and community organization: The provision and subsequent utilization of space in a rocky intertidal community. *Ecol. Mono.* 41:351–89.

Diamond, J. M. 1975. Assembly of species communities. In *Ecology and Evolution of Communities,* ed. M. L. Cody and J. M. Diamond, pp. 342–44. Harvard Univ. Press.

———. 1978. Niche shifts and the rediscovery of interspecific competition. *Am. Sci.* 66:322–31.

Dolby, G. R. 1982. The role of statistics in the methodology of the life sciences. *Biometrics* 38:1069–83.

Elton, C. S. 1946. Competition and the structure of animal communities. *J. Animal Ecol.* 15:54–68.

Fowler, N. 1981. Competition and coexistence in a North Carolina grassland. II. The effects of the experimental removal of species. *J. Ecol.* 69:843–54.

Gilpin, M. E., and J. M. Diamond. 1984. Are serious co-occurrences on islands non-random, and are null hypotheses useful in community ecology? In *Ecological Communities: Conceptual Issues and the Evidence,* ed. D. R. Strong et al., pp. 297–315. Princeton Univ. Press.

Gleason, H. A. 1939. The individualistic concept of the plant association. *Am. Midl. Nat.* 21:92–110.

Grant, P. R. 1972. Convergent and divergent character displacement. *Biol. J. Linn. Soc.* 4:39–68.

———. 1981. Speciation and adaptive radiation of Darwin's finches. *Am. Sci.* 69:653–63.

Grant, P. R., and I. Abbott. 1980. Interspecific competition, island biogeography, and null hypotheses. *Evolution* 34:332–41.

Haila, Y. 1982. Hypothetico-deductivism and the competition controversy in ecology. *Ann. Zool. Fennicia* 19:255–63.

Harvey, P. H., R. K. Colwell, J. W. Silvertown, and R. M. May. 1983. Null models in ecology. *Ann. Rev. Ecol. Syst.* 14:189–211.

Hilborn, R. C., and S. Stearns. 1982. On inference in ecology and evolutionary biology: The problem of multiple causes. *Acta Biotheoretica* 31:145–64.

Jackson, J. B. C. 1981. Interspecific competition and species distributions: The ghost of theories and data past. *Am. Zool.* 21:889–901.

James, F. C., and C. E. McCulloch. 1985. Data analysis and the design of experiments in ornithology. In *Current Ornithology,* ed. R. F. Johnston, vol. 2, pp. 1–63. Plenum Press.

Lack, D. 1983. *Darwin's Finches.* Cambridge Univ. Press

Lawton, J. H., and D. R. Strong. 1981. Community patterns and competition in folivorous insects. *Am. Nat.* 118:317–38.

Lewin, R. 1983. Santa Rosalia was a goat. *Science* 221:636–39.

Paine, R. T. 1969. The *Pisaster-Tegula* interaction: Prey patches, predator food preference, and the intertidal community structure. *Ecology* 50:950–61.

Palmgren, P. 1949. Some remarks on the short-term fluctuations in the numbers of northern birds and mammals. *Oikos* 1:114–21.

Pielou, E. C. 1982. The usefulness of ecological models: A stock-taking. *Quar. Rev. Biol.* 56:17–31.

Platt, J. R. 1964. Strong inference. *Science* 146:347–53.

Popper, K. R. 1959. *The Logic of Scientific Discovery.* Harper and Row.

Quinn, J. F., and A. E. Dunham. 1983. On hypothesis testing in ecology and evolution. *Am. Nat.* 122:602–17.

Salt, G., ed. 1984. *A Round Table on Research in Ecology and Evolutionary Biology.* Univ. Chicago Press.

Schoener, T. W. 1982. The controversy over interspecific competition. *Am. Sci.* 70:586–95.

———. 1983. Field experiments on interspecific competition. *Am. Nat.* 122:240–85.

Seifert, R. P., and F. H. Seifert. 1979. A *Heliconia* insect community in a Venezuelen cloud forest. *Ecology* 60:462–67.

Simberloff, D. 1970. Taxonomic diversity of island biotas. *Evolution* 24:23–47.

———. 1980. A succession of paradigms in ecology: Essentialism to materialism and probabilism. *Synthese* 43:3–39.

———. 1983a. Competition theory, hypothesis testing, and other community ecological buzzwords. *Am. Nat.* 122:626–35.

———. 1983b. Sizes of coexisting species. In *Coevolution,* ed. D. J. Futuyma and M. Slatkin, pp. 404–30. Sinauer.

Simberloff, D., and E. F. Connor. 1979. Q-mode and R-mode analyses of biogeographic distributions: Null hypotheses based on random colonization. In *Contemporary Ecology and Related Ecometrics,* ed. G. P. Patil and M. L. Rosenzweig, pp. 123–38. Burtonsville, Md: International Cooperative Publishing House.

Strong, D. R., L. A. Syszka, and D. S. Simberloff. 1979. Tests of community-wide character displacement against null hypotheses. *Evolution* 33:897–913.

Wiens, J. A. 1977. On competition and variable environments. *Am. Sci.* 65:590–97.

Williams, C. B. 1951. Intrageneric competition as illustrated by Moreau's records of East African birds. *J. Animal Ecol.* 20:246–53.

Intertidal Zonation of Marine Invertebrates in Sand and Mud

Communities on intertidal rocks are arranged in well defined horizontal bands. Is there an ecological analogue in soft sediments?

Charles H. Peterson

At the seashore, a dominant rhythm of plant and animal life is supplied by the rise and fall of the tides. The incoming tide, washing over rocks, sand or mud, arouses diverse marine creatures from their low-tide quiescence; as the tide falls, these organisms withdraw again into shells, burrows and other protective shelters. This retreat is not hard to understand. The residents of the intertidal zone, the area between the low- and high-tide marks, are almost exclusively marine plants and animals, and they are essentially hung out to dry for much of each day. The period of their exposure to air at low tide represents a time of physiological stress. Intertidal invertebrates generally cease feeding and, with the plants, risk overheating in summer, overcooling in winter, desiccation, damage from ultraviolet radiation from the sun, or osmotic shock after heavy rains. The higher the elevation on shore, the longer is the average duration of this exposure, and the greater are the physiological stresses.

The relationship between physiological stress and elevation in the intertidal zone contributes to the formation of distinct biotic zonation, which is readily noticeable to visitors to rocky shores. At low tide, one observes that intertidal organisms are arranged in horizontal bands that change with elevation on the face of exposed rock. On temperate rocky coasts of Europe and North America, for example, periwinkle snails commonly are seen at high elevations. Closer to the low-tide mark typically appear bands of barnacles, mussels, and ultimately seaweeds. These stratified communities have been the subject of much study; indeed, investigations of intertidal rocky-shore communities have supplied many unifying concepts in marine ecology (Connell 1972; Menge and Sutherland 1976; Sousa 1979; Gaines and Roughgarden 1985).

Along sandy or muddy shores, however, often the only obvious vertical zonation is that exhibited by large plants in protected embayments. Here salt-marsh plants or mangroves dominate the high ground; seagrasses begin to occur around the lower edge of the intertidal zone, forming the boundary between this zone and habitats that are continuously submerged, or subtidal. Separating these large and obvious plants are extensive expanses of sediments without large plants and usually without any animals obvious to the eye. The organization of intertidal invertebrate communities in expanses of soft sediment between the marsh and the seagrass, and along sedimentary shorelines that are less well protected and lack emergent vegetation, is not easy to observe. On intertidal sediments the animal assemblage is dominated not by attached creatures but by invertebrates that live beneath the surface of the sand or mud. The question of how the organization of these reclusive communities varies with elevation on intertidal shores has received relatively little attention from ecologists.

Still, there is growing evidence that some degree of vertical zonation of resident animals is a ubiquitous feature of intertidal sediments as well as of rocky shores (e. g. Johnson 1970; Remane and Schlieper 1971; Kneib 1984; Peterson and Black 1987). On both kinds of shoreline, zonation is produced by an interplay of physical and biological processes, but the critical forces at work in soft-sediment communities are not necessarily the same as those that are the prime players on the rocky shore. In this article I shall paint a picture of the vertical zonation of marine invertebrates on intertidal sand and mud flats by examining how switching from a rocky substrate to a soft-sediment environment alters the effectiveness of

Figure 1. Tide-washed sand flats are home to many species of marine invertebrates that often vary in abundance with elevation above the low-tide mark. Observations conducted on the broad intertidal flats near Monkey Mia on the shore of Shark Bay in Western Australia (facing page) have documented the zonation of several species of clams; an important factor influencing this zonation may be the physiological stresses that intertidal marine animals experience when exposed to air. Most inhabitants of tidal flats, including the clams, live beneath the surface sediments. Seen here are several plastic-mesh enclosures used in experiments to maintain the desired combinations of clams at fixed densities. The fences shown are located roughly midway between the low- and high-tide marks. Several ecological processes combine to influence biotic zonation on soft-sediment shores, and experimental manipulations can serve to separate the roles of various contributing factors. (Photograph courtesy of the author.)

Charles H. Peterson is professor of marine sciences, biology and ecology at the University of North Carolina at Chapel Hill. He received his bachelor's degree in biology from Princeton University in 1968 and his Ph.D., also in biology, from the University of California at Santa Barbara in 1972. He has conducted ecological experiments on benthic invertebrates of intertidal flats in coastal lagoons of the Pacific, Atlantic, Indian and Southern oceans. Address: University of North Carolina at Chapel Hill, Institute of Marine Sciences, Morehead City, NC 28557.

Figure 2. Marine invertebrates living beneath the sediments dominate the animal assemblage of a typical intertidal sand flat. Shown here are some of the major components of tidal-flat ecosystems: a burrowing animal, the ghost shrimp, and the inhabitants of its burrow *(center)*; tube-building polychaete worms *(right)*; a sea anemone; clams; shorebirds (including the marbled godwit, *foreground*), which are the principal "top-down" predators; and some mobile "bottom-up" predators (crabs, a ray and a bony fish). Although the zonation of invertebrates on such soft-sediment shores is not as well defined as that on rocky shores (*Figure 3*), factors such as physiological stress, predation, disturbance and competition for food may vary with elevation on the shore and help produce more gradual patterns of zonation. Wave action often creates a gradient in sediment size that is a potential additional influence on zonation; coarser sediments that tend to be found higher on shore are favored by suspension-feeding invertebrates, whereas deposit feeders are more common in finer sediments lower on the shore.

those biological and physical processes known to cause the distinct zonation on rocks. Specifically, I review how physiological stress, competition for space and food, disturbance of sediments, predation and settlement of larvae vary with elevation on the shore, and thereby produce vertical zonation.

Lessons from Rocky Shores

Models of zonation on rocky shores begin with the influence of physiological stress (Newell 1976). Each marine species is limited in its ability to withstand the stresses that accompany exposure to air; since the duration of exposure increases with elevation above the low-tide level, physiological tolerance commonly sets upper limits for the distributions of species of benthic, or bottom-dwelling, plants and animals on rocky shores.

Studies of rocky shores have also revealed that the physiological consequences of increasing elevation do not, by themselves, explain observed patterns of zonation. Instead, much of the explanation can be found in important local biological interactions that may be seen as indirect effects of elevation. The intensity of both interspecific competition and predation, in particular, varies with elevation on shore and contributes significantly to the creation of vertical distributions (Paine 1966; Dayton 1971; Connell 1972). At mid-intertidal levels, physiological stresses are low enough and the feeding time long enough that benthic organisms often settle and grow to cover all available space on the rocks. As a result, species commonly compete for limited attachment space, and some inferior competitors are excluded and thereby limited in their downward distribution. Still lower on the shore, predatory marine invertebrates have more time to feed because the rocks are covered with water most of the day; by this process, predators often set lower boundaries for the distribution of their preferred prey. Zonation of communities on rocky substrata also continues into the subtidal zone, where physical and biological processes also vary with water depth (Witman 1987).

No analogous model of the role of tidal elevation has emerged from studies of marine benthic communities on sandy and muddy shores, and the reasons for the absence of such a model deserve examination. Of primary importance are fundamental differences that make the rocky shore a much more tractable system for ecological modeling. Because of the underlying potential for competitive exclusion caused by local interference competition for space on rocky shores, one can model the influence of tidal elevation on the composition of benthic communities by examining how factors that can reduce the intensity of competition—including physiological stress, limitations on recruitment of new individuals, physical disturbance and predation—vary with elevation (Menge and Sutherland 1976). By contrast, residents of soft sediments do not generally compete for limited surface space; instead, they may compete for a mobile food source, the plankton and organic materials that are transported by tidal and other currents (Peterson 1979, 1982). This competition implies a larger scale than that comfortably employed in ecological experiments. Alternatively, many species in soft sediments may not be experiencing competition at all because densities are kept low by various disturbance agents.

Other characteristics of soft-sediment communities may have inhibited experimentation. The buried, or infaunal, benthic invertebrates of the sediments are often small and usually soft-bodied and require careful handling;

unearthing them can easily damage or kill the animals, disrupt their natural sedimentary environment and induce artifacts in experiments that require such intervention. Because of these concerns, soft-sediment systems are widely viewed as less amenable to experimental manipulation than rocky intertidal communities. This view needs qualification. The methodology for such experiments is necessarily different from that used to study animals attached to exposed rocks; nevertheless, the fact that these animals are not attached to substrata does allow one to create almost unlimited experimental combinations of species and densities.

The most significant inhibitor to ecological experimentation in soft sediments, then, may not be an inherent procedural constraint, but rather the lack of an overriding and compelling conceptual model to inspire such experiments. Fortunately, even in the absence of a simple, unifying model such as could be built upon established principles of local competitive exclusion, one can gain general insights into the effects of tidal elevation on soft-sediment communities from those experiments that have been performed and from what is known about ecological processes in the intertidal zone.

Physiological Stress

Tony Underwood and Liz Denley (1984) noted that one can distinguish two components in the physiological stress that marine invertebrates experience when they are exposed to air at low tide; this dichotomy is useful for examining responses of soft-sediment invertebrates to the tides. First, invertebrates exposed to air experience physical stresses: among these are possible temperature shock, desiccation, poisoning from ultraviolet radiation and osmotic imbalance. (The last stress is a consequence of the fact that saltwater creatures unable to adjust their internal salt concentration can be drowned by an influx of fresh water into their salty tissues through their external membranes.) The other component is biological stress; this includes the cessation of feeding and of aerobic respiration. Both components are likely to increase as elevation on shore and thus exposure times increase. On rocky shores the importance of physiological stress in setting upper distribution limits has been elegantly demonstrated for the barnacle *Semibalanus balanoides* by David Wethey (1984), who compared survivorship under shaded and translucent (control) roofs. The same process is likely to operate in soft-sediment benthic species, although few tests of this hypothesis have been attempted.

For organisms that live buried under sand and mud, the sediment blanket itself provides a buffer against physical stresses by blocking ultraviolet light, retaining moisture and slowing temperature changes. Except for the immediate surface layer of sediment, tidal-flat sands and muds remain moist during low tide. The residual pore waters not only prevent desiccation of buried organisms; they also inhibit temperature change through their high heat capacity. A sediment cover of 10 centimeters has been shown to buffer any detectable change in temperature or salinity during low tides on intertidal flats except for sites in the extreme high intertidal zone.

Sand- and mud-dwellers are not, however, buffered from the biological components of exposure stress. Only a small fraction of soft-sediment invertebrates is able to continue normal feeding and aerobic respiration at low tide; these are primarily very small animals that live in the interstices between coarse sand grains. Larger marine invertebrates on sand and mud flats are likely to follow a general pattern of di-

Figure 3. On rocky shores, discrete horizontal bands provide visual evidence of the zonation of intertidal species and the role played by competition among species for limited attachment space. On these wave-protected intertidal rocks near Monterey, California, are seen (*from top*) the supralittoral green alga *Prasiola meridionalis*, a "bare zone" of limpets and littorines, clumps of the goose barnacle *Pollicipes polymerus*, clumps of the red alga *Endocladia muricata*, the barnacle *Tetraclita squamosa rubescens*, several red algae, and finally surf grass (*Phyllospadix*). (Photograph courtesy of Michael Foster, Moss Landing Marine Laboratories.)

minishing feeding time and thus diminishing growth with increasing intertidal elevation, as exhibited in some experiments with clams.

One way to gather evidence about the effects of physiological stresses at low tide is by conducting transplant experiments, which examine the responses of organisms to environments where conditions differ. I have participated in transplant studies using sand-flat clams living on the shores of Shark Bay in Western Australia (Peterson and Black 1987, 1988) The tidal sand flats near Monkey Mia on the Peron Peninsula shore of Shark Bay's eastern gulf are extremely broad, extending nearly a kilometer from shore. Clams are both abundant and diverse on these flats, living on algae filtered from the waters. The six most abundant species at Monkey Mia exhibit clear patterns of vertical zonation on the intertidal flats (Figure 6). Three species—*Anomalocardia squamosa, Placamen berryi)* and *Pitar citrina*—are much more abundant at higher intertidal elevations, whereas two other species—*Circe lenticularis* and *Placamen gravescens*—are found only at low intertidal and shallow subtidal elevations. The sixth species, *Callista impar*, reaches maximal abundance between the mid-intertidal and shallow subtidal elevations.

Five of these species were experimentally transplanted to sites at a mid-intertidal and a shallow-subtidal elevation (Peterson and Black 1987, 1988). Growth rates of all five species declined as tidal elevation increased, but the two low-elevation species exhibited the largest reductions in growth with increasing elevation (Figure 7). One of the two low-elevation species also exhibited a much lower rate of survival at the higher experimental elevation; no difference in survivorship could be detected for the other species. The results suggest that the pool of physiological stresses associated with exposure to air may set the upper limits for the low-elevation clams at Shark Bay. Interestingly, this pattern of growth is precisely analogous to that found by Joe Connell (1961) between the rocky intertidal barnacles *Semibalanus balanoides* and *Chthamalus stellatus.*

It is not at all clear, however, that the physical component of physiological stress contributes to the differing responses of Shark Bay clams to exposure to air. The two low-elevation species, *Circe lenticularis* and *Placamen gravescens*, may simply differ from the others in their feeding and energetics, making them less able to exist with reduced feeding time or to tolerate respiration without oxygen. This could explain their failure to live as high on shore as the other three species of clams.

The effect of the physical component of low-tide physiological stress on rocky-shore zonation is well documented. Clonal marine invertebrates (those that grow in clones of genetically identical units through the asexual division of one to form many) are limited to subtidal elevations on the rocks by their sensitivity to aerial exposure (Jackson 1977), even though they are superior competitors for surface space. With no external skeleton to constrain their size, clonal invertebrates can spread rapidly by marginal growth. Barnacles and most other solitary rocky-shore invertebrates, although poorer competitors, have protective external skeletons, which enable them to survive at higher elevations than clonal forms. This tradeoff between competitive ability and external protection contributes to the vertical segregation of these groups on rocky shores (Jackson 1977).

If the physical component of physiological stress were the predominant factor limiting the upward spread of distributions of intertidal invertebrate species on sandy and muddy shores, one would expect to see clams and snails, tube-building worms and tube-building crustaceans higher on the tidal flats; nudibranchs, annelid worms without tubes, non-tube-building crustaceans and sea anemones should be restricted to lower levels on the shore. Unfortunately, there have been few surveys of such taxonomic distributions with elevation, and so this prediction has not yet been rigorously tested. Certain components of the prediction appear to hold true: sea anemones and nudibranch and opistobranch molluscs without external shells do seem confined to subtidal zones on sand and mud flats. Nevertheless, the buffering capacity of sediments suggests that this analogy to rocky-shore systems may not provide correct predictions.

One might also expect to see fewer resident species at higher tidal elevations as a general consequence of the gradient in physiological stress; the possibility that some other factor, such as intense predation, could eliminate enough species at the lower end to override this prediction seems unlikely. Ralph Gordon Johnson (1970) described an example of this pattern among the invertebrates in the soft sediments of Tomales Bay, California. As one moves deeper in the sediments new species appear, while those previously encountered higher on the shore continue to be found. At Shark Bay, there are some high-elevation "specialists" unlike the pattern Johnson described for Tomales

Figure 4. Physiological stress appears to limit the distribution of some species on sandy and muddy shores. Marine animals located higher on the shore must endure longer periods of exposure to air and reduced feeding time and the concomitant threats of desiccation, osmotic shock and damage from ultraviolet radiation. Some of the species whose relative distributions may be influenced by the gradient in stresses are seen here: sea anemones and mobile polychaete worms, which tend to be found lower on shore, and tube-building amphipods and clams, which possess protective "shells" that may enable them to better tolerate exposure. Burial under the sediments, however, buffers even soft-bodied infaunal invertebrates from physical environmental extremes.

Figure 5. Emergent plants on protected shorelines provide the most obvious evidence of zonation on intertidal flats. On the shore of Moreton Bay in Queensland, Australia *(above)*, the distribution of mangrove trees is clearly restricted to the higher elevations. Any zonation among the buried microalgae and invertebrates of the muddy sediments lower on the flat is not evident to the casual observer. (Photograph courtesy of the author.)

Bay—but here, too, a wider variety of species was found low on the shore, although we sampled three times as many clams at the higher, mid-intertidal elevation (Peterson and Black 1987).

Competition

As noted above, interspecific competition for space on the rocks is a strong force in the organization of many intertidal communities on rocky shores and has supplied the conceptual framework for several studies of intertidal zonation (Connell 1961, 1972; Paine 1966). Competition among species on sandy and muddy shores has been documented, but its consequences are entirely different. This contrast may explain much of the difference in the character of zonation seen in comparisons of hard and soft shores.

The intense competition among sessile species for limited surface space on many rocky shores contributes to the discrete nature of zonation there. If the vertical boundary of one species' distribution is set by competitive exclusion by another, that implies the generation of quite discrete, monospecific bands of dense coverage. There are certainly rocky shores where this is not true; on some gastropod-dominated rocky shores, for example—where species compete for food, rather than space-competitive exclusion is not a primary cause of the vertical boundaries of species distribution. Here, distribution patterns are not characterized by sharp, discrete zones of single species occupation. A similar pattern of more gradual change with elevation likewise characterizes the invertebrate communities on intertidal soft-sediment shores (Johnson 1970; Kneib 1984; Peterson and Black 1987)

Competition for space is not a dominant process in sand and mud flats (Peterson 1979; Black and Peterson 1988); as a result, this type of competition does not contribute significantly to zonation along these shores. In general, animals living in sand and mud are not subject to the competitive tactics of overgrowth, crushing and dislodgement used by aggressive competitors on hard surfaces. Sediment dwellers are not attached to the surface, and they can use a third dimension, depth of burial, to avoid contact (Peterson 1979). Some clear cases of local competitive exclusion through interference competition do exist; Herb Wilson (1980), for example, demonstrated in the laboratory that a tube-building worm, a terebellid polychaete, forcibly ejects any nearby individuals of a less aggressive nereid polychaete species. Nevertheless, such examples seem to be the exceptions, not the rule.

The competition that does take place among invertebrates of soft sediments more often involves reduction in food supplies (Levinton 1972; Peterson 1982). Whereas competition for space is effective only in the immediate vicinity of the competitors, competition for food occurs over a large spatial scale. For sessile organisms the implications

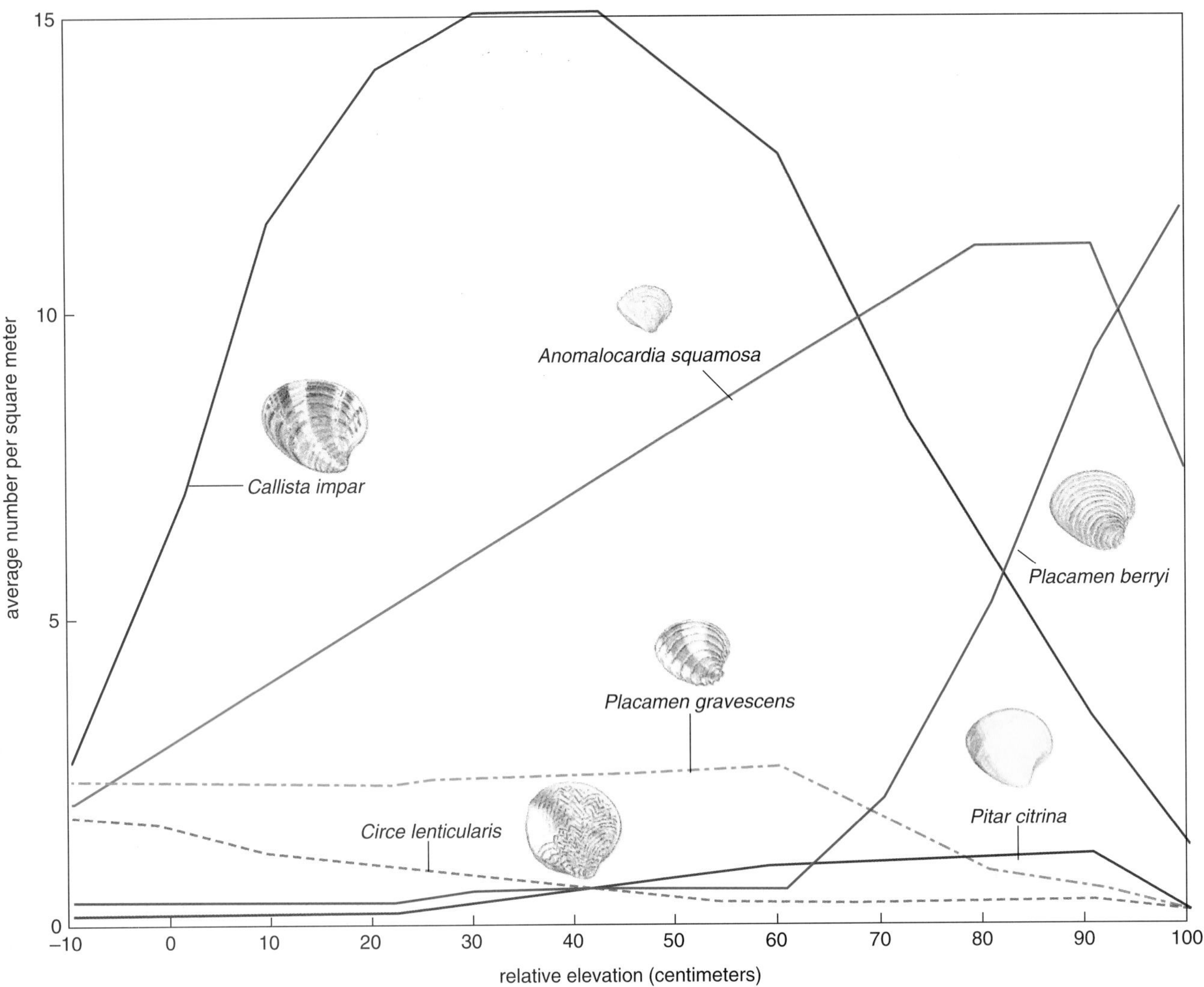

Figure 6. Abundance of each of the six most plentiful clam species at Monkey Mia in Shark Bay, Western Australia, changes with elevation in the intertidal zone. The smoothed density of each of the six most abundant species of clams is plotted here with respect to the approximate mean low-tide level (O centimeters). Sampling elevations ranged from a shallow subtidal level 10 centimeters below the low tide mark to a mid-intertidal location 100 centimeters above low tide. The high-tide mark is another 80 centimeters higher. These data suggest that three species are more common high on the shore; two "prefer" low elevations, and the most abundant (*Callista impar*) reaches maximum abundance between the extremes. The values for *Callista*, by far the most common species on the flat, have been reduced by one-third so that they could be included in the same graph with the other data. (Data from Black and Peterson, in preparation.)

of coming into interference competition for space are severe: displacement, overgrowth, crushing and death. Since the source of food for most soft-sediment benthic invertebrates is mobile and renewable, however, in competing for food even the poorer competitor gets some, usually enough to survive. The effects of food depletion (in all marine intertidal systems) are not localized because tidal and other currents spread the impact over a very large area. Even food particles on the sediment surface are continually being resuspended and redeposited.

Because incoming tides move directionally from low to high on the shore, progressive depletion of suspended foods by animals at low elevations could, by competition, set upper distributional limits for species higher on the shore (Peterson and Black 1987). No test of this hypothesis exists, and the distinction between the spatial scales of competition for space and competition for food is not yet fully explored in the ecological literature.

Disturbance

As the tides move in and out, the sediments in which infaunal invertebrates are buried become shifted continuously by wave action and currents. In addition, some animals process sediments for organic foods or continually build and modify burrows, and many predatory animals disturb the sediments by foraging, digging and moving. By these processes invertebrates in the sediment experience both physical and biological disturbance of their environment (Grant 1983).

On rocky shores an important interplay takes place between disturbance and competition; disturbance can serve to reduce the densities of animals and plants on the rocks and thereby diminish competition among species (Sousa 1979). Since densities high enough to create competitive exclusion are rarely observed on most sandy and muddy shores (Peterson 1979), one must ask in this context a different question: Could physical or biological disturbance of the sediments be severe enough to set upper or lower limits of species distributions? It seems likely that these factors may, indeed, contribute directly to zonation.

Wave action may help set patterns of zonation because its intensity is distributed unevenly, typically reaching a maximum at higher elevations on shore, where waves ultimately break. The water is shallower at these margins, so surface wave energy reaches the sea floor. The resulting overturning of sediments may limit the abundance of soft-sediment invertebrates higher on shore, but this hypothesis remains untested. Better understood is one consequence of the resultant variation in wave-energy intensity on sedimentology: creation of a gradient in average sediment size that may, in turn, affect the organization of these communities.

Marine sediments exist in dynamic equilibrium with the overlying water flow regime. Where wave action or rapid currents create strong forces of bottom disturbance (shear stresses), fine sediments are suspended, leaving on the sea floor only coarser sand particles, which are less subject to resuspension and erosion. Fine particles (muddy sediments) settle out onto the sea floor in areas less exposed to wave and current energy (see Butman 1987). By this process, the intertidal gradient in bottom shear stresses tends to sort sediments according to size, with coarser sediments higher on shore.

Sediment size has been shown to have a fundamental influence on soft-sediment invertebrates (Rhoads and Young 1970; Levinton 1972). Deposit feeders—animals that ingest deposited foods—are often more common in finer sediments, where the organic content is greater. Suspension feeders—which feed on phytoplankton and other organic matter suspended in the water—are more common in coarser sediments, where faster currents renew food supplies more quickly. By influencing sediment size, then, the differing severity of wave action with elevation may create a significant zonation of feeding types.

Biological disturbance, or bioturbation, of substrates is unique to the soft-bottom environments within the intertidal zone because few, if any, biological activities are forceful enough to overturn rocky substrates. In sediments, biological disruption can occur over a large scale; deposit feeders and burrow builders can often be sufficiently disruptive to influence zonation. Visual evidence of large-scale sediment disturbance is not hard to come by; one can often see mounds or fecal coils of freshly deposited sediment, burrow openings and honeycombs of subterranean burrows. Perhaps the best example is provided by the ghost shrimp, *Callianassa californiensis*, which inhabits California lagoons. Ghost shrimps live in dense subterranean colonies and turn over tremendous volumes of sediment during their almost constant burrow construction. This activity prevents colonization of the estuarine sea floor by many tube-building benthic animals and other sedentary species that cannot move easily to reestablish their required depth after they have been buried by *Callianassa* (Brenchley 1981; Posey 1986). *Callianassa's* burrowing helps maintain a striking zonation on the sand flats of Elkhorn Slough in northern California, where the tube-building invertebrate *Phoronopsis viridis* is excluded from higher intertidal flats; *Callianassa* breaks and buries the *Phonoropsis* tubes.

Similarly intense bioturbation is caused by several other species, including especially large enteropneust worms such as *Balanoglossus*. These bioturbators tend to occur at middle and upper levels on the shore, perhaps because they are limited in the downward direction by increasingly intense fish predation (Posey 1986). Bioturbators are especially susceptible to fish predation because they frequently expose various body parts above the bottom while transporting and depositing sediments.

Although bioturbators can limit the abundance and distribution of certain sedentary species, those that construct burrows also have positive influences on suites of (usually smaller) commensal species that depend on the provision of three-dimensional subterranean

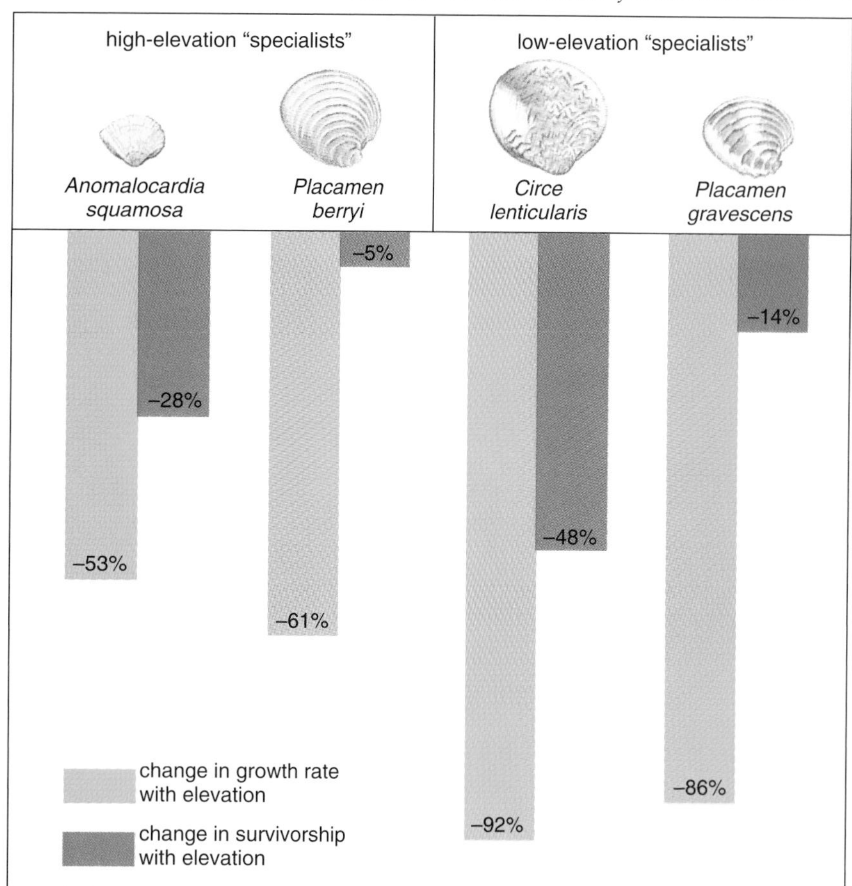

Figure 7. During transplant experiments, growth rates of four species of Monkey Mia clams (*see Figure 6*) were lower at the higher elevation, suggesting that the upper limits of these clam distributions in the intertidal zone may be set by relative tolerance to physiological stresses associated with increasing elevation and aerial exposure. Declines in growth with elevation were statistically significant for all species, but were larger for the low-elevation "specialists," a finding consistent with the hypothesis that physiological sensitivity to increasing stresses of higher elevation contributes to clam zonation at Monkey Mia. Survivorship declined significantly with elevation only for *Circe lenticularis*, a low elevation "specialist." (Data from Peterson and Black 1988.)

structure for their habitat (Reise 1985). This positive interaction has a direct rocky-shore analogue in the provision of three-dimensional habitat by mussel beds, which likewise promote the presence of an entire assemblage of smaller invertebrates otherwise absent.

The impact of disturbance by predatory animals on the distributions of species is less clear, since bioturbation effects have only rarely been separated from the direct effects of predation. Ron Kneib (1985) experimentally amputated the feeding appendages of grass shrimp to prevent consumption of prey while still allowing excavation activities. In this experiment densities of a gastropod, *Hydrobia*, declined due to disturbance alone. But disturbance as a by-product of predator activity seems unlikely to be intense enough over entire elevation zones to contribute significantly to maintenance of zonation patterns, except in special cases; large mobile, herding predators such as the horseshoe crab might be expected to cause intense enough local disruption of sediments, but the possible relationship of this activity to zonation has not been explored.

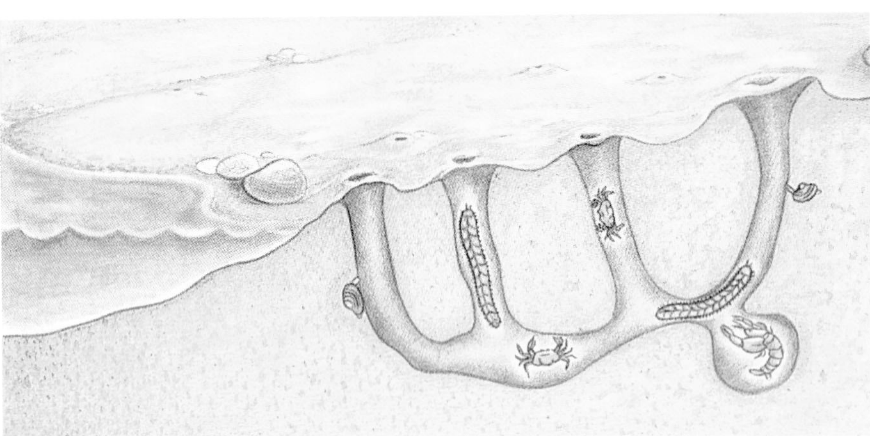

Figure 8. Ghost-shrimp burrows provide the three-dimensional structure that is home to several commensal species, including the pea crab, scale worm and *Cryptomya* clams. The almost constant burrowing of the ghost shrimp, *Callianassa californiensis*, also prevents colonization of the estuarine sea floor by tube-building benthic animals and other sedentary species that cannot survive burial by sediments excavated by *Callianassa*. Such intense biological disturbance, or bioturbation, can contribute to zonation patterns in the intertidal zone.

Figure 9. Sea-floor mounds and burrow openings on the floor of Mugu Lagoon in Southern California provide a clear indication that the ghost shrimp, *Callianassa californiensis*, has been at work. (Photograph courtesy of the author.)

Predation

Marine invertebrates on soft sediments face two kinds of predation: "top-down" predation from the land and "bottom-up" predation from the sea. Soft-sediment systems on tidal flats are extremely exposed to consumers from other habitats. On a flooding tide, it is quite common for swimming crabs and schools of predatory fishes, especially rays and flatfishes, to follow the tide in and to forage extensively on all elevations of the flat (see, for example, Virnstein 1977). The pits and holes they dig contribute to mortality of invertebrates and provide an opportunity for recolonization by other benthic organisms. Similarly, the flat topography of tidal flats makes them accessible during low tide to foraging shorebirds of many diverse groups; these predators find near the surface an abundance of infaunal prey (Quammen 1984). The consumers vary in their mobility; the activity of those that are less mobile is likely to be more constrained by the duration of tidal emersion and immersion, as well as by the spatial distances that they must travel from retreats that are permanently exposed or permanently inundated.

Predation by marine consumers can vary in intensity with intertidal elevation; the effect of this variation on zonation seems to depend on the physical characteristics of the shore and the dynamics of the local water regime. Marine predation is generally thought to become more intense with declining elevation on rocky intertidal shores (Connell 1972, 1975), a generalization that holds up best on shorelines exposed to waves (Underwood and Denley 1984). The sluggish predatory invertebrates (such as predatory gastropods and seastars) that are the dominant consumers on wave-exposed rocky shores probably do prey more intensely lower on the shore and thus set lower distributional boundaries of preferred prey. These predators are sufficiently slow afoot that their prey can escape them by locating high enough on the shore to be effectively out of reach during tidal immersion. Quiet waters, however, present opportunities for more mobile predators such as certain bottom-feeding fishes (including tautogs and sheepsheads) and mobile crabs (Kitching and Ebling 1967). High-elevation refuges may no longer exist for preferred prey in these waters; rock crevices, which probably do not vary in abundance with elevation, are more likely to provide refuges for invertebrate prey in quiet waters (Lubchenco et al. 1984).

Invertebrates residing in tidal flats are subject during high tides to bottom-up consumption by both mobile consumers such as bottom-feeding fishes and swimming crabs and sluggish predatory invertebrates such as gastropods and sand-bottom seastars. Predation by mobile marine consumers is potentially greater lower on the shore, but there is little evidence that it actually sets lower limits for species distributions on tidal flats. In contrast, sluggish predatory invertebrates such as moon snails and sea stars probably do exert greater pressure lower on the shore and could set lower distributional boundaries of preferred prey.

At low tide, the major predatory threat to invertebrates living in soft sediments comes from the top-down predators, the shorebirds; their role is much greater in soft sediments than on rocky shores, where their impact seems to be inhibited by the rough, vertical topography. In addition, many of the invertebrates of soft sediments lack protective outer shells and thus represent usable prey for a wider variety of shorebirds. As the tide ebbs on sandy or muddy flats, one witnesses the arrival of several guilds of shorebirds that have evolved bills of differing length for probing the sediments for prey (Baker and Baker 1973). Long-billed shorebirds such as dowitchers, whimbrels and godwits capture large, more deeply burrowing prey, whereas short-billed shorebirds such as dunlin and sandpipers forage on prey near or on the sediment surface. Herring gulls and other gulls often behave as shorebirds in extracting infaunal prey, and they also consume thick-shelled molluscs, crabs and urchins by using their relatively thick bills and the technique of dropping prey from the air to crack them open. Oystercatchers have strong, stout bills that afford them a similar ability to prey on larger molluscs, whose shells protect them from predation by other less-well-endowed shorebirds.

Like the fishes and swimming crabs, avian predators are mobile enough to be able to cover the entire intertidal zone. Benthic invertebrates living at higher elevations are clearly exposed to longer times of risk from consumption by top-down predators than are those invertebrates lower on shore. However, recent studies show that predation rates by gulls feeding on intertidal invertebrates on rocks and mud-flats are greater lower in the intertidal zone than they are higher (Ambrose 1986; Irons, Anthony and Estes 1986).

Shorebirds are migratory; their predation seems to exert substantial pressure on soft-sediment invertebrates, but only during the part of the year when they are present. These brief, intense periods of avian predation could create a temporary zonation between the intertidal zone and shallow subtidal areas. Study of these effects will require techniques such as the floating cage developed by Millicent Quammen (1984) to separate the effects of marine and avian predation and determine whether either changes across tidal elevations and is intense enough to cause zonation. To separate effects of top-down and bottom-up predators, one might be able to cage out birds by covering the top of the enclosure, fence out marine predators by covering the sides or exclude both or neither.

Figure 10. Large pit dug in the sand by a swimming crab (the blue manna, *Portunus pelagicus*) on a tidal flat at Careening Bay in Western Australia, illustrates the sediment disturbance that can be caused by mobile marine predators that operate when the tide is in over in the intertidal zone. Swimming crabs are among the marine predators that follow the tide in and, because of their mobility, forage at all elevations of tidal flats. Ripples on the sand provide evidence of the physical disturbance of sediments that is caused by wave action. (Photograph courtesy of the author.)

Larval Settlement

An additional biological process—the settlement of larvae—is a potentially important contributor to vertical zonation along rocky intertidal shores and merits consideration here, although its role in soft-sediment communities seems to be limited.

Propagules—the seeds or larvae from which most benthic organisms grow—are transported by ocean currents. Variations in the settlement of propagules onto the sediments could influence the zonation of adult benthic plants and animals (e. g., Eckman 1983; Underwood and Denley 1984; Gaines and Roughgarden 1985). For example, if propagules were well mixed throughout the entire water column over an intertidal shore, one might predict crudely (in the absence of any biological behaviors) that settlement intensity would be proportional to the amount of time that sediments spent under water. This would clearly produce a pattern of more intense settlement lower on the shore, although the precise relationship between elevation and immersion time is nonlinear and is a complex function of local tidal dynamics. Similarly, predictions of settlement patterns could be based on particle dynamics; propagules might be expected to settle preferentially at those locations where bottom shear stresses are least and where particle deposition is most intense (Butman 1987). In marine soft sediments, this process of passive deposition might be expected to enhance larval settlement in localities (again, primarily at lower elevations) where bottom shear stress is lower and to sort invertebrate species by elevation according to the physical characteristics of their larvae.

Studies of rocky shores have shown that biological interactions can produce or modify vertical patterns in the intensity of propagule settlement. One contributing factor, implicated in the vertical zonation of barnacles on shore, is the depths at which larvae swim in the water column immediately prior to settlement (Grosberg 1982). Competition for space also can contribute to differential settlement of propagules if most available space at a given elevation is preempted by species that colonized earlier (Underwood and Denley 1984). Invertebrate larvae may also have evolved vertical preferences for settlement locations, perhaps as a consequence of natural selection by predictable gradients in the intensity of predation, physical disturbance, competition, feeding time and physiological stress.

There are several reasons to expect larval settlement to be less important in creating zonation of soft-sediment communities than it may be on rocky shores. The benthic invertebrates of sand and mud flats usually possess an ability to move as juveniles and/or adults, so that settlement does not represent an irreversible commitment to a living place. Some soft-sediment ben-

Figure 11. Migrating shorebirds such as these western sandpipers find sand and mud flats far more accessible than rocky shores, which typically are topographically complex. Shorebirds forage intensively throughout the intertidal zone, exerting substantial pressure on soft-sediment invertebrates. The invertebrates on intertidal sand and mud flats are predominantly soft-bodied and thus more vulnerable to shorebird predation than the shelled invertebrates typical of rocky shores. The periods of intense predation, however, are usually brief, coinciding with migrations, and their influence on the zonation of intertidal animals may be temporary. (Photograph by T. Leeson/Photo Researchers)

Figure 12. Common predators on tidal flats of the East Coast of North America include both marine and terrestrial species. Marine predators coming from the bottom up include those that are highly mobile—fishes and swimming crabs—as well as those that are sluggish, including gastropods such as the moon snail. The effectiveness of sluggish marine predators may be limited to lower elevations by their inability to travel the large horizontal distances covered by the tides. The "top-down" predators are those that arrive largely from the land and air as the tide recedes and include many migratory shorebirds whose bills are adapted for probing the sediments to different depths for infaunal prey. Types of predators shown here include those represented in Figure 2, with the addition of a dunlin, or short-billed shorebird, and a sluggish marine predator, the moon snail *Polinices*.

thic invertebrates even reenter the water as juveniles to select a secondary living site. Furthermore, a general lack of competition for space implies that preemption of space is unlikely to prevent larval settlement on tidal flats (Peterson 1979). Nevertheless, it is possible that on intertidal flats with high invertebrate densities, resident adults may prevent successful settlement of larvae (Woodin 1976) and thereby contribute to vertical zonation.

Zonation in Sand and Mud

The intertidal zone is universally characterized by steep vertical gradients in the intensity of environmental processes. This variation over short distances provides ecologists with a valuable opportunity to discover lessons of community organization. The lessons learned from study of rocky intertidal seashores play a prominent role in the development of theory in community ecology. Marine scientists have only recently turned their attention to study of how invertebrate communities in tidal sand and mud flats change with elevation on shore.

On tidal sand and mud flats, the changing physical environment of the intertidal zone creates a gradient in the physiological stress for resident invertebrate animals; although the direct effects of exposure to air and solar radiation are buffered by the sediments, the limits on feeding and respiration time imposed by the withdrawal of the tide may have effects that are similar on invertebrates of both soft-sediment and rocky shores. These stresses make higher elevations a more rigorous environment, suitable only for those species most tolerant of them.

Another interesting effect of the changing environment, one without a parallel on rocky shores, is produced by the variation in the average intensity of wave action as one moves up the shore; this creates a gradation in the coarseness of sediments on tidal flats, sorting the grains from coarse to fine with decreasing elevation. Sediment size makes a direct contribution to zonation, since different types of sediment provide suitable habitats for different suites of species.

Biological activities also disturb the substrate in ways that may affect patterns of zonation. Many predatory fishes and crabs, as well as deposit-feeding invertebrates, dig burrows in sand and mud; this activity, particularly the large-scale digging of burrowing shrimps, can prevent cohabitation by a variety of sedentary species. On the other hand, the presence of the burrows themselves provides a new three-dimensional habitat for commensal species. Such creation of positive associations of species in soft sediments has an analogue on rocky shores, where mussel beds create new vertical habitat above the rocks and help enhance zonation by mutualistic biological interactions.

There are also interesting similarities and differences between patterns of predation on rocky shores and those observed in sand and mud flats. Predation on wave-exposed rocky shores is most intensely provided by sluggish marine invertebrates, limited in effectiveness to the lower slope by their lack of mobility and need for water coverage. Sandy and muddy shores, in contrast, are frequently utilized by swimming crabs, fishes and shore

Figure 13. Predator-exclusion cages such as these from Mugu Lagoon in Southern California make it possible to compare the effects of different types of predators on invertebrates buried in the sediments. Vertical fencing around the sides of a cage can keep out marine predators such as gastropods, whereas covering the top can exclude fishes and shorebirds. Cages with partial tops are intended to allow normal access by mobile marine consumers, such as crabs and fishes, while still including sufficient roofing material to provide a test of any artifacts of roofs (unintended effects of experimental intervention that need to be separated from the true effects of predator exclusion). Shown here are a full cage, a partial cage used as a control for cage structures, a fence and an uncaged area to control for fences. (Photograph courtesy of the author.)

birds—marine and terrestrial predators that tend to be highly mobile. Such mobility implies that available feeding times, as dictated by tides, will contribute less to creating patterns of changing predation intensity on soft-bottom intertidal shores. Still, the limited evidence available suggests that predation rates for both classes of predators may be more intense lower on shore, and may contribute to creating zonation of benthic invertebrates.

Variations in the settlement of larvae with elevation are likely in sand and mud as well as on intertidal rocks; however, because soft-bottom invertebrates are more mobile, larval settlement is likely to play a lesser role in creating vertical zonation patterns in soft substrates.

Perhaps the most significant contrast in community organization between rocky and soft shores lies in the relative importance of interspecific competition. Intense interference competition for space on many rocky intertidal shores produces sharp boundaries between competing sessile species and discrete bands of virtual monocultures.

Competitive exclusion is, by comparison, rare on sedimentary bottoms, and the competition for food that is common has more subtle effects. As a result, zonation is more gradual on sandy and muddy shores; it is, in general, less influenced by the indirect effects of physical environment on biological processes such as competition and predation and more strongly dictated by the direct effects of changes in the physical environment.

Dedication

This article is dedicated to the memory of Millicent Quammen, who inspired this work and who fell victim to breast cancer in 1990.

Bibliography

Ambrose, W. G., Jr. 1986. Estimate of removal rate of *Nereis virens* (Polychaeta: Nereidae) from an intertidal mudflat by gulls (*Larus* spp.). *Marine Biology* 90:243–247.

Baker, M. C., and E. A. M. Baker. 1973. Niche relationships among six species of shorebirds on their wintering and breeding grounds. *Ecological Monographs* 43:193–212.

Black, R,, and C. H. Peterson. 1988. Absence of preemption and interference competition for space between large suspension-feeding bivalves and smaller infaunal macroinvertebrates. *Journal of Experimental Marine Biology and Ecology* 120:183–198.

Brenchley, G. A. 1981. Disturbance and community structure: an experimental study of bioturbation in marine soft-bottom environments. *Journal of Marine Research* 39:767–790.

Butman, C. A. 1987. Larval settlement of soft-sediment invertebrates: the spatial scales of pattern explained by active habitat selection and the emerging role of hydrodynamical processes. *Oceanography and Marine Biology: Annual Reviews* 25:113–165.

Connell, J. H. 1961. Effects of competition, predation by *Thais lapillus*, and other factors on natural populations of the barnacle *Balanus balanoides*. *Ecological Monographs* 31:61–104.

Connell, J. H. 1972. Community interactions on marine rocky intertidal shores. *Annual Review of Ecology and Systematics* 3:169–192.

Connell, J. H. 1975. Some mechanisms producing structure in natural communities: a model and some evidence from field experiments. In *Ecology and Evolution of Communities*, ed. M. L. Cody and J. M. Diamond, 460–490. Cambridge, MA: Belknap Press.

Dayton, P. K. 1971. Competition, disturbance and community organization: the provision and subsequent utilization of space in a rocky intertidal community. *Ecological Monographs* 41:351–389.

Eckman, J. E. 1983. Hydrodynamic processes affecting benthic recruitment. *Limnology and Oceanography* 28:241–257.

Gaines, S., and J. Roughgarden. 1985. Larval settlement rate: a leading determinant of structure in an ecological community of the marine intertidal zone. *Proceedings of the National Academy of Sciences* 82:3707–3711.

Grant, J. 1983. The relative magnitude of biological and physical sediment reworking in an intertidal community. *Journal of Marine Research* 41:673–689.

Grosberg, R. K. 1982. Intertidal zonation of barnacles: the influence of planktonic zonation of larvae on vertical distribution of adults. *Ecology* 63:894–899.

Irons, D. B., R. G. Anthony and J. A. Estes. 1986. Foraging strategies of glaucous-winged gulls in a rocky intertidal community. *Ecology* 67:1460–1474.

Jackson, J. B. C. 1977. Competition on marine hard substrata: the adaptive significance of solitary and colonial strategies. *American Naturalist* 111:743–767.

Johnson, R. G. 1970. Variation in diversity within marine benthic communities. *American Naturalist* 104:285–300.

Kitching, J. A., and F. J. Ebling. 1967. Ecological studies at Lough Inc. *Advances in Ecological Research* 4:198–291.

Kneib, R. T. 1984. Patterns of invertebrate distribution and abundance in the intertidal salt marsh: causes and questions. *Estuaries* 7:392-412.

Kneib, R. T. 1985. Predation and disturbance by grass shrimp, *Palaemonetes pugio* Holthuis, in soft-substratum benthic invertebrate assemblages. *Journal of Experimental Marine Biology and Ecology* 93:91–102.

Levinton, J. S. 1972. Stability and trophic structure in deposit-feeding and suspension-feeding communities. *American Naturalist* 106:472–486.

Lubchenco, J., B. A. Menge, S. D. Garrity, P. J. Lubchenco, L. R. Ashkenas, S. D. Gaines, R. Emlet, J. Lucas and S. Strauss. 1984. Structure, persistence, and role of consumers in a tropical intertidal community (Taboguilla Island, Bay of Panama). *Journal of Experimental Marine Biology and Ecology* 78:23–73.

Menge, B. A., and J. P. Sutherland. 1976. Species diversity gradients: synthesis of the roles of predation, competition, and environmental stability. *American Naturalist* 110:351–369.

Newell, R. C. 1976. Adaptations to intertidal life. In *Adaptations to Environment: Essays on the Physiology of Marine Animals*, ed. R. C. Newell, 1–82. Sydney: Butterworth.

Paine, R. T. 1966. Food web complexity and species diversity. *American Naturalist* 100: 65-75.

Peterson, C. H.1979. Predation, competitive exclusion, and diversity in the soft–sediment benthic communities of estuaries and lagoons. In *Ecological Processes in Coastal and Marine Systems*, ed. R. J. Livingston, 233–264. New York: Plenum.

Peterson, C. H. 1982. The importance of predation and intra- and interspecific competition in the population biology of two infaunal suspension-feeding bivalves, *Protothaca staminea* and *Chione undatella*. *Ecological Monographs* 52:437–475.

Peterson, C. H., and R. Black. 1987. Resource depletion by active suspension feeders on tidal flats: influence of local density and .tidal elevation. *Limnology and Oceanography* 32:143-166.

Peterson, C. H., and R. Black. 1988. Responses of growth to elevation fail to explain vertical zonation of suspension-feeding bivalves on a tidal flat. *Oecologia* 76:423–429.

Posey, M. H. 1986. Changes in a benthic community associated with dense beds of a burrowing deposit feeder, *Callianassa californiensis*. *Marine Ecology Progress Series* 31:15–22.

Quammen, M. L. 1984. Predation by shorebirds, fish, and crabs on invertebrates on intertidal mudflats: an experimental test. *Ecology* 65:529–537.

Reise, K. 1985. *Tidal Flat Ecology*. Berlin: Springer-Verlag.

Remane, A., and C. Schlieper. 1971. *Biology of Brackish Water.* New York: Wiley Interscience.

Rhoads, D. C., and D. K. Young. 1970. The influence of deposit–feeding organisms on sediment stability and community trophic structure. *Journal of Marine Research* 28:143–161.

Sousa, W. P. 1979. Disturbance in marine intertidal boulder fields: the non-equilibrium maintenance of species diversity. *Ecology* 60:1225–1239.

Underwood, A. J., and E. J. Denley. 1984. Paradigms, explanations and generalizations in models for the structure of intertidal communities on rocky shores. In *Ecological Communities: Conceptual Issues and the Evidence*, ed. D. Simberloff et al., 151–180. Princeton, NJ: Princeton University Press.

Virnstein, R. W. 1977. The importance of predation by crabs and fishes on benthic infauna in Chesapeake Bay. *Ecology* 58:1199–1217.

Wethey, D. S. 1984. Sun and shade mediate competition in the barnacles *Chthamalus* and *Semibalanus:* a field experiment. *Biological Bulletin* 167:176–185.

Wilson, W. H., Jr. 1980. A laboratory investigation of the effect of a terebellid polychaete on the surivorship of nereid polychaete larvae. *Journal of Experimental Marine Biology and Ecology* 46:73–80.

Witman, J. D. 1987. Subtidal coexistence: storms, grazing, mutualism and the zonation of kelps and mussels. *Ecological Monographs* 57:167–187.

Woodin, S. A. 1976. Adult–larval interactions in dense infaunal assemblages: patterns of abundance. *Journal of Marine Research* 34:25–41.

Nonequilibrium Determinants of Biological Community Structure

Biological communities are always recovering from the last disturbance. Disturbance and heterogeneity, not equilibrium, generate biodiversity

Seth R. Reice

Not far from my home in the Piedmont region of North Carolina, the diversity of life is evident in New Hope Creek. More than 160 macroinvertebrates, most of them insects, thrive in the stream. Nearby Botany Pond, however, has only 15 species of macroinvertebrates. This pattern is far from unique; in fact, it is repeated among clean streams and ponds worldwide. Neighboring biological systems often include members of overlapping families and genera, but they tend to have distinctly different community structures.

This curious finding provokes questions: Why should neighboring systems be so dissimilar? What determines the differences in structure among similar communities?

The answers to these questions are at the heart of a long-standing debate in ecology—a debate that recently has taken a new turn. The model that has guided policies on ecosystem management and biodiversity issues may have ignored important aspects of community dynamics. Human efforts to stabilize ecosystems have resulted in the loss of biodiversity, when the opposite result was the goal. It is time to look at the response of biological communities to disturbance in a different way.

Historically, the question of what determines community structure has been approached from two directions. One group of theorists has supported environmental processes as determinants of community structure. In the view of H. Gleason, writing in the 1920s, and H. G. Andrewartha and L. C. Birch, in a 1954 book, a community's composition is determined by its environment, which sets the range of possibilities for colonization, reproduction, growth and survival. The presence or absence of a given taxonomic group results from random colonization processes and variability in the environment.

This view has not, however, been dominant in this century. Another paradigm, equilibrium theory, has dominated both ecological thought and public policy. This view has its origins in the writings of Charles Elton, who in 1927 emphasized the role of interspecific (between-species) interactions to explain why only a subset of the species that could colonize an area actually coexist there. His theory of limited membership argued that biotic interactions—competition and predation—determine the community structure we observe.

Equilibrium theory asserts that systems are at equilibrium—in a steady state, with overall species composition and relative abundances stable through time—as a result of biotic interactions among its members. Such systems return to their original structure after perturbation.

Equilibrium models presume a constant environment, excluding the significance of disturbance or any other environmental fluctuations. In such a worldview, it is a simple step to conclude that biotic interactions are the key determinants of community structure. R. H. MacArthur and E. O. Wilson's *The Theory of Island Biogeography* epitomizes this view. Under equilibrium conditions, the community is the direct result of the competitive and predator-prey relationships among species. Predator-mediated coexistence and spatial heterogeneity are important in maintaining high species diversity, but they still fall within the equilibrium.

The predictions of interspecific-interaction models, however, often do not appear to hold in nature. Indications of the shortcomings of equilibrium theories include the coexistence of quite similar species and the frequent lack of demonstrable effects both of suspected competitors on each other and of predators on their prey distributions. Such contradictions have been identified in a variety of ecosystems: Peter Sales's study of coral reef fishes, Joe Connell's work on tropical rain forest trees and coral reefs, Bob Peet's examinations of chalk grasslands and pine savannas, and my own work on benthic invertebrates in streams have all led to a new view of community dynamics.

Seth R. Reice is an associate professor of biology and chairman of the curriculum in ecology at the University of North Carolina at Chapel Hill. Growing up in New York City in no way impeded his development as an ecologist; he soon received his Ph.D. in zoology from Michigan State University. His particular area of research interest is in disturbance ecology as applied to stream communities and other systems. He also is interested in conservation and sustainable development, particularly in Latin America. Address: CB# 3275, Wilson Hall, University of North Carolina at Chapel Hill, Chapel Hill, NC 27599-3275.

Figure 1. Nonequilibrium theories of biological community structure counter the long-held notion that equilibrium in an ecosystem maximizes biodiversity. Instead, such theories emphasize environmental disturbance and spatial heterogeneity as factors that encourage colonization and species diversity. Here, in Yellowstone National Park, a surge of botanical diversity is obvious only months after the famous fires of 1988. (Photograph by Renee Lynn, Photo Researchers Inc.)

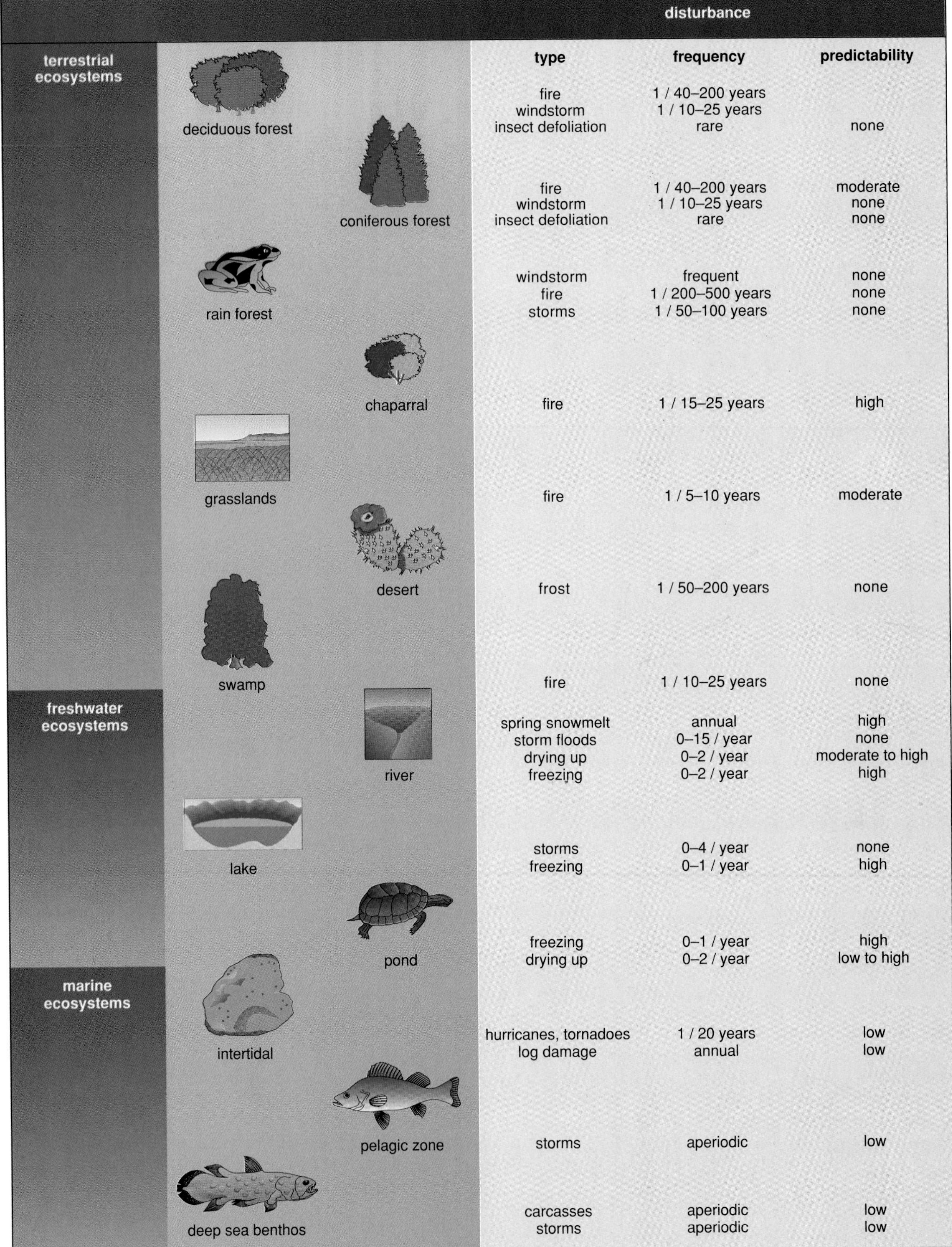

		disturbance		
		type	frequency	predictability
terrestrial ecosystems	deciduous forest	fire windstorm insect defoliation	1 / 40–200 years 1 / 10–25 years rare	none
	coniferous forest	fire windstorm insect defoliation	1 / 40–200 years 1 / 10–25 years rare	moderate none none
	rain forest	windstorm fire storms	frequent 1 / 200–500 years 1 / 50–100 years	none none none
	chaparral	fire	1 / 15–25 years	high
	grasslands	fire	1 / 5–10 years	moderate
	desert	frost	1 / 50–200 years	none
	swamp	fire	1 / 10–25 years	none
freshwater ecosystems	river	spring snowmelt storm floods drying up freezing	annual 0–15 / year 0–2 / year 0–2 / year	high none moderate to high high
	lake	storms freezing	0–4 / year 0–1 / year	none high
	pond	freezing drying up	0–1 / year 0–2 / year	high low to high
marine ecosystems	intertidal	hurricanes, tornadoes log damage	1 / 20 years annual	low low
	pelagic zone	storms	aperiodic	low
	deep sea benthos	carcasses storms	aperiodic aperiodic	low low

Figure 2. Frequency and predictability of disturbances affect characteristics of the biological community. For example, the number of years between disturbances is roughly proportionate to, and nearly always shorter than, the life span of the dominant species in the system. Biota adapt readily to predictable disturbances, such as spring snowmelt, but many more individuals are lost when disturbance is erratic.

To explain these apparent anomalies, nonequilibrium theories of community structure have recently been proposed. These include concepts such as disturbance theory, patch dynamics and supply-side ecology, which I shall discuss in detail later. In essence, nonequilibrium theories attribute the high diversity of species and the coexistence of similar species that we observe to processes of disturbance and recruitment.

I shall argue that community structure is primarily determined in a nonequilibrium fashion by the interactions of the heterogeneity of the physical-chemical environment, disturbance and recruitment. Homogeneity in natural systems is clearly illusory. Although I acknowledge that biotic interactions are certainly important in communities that are at or near equilibrium, I contend that equilibrium is an unusual state for natural ecosystems. In systems where the interval between disturbances is long relative to the generation time of the dominant species, biotic interactions may well have an important role. However, I suggest that the normal state of communities and ecosystems is to be recovering from the last disturbance. Natural systems are so frequently disturbed that equilibrium is rarely achieved.

Furthermore, it appears that disturbance not only creates patches (the environmental mosaic) but also is a product of the patches themselves. The underlying physical-chemical environment determines the frequency and magnitude of a disturbance, influences the time course of recovery and ultimately determines the community structure. The interaction between disturbance and patchiness in virtually all systems is the underlying basis for the control of community structure. Stochastic (random) recruitment in a heterogeneously disturbed, patchy environment results in high overall species diversity. These interdependent variables are the nonequilibrium determinants of community structure.

All Environments Are Heterogeneous

All environments in all ecosystems are patchy in space and variable in time. The environmental patchiness is based on physical and chemical gradients that are ubiquitous. In terrestrial ecosystems we encounter variations in elevation, slope, temperature and moisture, and a host of chemical gradients in the soil. In aquatic ecosystems we encounter variations in depth, temperature, current velocity, substrate particle size, salinity, pH, dissolved oxygen, nutrients and heavy metals. The limnologist and longtime *American Scientist* columnist G. Evelyn Hutchinson argued that even the pelagic (open-water) zones of lakes, which superficially appear homogeneously mixed, are patchy systems.

In spite of this obvious spatial heterogeneity and complexity, many ecological modelers assume the existence of a homogeneous environment. We would all agree that a superior experimental study manipulates or controls only a few environmental variables out of the thousands that are the parameters of the niche of every species. It is not surprising, then, that we tend to minimize the effects of environmental heterogeneity on community structure. We are simply overwhelmed by the complexity of the environment.

Spatial heterogeneity is frequently invoked as a mechanism for *maintaining* diversity. A heterogeneous environment allows species to subdivide resources. Thus the problem of diversity becomes the traditional issue of limiting similarity (dating back to Elton). In 1984, A. Shmida and Stephen Ellner (of North Carolina State University) modeled the coexistence of equivalent species in a spatially heterogeneous environment. They showed that dispersal from optimal microsites can support marginal populations in suboptimal microsites. This mechanism is important in all patchy environments whenever disturbances occur.

Ecologists acknowledge the importance of temporal variability. Tropical scientists, for example, are well aware of the seasonality of rainfall. Arctic and near-arctic ecologists, on the other hand, emphasize seasonal fluctuations in temperature. Temperate-zone ecologists deal with both. Even so, it is rare that temporal variability is seen as having an important role in structuring the community. Shmida and Ellner argue that temporal environmental variation can permit the coexistence of trophically equivalent species. Under one set of environmental conditions, one competitor is more fit; when things change, another species is favored. Coexistence results from fluctuations in the environment. This mechanism also has its greatest impact in disturbed systems. Fluctuations in the environment generate what John Weins in 1977 called "environmental crunches," which affect the relative importance of competition in communities through time. When resources are scarce, competition becomes much more important.

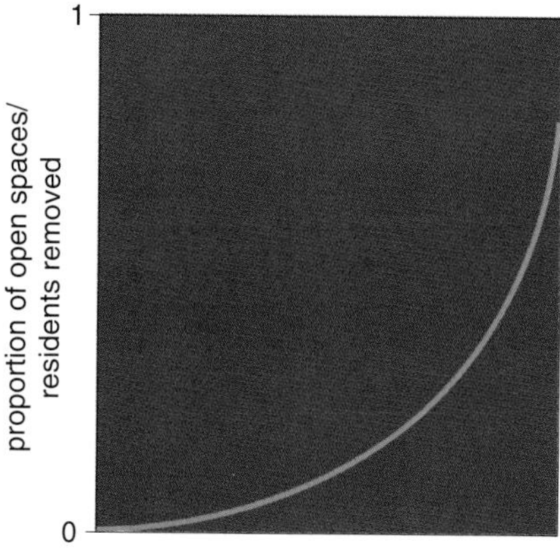

Figure 3. Magnitude of environmental disturbance determines the proportion of both open spaces created and individuals lost, creating opportunities for recolonization and greater diversity.

Disturbance versus Equilibrium

Ecologists have recently begun to understand the implications of disturbance for community structure. The theoretical basis for the role of disturbance in community organization has been articulated by many ecologists. I shall use the definition of Steward Pickett and Peter White, who wrote: "A disturbance is any relatively discrete event in time that disrupts ecosystem, community or population structure, and changes resources, availability of substratum, or the physical environment." This description defines disturbance as a physical event, not as the biotic outcome of the event. I view a disturbance as a physical force such as a fire, flood or tornado that damages natural systems and removes organisms. The first impact of a disturbance is always to remove organisms.

Disturbance is scale dependent. If the area studied is large enough or the period of observation long enough, all disturbances are predictable and "normal." At smaller scales or shorter durations, disturbances appear completely random. Vincent Resh and his colleagues have argued that if a disturbance is predictable, the biota can and will adapt to it; a disturbance that is unpredictable will have a greater impact. The *absence* of predictable disturbances (for example, spring snowmelt floods or the arrival of snow in the winter) often has a greater impact than their presence.

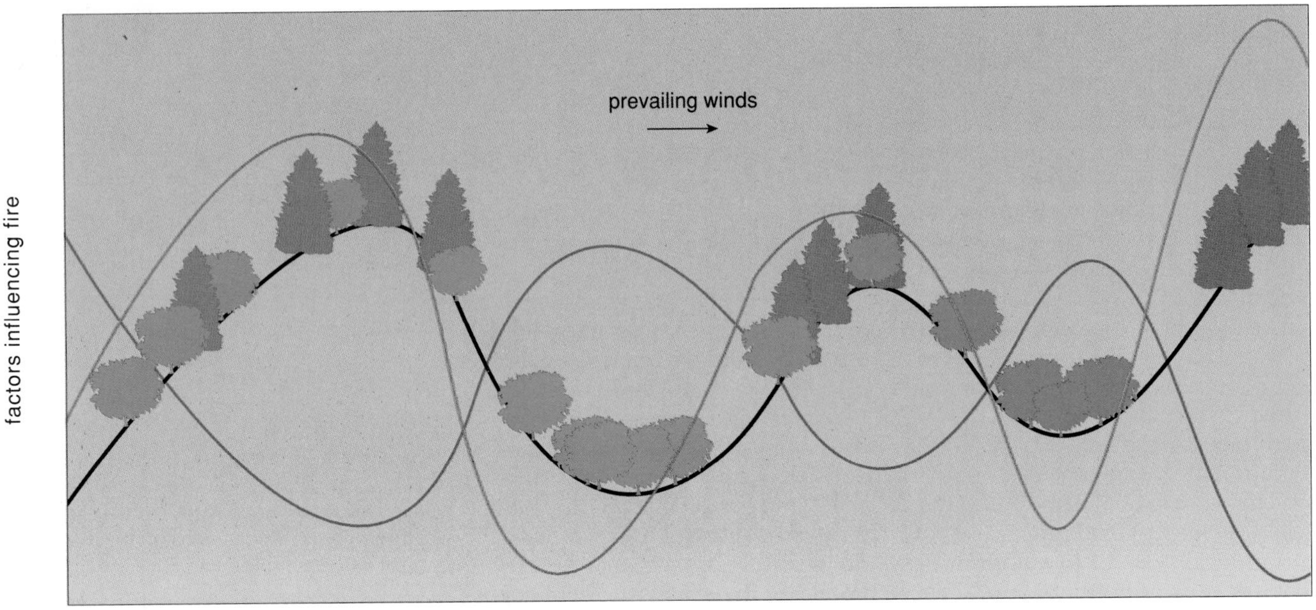

Figure 4. Susceptibility of a forest system to fire (*red*) is determined by the characteristics of the system itself. Its altitude, orientation to prevailing winds, the moisture content of the wood (*blue*) and the abundance of the fuel all affect the ability of fire to take hold and spread. Thus the system may influence both the likelihood and the severity of disturbance.

Contrast disturbance with predation. Predation is an important agent for the removal of prey individuals. It is not, however, a disturbance. Predation is intrinsic to the life of the prey species, which can and does adapt to it.

Enrichment events often constitute disturbances, although they appear not to fit the definition presented above. Fertilization often leads to the loss of species diversity by favoring particular species' growth and reproduction at the expense of others. The enrichment event can be viewed as the indirect cause of the disturbance. For example, a dam burst on a farm manure pond near Raleigh, North Carolina, and flooded the Neuse River with a million gallons of liquid pig manure. This material had a high concentration of organic matter, with high biochemical oxygen demand (BOD). The influx of the manure slurry was actually the indirect cause of the disturbance. The disturbance was the loss of dissolved oxygen, which killed thousands of fish.

The enrichment of lakes and streams that results from ordinary agricultural runoff is one step further removed from the disturbance. The nutrients in the runoff stimulate algal blooms, often shifting the community composition from green to blue-green algae. The BOD from the decay of the algae can alter community structure by selectively killing off organisms that have greater needs for dissolved oxygen. Again, it is the loss of dissolved oxygen that is the actual disturbance.

Disturbance typically is not a population process; it causes restructuring of the whole ecosystem (for example, following a fire or flood), wherein the habitat is altered or rearranged and, as a result, organisms are removed. Disturbance is a reinitializing step in the successional development of a community. Different disturbances remove different species to different degrees. Therefore, the particular response of the community is likely to be a unique successional sequence.

It is important to understand that community structure is made up of both species richness (the number of species) and evenness (the distribution of individuals among the species). (We ecologists combine these parameters into diversity indices, such as the Shannon-Weaver Diversity Index.) The effects of disturbance on richness and evenness can be quite different. In my experiments on disturbance in stream benthic communities (bottom dwellers), for example, species richness was rarely affected, because whole species were rarely eliminated or added. There are often, however, major shifts in relative dominance, which can dramatically affect evenness.

The Intermediate Disturbance Hypothesis (IDH) was presented by Joe Connell in 1978 to explain the high species diversity in tropical rain forests and coral reefs. The IDH presumes a competitive hierarchy of species. When disturbances are infrequent or small, superior competitors eliminate inferior competitors, reducing the species richness of the system. The model also presumes that the superior competitors are efficient occupiers of space (they are resident species).

Under a frequent or intense disturbance regime, by contrast, the dominant competitors are reduced or eliminated, and colonizing species (assumed to be inferior competitors) dominate the system. The absence of the resident species reduces species richness. Finally, under a disturbance regime that is intermediate in frequency, magnitude and intensity, some resident species persist in the system along with colonizing species, which exploit the disturbed areas. Thus intermediate disturbance leads to maximum species richness.

Michael Huston's 1979 "dynamic equilibrium" model showed that community structure can result from the trade-offs between population growth rates, rates of competitive exclusion and the frequency of population reductions. Huston demonstrated that if disturbance recurred at intervals shorter than the time necessary for competitive exclusion to take place, poorer competitors would persist in the system, increasing species richness. He wrote: "Diversity is determined not so much by the relative competitive abilities of the competing species as by the influence of the environment on the net outcome of their interactions."

It is my view that disturbance is the major source of these population reductions and is the key factor in determining community structure. Many systems are disturbed frequently enough to prevent equilibrium conditions from ever being established. And in a heterogeneous environment, when disturbance routinely recurs more quickly than equilibrium can be attained, high diversity is the result.

Nonequilibrium dynamics can best explain the temporal and spatial patchiness that characterizes nearly all communities. Disturbances are common and important in virtually all ecosystems (Figure 2). In terrestrial ecosystems, fire and windstorms are most important. Neither is predictable, and both can destroy plants and animals. Nevertheless, many terrestrial species are adapted to their disturbance regimes. In fact, many species require fire to flower or set seed. In chaparral vegetation, Norm Christensen and Cornelius Muller have shown that life histories are linked to the periodic disturbance of fire. As debris builds up during nonburn periods, the susceptibility to fire increases. The reproduction of these plants is keyed to fire, as the seeds will not germinate without it. This is true for serotinous pines, such as jack pine (*Pinus banksiana*) and pond pine (*Pinus serotina*). In the Australian bush, eucalyptus forests are fire adapted and fire maintained. This is not limited to a particular species but applies to the dominant genus of the entire continent. In Tasmania, Jackson showed that fire sets the boundaries and controls succession from rain forests to eucalyptus or moist sclerophyll (hard-foliage) forests. On poor soils, fire frequencies of more than once every 300 years will result in the transition from rain forest to moist sclerophyll forest. With fire frequencies of less than 1 per 500 years, the transition is to a diverse rain-forest system. This suggests that 1 per 500 years may be an intermediate disturbance frequency for these trees.

The role of disturbance in stream communities has been analyzed by myself, Resh and his colleagues, and J. David Yount and Gerald Niemi of the University of Minnesota. Benthic invertebrates in flowing waters that flood frequently have morphological features that allow them to cling to the substrate (for example, mayfly nymphs have claws). Others live in the boundary layer and filter above it (for example, black flies [Simuliidae] have collecting fans that project into the flow so their bodies are protected from the full force of the current). Simuliidae are excellent colonizers of newly opened space. Adaptations of species to disturbance regimes are common in many other systems.

Figure 5. Photographs of the same location in Yellowstone National Park taken immediately after the 1988 fire (*top*, November 4, 1988) and the next summer (*above*, July 26, 1989) show how disturbances are patchy in their distribution and how quickly systems respond to disturbance with a variety of new life. (Photographs by Jim Peaco, Yellowstone National Park.)

Not only are all ecosystem types disturbed, but most also are disturbed frequently relative to the life history of the dominant species. The return interval of disturbance is roughly proportionate to the life span of the dominant organisms (the space holders) in different systems (Figure 2). Forest trees can live hundreds of years, and the frequency of fire or windstorm disturbances is much shorter than that. Stream insects live up to four years, and fishes even longer. The flood cycle is typically annual or shorter. The return interval of disturbance may be a critical determinant of life-history characteristics such as the

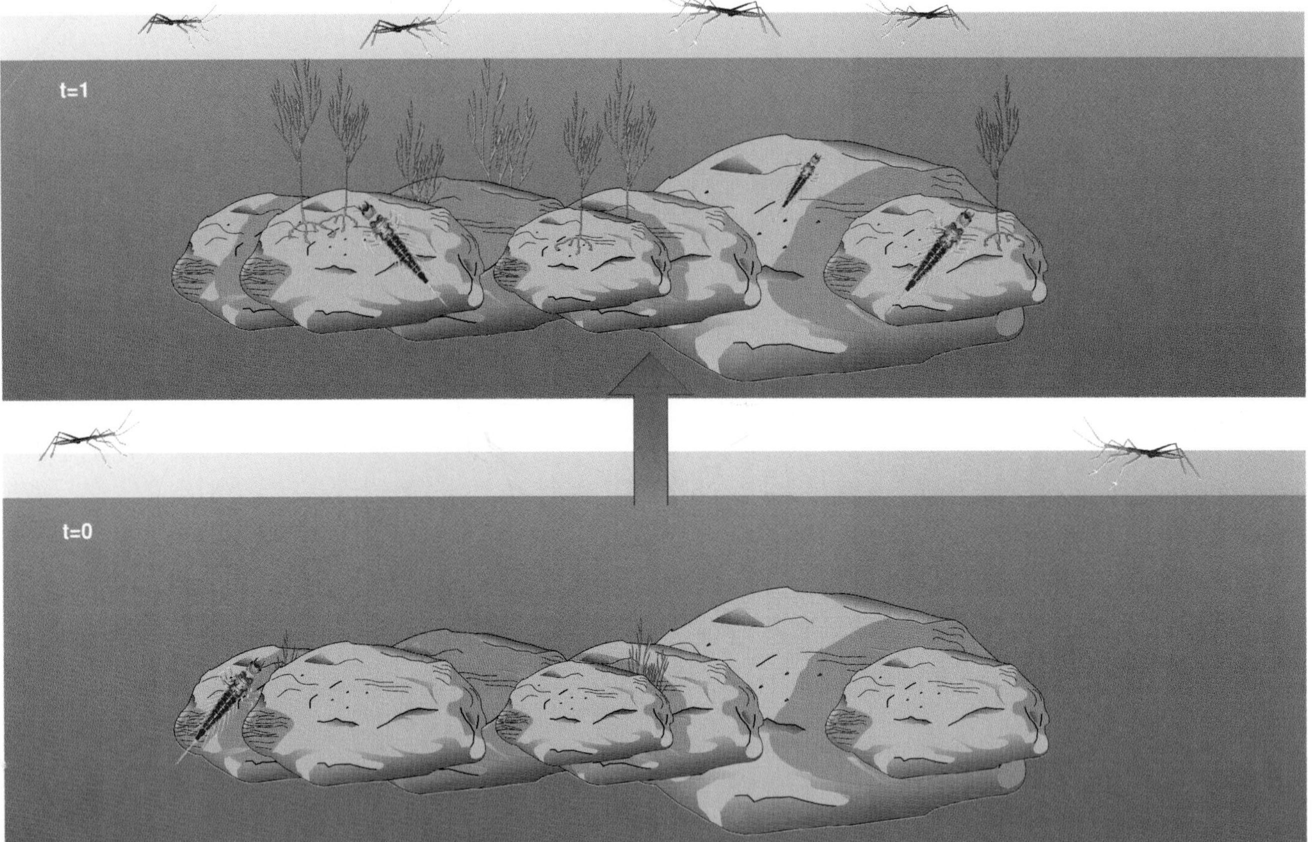

Figure 6. Recolonization takes place by only three mechanisms. Biota (typically plants) may regrow from surviving individuals; organisms that walk or crawl may migrate from adjacent patches; or individuals that swim, float or fly may be recruited from outside the proximate system.

generation time and the intrinsic rate of natural increase. The common condition for most communities is to be recovering from the last disturbance. Only when the return interval for the next disturbance is long, relative to the life span of the resident species, can something approaching equilibrium be attained. I argue that this nearly never happens.

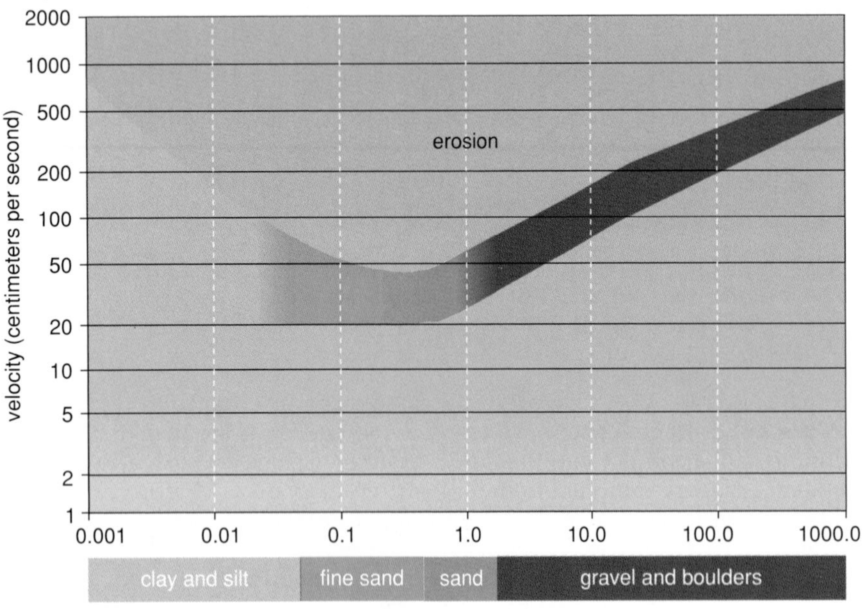

Figure 7. Grain size influences the water velocity at which material in a streambed begins to move. Adhesive properties cause the smallest particles, clays and silts (*left*), to respond at a wide range of threshold velocities. Thus consolidated clays and silts have as much resistance to disruption by moving water as boulders, whereas unconsolidated small particles move as readily as sand grains.

Disturbance Means Loss of Individuals

The initial impact of a physical disturbance in a community is to remove individuals, who may be either killed or displaced. Disturbance is the major mechanism of density-independent population reductions. The removal of individuals creates opportunities for new species to colonize, changing the community structure. Colonization by individuals of pre-existing species can transform the age structure of populations and change their interactions with the community. Whether fire or flood, the result of disturbance is that individuals and populations are lost from the system. As a result of the losses, new physical and niche space is created. The creation of opportunities for recolonization is the vehicle for generating high species diversity. Greater magnitude disturbances remove more individuals and create more open space (Figure 3).

Three Mechanisms of Recolonization

There are only three mechanisms by which a species can recolonize newly available niche space after disturbance: regrowth, migration or recruitment. Regrowth is repopulation by survivors of the disturbance. Trees can regrow from stump sprouts, and surviving coral polyps can regenerate an entire reef. Migration is recolonization from adjacent patches within the system. Mayflies can crawl from one stone to another, and crabgrass can spread into tilled gardens. Recruitment is recolonization by individuals from outside the proximate system. This includes windblown seeds in a forest, or egg laying or drift by stream insects. The germination of seeds from a seed pool is viewed as recruitment.

The relative importance of these three mechanisms varies widely among systems and taxa. Some systems, such as the rocky intertidal ecosystem, are dominated by recruitment. Migration is less important in others (temperate forest trees, for example), where regrowth and recruitment have important roles. Regrowth in sexually reproducing animals is rare, but migration is common. Regrowth is strictly fixed: Only survivors of disturbances can regrow, and they replace themselves.

Migration blends random and fixed processes. Immigrants are a random sample but only of the residents of the neighboring patches (a limited subset of potential colonists). Migration results from crowding. In nature recruitment is largely random: Open space is colonized by a random sample of potential occupants.

Disturbance *per se* does not determine diversity. It creates opportunities for colonization of vacated spaces by new species. If regrowth dominates, there is no change. Disturbance leads to change in community structure only if there is an ample supply of recruits to the community.

Patchiness and Disturbance

Most of the literature focuses on disturbance as the agent causing or maintaining the mosaic of species distribution. A tornado or fire destroys one patch of trees but not another. A flood scours one riffle, opening some patches, and transports other animals into it. It is easy to see that disturbance is a source of environmental patchiness. A complementary perspective is that the patches influence the disturbances. Patches can mitigate or intensify a disturbance event. Various patches have differential susceptibility to a unit disturbance (one of uniform strength or magnitude across the range of patches). The result of the disturbance is to enhance the patchiness, in that one patch sustains far greater population losses than another. As a result, different patches will be differentially open to recolonization. Two examples will illustrate this idea.

Fire susceptibility in forests depends on many variables, but the two main factors are fuel quality and fuel quantity. Soil moisture influences fuel *quality*, because wet wood does not burn well. Furthermore, decomposition on the forest floor is more rapid where the soil is wetter. So soil moisture affects the fuel *quantity* as well. Wind is critical, too. It can create fire danger by blowing down trees and drying the wood, and it can also help spread flames. Fire vulnerability is greatest on the mountaintops and lowest on the moist valley floors. The risk of fire is not uniform across the forest (Figure 4). In the Yellowstone National Park fire of 1988, whole forest stands were burned to the ground while others were barely singed. Mark Harman, Susan Bratton and Peter White showed that fire frequency in the Smoky Mountains is dependent on elevation and soil moisture. Variability in distribution and intensity of burns is common in all terrestrial systems.

Stream beds are patchy in their responses to floods. Sedimentologists, such as R. W. Newbury, have shown that particles of different sizes begin to move at various ranges of critical stream velocities. In Figure 7 we see that sand moves at lower velocities than all other particle sizes. Boulders (because of their mass) and clays (because of their adhesive properties) are hardest to mobilize. The higher the critical velocity for a particle, the less frequently will that velocity be exceeded. Sediments in streams are distributed in a patchy fashion. Patches of different size sediments are disturbed at different frequencies, based on their critical velocities (Figure 8). Sands are disturbed most frequently; they move even in small spates. But only major storms generate the velocities necessary to disturb the clays or boulders. Shifting sediments can be a significant disturbance to the benthic community. Invertebrates, fish eggs and algae can be dislodged, buried or crushed, resulting in injury or death. Consequently, space is opened up, creating opportunities for recolonization. So the frequency of disturbance to the biota is a result of the physical patchiness of the environment.

The relation between sediment particle size, frequency of disturbance and community structure has not yet been thoroughly investigated. The resilience of community structure in a stream with a particular sediment type should

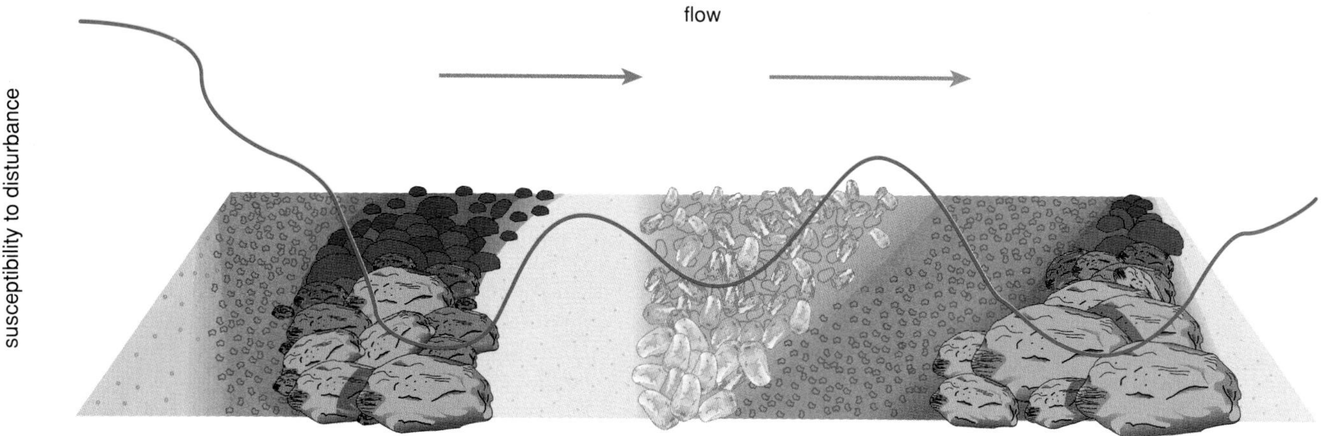

Figure 8. Stream bottoms are rarely composed of only one grain size. More typically they contain grains ranging from clay to boulders, materials that have great differences in susceptibility to disturbance. In flood conditions, biota located in sandy areas are much more likely to be disturbed than those that inhabit boulders.

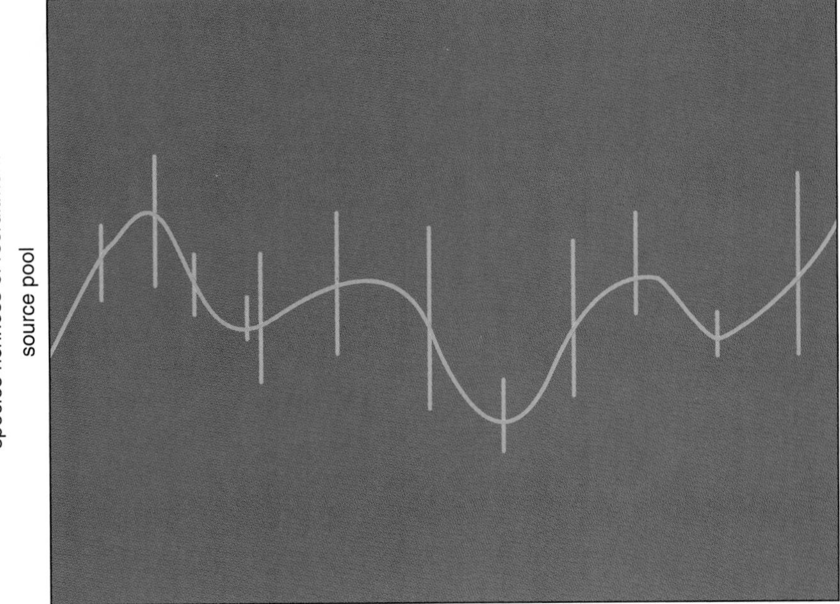

Figure 9. Richness of the recruitment pool—the individuals available from outside to replenish a system depleted by a disturbance—varies significantly in time and space. Thus the conditions in neighboring or even remote systems may influence the diversity of recruitment after a disturbance in another system.

be directly related to the frequency of disturbance in that sediment type. A higher frequency of disturbance should produce communities with higher rates of recovery. My students and I imposed unit tumbling disturbances on patches of cobbles and patches of sand mixed with gravel. We assessed sediment-specific recovery in terms of changes in species richness and population sizes in disturbed and undisturbed patches of both sediment sizes. Following a disturbance, species richness was reduced 24 percent in cobbles, and 40 percent in sand/gravel. Total animal densities were reduced 48 percent in cobbles and 79 percent in sand/gravel. Within two weeks these animal densities rebounded 240 percent in cobbles and 510 percent in sand/gravel. These data show that, in response to a unit disturbance, there are dramatic differences in both the space available and the recolonization between two distinct stream sediment patches.

These examples demonstrate that a given disturbance has different impacts on various patches in the same system. Whether the disturbance is a fire or a flood, each distinct patch type in each system has a unique response. This results in differential loss rates of organisms from the different patch types, intensifying the patchiness of the system. The result of the disturbance is to create a mosaic of open spaces in the system and a range of opportunities for recolonization. Such varied disturbance regimes have direct consequences for the structure of biotic communities in all ecosystems.

Recruitment Varies in Space and Time

In streams colonists drift, crawl or swim into available places. Egg laying supplies recruits in lakes, streams and terrestrial systems. In marine systems (the rocky intertidal zone, for example) pelagic larvae settle. In terrestrial plant systems the recruits are seeds that are either already in the seed bank or are part of the seed "rain." Plant ecologists have long focused on the importance of dispersal and recruitment to plant community structure.

Recruitment has a major impact on the outcome of competition and, consequently, on community structure. This idea, dubbed supply-side ecology by R. Lewin, asserts that physical transport processes are important determinants of community structure because they set the supply of colonists to habitat patches. Steven Gaines and J. Roughgarden stated that, "Settlement in the high intertidal community … appears to play as important a role as postsettlement processes such as predation and competition." They also show that a tacit assumption of high settlement rate underlies the Intermediate Disturbance Hypothesis. They suggest that without high recruitment, diversity may decrease with disturbance.

Shmida and Ellner showed how recruitment promotes the coexistence of even very similar species. The abundance of recruits reaching a site can offset the effect of competition. As individuals of the poorer competitor are killed, they are replaced by more individuals of that species. With a big supply of recruits of the weaker competitor, competitively dominant and inferior species can coexist indefinitely.

What determines the identity and variety of the new recruits? At any point in space and time, the pool of potential recruits is highly variable (Figure 9). This is a venerable concept, illustrated by Louis Pasteur's experiment in which he exposed a series of identical sterile flasks of "sugared yeast water" to air. He found different bacterial and fungal "species" in the identical flasks. A century later, Ruth Patrick exposed identical glass slides in a stream and found that they were colonized by a wide and unpredictable array of diatoms. Studies of seed rain and the seed pool show temporally and spatially variable arrays of potential recruits. In Figure 10, the recruits available to colonize an opening at a given instant are shown as random subsets of the source pool. Those individuals and species from the source pool who happen to be in the right place at the right time become the recruits to that newly available patch.

I compared the stream macrobenthic community structure in patches of clean cobbles and pebbles, half with fish-excluding mesh tops. Large differences were found between cobble and pebble patches, but the exclusion of fish had only minor effects on the abundance of benthic invertebrates. Particular patches were "hot spots" for recruitment. Highly aggregated populations of several species occurred in individual patches. One patch (cobbles, fish excluded) had the highest density observed for seven of the nine species common in cobbles. Another patch was similarly popular. This shows that among initially identical substrate patches, set out randomly, there is still differential recruitment, leading to diverse communities.

When openings in the habitat are created, who settles in is unpredictable.

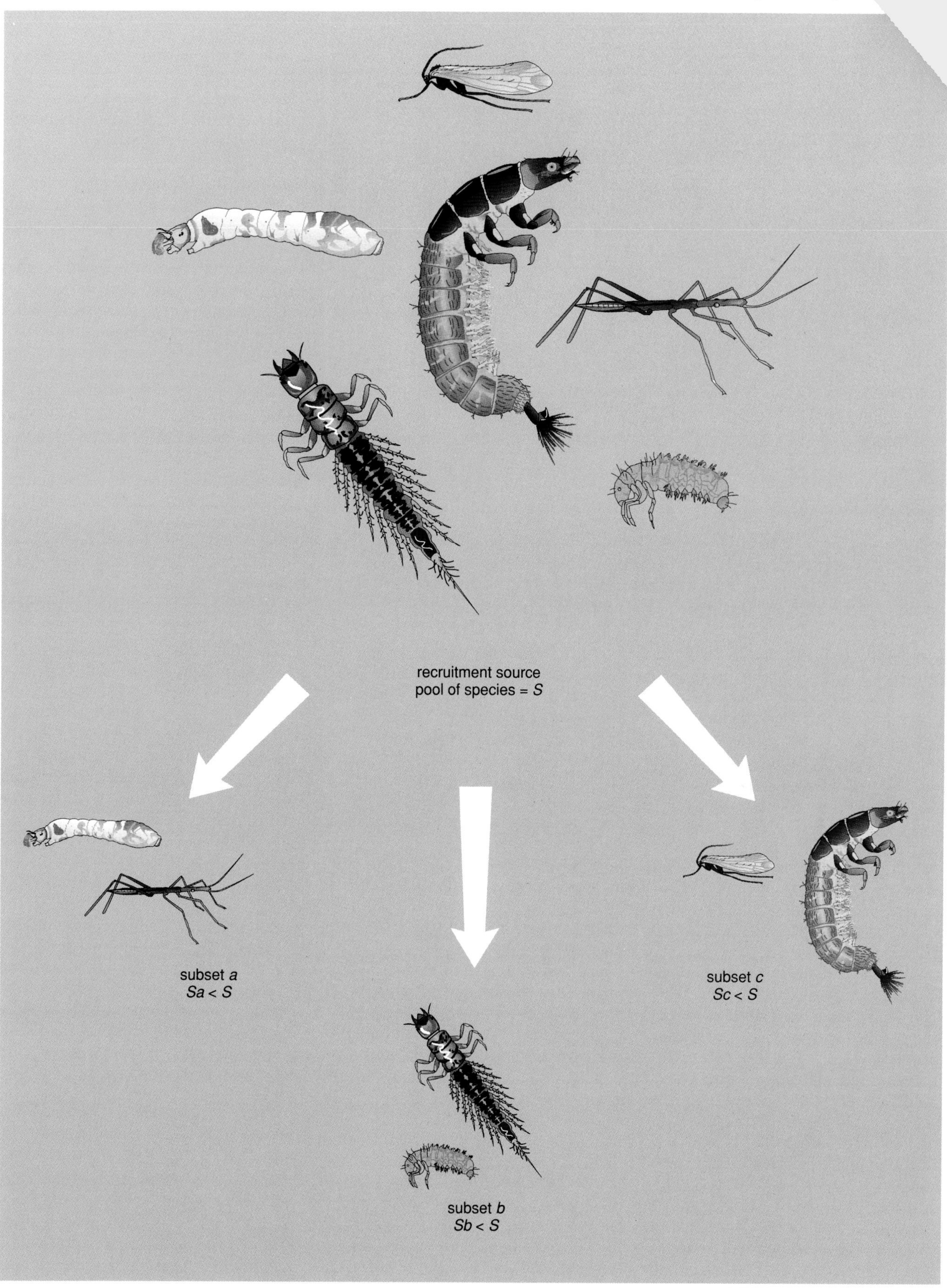

Figure 10. Nearby patches opened by disturbance may develop distinctly different communities because only a portion—not necessarily the same portion—of the recruitment pool reaches each patch. Each patch's community consists of a random subset of the populations available for recruitment.

ture 165

Figure 11. New Hope Creek, the subject of much study by the author and his students, is disturbed frequently by flooding. It contains more than 10 times as many different macroinvertebrate species as nearby Botany Pond. The key to this difference can be found in nonequilibrium theory. (Photograph by the author.)

It is highly unlikely that the new colonists will exactly replicate the lost taxa. What determines which colonists stay in the patch? The suitability of the patch for the survival and growth of the colonists is primary. If a colonist can live in the patch, it remains. However, if there are surviving species remaining in the patch, the interactions between the residents and the colonists determine the outcome. If the interactions are weakly negative (they include limited competition or predation), the species can coexist, and the colonists are recruited into the community of that patch. Mutualistic interactions can facilitate the persistence of the recruits. The result is that even identical patches can support distinct local assemblages of species.

Even identical patches left open for colonization can generate different species assemblages as a result of disturbance and stochastic recruitment; consider, then, the diversity that can be supported by the normal levels of heterogeneity in nature. Various patches may differ in dozens of microhabitat variables. They will support very different communities because of the adaptations of particular colonists to distinct microhabitats.

Policy Implications

Disturbance should be viewed as both natural and beneficial to the world's biodiversity. The most diverse systems are ones that are frequently disturbed—such as streams, pine savannas, prairies and grasslands, tropical rain forests and coral reefs. The positive impact of disturbance on biodiversity is apparent.

Still, when natural disturbance conflicts with economic interests, there is a drive to suppress disturbance. Human efforts to minimize disturbance have resulted in loss of biodiversity. James Ward and Jack Stanford have demonstrated loss of fish and insect diversity in streams dammed or channelized for flood control, and Norm Christensen has proved loss of plant and animal diversity in forests where fire has been suppressed.

Policy makers and environmental managers need to gain an appreciation of the value of disturbance, heterogeneity and recolonization to the maintenance of biodiversity. That insight can be applied as a test of a policy. We need to ask: Will the proposed change homogenize the system? Will it reduce the frequency or magnitude of disturbance? Or will it cut off routes of recolonization by the flora and fauna? If the answer to any of these questions is yes, we ought not do it. We need to value, nurture and preserve our planet's biodiversity. Understanding that heterogeneity and disturbance are important contributors to biodiversity will help us achieve these goals.

Summary and Conclusions

Spatial heterogeneity and disturbance are interdependent. Disturbance creates patches, but patchiness modifies and sets the extent of the disturbance—that is, disturbance responds to the underlying heterogeneity of the environment. Different patch types generate different frequencies and degrees of disturbance and provide numerous opportunities for recruitment. The combination of these factors can be invoked to explain high diversity in disturbed systems, which include most natural ecosystems. The interplay between spatial heterogeneity and disturbance creates the opportunities for recolonization. All these factors—heterogeneity, disturbance and recolonization—acting together, determine community structure.

This theory deemphasizes competition and predation as determinants of community structure by focusing on the nonequilibrium nature of the environment. Disturbance is viewed as the environmental cause of biological change, rather than as the direct changes in the biota. Defining disturbance by its effects is inherently circular and limits our insight into the interplay of disturbance and community structure.

In some systems the return frequency of disturbance is so long that the impression of equilibrium conditions develops. This is what underlies the traditional idea of climax communities. However, careful observation reveals that disturbance is ubiquitous and frequent relative to the life spans of the dominant taxa. Thus communities are always recovering from the last disturbance. It is the process of that recovery that produces the high diversity we find in disturbed systems.

There is, of course, an upper limit to the positive effects of disturbance on diversity. Diversity is bounded by the richness of the source pool for recolonization. As Joe Connell pointed out, at some point the disturbance regime becomes so severe that it makes the environment uninhabitable except for the hardiest colonizing species.

So, now to answer the original question I posed about New Hope Creek and Botany Pond. Why are the bottom dwellers of the stream so much more diverse than those of the pond? It is because streams are both more patchy and more frequently disturbed. The opportunities created by the disturbances permit the recruitment of taxa into the various patches. Thus stream communities maintain a far more diverse fauna than do the less heterogeneous and less disturbed pond communities. This result holds when one compares naturally disturbed to undisturbed systems of all types.

Acknowledgments

This research was supported in part by a Kenan Research Leave from the University of North Carolina. The ecology groups at the University of North Carolina at Chapel Hill and Monash University (Melbourne, Australia) provided stimulating discussions that helped me develop these ideas. Special thanks are due to Dr. Robert Peet, Dr. Alan Stiven, Dr. P. Sam Lake, Jonathan Parkinson, Janette Schue and Merri White.

Bibliography

Andrewartha, H. G., and L. C. Birch. 1954. *The Distribution and Abundance of Animals*. Chicago: University of Chicago Press.

Christensen, N. L., and C. H. Muller. 1975. Effects of fire on factors controlling plant growth in Adenostoma Chaparral. *Ecological Monographs* 45:29–55.

Christensen, N. L. 1988. Succession and natural disturbance: Paradigms, problems and the preservation of natural ecosystems. In *Ecosystem management for Parks and Wilderness*, ed. J. K. Agee and D. R. Johnson. Seattle and London: University of Washington Press.

Clements, F. E. 1916. *Plant Succession*. Carnegie Institute of Washington Publication 242.

Connell, J. H. 1978. Diversity in tropical rainforests and coral reefs. *Science* 199:1302–1310.

Elton, C. 1927. *Animal Ecology*. London: Sidgwick and Jackson.

Gaines, S. D., and J. Roughgarden. 1985. Larval settlement rate: A leading determinant structure in an ecological community of the marine intertidal zone. *Proceedings of the National Academy of Sciences* (USA) 82:3707–3711.

Gleason, H. 1926. The individualistic concept of the plant association. *Bulletin of the Torrey Botany Club* 53:1–20.

Harmon, M. E., S. P. Bratton and P. S. White. 1983. Disturbance and vegetation response in relation to environmental gradients in the Great Smoky Mountains. *Vegetatio* 55:129–139.

Harper, J. L. 1977. *Population Biology of Plants*. New York: Academic Press.

Huston, M. 1979. A general hypothesis of species diversity. *American Naturalist* 113:81–101.

Hutchinson, G. E. 1961. The paradox of the plankton. *American Naturalist* 95:137–145.

Lewin, R. 1986. Supply-side ecology. *Science* 234:25–27.

MacArthur, R. H. 1972. *Geographical Ecology*. New York: Harper and Row.

MacArthur, R. H., and E. O. Wilson. 1967. *The Theory of Island Biogeography*. Princeton, N.J.: Princeton University Press.

McIntosh, R. P. 1985. *The Background of Ecology: Concept and Theory*. Cambridge, England: Cambridge University Press.

Paine, R. T. 1966. Food web complexity and species diversity. *American Naturalist* 100:65–75.

Peet, R. K., D. C. Glenn-Lewin and J. Walker Wolf. 1983. Prediction of man's impact on plant species diversity. In *Man's Impact on Vegetation*, ed. W. Holzner, M. J. A. Werger and I. Ikusuma. The Hague: Junk.

Pickett, S. T. A., and P. S. White, eds. 1985. *The Ecology of Natural Disturbance and Patch Dynamics*. New York: Academic Press.

Reice, S. R. 1985. Experimental disturbance and the maintenance of species diversity in a stream community. *Oecologia* 67:90–97.

Reice, S. R., R. C. Wissmar and R. J. Naiman. 1990. Disturbance regimes, resilience, and recovery of animal communities and habitats in lotic ecosystems. *Environmental Management* 14:647–659.

Resh, V. H., A. V. Brown, A. P. Covich, M. E. Gurtz, H. W. Li, G. W. Minshall, S. R. Reice, A. L. Sheldon, J. B. Wallace and R. C. Wissmar. 1988. The role of disturbance in stream ecology. *Journal of the North American Benthological Society* 7:433–455.

Roughgarden, J. 1987. The structure and assembly of communities. In *Perspectives in Ecological Theory*, ed. J. Roughgarden, R. M. May, and S. A. Levin. Princeton, N.J.: Princeton University Press.

Sale, P. 1977. Maintenance of high diversity in coral reef fish communities. *American Naturalist* 111:337–359.

Shmida, A., and S. Ellner. 1984. Coexistence of plant species with similar niches. *Vegetatio* 58:29–55.

Sousa, W. P. 1984. The role of disturbance in natural communities. *Annual Review of Ecology and Systematics* 15:353–392.

Ward, J. V., and J. A. Stanford. 1983. The intermediate-disturbance hypothesis: an explanation for biotic diversity patterns in lotic ecosystems. In *Dynamics of Lotic Ecosystems*, ed. S. Bartell and T. Fontaine. Ann Arbor, Mich.: Ann Arbor Science.

Weins, J. 1977. On competition and variable environments. *American Scientist* 65:590–597.

White, P. S., and S. T. A. Pickett. 1985. Natural disturbance and patch dynamics: an introduction. In *The Ecology of Natural Disturbance and Patch Dynamics*, ed. P. S. White, S. T. A. Pickett. New York: Academic Press.

Tadpole Communities

Pond permanence and predation are powerful forces shaping the structure of tadpole communities

David K. Skelly

Consider life in a freshwater pond. There are clam shrimp, tadpoles, beetles, dragonfly larvae, mosquito larvae, backswimmers, salamanders and dozens of other species, all packed into an area smaller than your living room. Yet this pond is temporary, and in the space of several weeks the entire basin will dry, leaving little indication of a once and future thriving community.

Ponds with alternating wet and dry phases have long fascinated biologists because of the peculiar challenge they present to their inhabitants. The puzzle, of course, is how the so-called aquatic species manage to survive when the ponds dry up. Scientists have observed a number of ways in which different species deal with this paradox. Some insects and amphibians persevere by metamorphosing into terrestrial forms before the pond dries. Other groups lay resting eggs or burrow into the sediment and wait for the pond to refill. Still other species have no resistance to drying at all. Populations of these species become locally extinct with each drying event and must recolonize the pond when it refills.

The initial studies to provide answers to the survival puzzle posed by temporary-pond dwellers also led scientists to explore a more subtle, but no less important, pattern. The composition of the *communities* within temporary ponds is distinct from community composition within more permanent ponds. In fact, ecologists have realized that the permanence of a body of water exerts an enormous influence on the kind of animal communities that form within it.

In order to understand this pattern, it helps to think about arranging all freshwater, from the smallest puddle to the Great Lakes, along an axis of habitat permanence. When viewed this way, the distribution of aquatic organisms becomes surprisingly ordered. Most species are able to live on just a small section of the gradient using, for example, only ponds that dry up once a year to the exclusion of all others. There is, in fact, an entire suite of species that appear to have comparable limits on their distributions. Why can't these species live in more temporary ponds? Why can't they live in ponds that never dry up? Why do some species coexist while others are segregated in their distributions? Such questions of distribution and community composition are some of the fundamental problems tackled by ecologists.

To get at some of these fundamental questions, I have, for the past 9 years, been studying the distribution patterns of tadpoles. Research on tadpole assemblages has given ecologists some of the clearest indications of how the physical and living components of the world interact to determine the lot of individual species. That lot, we are coming to realize, is strongly affected by both habitat permanence and predation. Through their impacts on individual species, these factors may indirectly determine the composition of communities.

Tadpoles and the Permanence Gradient

Like many aquatic animals, tadpoles are faced with a time problem. They must reach a critical stage—metamorphosis, in this case—before their pond dries up. Among North American tadpole species, the developmental period can last anywhere from 2 weeks to 3 years. Obviously, species with long larval periods cannot successfully breed in ponds with short hydroperiods, and a number of studies have shown that species with longer development times tend to breed in more permanent habitats. If maximum development time were the only constraint on the composition of tadpole assemblages, we might expect a gradual increase in species richness as we proceed from ephemeral to permanent habitats, because permanent ponds would be able to include both temporary and permanent pond species. In fact, this is not the pattern that is observed. The most permanent ponds may have just a single tadpole species compared with up to a half dozen in some temporary ponds. This striking difference in tadpole assemblages requires another explanation.

The exclusion of many species from more permanent ponds may be the result of a second relationship with pond permanence: Predators, like their prey, are also distributed along the permanence gradient. In the most temporary habitats there may be few or no predators at all. Those that are present tend to be small. As the time constraint relaxes in more permanent ponds, more and larger predators are found. These include several types of insects, such as larval beetles and dragonflies, as well as salamanders. In fully permanent waters, animals that require water year round—notably fish—become important predators.

David Skelly is assistant professor of ecology at Yale University's School of Forestry and Environmental Studies. Following his doctoral work at the University of Michigan, he completed postdoctoral fellowships at the University of Wollongong in Australia and the University of Washington in Seattle. His research centers on understanding the distributions of amphibians and the influence of large-scale processes on ecological patterns. Address: School of Forestry and Environmental Studies, Yale University, 370 Prospect Street, New Haven, CT 06511. Internet: dskelly@minerva.cis.yale.edu.

Figure 1. Timing is everything for amphibians living in temporary ponds. These Australian tadpoles of the genus *Crinia* did not complete metamorphosis before their pond dried up and have suffered the consequences. The relative permanence of a body of water is but one of the factors shaping the life histories of aquatic animals. Ecologists are coming to understand how the interplay between such forces helps determine which animals dwell in which habitats, and by extension dictates community structure. (Except where noted, all photographs are courtesy of the author.)

Based on the patterns of distributions of tadpoles and their predators, ecologists formulated a model of amphibian-community structure that emphasizes the interplay between competition and predation. Because early drying prevents many predators from living in temporary ponds, it was reasoned that competition between species would be intense and would exert a greater influence on community structure than predation in these habitats. According to this logic, the tadpoles that persist in temporary ponds are the ones that can endure the contest to secure resources and still develop fast enough to metamorphose before the pond dries. In more permanent ponds, predators would reduce tadpole densities, mediating a release from interspecific competition. Species that succeeded in permanent habitats would, according to the hypothesis, be those able to survive in the presence of the predators; competitive ability becomes less important.

Experiments in artificial ponds confirmed the predictions of the competition–predation-gradient hypothesis and suggested that good competitors were poor at surviving with predators. Investigators reasoned that tadpoles, like the invertebrates of the rocky shore, are subject to a trade-off between competitive ability and susceptibility to predators that may drive the pattern of distribution. These ideas made intuitive sense, but they had not been evaluated in natural tadpole populations.

I set out to test the competition-predation hypothesis in a more natural setting. I focused on two species that live in southeastern Michigan: the striped chorus frog *(Pseudacris triseriata)* and the spring peeper *(P. crucifer)*. These species are quite similar in most aspects of their life history but differ in their distribution among ponds. Chorus-frog tadpoles tend to be found in temporary ponds that dry each summer as well as in intermediate ponds that dry some years, but not others. Chorus-frog tadpoles are rarely, if ever, found in ponds that remain consistently filled. By contrast, spring-peeper tadpoles are found in all three types of ponds and can be common even in the permanent ponds where chorus-frog tadpoles are absent. Using the logic of the competition-predation hypothesis I would expect chorus frogs, which dwell in temporary ponds, to be superior competitors, whereas spring peepers would be expected to dominate in more permanent ponds because of their superior resistance to predation.

tadpoles of species such as the chorus frog, commonly found in temporary waters, are rarely found in permanent habitats. Part of the answer is that the tadpoles' predators—insect larvae and, in the most permanent habitats, fish—are also distributed along a permanence gradient. Ecologists have concluded that habitat permanence and predation help shape the structure of aquatic communities. Early theories posited a trade-off between competition for resources, which would be more intense in temporary ponds, and predation.

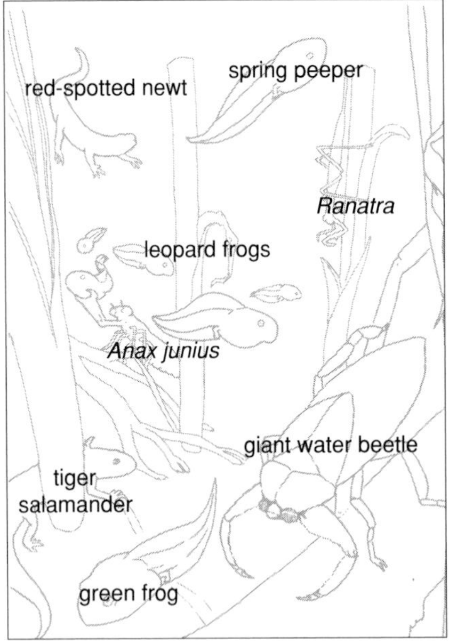

cages. In all, I installed more than 70 cages in six different ponds—two permanent, two temporary and two intermediate. This experiment allowed me to assess the effects of drying, locally present predators and potential competitors on each tadpole species at various points along the permanence gradient.

Two months after I set up the experiment, the results emerged in the form of metamorphosing froglets. I recorded the numbers that survived in each enclosure, their size and the time it took them to reach metamorphosis. These variables are measures of ecological performance. Tadpoles that survive, grow rapidly to a large size and develop rapidly to metamorphosis are more likely to contribute to the next generation than are tadpoles that lag in any of these dimensions. Performance can be compared within a species to determine variation across the permanence gradient or to determine the influence of predators. Measures can also be compared between species to assess changes in their relative abilities to exploit their habitat.

I found that the permanence gradient most definitely affects species performance. Chorus frogs grew larger and survived better at the temporary end,

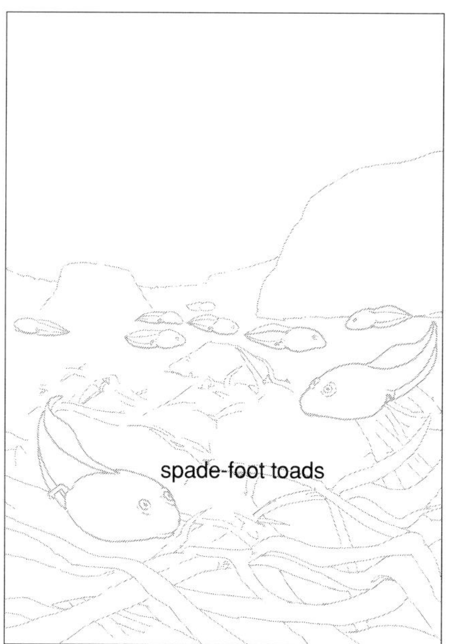

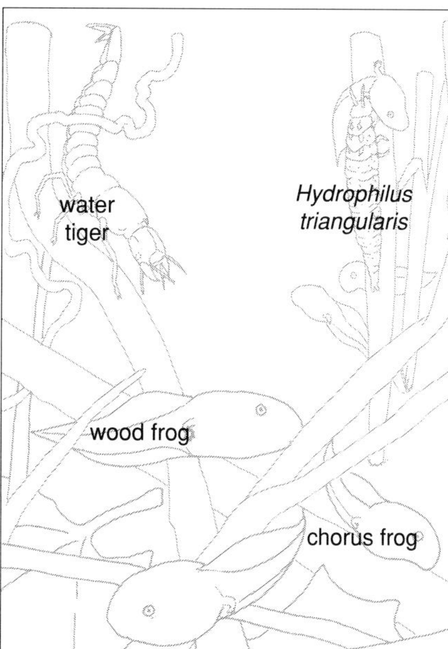

Figure 2. All freshwater habitats, from the smallest puddle to the Great Lakes, can be arranged along a gradient of permanence. Representative communities at four points along this continuum are shown here, beginning at left with the most temporary pond. Most aquatic species are able to occupy only a limited segment of the gradient. Bullfrog tadpoles, for instance, are associated with permanent habitats and never found in temporary ones, where they would meet the fate of the unfortunate tadpoles in Figure 1. It is harder to understand why the

In order to assess the potential roles of pond drying, predation and competition in the segregated distributions of *Pseudacris* tadpoles, I performed a field-transplant experiment in southeastern Michigan. I first identified several natural ponds that lie along different points of the permanence gradient. I then assessed the densities of tadpoles and their predators in those ponds. Even though I was performing these experiments in a natural body of water, I wanted to have some way to know exactly how many tadpoles I was starting with so I could determine how they fared by the end of the season. To do that, I placed several fine-mesh cages inside each of the ponds I selected. The mesh of the screen allowed the inhabitants of the cage to experience all of the environmental changes that affected the pond, but was fine enough to prevent most organisms from entering or exiting the cage.

Each cage contained just chorus-frog tadpoles, just spring-peeper tadpoles or both species together. In addition, each predator-treatment enclosure was stocked with the major predators, such as beetle larvae, salamanders and fish, present within the pond at their natural density. I added no predators to control

and spring peepers greatly outsurvived and outgrew chorus frogs at the permanent end. By itself this result is important. It could have been that observed distributional patterns were predominantly an effect of adult breeding-site choice. In that case tadpoles from the two species might have been equally adept at living in the different types of ponds. By contrast, my result suggests that, whatever the adult preferences are, sorting takes place during the larval period. In addition, I learned that the influence of predators becomes increasingly important in more permanent ponds. Predators found in permanent ponds have much larger affects on the survival rate of tadpoles compared with the influence of predators in more temporary ponds.

One finding, however, came as a particular surprise. My research showed that interspecific competition exerted little or no effect on the outcome of the experiment. Previous studies in artificial ponds had suggested that competitive effects should be important, particularly where predation is weak. I found that each species segregated into particular ponds even in the absence of interspecific competition. Finally, I had to conclude that for *Pseudacris* tadpoles, at least, the interaction between competition and predation did not adequately explain the distribution patterns I observed.

A Behavioral Trade-off

My results required an alternative hypothesis. It was clear from my experiment that both pond drying and predation had strong impacts on tadpole performance. Importantly, these impacts fell unequally on the two species. I wanted to determine whether a species was constrained in its abilities to survive and grow such that good performance at one end of the gradient would mean poor performance at the other end. I knew that chorus frogs survived better in temporary ponds

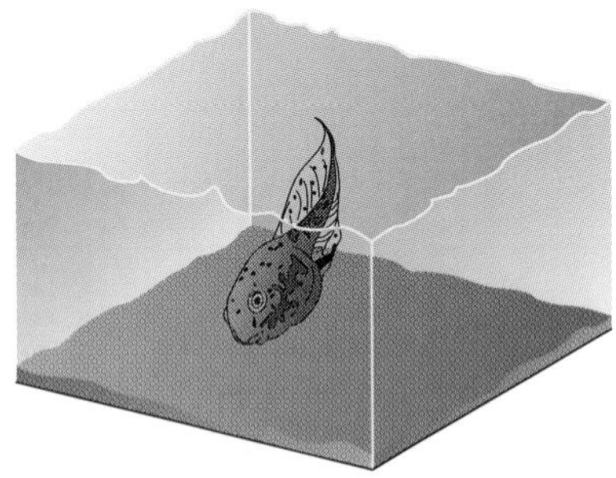

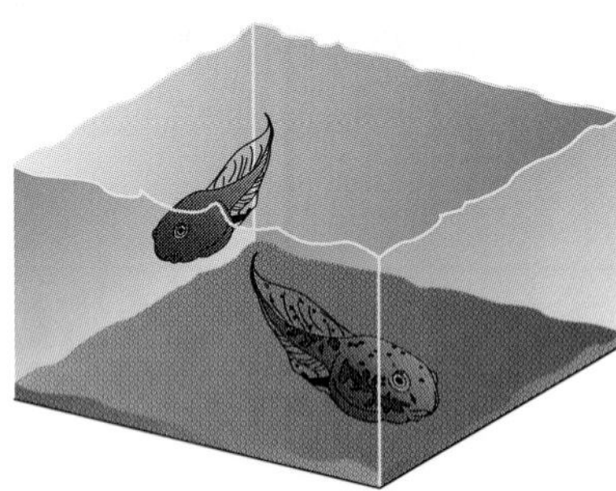

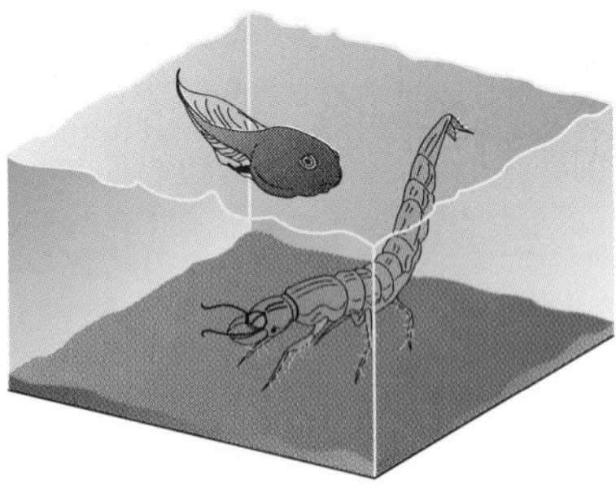

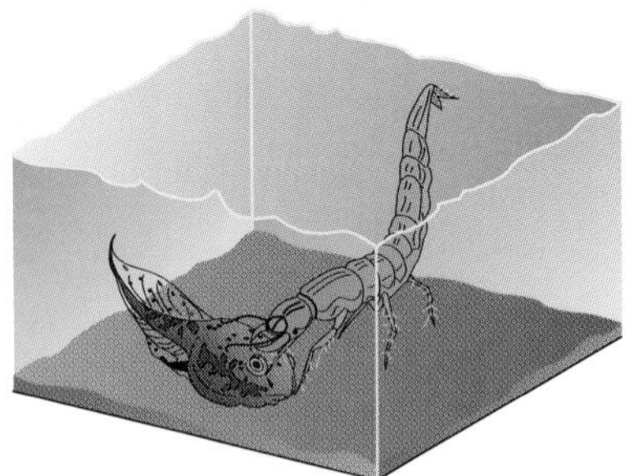

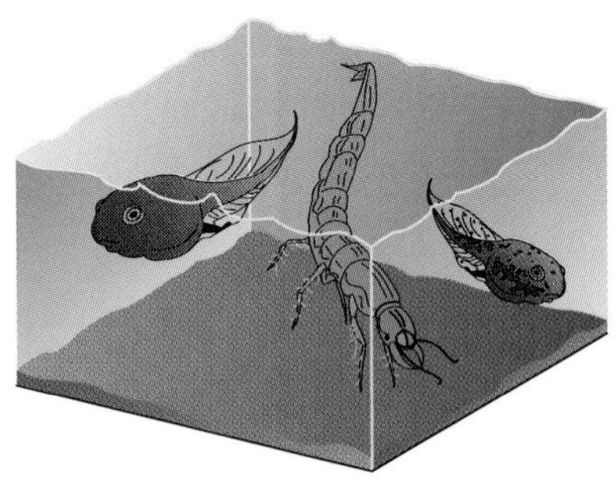

Figure 3. Effects of habitat permanence, competition and predation on tadpole-community structures were tested in the author's field-transplant experiments. Fine-mesh enclosures *(photograph)* were placed in natural ponds of varying degrees of permanence. Each enclosure contained just chorus-frog tadpoles, just spring-peeper tadpoles (found in all but the most permanent habitats) or both species. In the control cages *(top row)* there were no predators. Other cages were stocked with the tadpoles' major predators present within the pond at their natural densities. In this schematic, depicting the experiment as performed in a temporary pond with only invertebrate predators, the beetle larva *Dytiscus*, commonly known as the water tiger, has been added. More than 70 cages were installed in six ponds—two permanent, two temporary and two intermediate.

partly because they develop more rapidly, reaching metamorphosis up to two weeks sooner than spring peepers. Spring-peeper tadpoles, on the other hand, were not equipped to deal with the timetable of a temporary pond, and I found many of these tadpoles stranded in the bottoms of enclosures in temporary ponds at the end of the season. Chorus frogs fared relatively poorly in the presence of predators.

It appeared as if there were some sort of trade-off between rapid development, at the risk of being eaten by a predator, or slower growth to stand a chance of avoiding predation. Such a trade-off could explain both the natural distributions and the results from the transplant experiment. It would also eliminate the need to invoke competition.

Why might tadpoles be subjected to such a trade-off? As it turns out, there are good reasons to suspect that movement is positively related to feeding rate as well as to risk of predation. The behavior of a typical tadpole includes periods of inactivity interspersed with bouts of rasping and swimming. Rasping is a feeding behavior in which the tadpole uses its tail to steady itself against an object, say a decaying leaf, in order to scrape off food from the object's surface. Since food fuels growth, any increase in the time the tadpole spends moving to feed should translate into higher rates of growth and development.

Unfortunately for the tadpole, activity may also increase the risk of predation for at least two reasons. First, many aquatic predators sit and wait for their prey. Rather than hunting for food, these predators depend on having their prey stumble into them. The more a tadpole moves around, the more likely it is to encounter a waiting predator. Second, most aquatic predators actually use movement to identify potential prey. Salamander larvae, for example, use their vision to detect moving prey. Dragonfly larvae have mechanosensory hairs that detect pressure waves from objects moving nearby, which means they can strike at a prey without having to see it. Finally, I already knew from earlier work with Earl Werner at the University of Michigan that American toad tadpoles *(Bufo americanus)* reduce their activity in the presence of a predator.

If this hypothesis is correct, then, the tadpole finds itself in something of a bind. Eating engenders the risk of being eaten. I evaluated in a series of lab ex-

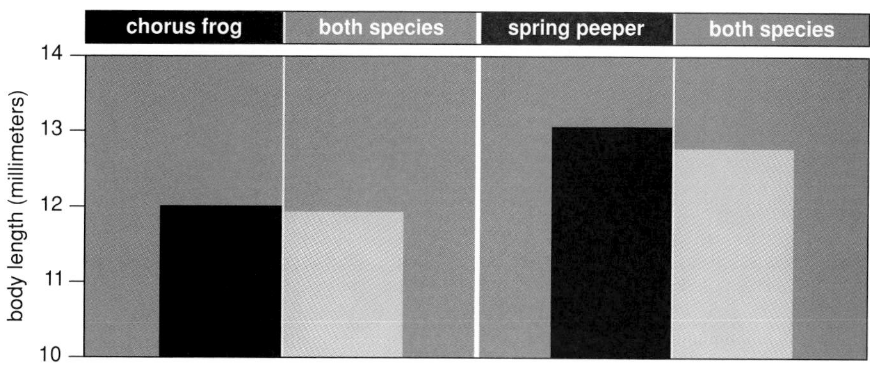

Figure 4. Results from the experiment shown in Figure 3 emerged in the form of metamorphosing froglets. The number surviving in each enclosure, their size and the time it took them to reach metamorphosis were assessed. In each habitat, froglets reached approximately the same size and survived about equally well in the mixed-species enclosures as they did in the enclosures containing just one or the other species alone. This result suggests that competition between species is of minor importance.

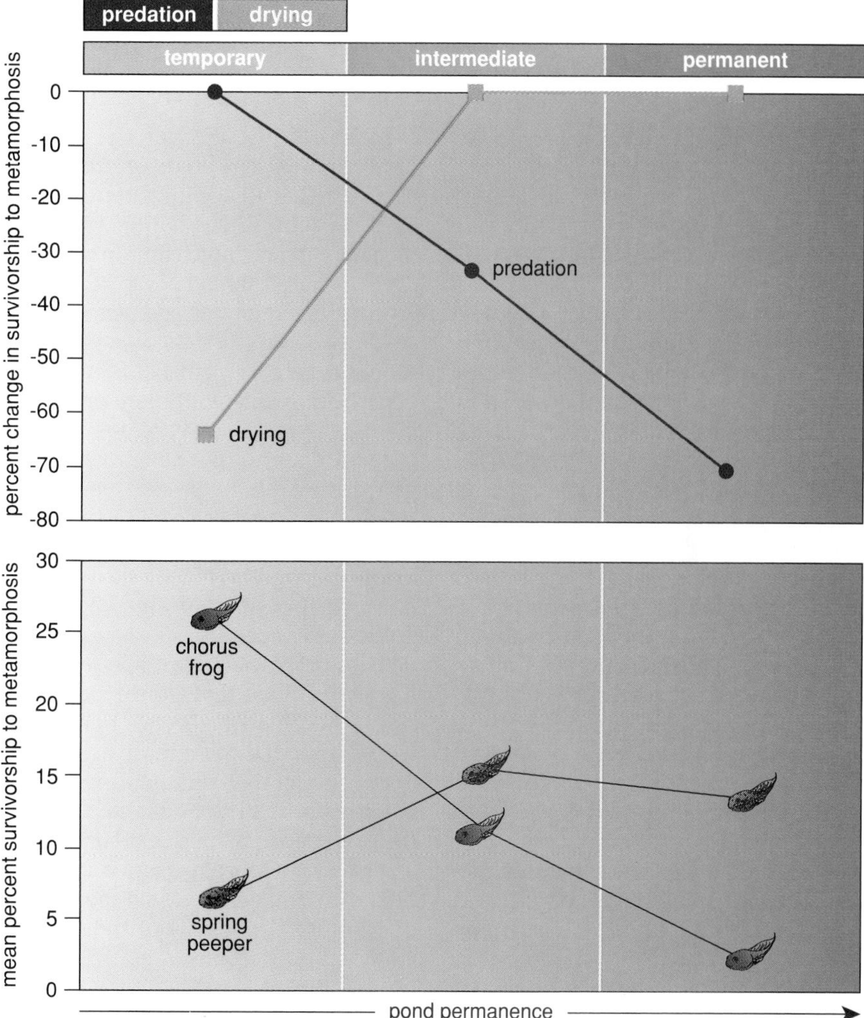

Figure 5. More important influences on tadpole survival were predation and habitat permanence. Pond drying accounted for almost all of the mortality in the most temporary habitats tested in the experiment shown in Figure 3, whereas predation became more important in more permanent habitats *(top graph)*. Chorus frogs grew larger and survived better in the more temporary habitats, and spring peepers greatly outsurvived and outgrew chorus frogs in more permanent habitats *(bottom graph)*. These results suggest that a species' distribution along the permanence gradient is not merely a matter of adult breeding-site choice. Rather, the life history of a species somehow makes it better suited to live in one habitat over another.

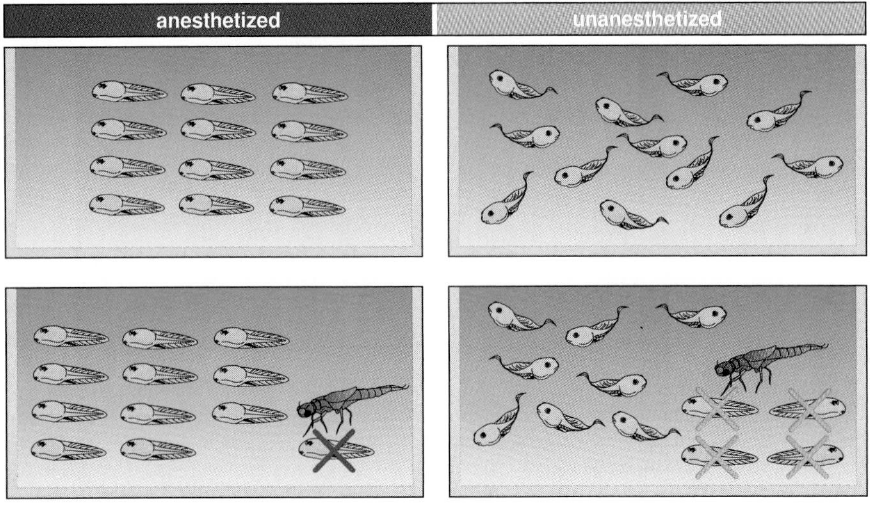

Figure 6. Movement attracts predation, as demonstrated by this experiment. Tadpoles were rendered inactive with anesthesia. Only one of 12 anesthetized wood-frog tadpoles was killed when a predator, the dragonfly larva *Anax junius*, was introduced into the container. In contrast, four of the 12 active tadpoles were killed by the predator. Less active species are predicted to be more successful in avoiding predation.

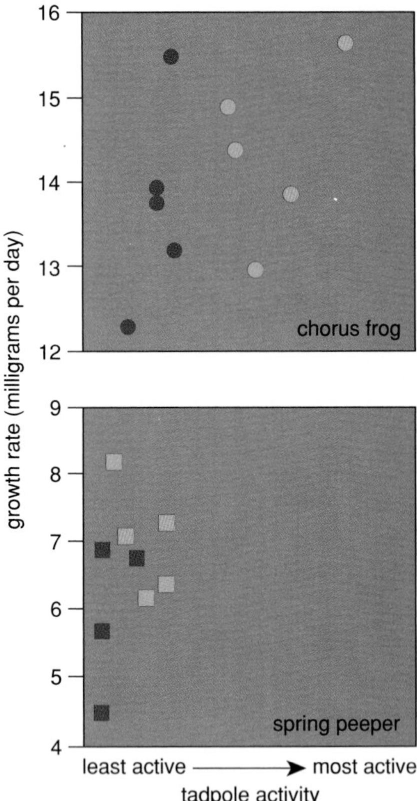

Figure 7. Different species do, in fact, exhibit different activity levels. The author raised chorus-frog and spring-peeper tadpoles in the presence and absence of predators in the laboratory and found that both species limit their activity in the presence of predators *(purple markers)*. Nevertheless, chorus frogs *(top)* are almost twice as active as are spring peepers *(bottom)*, which cope more successfully with predation. Even in the absence of predators (*orange markers*), spring peepers are far less active than are chorus frogs. This experiment shows that the lowered activity level has an additional consequence for both species: Tadpoles raised in the presence of predators on average grew more slowly than tadpoles raised free of predators. The need to limit activity presumably makes it more difficult for the tadpoles to forage for food, and their reduced intake slows growth.

periments whether tadpoles are in fact subject to a behavioral trade-off. To determine the relation between activity and predation risk, I performed a simple experiment in which I exposed different groups of tadpoles to predators. I set up two kinds of situations. In one set of containers, dragonfly larvae were placed in the company of anesthetized tadpoles. The tadpoles in the second set of containers had not been anesthetized and could move around normally. The result was consistent with the hypothesis. Mobile tadpoles were consumed at over four times the rate of the anesthetized animals. In fact, the hungry predators did not even attack tadpoles in half of the anesthetized-treatment containers, underscoring the importance of movement as a cue.

I next evaluated the relation between activity and growth rate for *Pseudacris* tadpoles. I placed each species in containers that held either a caged predator or an empty cage, for a control. I observed the same thing for each species. In the presence of predators, tadpoles were less than half as active as control tadpoles. As predicted, the growth rates of the more stationary tadpoles dropped correspondingly. Taken together, the results from the two sets of experiments suggest that *Pseudacris* tadpoles are confronted with a behavioral trade-off pitting growth against risk from predators. Increased activity allows tadpoles to grow faster, but it also invites attacks from predators.

Because the two *Pseudacris* species tend to live in different kinds of ponds, the trade-off should weigh differently on each. In temporary ponds with few predators and not much time to complete development, it should pay to be relatively active. Rapid growth and development are crucial, and the risks of being active should be relatively small. In contrast, in permanent ponds with more predators, the risks of predation become more important, whereas the benefits of rapid growth diminish. Tadpoles in these ponds should profit from being less active.

I performed a third laboratory experiment to compare directly the behavior of chorus-frog and spring-peeper tadpoles. I gave either high or low food rations to tadpoles of each species, starting with animals that were matched for size. At both food levels, chorus-frog tadpoles tended to be more active and to grow faster than did the spring-peeper tadpoles. The dissimilar behaviors of the two species helps to explain the observed differences in performance in the field experiment as well as the natural distributional patterns of the species. Chorus-frog tadpoles seem to be specialized for life at the temporary end of the gradient, where there is a strong time constraint on development, but where predation pressures are relatively weak. By contrast, slower development and greater resistance to predation make the spring-peeper tadpoles better suited to life in more permanent ponds.

I was encouraged by the internal consistency of the results from the three sets of experiments. In addition, David Smith of Williams College and Josh Van Buskirk of University of Zurich, Switzerland, independently performed similar experiments in rock pools on the shore of Lake Superior. Results from the two experiments corroborated each other well and suggest that we are developing a good understanding of the mechanisms of distribution for these species.

At this point I was curious to determine whether the behavioral trade-off shaping the distribution of species of *Pseudacris* might also influence the distributional patterns of other species. I tested this hypothesis by comparing the behavior of a large number of tadpole species that live in a wide variety of habitats. Because of its high amphibian diversity, Australia was a natural choice for this research. I performed my studies in the region surrounding Wollongong, New South Wales, where there are over a dozen species within the single genus *Litoria* alone.

Each *Litoria* species I studied has a relatively narrow distribution among breeding habitats; some species are found exclusively in temporary ponds, whereas others are found in permanent ponds or in permanent streams. As is the case in North America, the diversity and size of tadpole predators increases with habitat permanence.

I brought tadpoles from seven *Litoria* species into the lab and assessed their activity. Of these species, three breed in temporary ponds *(L. caerulea, L. chloris* and *L. freycineti),* one breeds in permanent ponds *(L. peroni),* and three breed in streams and rivers, which are the most predator-rich habitats of the habitats I tested *(L. citropa, L. leseueri* and *L. phyllochroa).* Based on the *Pseudacris* results, I predicted that tadpoles from the more permanent, more predator-rich habitats should be less active than tadpoles from the temporary ponds. And this is what I saw. In fact, I found that tadpole species from temporary ponds were, on average, more than twice as active as tadpoles from permanent ponds and streams. As with the *Pseudacris* species, *Litoria* tadpoles exhibit behavioral patterns consistent with the trade-off.

Plastic Tadpoles

These studies suggest that tadpoles are caught in an evolutionary balancing act driven by the covariation between pond permanence and predation, and expressed through traits, such as activity. At one time, some ecologists believed that the force of natural selection would eventually hone a species until it fit perfectly into its environment, whereupon evolution would cease. Today, the view is much different.

Biologists now know that an attribute of an individual—its phenotype—such as the activity of a tadpole, can greatly affect how well it will perform under different environmental conditions. It is also apparent that, from the perspective of this individual, the environment can vary tremendously. No single phenotype could possibly prepare an individual to cope with all of the conditions that it could face. Tadpoles living in ponds provide a good example. A particular pond can dry early in the summer one year and not at all the next. To deal with this variation, individuals can improve performance by assessing the conditions and altering a crucial trait—a phenomenon known as phenotypic plasticity. Plasticity could help tadpoles maintain their performance as the environment varies within a pond and also help a species maintain populations across a wider swath of the permanence gradient.

There is abundant evidence that tadpoles adaptively modify activity. Several species, including some *Pseudacris* species, are known to reduce their activity in the presence of predators. In a number of cases, the mere chemical trace of a predator is enough to limit tadpole activity, even if the predator has not recently eaten any tadpoles. Behavioral responses can also influence the life-history structure of amphibians. The activity levels, growth rates and the size at metamorphosis of at least two species are reduced in the presence of predators. Tadpoles are faced with two alternatives. They can remain in the water

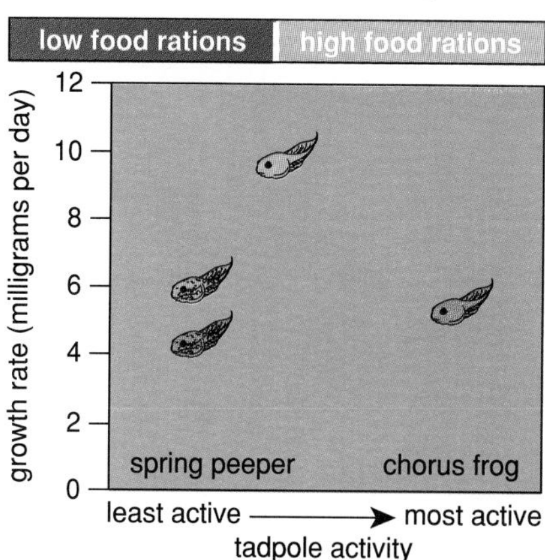

Figure 8. Reduced food intake more profoundly affected chorus-frog than spring-peeper tadpoles. When food rations were low, growth rates were more dramatically reduced for chorus-frog tadpoles than for spring peepers. Surprisingly, chorus-frog tadpoles increased their activity when their rations were low, whereas spring-peeper activity did not change at all. The consequences of activity changes are probably different for tadpoles adapted to permanent versus temporary ponds. In more permanent ponds, the major cost of low activity is a reduced growth rate, delayed metamorphosis and potentially reduced size at metamorphosis. In ephemeral ponds the cost of low activity is death, if metamorphosis cannot be initiated before the pond dries.

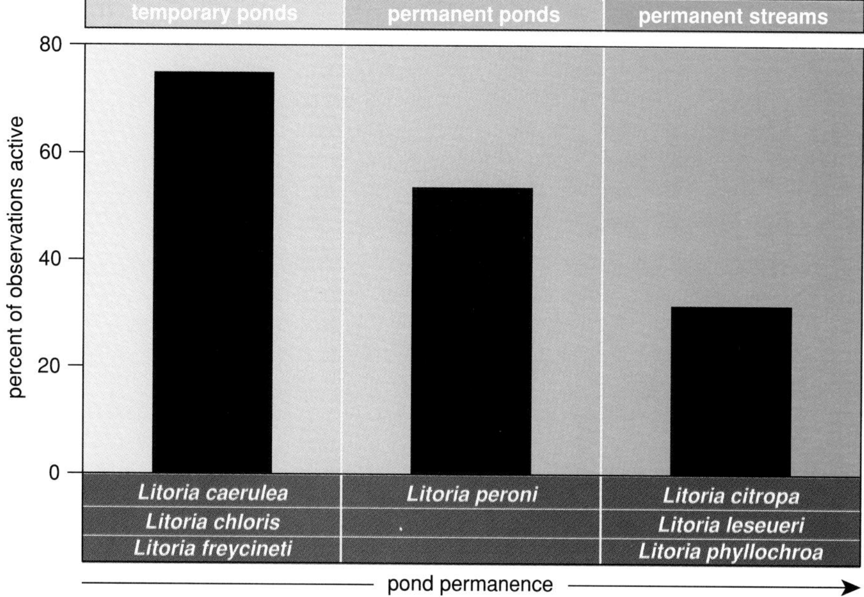

Figure 9. Connection between activity level and pond permanence has been observed for additional species. Various members of the Australian frog genus *Litoria* were collected from their respective habitats, ranging over almost the entire permanence gradient, and their activity level was measured. *Litoria* tadpoles exhibited, on average, the activity levels predicted by the author's model. The three species *Litoria caerulea, Litoria chloris* and *Litoria freycineti,* which live in temporary ponds, are the most active. *Litoria peroni,* which dwells in permanent ponds, shows intermediate levels of activity. *Litoria citropa, Litoria leseueri* and *Litoria phyllochroa* live in streams, the most predator-rich habitats, and were the least active. Tadpoles, it seems, are faced with a trade-off between maintaining a high activity level, with its risk of predation, in order to grow rapidly and metamorphose sooner, versus reduced activity to avoid predation, at the cost of slow growth. The way in which each species resolves this trade-off influences the kind of community it can live in.

Figure 10. Tadpoles can modify activity in varying conditions, but some species can also alter their body shape and coloration. Gray-treefrog tadpoles, *Hyla versicolor*, look completely different when raised in the presence of predators *(right)*. (Photographs courtesy of Josh Van Buskirk.)

and continue to increase in size before metamorphosing, or metamorphose at a smaller size and move onto land. Life-history theory suggests that a poor-quality aquatic environment makes it more advantageous for the tadpole to undergo metamorphosis instead of continuing to grow.

The array of responses exhibited by tadpoles has surprised the biologists who study them, but perhaps nothing has been as surprising as the changes in body shape and color exhibited by some species. A growing catalogue of tadpole species actually alters morphological features in response to predators. Andy McCollum of the University of Michigan and Josh Van Buskirk showed that in the presence of predators, the tail fin of the gray treefrog, *Hyla chrysoscelis*, becomes larger and turns brilliant red. These changes appear to reduce the risk of predation from dragonfly larvae by two means. First, by increasing the tadpole's ability to swim quickly away from a striking predator, and second, by directing predator strikes away from the tadpole's vulnerable head region and toward the tadpole's flashy tail. The costs associated with different tails are still being worked out. However, preliminary evidence suggests that tail shape and color are also traits molded by trade-offs observed in animals living along the permanence gradient.

The variety and striking degree of phenotypic plasticity exhibited by tadpoles might lead one to conclude that tadpoles are able to modify themselves to perform well in any situation. However, despite often parallel responses to the same situations, decreased activity in response to a predator for example, two tadpole species usually also retain differences. Such differences in spite of plastic responses may explain why most tadpoles perform well over a fairly narrow range of the permanence gradient.

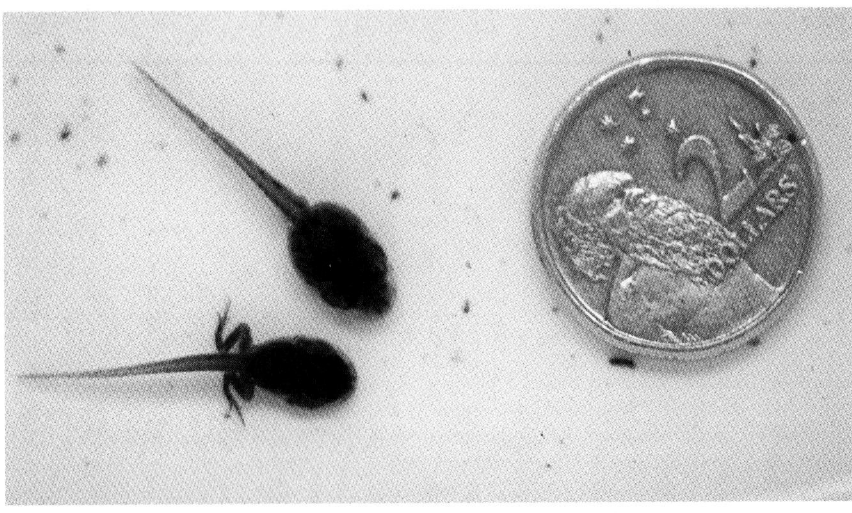

Figure 11. Most tadpoles live on a narrow segment of the permanence gradient, but generalists like *Crinia signifera*, shown here, are the exceptions that prove the rule. Their eggs can be found in everything from horse hoofprints to large ponds and streams. *Crinia* manages to maintain its broad distribution by being extremely inactive and extremely small, enabling it to reach metamorphic size quickly. The two tadpoles here are shown next to an Australian coin about the size of a dime.

Generalists

There are a few exceptions to the rule. A small number of amphibian species breed in a wide variety of habitats. These generalists, by doing what most species cannot, provide important tests of the behavioral-trade-off hypothesis. One generalist is an Australian species, a small and otherwise undistinguished animal known as the common froglet, *Crinia signifera*. This species is one of the most widespread and numerous frogs in southeastern Australia. Its call is so common that it can frequently be heard in the background of the soundtracks to Australian films.

The breeding habits of *Crinia* are remarkable. Its eggs are found in everything from horse hoofprints to sizable ponds and streams. In order to determine how *Crinia* maintains its unusual breeding distribution, I performed a set of experiments in conjunction with Mick Gregory of the University of Wollongong to examine characteristics of its

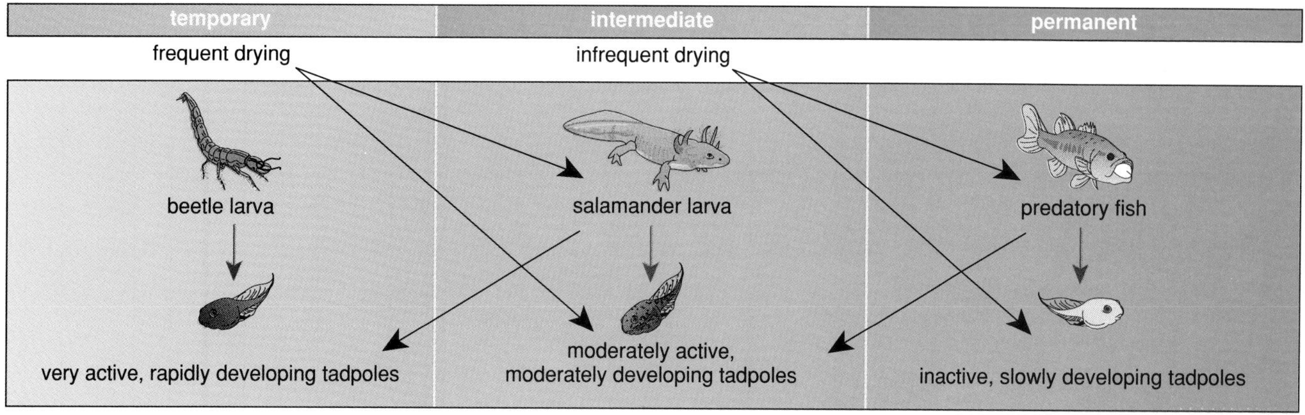

Figure 12. Mechanisms generating community structure along the freshwater permanence gradient are summarized in this schematic. Black arrows indicate factors that act strongly to constrain the distribution of affected species. For example, the inactive, slow-developing tadpoles found in intermediate ponds are intolerant of infrequent drying. Conversely, moderately active species from intermediate ponds are unable to cope with predatory fish. Purple arrows indicate more weakly acting factors that do not exclude a species from living in a community. According to this model, the distribution of tadpole species is limited by the frequency of drying at the temporary end and by the intensity of predation at the permanent end.

behavior and life history. This species's ability to coexist with predators in more permanent habitats stems in large part from its extreme inactivity. *Crinia* tadpoles typically spend less than 10 percent of their time being active. Most other tadpoles are several times as active. *Crinia* manages to survive in extremely small, ephemeral ponds by developing rapidly, completing larval development in as little as three weeks.

According to the behavioral trade-off hypothesis, inactivity and rapid development should be mutually exclusive. *Crinia* circumvents the trade-off by growing very little and metamorphosing at a small size. In fact, the smallest *Crinia* metamorphs are among the most diminutive frogs, emerging when they are as small as 20 milligrams—a size at which two or three could sit on your smallest fingernail. In essence, *Crinia* has reduced its dependence on the aquatic habitat to the point that almost any water will do. For some reason, most other species metamorphose at a much larger size. Perhaps this is related to the difficulties of being an insect-sized vertebrate.

Conclusions

The distributional pattern of an organism is one of its most basic ecological attributes, but the mechanisms that shape distributions are far from obvious. Previous concepts of tadpole distributions had placed a heavy emphasis on the role of interspecific competition. It now appears that tadpoles can maintain their characteristic distributional patterns in the absence of that competition. This is an important insight. However, if amphibian ecologists had stopped with the standard type of field experiments there would have been little success in developing an alternative hypothesis. Evaluating the roles of behavior and other mechanisms have allowed scientists to come to a better understanding of why tadpoles are distributed as they are. In fact, experiments by Earl Werner demonstrate how behavior may contribute to competitive ability. In situations where interspecific competition is important, more active tadpoles from more temporary ponds may be superior competitors compared with less active tadpoles from more permanent ponds.

The results from research on tadpole distributions are likely to apply in other contexts as well. Many freshwater organisms are probably subject to trade-offs comparable to those experienced by larval *Pseudacris*. And preliminary evidence indicates that several invertebrate taxa show a similar correspondence between behavior and distribution. These findings offer some hope of a more unified picture of ecological forces to ecologists wearied by ever more detailed studies of particular species in particular places and times. The search for mechanisms may yet yield answers in the quest to find general explanations for complex ecological phenomena.

Acknowledgments

This article has benefited from discussions and collaborations with a number of colleagues, including Earl Werner, Gary Wellborn, Mark McPeek and Spencer Cortwright. Harry Ehmann and Garry Daly provided valuable advice during the Australian research. My Australian research was supported by Australian Flora and Fauna Research Program of the University of Wollongong. Comments from Lauri Freidenburg and Michelle Hoffman improved the manuscript.

Bibliography

McCollum, S. A., and J. Van Buskirk. 1996. Costs and benefits of a predator-induced polyphenism in the gray treefrog *Hyla chrysoscelis*. *Evolution* 50:583–593.

Schneider, D. W., and T. M. Frost. 1996. Habitat duration and community structure in temporary ponds. *Journal of the North American Benthological Society* 15:64–86.

Skelly, D. K. 1994. Activity level and the susceptibility of anuran larvae to predation. *Animal Behaviour* 47:465–468.

Skelly, D. K. 1995. A behavioral trade-off and its consequences for the distribution of *Pseudacris* treefrog larvae. *Ecology* 76:150–164.

Skelly, D. K. 1996. Pond drying, predators and the distribution of *Pseudacris* tadpoles. *Copeia* 1996:599–605.

Skelly, D. K., and E. E. Werner. 1990. Behavioral and life-historical responses of larval American toads to an odonate predator. *Ecology* 71:2313–2320.

Smith, D. C., and J. Van Buskirk. 1995. Phenotypic design, plasticity, and ecological performance in two tadpole species. *American Naturalist* 145:211–233.

Wellborn, G. A., D. K. Skelly and E. E. Werner. In press. Mechanisms creating structure across a freshwater habitat gradient. *Annual Review of Ecology and Systematics*.

Werner, E. E., and M. A. McPeek. 1994. Direct and indirect effects of predators on two anuran species along an environmental gradient. *Ecology* 75:1368–1382.

The Recovery of Spirit Lake

A natural calamity provides scientists with a rare opportunity to study the rejuvenation of a once-pristine lake

Douglas Larson

On May 18, 1980, Mount St. Helens, a volcano in southwest Washington State, erupted with the force of a 10-megaton nuclear explosion, tearing open the north flank of the mountain and sweeping away all visible signs of life in a 500-square-kilometer area called, appropriately enough, the blast zone. The eruption obliterated whole forests, triggered massive landslides and scalding mudflows, and filled nearby lakes and rivers with downed timber and volcanic debris.

But a disaster in nature may be a boon to scientists. Limnologists and other aquatic experts quickly saw in the mass destruction a rare opportunity to study the succession of organisms that colonized the devastated lakes. The scientists focused their attention on the rebirth of several heavily damaged lakes. Also included in the studies were two new lakes formed by giant mudflows, or lahars, which had dammed major tributaries of the North Toutle River.

The largest and most prominent lake in the blast zone, at a distance of 8 kilometers from the volcano, was Spirit Lake. Before the blast, Spirit Lake was surrounded almost entirely by steep, rugged, densely forested mountain slopes in a relatively undisturbed province of the Gifford Pinchot National Forest.

Within minutes of the eruption, the lake received prodigious quantities of avalanche debris, tens of thousands of mostly old-growth coniferous trees, burnt forest vegetation and minerals deriving from molten and sedimentary rock. Superheated volcanic rock and mudflows, ashfall and geothermal waters also entered the lake, causing lake-water temperatures to rise abruptly, from around 10 degrees to more than 30 degrees Celsius.

The first limnologists to reach Spirit Lake were awestruck by the lake's greatly deteriorated condition. This once-pristine lake had been transformed into a roiling, steaming body of degraded water choked with logs and mud. At the time, the scientists predicted that the lake's recovery—broadly defined as the return to pre-eruption chemical and biological conditions—would require 10 to 20 years (Wissmar et al. 1982a, 1982b).

The daunting task of obtaining initial (baseline) post-eruption limnological data for Spirit Lake fell to a team of investigators from the University of Washington, Oregon State University and the U.S. Forest Service. The team airlifted to the lake during the summer of 1980 included Cliff Dahm, Marvin Lilley, Rick Kepler, Amy Ward, John Baross, James Sedell and Robert Wissmar.

Perhaps only once before in this century had scientists embarked on a similar exploration to track a lake recovering from the impact of a volcano. The volcano was Mount Besymjanny on the Kamchatka Peninsula along Russia's Pacific coast, which erupted violently in 1956. Ash from the eruption reached Lake Asabatchye, 80 kilometers northeast of the volcano, where average accumulations were as high as 20 kilograms per square meter across the 340-square-kilometer lake basin (Kurenkov 1966). Soviet scientists monitored the lake before and after the eruption, but they limited their study to plankton, the microscopic plant and animal life freely suspended in the water, and to a commercially important fish, sockeye salmon. At the time, limnologist I. I. Kurenkov, then with the Soviet Union's Pacific Research Institute of Marine Fisheries and Oceanography, stated, with a hint of regret, that "no detailed hydrochemical investigations were carried out."

Fortunately, Spirit Lake provided another chance for studies elucidating the changes in water chemistry following a volcanic eruption and their effects on biological communities. Spirit Lake had not supported much biological activity, at least not in recent years, as evidenced by the lake's exceptional clarity. A 1974 study found extremely low concentrations of most dissolved and particulate organic and inorganic substances, which are required to support life, and

Figure 1. Gray ash blankets the area around once-pristine Spirit Lake, imparting a ghostly calm, which even in spring 1985 was the legacy of the violent eruption on May 18, 1980, of Mount St. Helens (*background*), a volcano in southwest Washington State. The lake, 8 kilometers from the mouth of the volcano, is the largest of those found in the "blast zone," a 500-square-kilometer area demolished by the volcano. Avalanche debris and felled trees (*foreground*) blocked lake drainage, contributing to lake-volume instability and the subsequent rise and fall of many unstable biological communities throughout the years of recovery. By the end of the 1980s, Spirit Lake had regained a stable biological structure, but the lake's ecology has not returned to its pre-eruption composition.

Douglas Larson is an adjunct professor in the Department of Biology at Portland State University in Oregon. He is also coordinator of the Clean-Lakes Program for the Oregon Department of Environmental Quality. He was engaged in limnological research at Spirit Lake between 1980 and 1989, making more than 100 field trips by helicopter. Larson holds a Ph.D. in limnology from Oregon State University and a B.S. in biology from Jamestown College. He is currently preparing a book on the limnological recovery of Spirit Lake and other lakes in the blast zone of Mount St. Helens. Address: 10325 N.W. Flotoma Drive, Portland, OR 97229.

Kraft-Explorer/Science Source (Photo Researchers, Inc.)

Figure 2. Equivalent to a 10-megaton nuclear explosion, the force of the volcano's blast swept away all visible signs of life, felling trees and creating mud and avalanches that filled surrounding lakes with debris. Some of that debris blocked the North Toutle River, Spirit Lake's outlet, causing the lake to eventually fill with water to above its capacity. Spirit Lake is located to the north *(left)* of the volcano, in the path of the greatest destruction.

trace amounts of chlorophyll *a* from phytoplankton, the microscopic plant life suspended throughout the lake (Bortelson et al. 1976). Over the years following the eruption, limnologists would note that life not only returned to Spirit Lake, but that it became more abundant than it had ever been. The lake would eventually be reborn, but its ecology would be vastly changed from what it was before the blast.

An ecosystem approach was required in tracking the limnological recovery of Spirit Lake. Limnologists are basically ecologists, and as such they set out at Spirit Lake to understand key relationships among the lake's many physical, chemical and biological properties. The foundation of this understanding was a wide assortment of scientific data, systematically collected and interpreted over the course of the study. During a typical expedition to Spirit Lake, various instruments were deployed to determine vertical (surface to bottom) temperature, *p*H, oxygen, salinity and solar-transmission gradients in the water column. Water samples were routinely collected and analyzed for dissolved gases (oxygen, carbon dioxide, hydrogen sulfide, methane), dissolved metals (iron, manganese, sodium, calcium, potassium, magnesium), dissolved organic compounds, nutrients (nitrogen, phosphorus, silica) and other chemical constituents. Water samples were also routinely collected for bacteriological and phytoplankton analyses, including determinations of relative abundance, spatial distribution and species composition. Various microbial metabolic activities and phytoplankton photosynthetic rates were measured *in situ*. Very fine nets and other devices were used to capture aquatic invertebrates, including zooplankton. By accumulating these data year after year, investigators began to observe trends in the lake's physical, chemical and biological properties, indicating the rate and direction of lake recovery as well as the changing populations and communities.

Pre-Eruption Limnology of Spirit Lake

Unfortunately, Spirit Lake had not been studied in detail prior to 1980. Consequently, scientists were left with only a sketchy description of pre-1980 Spirit Lake to consider in assessing the degree of lake recovery. Especially sparse was information regarding the lake's biological communities. Pre-eruption information about phytoplankton and other aquatic plants was virtually nonexistent. The literature contained only a brief mention of single-celled diatoms and green algae collected during the first limnological survey of the lake in 1937 (Crawford 1986). Information about aquatic invertebrates was equally scant: The 1937 survey referred to "one species of rotifer and a few copepods" obtained with a plankton net towed for five minutes (Crawford

1986). A survey conducted only six weeks before the eruption reported more diverse microscopic life, including *Bosmina*, *Daphnia*, *Diaptomus*, cyclopoid copepods, and the rotifers *Asplanchna* and *Kellicottia* (Wissmar et al. 1982b). These organisms are all forms of zooplankton, the microscopic animal life suspended in lake water. These species exist typically in so-called oligotrophic lakes, which are well supplied with oxygen but are generally deficient in essential nutrients such as nitrogen, phosphorus and, perhaps, silica.

The 1937 survey focused on the lake's capacity to support fish, chiefly rainbow and brook trout stocked by the state of Washington's sport-fisheries agencies beginning in 1913. Fisheries biologists found that the lake's plankton populations were indeed sparse. Samples taken from the lake's sediments yielded no bottom-dwelling, or benthic, animals. Noting the shortage of organisms providing a food source for fish, the biologists concluded that the lake's potential for sustaining fish was relatively small, probably less than one kilogram of fish per hectare per year. Nevertheless, to promote fishing and other recreational activities, the lake was stocked with 40,000 rainbow trout annually between 1951 and 1979.

Chemical analyses before the eruption revealed very low levels of dissolved chemicals in the lake, which helps explain why it was so poor at sustaining life. Robert Wissmar and his colleagues found no detectable amounts of iron, manganese, copper, lead, zinc, arsenic and antimony, and they found that other dissolved metals, although measurable, were present in concentrations only slightly above the minimum detection limits of conventional analytical techniques. Extremely low were concentrations of the major positively charged chemical species, or cations—calcium, sodium, potassium and magnesium—as well as the major negatively charged chemical species, or anions—sulfate and chloride (Wissmar et al. 1982a).

Post-Eruption Limnology 1980–1981

The volcanic debris left by the eruption raised the lake's surface elevation by about 60 meters. In a pre-eruption survey the lake's depth was recorded at 58 meters, and its surface area was 5.3 square kilometers. The deposition produced a shallower, expanded basin, with its storage capacity reduced by 10 percent or more, but its surface area increased by more than 80 percent from pre-eruption conditions. Much of the deposited organic matter was incorporated in the lake's new sediment layer, now 50 to 60 meters thick. Thousands of shattered trees and other forest debris floated across the lake, eventually forming an immense log-raft that covered roughly 40 percent of the lake's entire surface area of 9 to 10 square kilometers.

Volcanic deposits also blocked the Spirit Lake outlet, the headwaters of the North Toutle River, which held the lake at a fairly constant volume year-round, a condition referred to as hydrologic balance. Spirit Lake was thereafter impounded in a closed, hydrologically unstable basin by a debris dam 150 to 180 meters thick. As the lake continued to fill with water from inflowing streams, the probability increased that the lake would eventually breach or overtop the debris dam. Either of these possibilities might, in turn, have caused the dam to fail entirely, in which case a substantial portion of the lake would have escaped, perhaps suddenly, downstream into the Toutle River valley. According to a worst-case scenario developed by the U.S. Geological Survey in 1983, a flood of this magnitude would generate a titanic mudflow, which in turn would inundate the downstream communities of Castle Rock, Kelso and Longview (having a combined population of roughly 50,000) with 40 to 60 feet (12 to 18 meters) of water and mud.

Also on the rise were concentrations of inorganic chemical constituents, as well as dissolved and particulate organic matter. The influx of mineral and organic matter, combined with the unusually high water temperatures, made Spirit Lake ripe for bacterial life. The lake's post-eruption concentrations of iron and manganese were, respectively, more than 1,000 times and more than 5,000 times their pre-eruption levels. The concentration of total organic carbon rose from 1.3 milligrams per liter to 41 milligrams per liter after the eruption. The concentration of dissolved calcium showed a spectacular increase, rising from a pre-eruption level of 2.15 milligrams to 66.9 milligrams per liter. The rise in concentrations of other inorganic chemical constituents, including sodium, potassium, magnesium, sulfate and chloride were similarly dramatic. The lake contained 33 times more phosphorus after the eruption than before, which greatly enhanced lake-water fertility. The additional phosphorus, as phosphate, combined with an abundance of bicarbonate, produced a substantial increase in lake-water alkalinity: The alkalinity of Spirit Lake in October 1980 was 248 milligrams per liter, or 35 times pre-eruption levels.

Curiously, the concentrations of nitrogen remained relatively unchanged following the eruption. Several investigators reported that most of the volcanic debris in the lake, particularly the ash and forest vegetation, was nitrogen-de-

Figure 3. Spirit Lake before the eruption was not rich in minerals and therefore supported very little life. Especially scarce were algae, accounting for the lake's exceptional clarity. Mount Rainier appears in the background. (U. S. Forest Service photograph.)

ficient. So even though concentrations of particulate organic nitrogen rose from 47 to 70 micrograms per liter, this was still relatively low. The scarcity of nitrogen limits biological productivity, in spite of the large increase in other life-sustaining chemicals and organic compounds. Without sufficient nitrogen, the organisms living in the lake would be less able to manufacture the building blocks for important biological molecules. For example, nitrogen is required in the synthesis of the amino acids that are needed to build proteins and nucleotides, from which is built DNA.

The Stirrings of Life

Shortly after the eruption, microbial populations in Spirit Lake became extremely active and prolific. By late June 1980, bacteria in surface waters numbered about 5 million cells per milliliter. A portion of these were heterotrophs, which derive energy from the decomposition of organic materials in either aerobic or anaerobic environments. They quickly decomposed and oxidized dissolved and particulate organic matter found abundantly in the lake's water column and sediments. Consequently, the lake's supply of dissolved oxygen was soon depleted, except near the surface, which was aerated by the wind and maintained a relatively small concentration of dissolved oxygen (2 to 3 milligrams of oxygen per liter of water). Not long afterward this small amount of oxygen was completely consumed by the bacteria, and by July 1980 the lake had become devoid of oxygen. It remained so until late October 1980. Thus throughout the summer and early fall of 1980, the only organisms remaining in Spirit Lake were those that could function fully in anoxic waters, chiefly anaerobic bacteria, protozoans and other primitive life forms.

In the absence of oxygen, biochemical conditions developed that generated extraordinarily high concentrations of reduced elements—principally iron, manganese and sulfur—and various gases, including carbon dioxide, hydrogen sulfide and methane. Chemosynthetic bacteria, which are able to derive energy by oxidizing reduced metals such as iron or manganese, reduced sulfur or reduced organic compounds, proliferated throughout the water column and in sediments of Spirit Lake during the summer of 1980. The manganese- and sulfur-oxidizing bacteria alone reached more than 5 million cells per milliliter and were distributed abundantly throughout the anaerobic water column. But even this number pales when compared with the estimated density of all bacteria in Spirit Lake, nearly one-half billion cells per milliliter, a concentration that is possibly unprecedented in the annals of environmental microbiology.

Microbial and chemical activity intensified during the summer and fall of 1980, increasingly enriching Spirit Lake with various nutrients and metals. This enrichment did not include nitrogen, however, which remained in short supply and limited the growth of organisms that required larger amounts of nitrogen, such as cyanobacteria. As the summer wore on, nitrogen supplies grew increasingly small, as the demand became increasingly greater from the burgeoning bacterial populations already in the lake. As a consequence organisms capable of fixing atmospheric nitrogen dominated and then overtook other life forms in the lake.

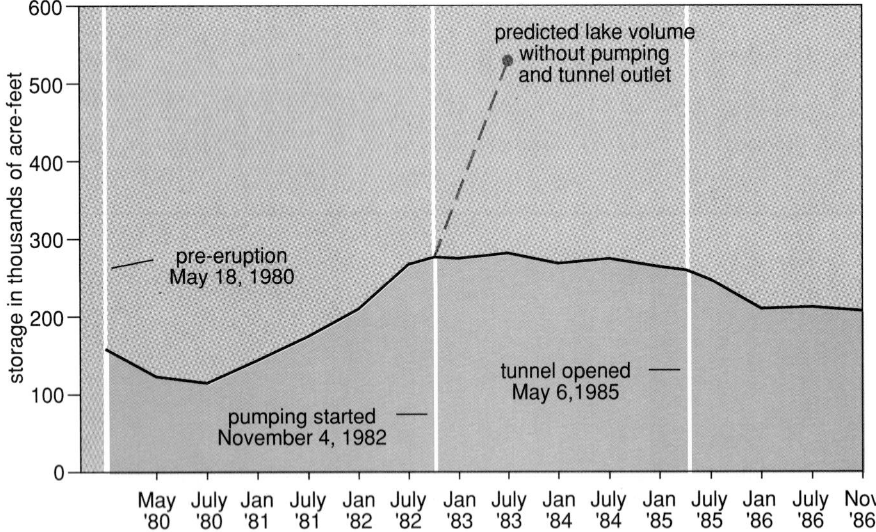

Figure 4. Volcanic debris raised the lake's surface elevation by nearly 60 meters over the pre-eruption elevation of 975 meters (above mean sea level). Lake volumes continued to rise for the next two years, as precipitation and inflowing streams brought water into the lake, and volcanic deposits blocked the lake's only outlet, the headwaters of the North Toutle River. By November 1982, as the lake's volume approached 300,000 acre-feet, the Army Corps of Engineers warned that water volume could top 500,000 acre-feet by July 1983, and breach the debris dam. Trying to avert a potential flood into the communities of the Toutle River valley, with a population of 50,000 people, the Corps intervened. They first pumped the lake to stabilize its surface elevation and then constructed an outlet tunnel connecting Spirit Lake to the North Toutle River. The tunnel now drains excess water flowing into Spirit Lake, although these same waters have contributed to recovery by diluting the chemically enriched lake-waters.

Fish and Plants

The fate of macroscopic creatures living in the lake before the eruption was not well documented. Most scientists involved in the Spirit Lake study have speculated that whatever biological communities had been there were simply buried under the overwhelming deluge of volcanic deposits. Fish and other vertebrates that happened to survive the initial impact were killed shortly afterward by the abrupt loss of oxygen and perhaps by toxic compounds building in the lake.

After the eruption, plants seemed to have essentially disappeared. Only a few remnant phytoplankton species were found during August and October 1980. These included 13 species of diatoms and an unidentified blue-green colonial form of algae. Although there is little information on phytoplanktonic life in Spirit Lake before the blast, limnologists can estimate pre-eruption species composition and abundance based on data from similar Cascade Range lakes whose biological communities have been better documented. Such estimates would suggest that the phytoplanktonic assemblage in Spirit Lake consisted of 100 or more species. The post-eruption numbers were very low by comparison.

The loss of green plants effectively put an end to photosynthesis. Without this principal means of converting solar energy into fuel for growth and production of organisms higher in the food chain, life forms remained very simple for a long time after the eruption. For two years, chemosynthetic bacteria accounted for nearly all of the

Figure 5. Stabilization of lake volumes made it possible in 1983 to construct this field laboratory at the south end of the lake. Prior to this time, the area had been too remote and too hazardous for scientists to remain for any substantial period. The laboratory served as a base of operations for scientists, who now could venture onto the lake about once every two weeks between June 1983 and October 1986, and greatly facilitated studies on the lake's limnological recovery. (Photo by the author, June 1984.)

organic production in Spirit Lake.

Further preventing the emergence of higher life forms were toxic chemicals emanating from the lake. Hazardous gases such as methane and hydrogen sulfide were found in lake sediments, as were toxic organic compounds that contain phenol. In addition, there were high concentrations of the potentially toxic metals mercury, lead, arsenic and cadmium. The excessive organic decomposition and sulfur turned the once-clear lake waters black and opaque. But most ominous were the pathogenic bacteria populating the lake, including the species *Klebsiella pneumoniae* and *Legionella*. In fact, two new strains of *Legionella* were discovered there and named *Legionella spiritensis* and *Legionella sainthelensi*, to commemorate the event. On a more troubling note, one or more strains of *Legionella* were suggested as the cause of a puzzling respiratory illness contracted by several scientists and others working in the vicinity of Spirit Lake in the months following the eruption; some scientists believed this illness to be a mild type of Legionnaires' disease (Baross et al. 1981, Campbell et al. 1984, Larson and Glass 1987).

Flood

Spirit Lake was headed for a stagnant end until nature once more intervened. Beginning in late October, winter storms typical for the Pacific Northwest brought heavy precipitation to the Spirit Lake watershed. Between November 1980 and April 1981, precipitation runoff into the lake increased its volume by almost 30 percent. Substantial inflows diluted the chemically enriched lake waters and dramatically altered the course of recovery.

By April 1981 concentrations of

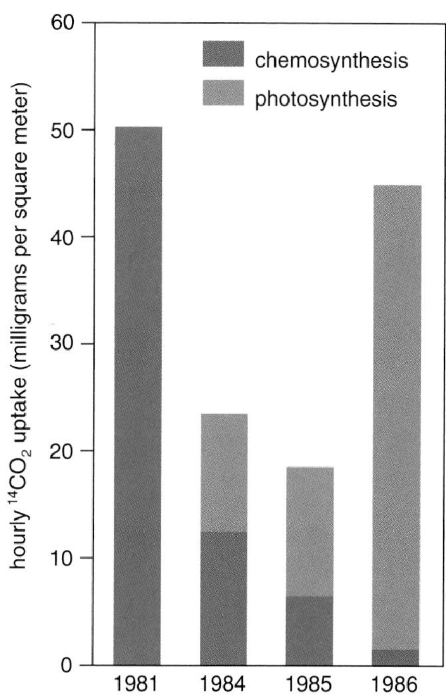

Figure 6. Shift of the lake populations from chemosynthetic to photosynthetic organisms is one indication that recovery is under way at Spirit Lake. Both groups use carbon dioxide, but among the compounds chemosynthesis produces are carbohydrates and hydrogen sulfide, whereas photosynthesis produces carbohydrates, oxygen and water. Only photosynthetic organisms provide fuel usable by more complex plants and animals, but only chemosynthetic bacteria could endure the harsh and toxic conditions in Spirit Lake in the years following the eruption. Immediately after the eruption in the summer and fall of 1980 rates of chemosynthetic uptake of carbon dioxide probably exceeded 1981 levels, but precise measurements were not made. As its chemistry changed, the lake became more hospitable to photosynthesizing organisms, which in turn paved the way for more complex life. Radioactive carbon dioxide ($^{14}CO_2$) provides a tracer for determining the relative metabolic activities, and hence the relative production of either group. (Graph adapted from unpublished data supplied by Cliff Dahm.)

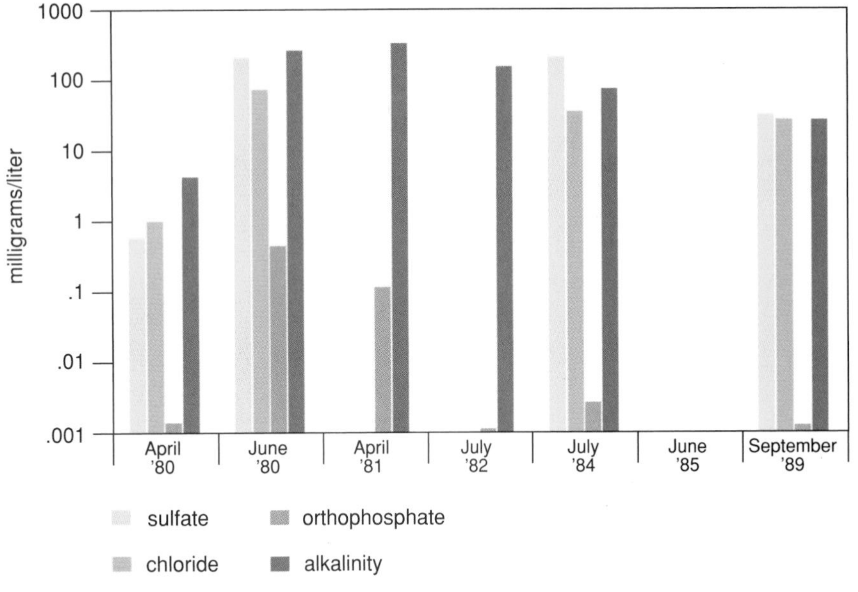

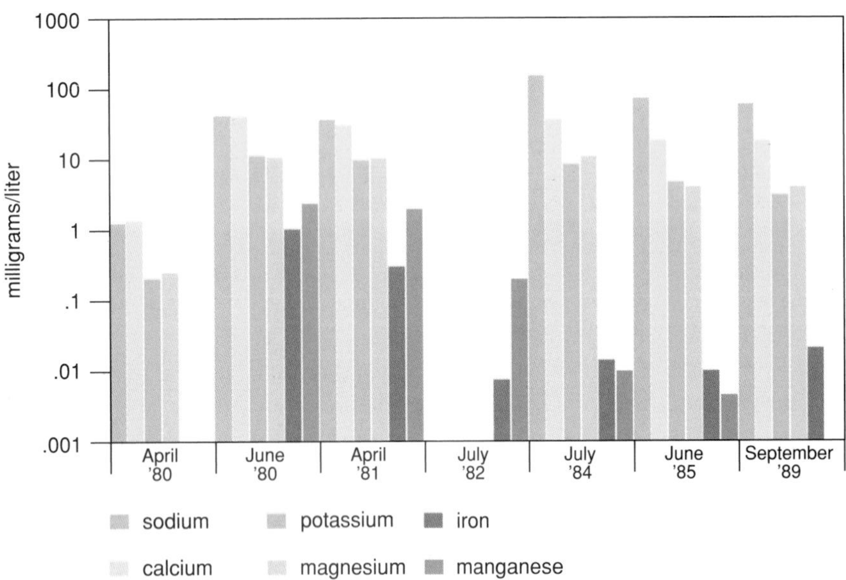

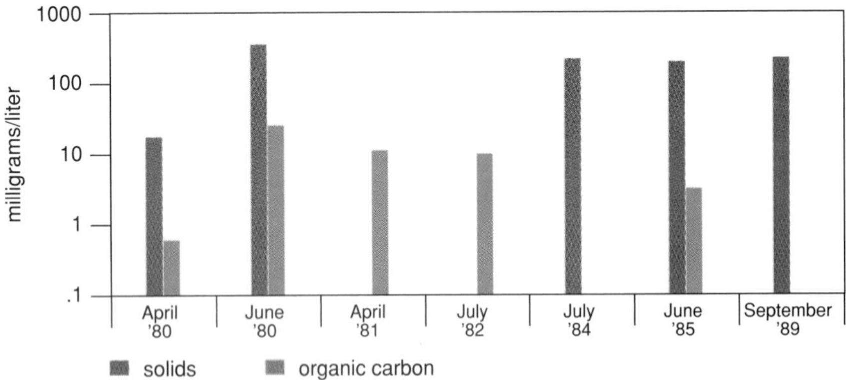

Figure 7. Trends in water chemistry at Spirit Lake indicate that concentrations of all ionic constituents rose dramatically as a result of debris and mudflows into the lake following the eruption. Over time, concentrations of all constituents gradually fell, with the exception of sodium, as precipitation and inflowing waters from streams diluted them. Nevertheless, all constituents remain at levels considerably higher than their pre-eruption values. The sustained high levels of sodium have not yet been explained; the dissolution of sodium-bearing mineral deposits has been proposed as one possible source.

most of the ionic constituents, particularly sodium, potassium, iron, magnesium, calcium, manganese and sulfate, had decreased moderately from the levels they had reached in May 1980, immediately following the eruption. The lake's dilution also diminished the bacterial populations by about two orders of magnitude.

While levels of other components were falling, the oxygen content of Spirit Lake was on the rise. The restoration of oxygen is attributable largely to autumnal lake turnover and to the more or less continuous vertical mixing throughout the water column during winter. Other factors possibly contributed to lake reoxygenation. For example, seasonally cooler lake waters ensured that more oxygen could be held in solution. In addition, lake waters were more dilute and contained far fewer bacteria, hence less biological and chemical oxygen demand.

In spite of the improving water quality, Spirit Lake was not yet stabilized. It still retained substantial quantities of dissolved inorganic and organic materials. Concentrations of dissolved organic carbon, for example, which had reached more than 50 milligrams per liter in August 1980 (up from pre-eruption levels of around one milligram per liter), were still 16 to 18 milligrams per liter during the spring and summer of 1981. These materials provided a rich energy source for oxygen-consuming bacteria, which proliferated anew. Consequently, oxygen concentrations began to diminish rapidly between April and June 1981. By August, the lake was once more without oxygen, except in near-surface waters. Spirit Lake was in danger of repeating the cycle.

Stabilization

The volume of Spirit Lake continued to increase during 1982, reaching 265,000 acre-feet by August, further diluting the waters and reducing the alkalinity as well as the concentrations of dissolved organic carbon, iron and manganese. Nevertheless, the lake remained chemically enriched, with all measured chemical variables well above their pre-eruption levels. Also, the summertime oxygen depletion appeared to have slackened somewhat: Relatively large concentrations of dissolved oxygen were still present in the lake's upper 10 meters as late as August 1982. This represented a significant turning point in the recovery of Spirit Lake. The return to oxygenated

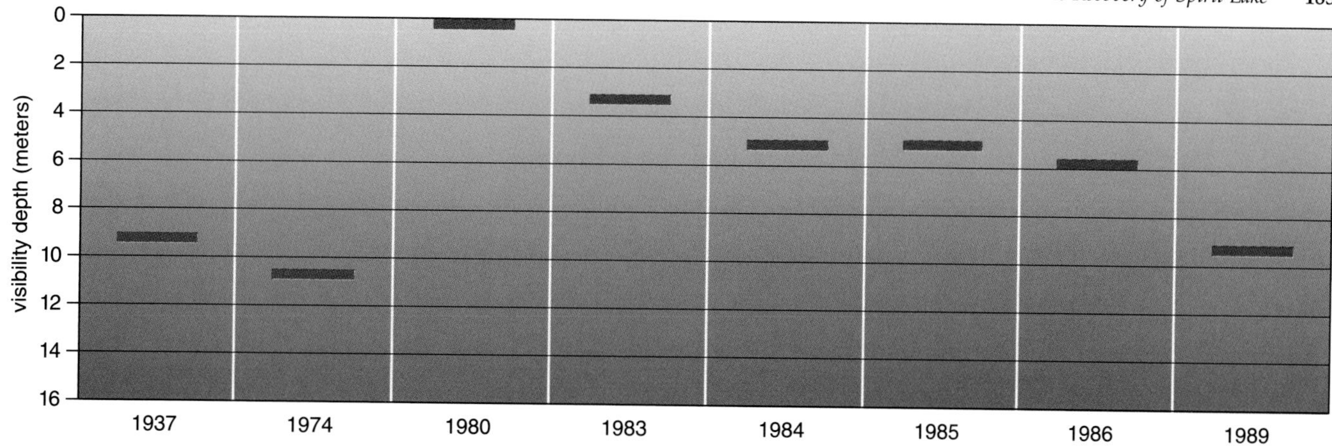

Figure 8. Water clarity of Spirit Lake before and after the eruption was determined using Secchi disk measurements. The Secchi disk, 20 centimeters in diameter, is lowered through the water column until it disappears from sight. The disk is then carefully raised until it reappears. The mean of the distance between the disk's disappearance and reappearance is called the Secchi depth. Soon after the eruption of Mount St. Helens, Spirit Lake waters were blackened by extremely large concentrations of organic compounds that leached from forest debris deposited in the lake. In addition sulfides, chemically reduced metals and particulate matter all contributed to the water's opacity. As a consequence, the Secchi disk on June 30, 1980, was visible to a depth of only a few centimeters. Since then, the lake's water clarity has been restored to nearly pre-eruption levels as water quality continues to improve.

water year-round at all depths meant that the lake's recovery was well under way after only two years.

It was clear by 1982 that the lake was in imminent danger of breaching the debris dam if it continued filling. At that point, the Army Corps of Engineers intervened and pumped the lake to stabilize its surface elevation below the critical level of 1,055 meters. In addition, the engineers constructed an outlet tunnel, which was completed by 1985 and connected Spirit Lake to the North Toutle River. The tunnel now permits excess water to be drained safely from the lake basin.

The participation of the Corps ensured the stabilization of the lake and facilitated access to the lake for limnologists. Before May 1983 the area had been too remote and too hazardous for scientists to remain for any substantial period. But in the latter months of 1983, work was accelerated when a

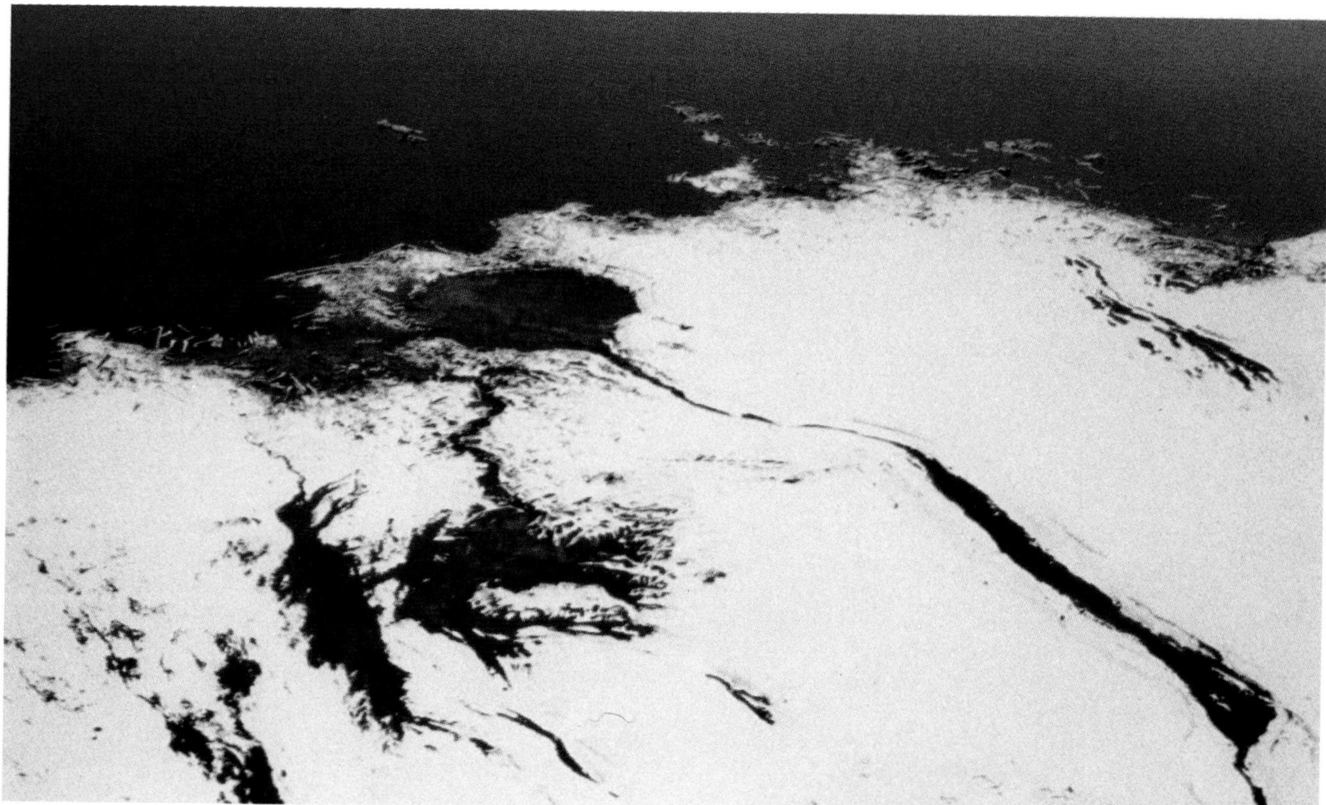

Figure 9. Geothermal "hot spots" and geothermal seeps create snow-free areas along the south shore of Spirit Lake. To stabilize its surface elevation, the drawdown of the lake has exposed a kettle hole, formed by glacial ice blown from the mountain; steam can be seen rising from this nearly circular basin, which is being fed by a stream of high-temperature water emanating from the crater of Mount St. Helens in this photograph taken by the author in March 1985.

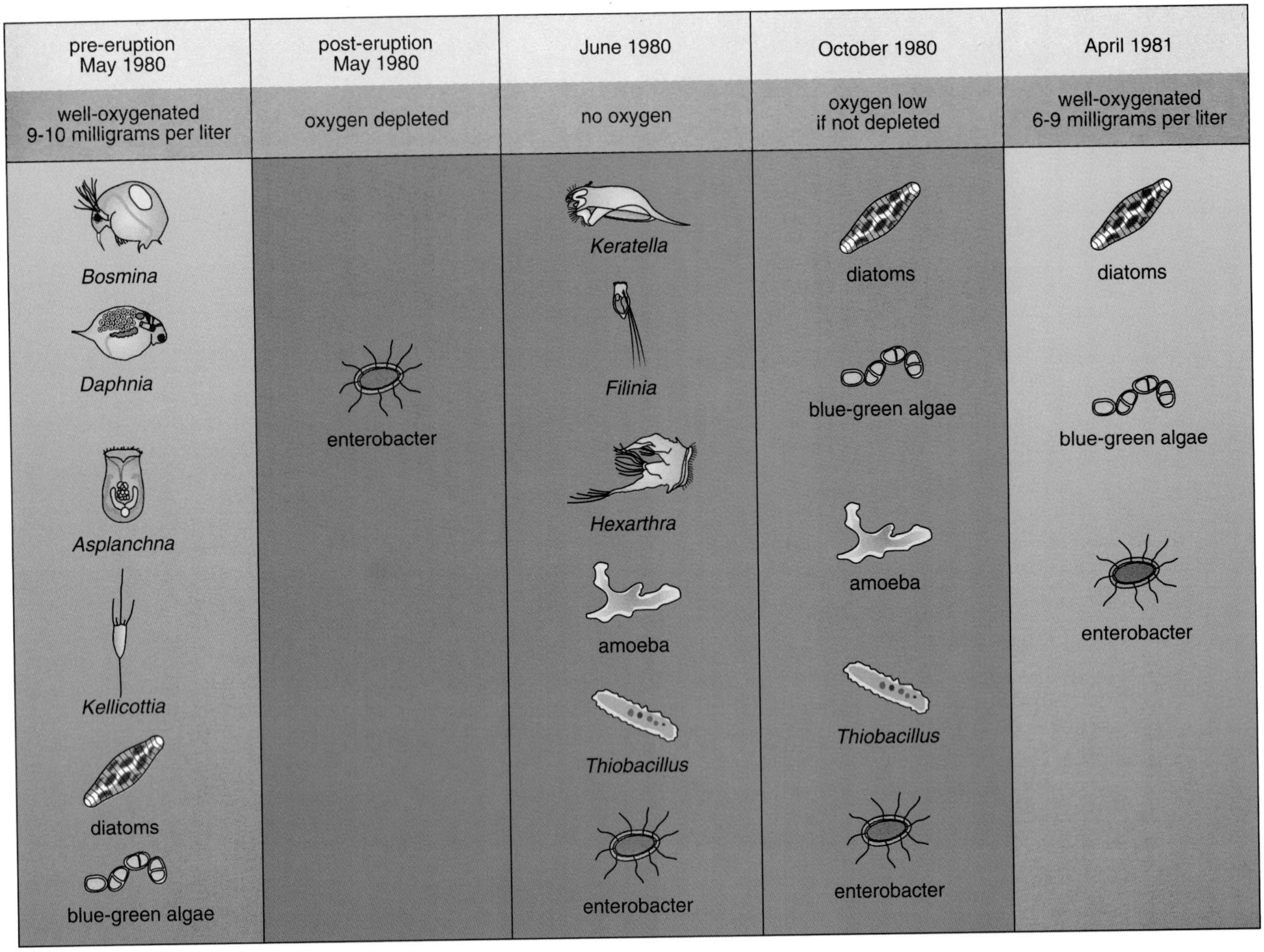

pre-eruption May 1980	post-eruption May 1980	June 1980	October 1980	April 1981
well-oxygenated 9-10 milligrams per liter	oxygen depleted	no oxygen	oxygen low if not depleted	well-oxygenated 6-9 milligrams per liter
Bosmina	enterobacter	*Keratella*	diatoms	diatoms
Daphnia		*Filinia*	blue-green algae	blue-green algae
Asplanchna		*Hexarthra*	amoeba	enterobacter
Kellicottia		amoeba	*Thiobacillus*	
diatoms		*Thiobacillus*	enterobacter	
blue-green algae		enterobacter		

Figure 10. Biological communities changed quickly and dramatically while Spirit Lake was regaining its stability. Often the changes in the composition of communities related to the changing concentrations of constituents dissolved in the water. Oxygen levels had the most dramatic effects on the kinds of organisms found in the lake. Examples of some of the most dominant species in each time period are shown. Before the eruption, the lake was well oxygenated and supported several microscopic animal and plant species, called zooplankton and phytoplankton, respectively, suspended throughout the lake. Conditions and oxygen levels changed drastically after the eruption, which meant that only the most primitive life forms could be found in the lake for a long time. Immediately after the blast, the only organisms that thrived in the lake were oxygen-using, or aerobic, bacteria. But these consumed so much oxygen that the lake's supply was very soon depleted. By late May 1980, the bacteria found in the oxygen-poor waters were either obligate anaerobes, which can survive in oxygen-poor environments, or facultative anaerobes such as the enterobacteria, which can

field laboratory was established at the south end of the lake. This base of operations enabled scientists to venture onto Spirit Lake about once every two weeks between June 1983 and October 1986. Only two visits were attempted after 1986, in August and September 1989, to update information on the lake's changing condition.

The period between 1983 and 1986 marked a biological renaissance for Spirit Lake. By 1986, we had identified a total of 138 species of phytoplankton, many of which were species typical for lakes like Spirit Lake before the eruption. Almost 65 percent of all species identified were diatoms, with the green algae representing an additional 15 percent. *Rhodomonas minuta*, a small, single-celled, motile cryptophyte, was easily the predominant planktonic plant in 1985 and 1986, regularly accounting for more than 90 percent of all phytoplankton present. Other dominant species collected during this period included the diatom *Asterionella formosa*, followed by the diatoms *Cyclotella meneghiniana*, *Achnanthes minutissima*, *Synedra rumpens* and *Diatoma tenue*. A bloom of *Asterionella formosa* was observed during April and May 1986; population densities ranged from 1.5 to 3.7 million cells per liter, or about 95 percent of the lake's entire phytoplankton assemblage. During our two visits to Spirit Lake in 1989, the diatom *Cyclotella comta* was inexplicably the most abundant planktonic plant throughout the water column, representing about 65 to 96 percent of all phytoplankton collected then. The renewal of the phytoplanktonic community carried important implications for Spirit Lake. The expanding population of photosynthesizing organisms overtook and eventually replaced the chemosynthetic bacteria that had for so long dominated the lake's living communities. Gradually, rates of carbon fixation and organic production by phytoplankton in Spirit Lake approached those reported for Crater Lake in Oregon, whose phytoplankton productivity closely resembles that of pre-eruption Spirit Lake.

The zooplankton community also started showing signs of recovery as early as 1982, when relatively large numbers of cladocerans, copepods and rotifers were found. By 1986 microcrustaceans (*Daphnia pulex* and *Bosmina longirostris*) could be found at densities that ranged from 2,700 to 23,500 individuals per cubic meter.

August 1981	October 1986	September 1989
oxygen at surface 3.6 milligrams per liter	well-oxygenated	well-oxygenated some deficiency on bottom 2-3 milligrams per liter
Thiobacillus enterobacter	*Rhodomonas minuta* (cryptophyte) green algae *Asterionella formosa* *Cyclotella meneghiniana* *Diatoma tenue* *Synedra rumpens* *Achanthes minutissima* *Bosmina longirostris* *Daphnia pulex* *Camptocercus rectirostris* *Chydorus sphaericus*	*Cyclotella comta* *Rhodomonas minuta* *Keratella* snail charophyte algae milfoil

survive with or without oxygen. As the months wore on, chemosynthetic bacteria such as *Thiobacillus* emerged as the dominant population, along with a few zooplankton and protozoa, such as the amoebas. By April 1981, precipitation had brought fresh water and new oxygen into the lake, and the bacterial populations once again were dominated by aerobic bacteria and facultative anaerobes. These repeated the earlier cycle, consuming all of the available oxygen and making the lake habitable only for anaerobes and chemosynthetic bacteria. With the intervention of the Army Corps of Engineers, lake-water levels had stabilized by October 1986 and so, too, had oxygen levels. The lake was well oxygenated at all depths. At this point, it was finally possible for the lake to once again support a variety of zooplankton and phytoplankton. By September 1989, lake-life diversity had advanced to include macroscopic plants and animals such as milfoil, *Chara* and freshwater snails. Oxygen levels in the lake are indicated by color, with brown and dark blue indicating low levels of oxygen, and lighter, brighter blue indicating higher oxygen levels.

Additional data collected between 1983 and 1986 demonstrated further the lake's continued recovery. Organic matter and other oxygen-consuming agents had diminished considerably. As a result the lake remained well-oxygenated year-round (except during summer in waters near the bottom, when oxygen consumption along the water-sediment interface was persistently high), and concentrations of most lake-water ions, notably iron and manganese, had decreased significantly by 1986. In addition bacterial levels had returned to near pre-eruption levels, dropping down a thousand times from the post-blast levels reported during 1980 and 1981. All of these changes were visible in the lake's improved appearance. The lake was once again transparent and clean, and freshwater

Figure 11. Freshwater snails enjoy the peace of a summer's day at Spirit Lake in August 1989, surrounded by milfoil. (Photo courtesy of Joe Liburdi, Orca Publications, San Diego.)

Figure 12. Spirit Lake in 1991 was rejuvenated and supporting a rich diversity of life. *Rhizoclonium hieroglyphicum*, lime-green vegetation, grows abundantly on the lake's shore (*lower right*), and swarms of damselflies engage in a mating frenzy across the surface of the lake. Although the lake is once again stable and life-supporting, its ecology is quite different from what it was before the eruption. Eurasian milfoil (*Myriophyllum spicatum*), which grows profusely in shallow waters along the lake's south shore (*left side of photo*), had never before been seen at Spirit Lake. The milfoil and the yellowish pebbles of pumice (*right center*) stand as permanent reminders of the drama of the Mount St. Helens eruption. While all is now calm, the possibility remains that Mount St. Helens will someday reawaken, once again bringing devastation and destruction to the area. (Photograph taken by the author in September 1991.)

snails started to thrive in the well-oxygenated waters near the lake's shore.

Spirit Lake Today

Despite substantial improvements in water quality and the reestablishment of more typical lake biological communities, Spirit Lake is today a very different place from what it was before the eruption of Mount St. Helens. And limnologists predict that Spirit Lake will probably never return to its pre-eruption condition. In September 1991 we observed vast beds of rooted aquatic vegetation at the south end of the lake, an area shoaled by organically laden, chemically reduced muds and fed by warm, chemically enriched geothermal seepwaters originating in the debris-avalanche materials at the base of Mount St. Helens. In this part of the lake, at least, shallower, more fertile waters now support prolific growths of Eurasian water milfoil and other plants, which could not be found in the lake before.

The appearance of the milfoil is not entirely a good thing, as milfoil is regarded as something of a nuisance, and it is almost impossible to remove from waterways once it has taken hold. It is likely that the plant was brought by waterfowl from neighboring lakes and ponds and invaded Spirit Lake after the May 1980 eruption. The end result could be an overabundance of vegetation infesting large portions of the lake and consequently accelerating the lake's aging process.

Other permanent changes in the lake's chemistry can now be detected as well. Concentrations of various inorganic chemicals remain well above their pre-eruption levels, probably because of continued geothermal inflows and runoff from the lake's denuded, erosion-prone, geochemically rich watershed. Indeed, centuries will pass before the lake's watershed is restored to its pre-eruption status as an old-growth forest.

Spirit Lake is now part of the Mount St. Helens National Volcanic Monument, established by Congress in 1982 for scientific research and education. Viewing the lake today, one is deeply impressed with nature's resilience and vigor. But one is also mindful of nature's power, unpredictability and hazards. The history of Mount St. Helens indicates that the volcano will someday reawaken, perhaps abruptly, and once more burst destructively across the lands around its base. And considering that waters throughout the blast zone now harbor pathogenic bacteria and prolific noxious plants, even life renewed in Spirit Lake warrants caution.

References

Baross, J. A., J. R. Sedell and C. N. Dahm. 1981. Letter to the Editor and Editor's Reply. *National Geographic* 160:3.

Baross, J. A., C. N. Dahm, A. K. Ward, M. D. Lilley and J. R. Sedell. 1982. Initial microbiological response in lakes to the Mount St. Helens eruption. *Nature* 296:49–52.

Bortleson, G. C., N. P. Dion, J. B. McConnell and L. M. Nelson. 1976. *Reconnaissance data on lakes in Washington.* Vol. 4. Washington Department of Ecology Water-Supply Bulletin 43.

Campbell, J., W. F. Bibb, M. A. Lambert, S. Eng, A.G. Steigerwalt, J. Allard, C. W. Moss and D. J. Brenner. 1984. *Legionella sainthelensi:* A new species of *Legionella* isolated from water near Mount St. Helens. *Applied and Environmental Microbiology* 47:369–373.

Crawford, B. A. 1986. *Recovery of game fish populations impacted by the May 18, 1980 eruption of Mount St. Helens. Part II. Recovery of surviving fish populations within the lakes in the Mount St. Helens National Volcanic Monument and adjacent areas.* Fishery Management Report 85-9B. Vancouver, Wash.: State of Washington Department of Game.

Dahm, C. N., J. A. Baross, A. K. Ward, M. D. Lilley, R. C. Wissmar and A. Devol. 1981. North Coldwater Lake and vicinity: Limnology, chemistry and microbiology. Unpublished report, U. S. Army Corps of Engineers, Portland District.

Dahm, C. N., J. A. Baross, A. K. Ward, M. D. Lilley and J. R. Sedell. 1982. Lakes in the blast zone of Mount St. Helens: Chemical and microbial responses following the May 18, 1980 eruption. In *Conference on Mount St. Helens: Effects on Water Resources, Symposium Proceedings*, Report Number 41, pp. 98–137.

Dahm, C. N., J. A. Baross, A. K. Ward, M. D. Lilley and J. R. Sedell. 1983. Initial effects of the Mount St. Helens eruption on nitrogen cycle and related chemical processes in Ryan Lake. *Applied and Environmental Microbiology* 45(5):1633–1645.

Dethier, D. P., D. Frank and D. R. Pevear. 1980. *Chemistry of thermal waters and mineralogy of the new deposits at Mount St. Helens—A preliminary report.* Open-File Report, U. S. Geological Survey, Number 81-80.

Dion, N. P., and S. S. Embrey. 1981. *Effects of Mount St. Helens eruption on selected lakes in Washington.* U. S. Geological Survey Circular 850-G.

Jennings, M. E., V. R. Schneider and P. E. Smith. 1981. *Emergency assessment of Mount St. Helens: Post-eruption flood hazards, Toutle and Cowlitz rivers, Washington.* U. S. Geological Survey Circular 850-I.

Kurenkov, I. I. 1966. The influence of volcanic ashfall on biological processes in a lake. *Limnology and Oceanography* 11(3):426–429.

Larson, D. W. 1972. Temperature, transparency, and phytoplankton productivity in Crater Lake, Oregon. *Limnology and Oceanography* 17(3):410–417.

Larson, D. W., and N. S. Geiger. 1982. Existence of phytoplankton in Spirit Lake near active volcano Mount St. Helens, Washington, U. S. A.: Post-eruption findings. *Archives of Hydrobiology* 93(3):375–380.

Larson, D. W., and M. W. Glass. 1987. *Spirit Lake, Mount St. Helens, Washington: Limnological and bacteriological investigations.* Final Report. Portland, Ore.: U. S. Army Corps of Engineers, Portland District.

Larson, D. W., C. N. Dahm and N. S. Geiger. 1987. Vertical partitioning of the phytoplankton assemblage in ultraoligotrophic Crater Lake, Oregon, U. S. A. *Freshwater Biology* 18:429–442.

Meyer, W., and P. J. Carpenter. 1983. *Filling of Spirit Lake, Washington, May 18, 1980, to July 31, 1982.* Open-File Report, U. S. Geological Survey, Number 82–771.

Rosenfeld, C. L. 1980. Observations on the Mount St. Helens eruption. *American Scientist* 68:494–509.

Staley, J. T., L. G. Lehmicke, F. E. Palmer, R. W. Peet and R. C. Wissmar. 1982. Impact of Mount St. Helens eruption on bacteriology of lakes in the blast zone. *Applied and Environmental Microbiology* 43(3):664–670.

U. S. Army Corps of Engineers. 1984. *Alternative strategies for a permanent outlet for Spirit Lake near Mount St. Helens, Washington.* Final Environmental Impact Statement. Portland, Ore.: U. S. Army Corps of Engineers, Portland District.

U. S. Geological Survey. 1983. *Mudflow hazards along the Toutle and Cowlitz rivers from a hypothetical failure of the Spirit Lake blockage.* U. S. Geological Survey Water Resources Investigation Report 82–4125. Tacoma, Wash.: U. S. Geological Survey.

Ward, A. K., J. A. Baross, C. N. Dahm, M. D. Lilley and J. R. Sedell. 1983. Qualitative and quantitative observations on aquatic algal communities and recolonization within the blast zone of Mount St. Helens, 1980 and 1981. *Journal of Phycology,* 238–247.

Wissmar, R. C., A. H. Devol, A. N. Nevissi and J. R. Sedell. 1982a. Chemical changes of lakes within the Mount St. Helens blast zone. *Science* 216:175–178.

Wissmar, R. C., A. H. Devol, J. T. Staley and J. R. Sedell. 1982b. Biological responses of lakes in the Mount St. Helens blast zone. *Science* 216:178–181.

Wolcott, E. E. 1973. *Lakes of Washington, Volume 1: Western Washington.* Washington Department of Ecology Water-Supply Bulletin 14. Olympia, Wash.: Washington Department of Ecology.

The Ecology of the Southern Ocean

Richard M. Laws

The Antarctic ecosystem, based on krill, appears to be moving toward a new balance of species in its recovery from the inroads of whaling

Almost every aspect of the Antarctic—the southernmost tenth of our planet—is on a grand scale. It is the coldest, highest, and most isolated continent, surrounded by the windiest and roughest body of water, the Southern Ocean. Almost all life there is dependent on the sea, and only 1% of the continent is not covered by ice. The Southern Ocean is the site of a vast uncontrolled experiment that began when commercial sealing and whaling activities in the nineteenth and twentieth centuries brought some seal and whale species near extinction (Laws 1977). The resulting perturbations set off a series of large-scale interactions centered on the staple food of these species, krill—a small, shrimplike animal 5 to 7 cm in length that occurs in remarkable abundance in these waters. There are now in existence several farsighted international conservation and management agreements containing some unique provisions that demand further research into the workings of the Antarctic ecosystem as a whole for their implementation. This article is concerned with present knowledge of the oceanic environment, the ecology of the components of the Antarctic food web, and the interactions between the two (Fig. 1).

The Antarctic continent is isolated from the rest of the world by a circumpolar ocean (Fig. 2). An important feature of this Southern Ocean is the Antarctic Convergence, a front that marks the point at which cold Antarctic surface water flowing northward meets warmer sub-Antarctic water flowing southward from the Atlantic, Pacific, and Indian oceans. Although this front is a mobile one, with eddies and loops that span a zone as wide as 150 km, it has a fairly constant mean position from year to year, appearing at an average latitude of about 50°S. The seas south of the convergence extend over 35 million km^2 and comprise about a tenth of the World Ocean. The Southern Ocean is not contained within the circle of the convergence; it extends farther north and is larger than this, but its northern boundary is undefined. Dynamic aspects of this immense ocean

Richard M. Laws is Director of the British Antarctic Survey and of the Sea Mammal Research Unit of the Natural Environment Research Council. He received a B.A. in natural sciences and a Ph.D. in zoology from Cambridge University, and was elected to the Royal Society in 1980. Since 1947 he has spent three southern winters and fourteen summers in the Antarctic, including a season in which he served as inspector/biologist on a whaling ship. His research has focused on the ecology and physiology of seals and whales, but as a result of eight years' research in East Africa he is also an authority on the elephant and the hippopotamus. In 1966 he received the Scientific Medal of the Zoological Society of London, of which he is now Secretary. Address: British Antarctic Survey, NERC, High Cross Madingley Road, Cambridge CB3 OET, England.

have recently been described in detail by Foster (1984).

The surface currents of the Southern Ocean are determined mainly by the driving force of the winds, which are predominantly easterly at higher latitudes, creating a current known as the East Wind Drift close to the continent. Both surface and deeper currents are deflected northward by the Antarctic Peninsula and are then driven clockwise around the continent by the westerly winds north of about 60°S. This West Wind Drift, or Antarctic Circumpolar Current, originated with the separation of the Antarctic Continent from South

Figure 1. A more precise understanding of how members of the Antarctic food chain interact with each other and with the vast Southern Ocean that supports them is becoming increasingly important in light of recent proposals that man harvest "surplus" krill. In this typical summer assemblage of species sharing the ocean's edge at Right Whale Bay, South Georgia, a king penguin (*left*) prepares to lunge at a fur seal. Fur seals kill and eat numerous macaroni penguins at South Georgia, but the larger king penguin is more than a match for them. The birds in the background are dominican gulls; normally intertidal feeders that consume limpets, they will also eat carrion and even small petrels. A number of elephant seals are visible in the distance. (Unless otherwise noted, all photographs are by the author.)

America and Australia by continental drift, a process completed about 37 million years ago. The West Wind Drift now transports more water than any other current system in the world and is very broad, ranging from 200 to 1,000 km in width, the only constriction being the Drake Passage at the tip of South America. Concentrations of nutrients are much higher south of the Antarctic Convergence than for oceanic surface waters in general. The warmer Circumpolar Deep Water, flowing from the north, rises over the Antarctic Bottom Water (Fig. 3), producing an upwelling of nutrients especially in the Antarctic Divergence, a region between the East Wind Drift and the West Wind Drift where the deep water rises to within about 100 m of the surface and diverges north and south. This warmer water eventually mixes with the Antarctic surface water and may flow onto the continental shelf.

Because of the weight of the vast continental ice sheet, the continental shelf has been depressed to a depth of 400 to 500 m and is very narrow, except in the huge embayments of the Weddell and Ross seas. Fringing the thick land-based Antarctic ice cap are extensive floating ice shelves, which are particularly evident to the south of the Weddell and Ross seas, but which also occur as narrower zones along two-thirds of the coastline. These ice shelves are usually about 200 m thick, and are formed mainly by the extrusion of continental ice and snow accumulation. Salt is expressed when sea water under the ice shelves freezes, and the cold saline water sinks to become the Antarctic Bottom Water. The major source of this frigid water is the western portion of the Weddell Sea; flowing northward, it influences the oceanography and hence the climate as far away as the North Atlantic region. At their peripheries the ice shelves crack when exposed to the ocean swell, forming tabular icebergs that are carried northward by the currents and eventually melt in latitudes of 50 to 60°S.

Sea ice—that is, pack ice and the fast ice from which it is formed—is the dominant influence in the Southern Ocean. It is subject to freezing and melting each year, expanding to form an area of pack ice as large as 20 million km^2 in winter and contracting to an area of about 4 million km^2 by the end of the Antarctic summer, which extends from December to March. This is the largest seasonal process in the World Ocean. The ecosystem is synchronized with this annual pulse, which effectively doubles the area of the continent from 18 to 34 million km^2, leading to increased reflection of solar radiation and thus reducing the penetration of radiant heat into the sea. The edge of the pack ice is of great importance both as a productive region and as the habitat of large populations of birds and marine mammals.

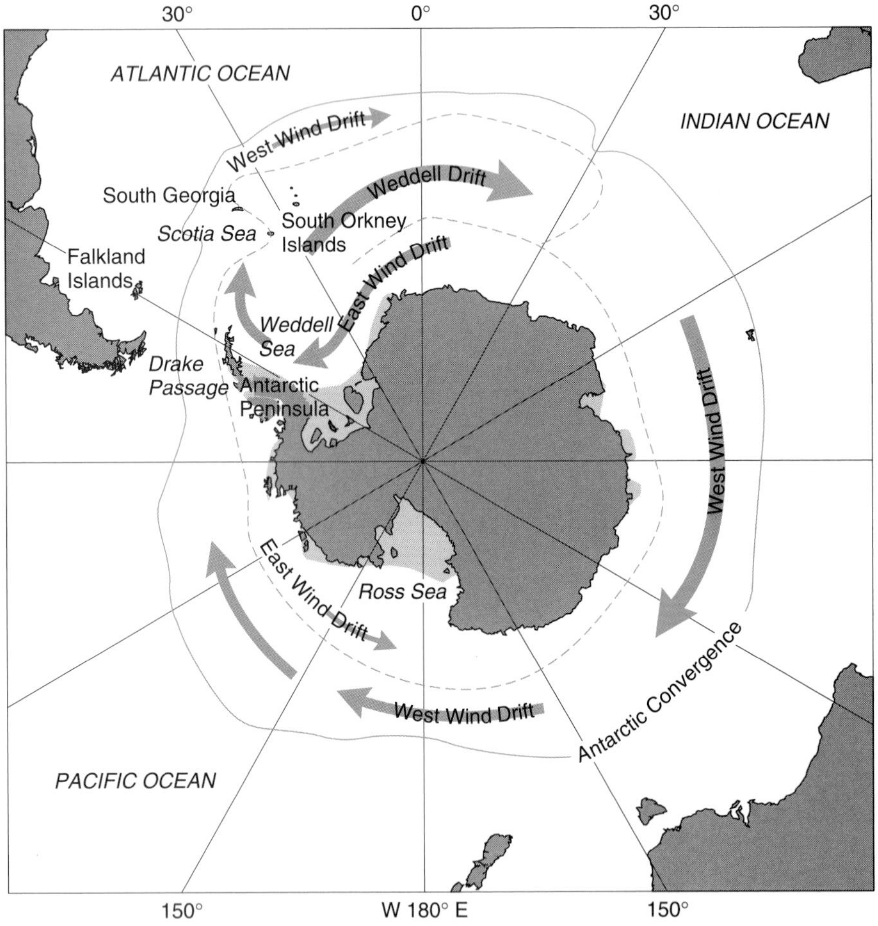

Phytoplankton and microplankton

The food web of the Southern Ocean (Fig. 4) reflects the fact that incoming energy from the sun, nutrients, bacteria, and phyto- and microplankton is the basis of all higher life in the sea. Although the shading effect of the ice cover prevents significant primary production in the underlying water column, a layer of ice-associated algae seeds the water when the sea ice melts, a process which may contribute more than a tenth of annual phytoplankton production. Levels of chlorophyll *a* are enhanced up to 250 km away from the edge of the ice—an indication of recent ice cover. These layers of tiny brown pigmented organisms, by absorbing solar radiant energy and thus accelerating the melting of the pack ice, probably have a disproportionately large effect on the global environ-

Figure 2. A distinctive feature of the Southern Ocean is the Antarctic Convergence, a front at which warm water flowing southward meets cold water flowing to the north. The convergence marks an abrupt change of temperature. To the south, Antarctic conditions essentially prevail; in the sub-Antarctic region to the north the climate is more temperate, with corresponding changes in the flora and fauna. In summer the temperature of the Antarctic surface water ranges from −1°C near the continent to 3.5°C farther north, rising rapidly by 2 or 3 degrees at the Antarctic Convergence; in winter it varies from −1.8°C to 0.5°C over the same latitudinal spread. Two main surface currents – the East Wind Drift near the continent and beyond it the clockwise West Wind Drift – dominate the ocean. The lighter areas along the coast represent major ice shelves. (After Mackintosh 1973.)

ment (Heywood and Whitaker 1984; Siegfried et al., in press).

Different phytoplankton species show different seasonal timing of their maximal populations, with peaks reached later and maintained for shorter periods of time in higher latitudes. Like most of the Antarctic life forms, the most important phytoplankton species are circumpolar in distribution. High values of nutrients, chlorophyll, and primary production are recorded at oceanic fronts (Lutjeharms et al., in press), owing either to vertical upwelling or horizontal transport. However, nutrients—especially nitrate and phosphate—are unlikely to be a limiting factor for primary production, because concentrations remain high after the period of maximum production. In general the availability of trace elements, differences in incident radiation with latitude, sea-ice cover, and turbulence are likely to be much more significant factors. The westerly winds in high southern latitudes are consistently stronger than winds in any other oceanic zone, and wave heights are correspondingly large (Tranter 1982). The wind produces turbulence and vertical mixing of the surface layers to depths of 60 to 80 m or more, a phenomenon that probably depresses primary production by preventing phytoplankton from maintaining an optimum depth near the surface.

There is an apparent seasonal progression of maximum primary production southward, but this overall pattern is complicated by storms, changes in pack-ice cover, and varying intensities of zooplankton grazing—factors which probably produce a mosaic of areas in different states. This makes the estimation of standing crop and production for the Southern Ocean as a whole quite complex. However, current thinking is that the region is no more productive per unit of area than other oceans (Heywood and Whitaker 1984). The estimated rate of primary production varies widely, with a mean figure probably lying between 20 and 100 grams of carbon/m^2/yr in the summer in the ice-free areas. Although other living organisms ultimately depend on these levels of productivity, there is as yet no reliable figure for total annual primary production, and this is a priority for research. Estimates range between 260 and 6,400 million metric tons (Heywood and Whitaker 1984).

Recent research has revealed a significant microplankton biomass (Holm-Hansen, in press). A large proportion of the total chlorophyll—possibly as much as three-quarters of primary production—derives from this source, and some organisms feed on bacteria, picoplankton, and the debris of phytoplankton. Hewes and his colleagues (in press) suggest that on the basis of this new information we should revise the classical concept of the Antarctic "diatom/krill" food chain to one which incorporates the feeding by krill on protozoans as well as diatoms, recognizing the role of protozoans as an efficient link that couples pico- and nanoplankton production to the higher trophic levels. However, the short annual burst of primary production may be in excess of the needs of pelagic standing stocks, and much of it may go to feed detrital feeders in the benthos (Fig. 4).

The effects of bacteria have also been much neglected, but they are important for their interactions with other animals, for their role in facilitating the flow of carbon and nitrogen and the regeneration of nutrients, as enzyme producers in other organisms, and as a food base for grazing organisms. In general the detrital food chain plays a crucial part in transforming particles in transit in the water column into dissolved nutrient chemicals.

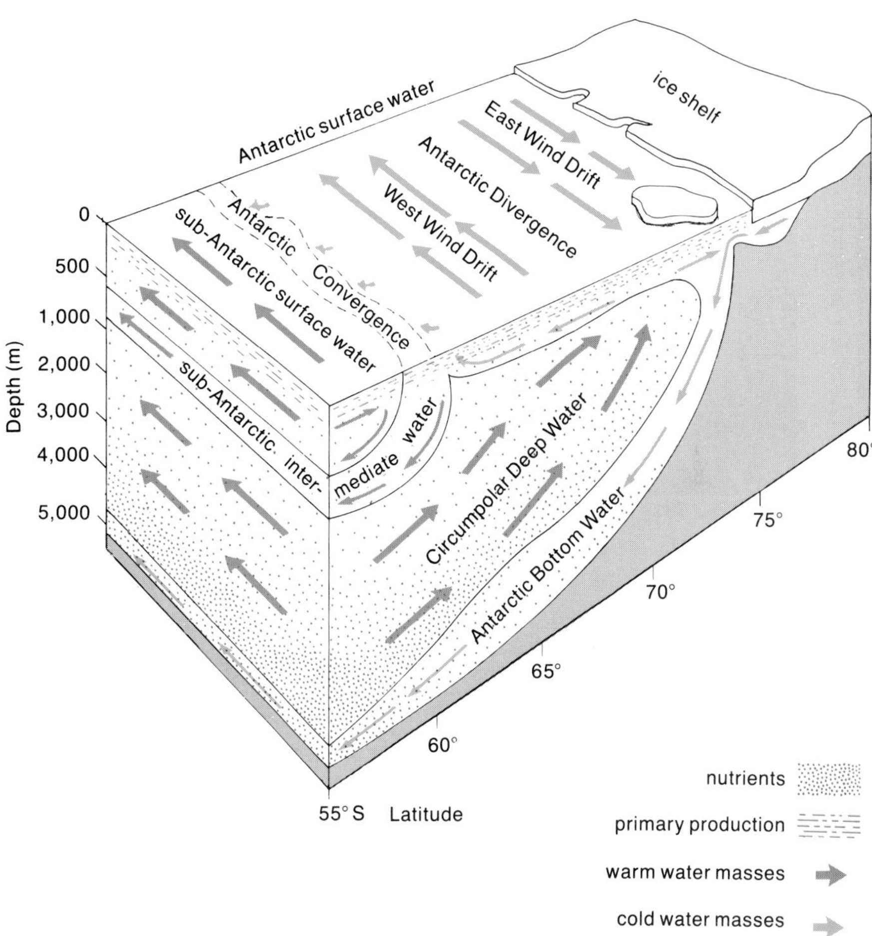

Figure 3. The unusual concentration of nutrients carried to the surface when the warmer Circumpolar Deep Water rises over the Antarctic Bottom Water has important consequences for the food web of Antarctica. The presence of this rich concentration means that the availability of nutrients does not constrain the plant primary production on which all higher life forms depend. Rather, a limit may be imposed by the wind-induced turbulence of the water, which prevents the primary producers from remaining at the optimum depth near the surface for utilizing incoming solar radiation. Many zooplankton organisms including krill migrate vertically between the northward- and southward-flowing water masses either in a daily or seasonal cycle or at different times during their life cycles.

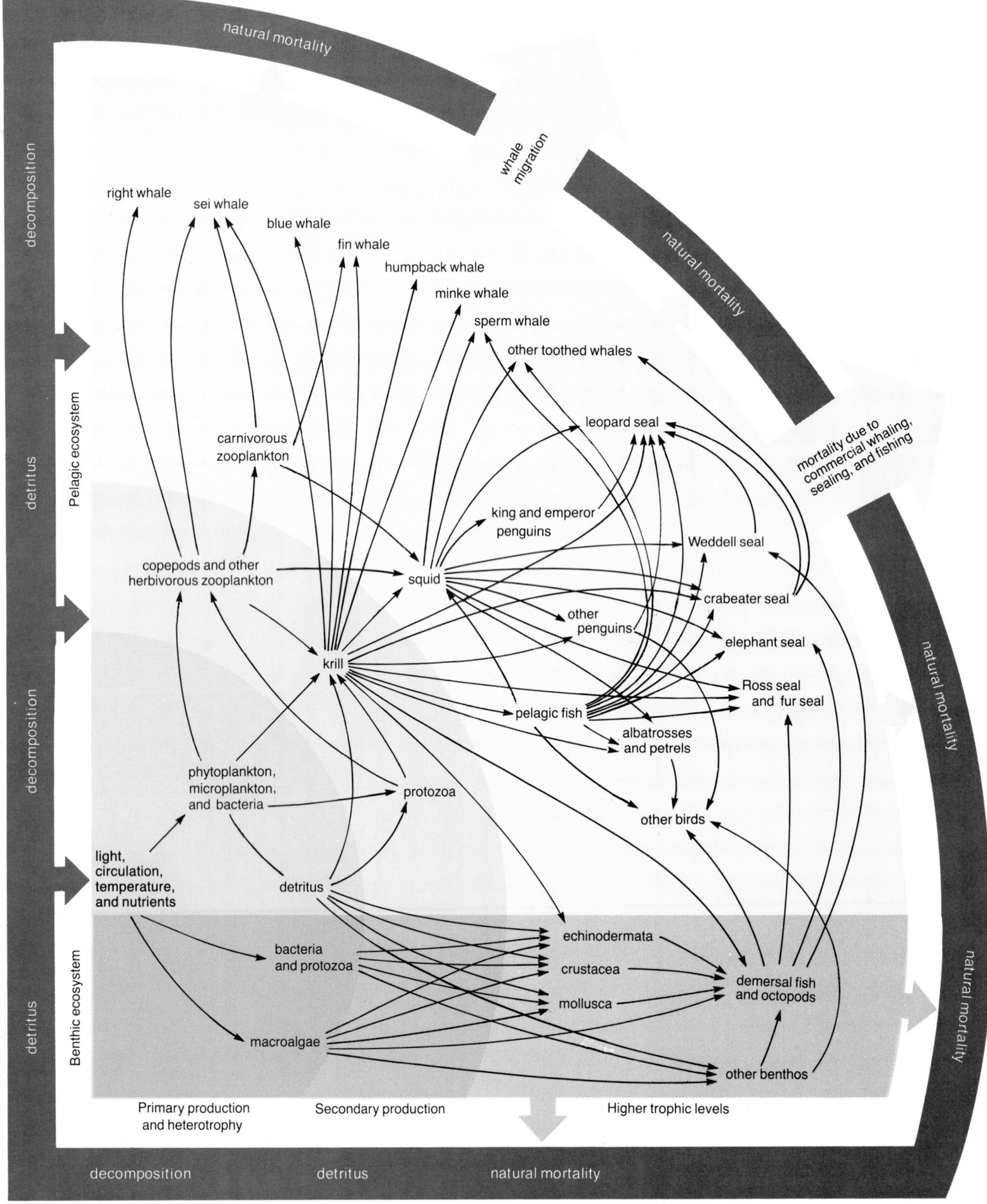

Figure 4. All life in the Antarctic is linked within a great food web. Primary production by phytoplankton and microplankton supports bacteria, protozoa, and zooplankton, including a krill population estimated at 500 to 750 million metric tons. Other key groups in the food web are copepods, carnivorous zooplankton such as chaetognaths, and some crustaceans, squid, and fish. Together with krill, these organisms directly or indirectly form the food base for the more conspicuous birds, seals, and whales. Benthic life is dependent on the rain of debris from above, and in near-shore waters on primary production and detritus from seaweeds. Demersal fish living near the bottom feed on molluscs, crustacea, and echinoderms such as sea urchins and starfish. Although the whales were once the principal consumers of krill, eating an estimated 190 million tons annually, their decline in numbers with whaling may mean that the available krill has been reapportioned, with birds and seals now taking the largest shares (see Table 1). (After Everson 1977.)

Krill and other zooplankton

Zooplankton species also tend to be circumpolar in distribution, occupying broad latitudinal zones in which they move vertically between the northward- and southward-flowing water masses daily, seasonally, or at different times during their life cycle. The zooplankton standing crop in Antarctic waters is significantly higher than that in tropical and temperate regions, averaging about 105 mg/m^3. A single zooplankton species, *Euphausia superba*, the Antarctic krill (Fig. 5), probably represents half of this total; it is the key organism in the Antarctic marine ecosystem, although there are several other species known collectively as krill. Krill is predominant among the zooplankton of the East Wind Drift and the Weddell Drift, but is almost without exception absent north of the Antarctic Convergence.

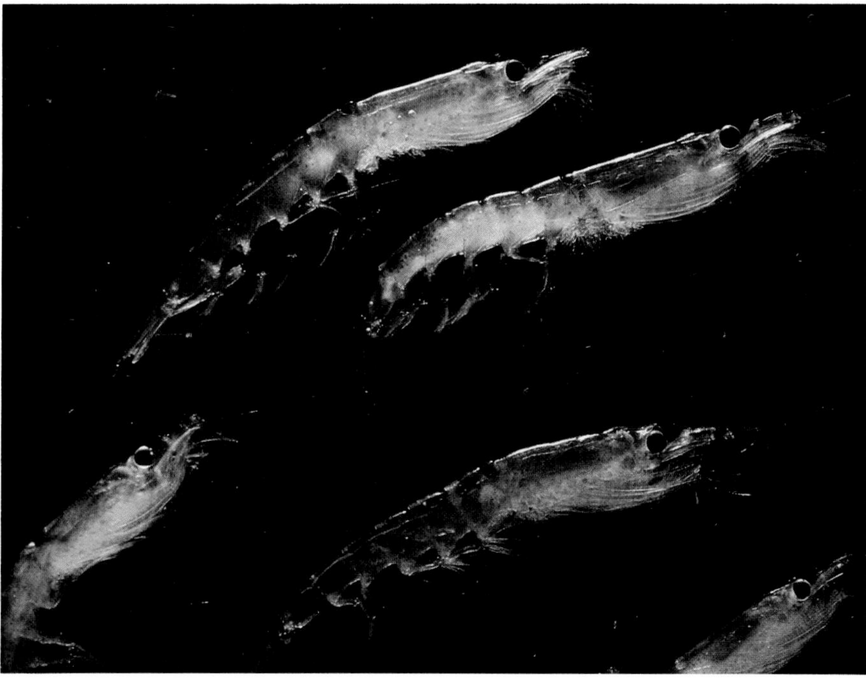

Figure 5. The abundant Antarctic krill, *Euphausia superba*, is central to the Antarctic ecosystem. Visible in individuals in this portion of a krill swarm is the krill's "filtering basket," a specialized arrangement of progressively finer filters designed to extract a wide variety of food organisms from the water; the green phytoplankton-filled gut can also be seen. It appears that the large body size of the adult krill as opposed to other euphausiids is the key factor in its success in utilizing the seasonally fluctuating primary production of the Antarctic. However, the large body size also means that krill must swim actively to avoid sinking; forward speeds of up to 46 cm/sec have been recorded. Because of this need for activity krill can live only in well-oxygenated waters. (Photograph courtesy of C. Gilbert.)

Other zooplankton in the food web include copepods, amphipods, other euphausiids, chaetognaths, salps, siphonophores, pelagic gastropods, and fish larvae, but these groups have received little attention. In the West Wind Drift the major herbivores are not krill but copepods, especially *Calanoides acutus*, which maintains populations in this zone by seasonal vertical migrations, deeper in winter, between northward- and southward-moving water masses. In winter two-thirds of the total zooplankton biomass is located below 200 m, but in summer there is a marked increase in zooplankton in the top 100 m of the ocean because of seasonal vertical migration and growth from grazing on phytoplankton.

In winter the extent of the krill range not covered by ice is only about 1 million km^2, but the contraction of the pack-ice zone southward in summer enlarges this area to a peak of about 13 million km^2 (Laws 1977). Together with vertical migration and high summer production, the retreat of the pack ice in summer is associated with a great increase in the amount of krill available to predators in ice-free surface waters. Although krill is circumpolar in distribution (Fig. 6), there appear to be several areas of high concentrations of krill which may or may not represent separate stocks (Mackintosh 1973), an issue of crucial importance to the way this resource is managed. In general these concentrations correspond to a series of eddies or gyres in the oceanic circulation. The largest such feature, the Weddell Gyre located between 0° and 60°W, is associated with krill populations much greater than those in other sectors, and mainly for this reason populations of whales were also much greater there formerly, when whale stocks were large. Fish stocks, seal populations, and breeding colonies of birds are also more abundant in the Weddell Gyre than elsewhere.

The main spawning season for krill extends from January to March and is most active at fronts such as the Antarctic Divergence. The eggs are released at the surface and sink to the warmer Circumpolar Deep Water, where they hatch at a depth of about 750 m. The larvae develop as they rise to the surface, growing over the next two to three years to a length of about 6 cm (Marr 1962). During this time they are found mainly in the top 200 m of the water column, although concentrations exist down to a depth of 600 m. Early workers believed that krill had a two-year life cycle punctuated by molts, but it is now known that they live longer. Studies of concentrations of lipofuscin, a pigment that accumulates with age and thus provides a measure of metabolic time, suggest that there may be up to seven age groups in krill, including larval and juvenile stages (Ettershank 1983). Such longevity is unusual if not unique in plankton organisms—an exciting aspect of krill biology. Molting represents a large loss of organic material, and the growth processes of krill have been shown to be remarkably plastic. In the waters around South Georgia mature male krill grow from 0.4 to 1.6 g in a 190-day summer period (Clarke and Morris 1983). However, it is clear that in vitro studies may show very different results; growth between molts can vary between −15 and +20%, depending on experimental conditions, and molting may be less frequent or may stop altogether in the winter (Siegfried et al., in press).

E. superba is a fast swimmer with a high metabolic rate; it is calculated that 30% of its energy output is allocated to active feeding. Krill feed on the most abundant food particles over an enormous size range, from microorganisms to other crustacea, and at high rates. Huge swarms comprising up to 2.1 million tons of krill in an area of 450 km^2 have been described. The ammonia output of krill is high compared with that of birds and mammals—a significant fact because some 50 to 90% of all nitrogen assimilated by phytoplankton is in the form of ammonia (Siegfried et al., in press).

It appears that in the open ocean sexually mature krill undertake daily vertical migrations within the surface layers, rising to the surface and dispersing when feeding actively. Growth of zooplankton tends to be confined to the summer, and survival through the winter depends on the presence of lipid reserves, a switch of diet, or reduced metabolic demands. In winter krill may descend to deeper water and feed on plankton detritus. With primary production virtually at a standstill in the surface water because of increased turbulence, reduced light, and pack-ice cover, this is likely to be their only source of food at that time (Everson 1984a). They may then fast, shrinking in body size, as in laboratory experiments (Ikeda and Dixon 1982).

All these mechanisms and behaviors, including a very efficient filter-feeding apparatus and a very wide spectrum of food, ranging from microplankton to other krill, suggest that the krill-centered system is one of the most efficient in the animal kingdom, and this may be the key to understanding the apparently large secondary and tertiary production in the Southern Ocean. The total krill biomass has been estimated at 500 to 750 million tons; annual production may be as high as 750 to 1,350 million tons. Although little confidence can be attached to these figures, further sonar surveys, calibrated by net sampling, will improve the accuracy of estimates (Everson 1984a).

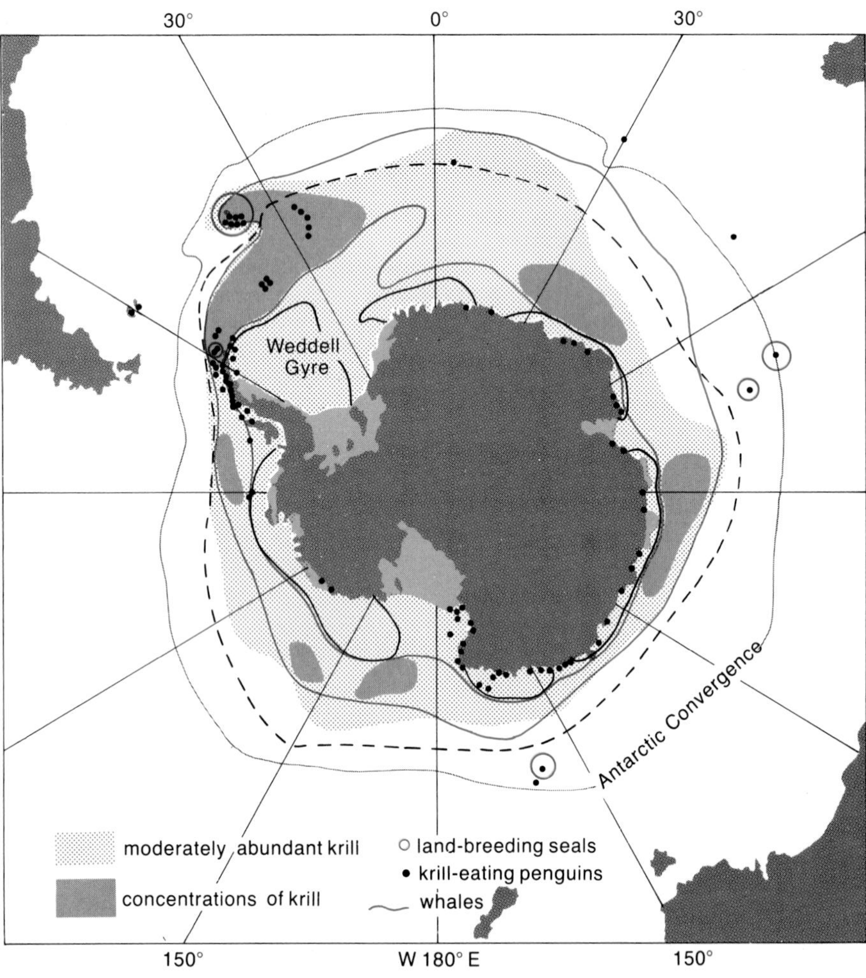

Figure 6. The distribution of species within the Antarctic ecosystem appears to be related not only to major environmental boundaries such as the Antarctic Convergence and the mean outer limit of pack ice in winter (*dashed black line*) and summer (*black line*), but to the distribution of krill. The area of moderately abundant krill is shown in red stippling; supposed concentrations of krill (*solid red*) occur at several points, the largest in the Weddell Gyre. The distribution of breeding colonies of krill-feeding penguins (*black dots*) is similar, with concentrations in the Scotia Sea and the Antarctic Peninsula. Krill-feeding petrels and albatrosses, not shown here, follow a comparable pattern. Distributions of ice-breeding seals are related to the shifting boundaries of the pack ice indicated here, contracting in summer so that they are complementary to whale distributions; the distribution of land-breeding seals on sub-Antarctic islands is shown in green. The distribution of whales (*blue lines*) is based on earlier information from whaling catches. The major concentrations of krill, birds, seals, and whales occur in the Atlantic sector, and it is here that the overlap in distributions is greatest. (After Laws 1984; Mackintosh 1973; Wilson 1983.)

Oceanic squid and fishes

Another important invertebrate group in this ecosystem is the Cephalopoda, mainly the oceanic squid. The remains of squid are frequently found in the stomachs of whales, seals, and birds, but living squid are very difficult to sample in the ocean. There may be more than 20 species occurring in large numbers, most ranging in size from 1 to 4 m, with some reaching greater sizes. Very little is known about these organisms, although two species—*Kondakovia longimanna* and *Moroteuthis knipovitchi*—are predominant in pelagic catches by research trawlers. These species and the genus *Architeuthis* are commonly cannibalistic, but their most important food appears to be Antarctic krill. *Themisto gaudichaudii*, an amphipod, is also a frequent item in their diet, and they feed on other zooplankton and microplankton as well.

We know that squid perform vertical migrations, the krill-feeders probably diurnally and perhaps on a lunar cycle. Squid seem to select the same size range of krill as baleen whales, and indeed there is increasing evidence of considerable overlap among all the krill predator groups, since their diurnal and seasonal cycles are locked in on the krill cycle. In general the size of squid seems to increase with the depth at which they are captured, and it has been suggested that the larger squid feed on fish in deeper waters (Nemoto et al., in press). The total annual consumption of squid by whales, birds, and seals is probably about 35 million tons (Clarke 1983), but their total biomass may be much higher, and even the estimated 35 million tons of squid eaten by predators may in turn have consumed about 100 million tons of krill.

The Southern Ocean is clearly a very special environment: of the 20,000 species of modern fishes only about 120, of 29 families, occur south of the Antarctic Convergence (Everson 1984b). The dominant group is the Nototheniiformes, which make up nearly three-quarters of the coastal fish species. Most Antarctic fish have slow growth rates and do not mature until they are several years old. Some species—for example, the South Georgia cod (*Notothenia rossii*)—spend their early years in the waters near shore before dispersing more widely to feed on krill, when they show an increased growth rate. Others, such as the southern blue whiting (*Micromesistius australis*), spawn on continental shelves in temperate waters and migrate into the Antarctic each summer to feed on the high concentrations of krill. It has been shown that the commercially important *Champsocephalus gunnari* feeds on krill in surface waters at night, remaining quiescent on the bottom during the day.

Figure 7. Among the best-adapted Antarctic seabirds are the petrels, which contribute 18 of the 35 species that breed south of the convergence. The Antarctic petrel, seen here in flight over pack ice, feeds on krill, other zooplankton, squid, and fish. Together with the snow petrel it is the most southerly of the petrels, and is usually found in regions of pack ice or in upwelling water at the edge of ice shelves. These two species nest up to 300 km inland along the eastern coast of the Weddell Sea, seeking out isolated nunataks, exposed rocky outcroppings which provide the only suitable nesting sites along a coast that is continuously fringed by ice shelves. This location enables them to exploit krill populations in the East Wind Drift.

In contrast to other oceans, the Southern Ocean does not contain dense shoals of pelagic fish, although small bathypelagic fish, mainly Myctophidae, may well be abundant in the open ocean. A recent study of the evolution of neutrally buoyant fish (Eastman, in press) has indicated the dominant role of one pelagic fish, the Antarctic silverfish (*Pleurogramma antarcticum*), in the Ross and Weddell seas and its importance in turn to the giant *Dissostichus*, also a midwater species. These fish are pelagic seasonally, when there is an abundant supply of food at these depths, but appear to rest on the bottom during the rest of the year, paralleling the diurnal pattern of *Champsocephalus* on a longer time scale.

The feeding ecology and diet of fishes in western Antarctic waters and off South Georgia have recently been investigated by Kock (in press). In these waters the seven species studied consumed about 5 million tons of krill a year; the two most important species, *C. gunnari* and *N. rossii*, shared 4 million tons of this total, although their stocks have been greatly depleted by commercial trawling. It has been estimated that all Antarctic fish combined consume on the order of 100 million tons of krill annually (Everson 1984b; Kock, in press; Siegfried et al., in press).

Seabirds

The vast stretches of ocean and small extent of ice-free land mean that almost all Antarctic birds are seabirds of the two best-adapted marine groups, the penguins and the albatrosses and petrels (Fig. 7). The ecology of these groups has recently been reviewed by Croxall (1984). The range of size is great, with the weight ratio between the emperor penguin and the smallest petrel over 1,000 to 1. South of the Antarctic Convergence only 35 species breed: 6 penguins, 6 albatrosses, 18 petrels, 1 shag, and 4 skuas, gulls, and terns. Some of these breed in the sub-Antarctic zone north of the convergence as well, and an additional 29 species, including 4 penguins, 2 albatrosses, 13 petrels, 4 shags, and 6 skuas, gulls, and terns, are also found there. The distributions of these birds are greatly influenced by oceanographical and environmental features such as islands, fronts where waters diverge or converge, pack-ice zones, ice shelves, or ice-free ground suitable for breeding (Fig. 6). Although the species are few, populations are usually very large; most of the species are colonial breeders, with some colonies containing millions of pairs. Some petrels occupy burrows, especially in the sub-Antarctic. Other birds—petrels, albatrosses, and penguins—nest in the open (Figs. 8 and 9), but the snow petrel seeks out crevices. These differing preferences in nest sites help to promote ecological separation.

The climate determines the breeding season, which is about two to three weeks earlier in northerly than in southerly populations of the same species. In most species egg-laying and the fledging of chicks occur on land in the spring and summer, during the main period of zooplankton abundance; chicks have long fledging periods, extending to more than a year in the wandering and gray-headed albatrosses and the king penguin, which are all sub-Antarctic species. The emperor penguin breeds on the fast ice in winter under the most ex-

Figure 8. A nesting colony of gray-headed albatrosses at Bird Island off South Georgia occupies raised nests constructed of mud and tussac grass. Used seasonally by each pair for many years, the nests keep eggs and chicks above the cold surface-water runnels. The wet but relatively warm climate of this sub-Antarctic island supports a lush growth of tussac grass at low altitudes near the sea, where it is exposed to spray. Gray-headed albatrosses feed primarily on krill and squid. The relatively low calcium content of squid means that the young grow more slowly and take longer to fledge, resulting in a biennial breeding cycle rather than an annual one like that of the black-browed albatross.

Figure 9. Gentoo penguins nest in the open at South Georgia. Females lay two eggs, and the sexes exchange incubation duties daily. One of the least abundant of the 16 species of penguins, the gentoo is found mainly in the sub-Antarctic, particularly in the western portion of the Scotia Sea and the Antarctic Peninsula. Its feeding range is more restricted than those of other Antarctic penguins, rarely exceeding 30 km, and it supplements a diet of krill by taking fish from kelp beds near shore.

treme environmental conditions experienced by any vertebrate (Le Maho 1977), with temperatures as low as −60°C and winds that may reach over 150 km/hr. The eggs are laid on the fast ice formed in late summer and incubated on the penguin's feet, protected by a brood pouch; after the chick hatches it is reared through the winter, fledging in midsummer. Male emperor penguins fast for three to four months and lose almost half their body weight during courtship and incubation. So harsh are the conditions that only 19% of the emperor chicks survive their first year.

The related sub-Antarctic king penguin generally lays its eggs in early December; the chicks fledge a year later, and 93% survive the first year. However, the shorter fledging period of the emperor penguin means that it can produce a chick every year, whereas the king penguin averages two chicks in three years. Male and female penguins take turns sheltering the egg and foraging, and the length of their incubation shifts is correlated with varying foraging ranges. These ranges are extensive in the Adelie penguin and even longer in the emperor, perhaps because in the early part of some breeding seasons the penguins must cross a zone of fast ice to reach their feeding areas.

The extent of foraging ranges during the breeding season varies from relatively small in the penguins (gentoo, 30 km; chinstrap, 100 km; macaroni, 120 km; king, 500 km) to larger in the flying birds (petrels, 300 to 900 km; albatrosses, up to 2,650 km—probably the most extensive range of any bird). Although penguins have a relatively short horizontal range, their vertical feeding range is much greater (emperor penguins, up to a depth of 265 m; king, 235 m; gentoo, 100 m; chinstrap, 70 m). Both penguins and shags are pursuit divers, with shags reaching depths of up to 25 m. Time-depth recorders attached to chinstrap penguins indicated that 40% of their dives were to depths of less than 10 m (Croxall 1984); the large krill they pursue in these dives are probably caught individually, possibly by a combination of simple echolocation and bioluminescence. It has also been shown that king penguins on four- to eight-day feeding trips make from 500 to 1,200 dives, more than half to depths below 50 m, with only one dive in ten resulting in the capture of prey (Kooyman et al. 1982).

The flying birds are much more closely tied to the surface, with possible depth ranges of about 0.1 to 3 m, and show a variety of feeding methods: storm petrels and terns feed by shallow dipping through the surface film to catch individual prey, prions filter-feed at the surface with their specialized bills, and albatrosses, petrels, and diving petrels use shallow surface dives or plunges. Activity meters attached to gray-headed albatrosses have shown that they spend 75% of their time away from the nest in flight and 25% on the sea, mostly at night, when food is near the surface (Croxall and Prince 1983). Thus ecological separation is promoted both by differences in feeding preferences and by morphological and behavioral adaptations. The range and depth of the foraging area, the type of food, the method and timing of feeding, and the timing of the breeding season all vary, reducing direct competition for food. When the preferred food is scarce, however, birds may switch to other prey and the overlapping increases.

The birds of Antarctica feed mainly on crustacea (largely krill, copepods, and amphipods), squid, fish, and carrion, with krill amounting to 78% of all the food they eat. According to recent estimates (Croxall 1984), about 115 million tons of krill are taken annually by birds either directly or indirectly. Flying birds that feed on crustaceans in summer must find another source of food in winter, when crustaceans are located at greater depths or covered by pack ice. They either switch to a diet of squid or fish or migrate northward like the storm petrels and terns. There are important differences in the diets of quite closely related species, for example the common and the South Georgia diving petrels, the dove prion and the blue petrel, and the black-browed and gray-headed albatrosses. Thus the two species of diving petrels nesting on Bird Island off South Georgia take food items in different proportions: whereas the common diving petrel selects 76% krill, 4% amphipods, and 20% copepods, the South Georgia diving petrel consumes respectively 15%, 17%, and 68% of the same foods (Payne and Prince 1979).

Figure 10. The crabeater seal is the most abundant species of seal in the world, with a population estimated at over 30 million – more than the total population of all other seals combined. It is confined to the outer fringes of the pack ice, although occasional stragglers are seen on the southern temperate shores of the surrounding continents. An obligate krill-feeder, it takes its prey mainly at night, when krill is nearer the surface. Its numbers appear to have increased with release from competition with the baleen whales, but the rate of increase has now probably leveled off or declined.

Figure 11. This pod of southern elephant seals is undergoing a period of molting, during which both sexes fast and hair and epidermis are shed. This species is predominantly sub-Antarctic in distribution, with a world population of about 750,000. A land-breeder, the elephant seal exhibits very conspicuous sexual dimorphism; adult males average six to seven times the weight of adult females – the largest ratio of all seals. The related northern elephant seal breeds off the coast of California in subtropical conditions. In contrast, at the southern limit of its range some small colonies of the southern elephant seal breed on coastal fast ice.

Penguins account for virtually all avian biomass and food consumption in the Antarctic, and two-thirds of their populations consist of Adelie penguins; in the sub-Antarctic, penguins constitute effectively 80% of the avian biomass, and half of their numbers are macaroni penguins. Most penguin species feed overwhelmingly on euphausiid crustaceans, supplemented in some cases by fish. The two largest, however—the emperor and king penguins—take fish and squid almost exclusively. Krill has a higher nutritive value than squid, and the chicks of krill-feeding species such as the macaroni penguin and the black-browed albatross therefore grow faster and have a shorter fledging period than the offspring of squid-feeders such as the king penguin and the gray-headed albatross.

Adult survival rates are high, up to 95% yearly in petrels, albatrosses, and the emperor penguin, but somewhat lower—82 to 87%—in the other penguins. Life spans are correspondingly long, reaching an estimated 70 to 80 years in the wandering albatross and the emperor penguin. The age at first breeding ranges from two years in small species like the diving petrel to ten years in albatrosses. Breeding success may vary from year to year. In the 1976/77 season, gentoo penguins at Bird Island fledged 1.2 chicks per pair; in 1977/78, when there is known to have been a dearth of krill swarms, they fledged less than 0.1 chick per pair; in 1978/79 the average number fledged was up again to 0.6 (Croxall and Prince 1979). The season of 1983/84 has been marked by a recurrence of krill shortages and reduced breeding success. Macaroni penguins were less affected in 1977/78 than the gentoo, probably because their foraging range is greater, and gray-headed albatrosses were more successful than black-browed, probably because they depend primarily on squid rather than krill.

Antarctic seals

Seven species of seals are found in the Southern Ocean, and their ecology has been explored in some detail (see Laws 1984). Their breeding populations are confined to the Antarctic, with three exceptions. The southern elephant seal also occurs less abundantly on sub-Antarctic islands, the Falkland Islands, and South American coasts, and the sub-Antarctic fur seal is rarely seen south of the Antarctic Convergence. A third species, the Antarctic fur seal, breeds in small numbers north of the convergence. The four true Antarctic species, virtually confined to zones of pack and fast ice, are the crabeater, leopard, Ross, and Weddell seals; in contrast, the elephant and fur seals are land-breeders and are rarely found in areas of pack ice. There is a well-defined ecological separation of the species. The crabeater and leopard seals have similar distributions in the pack-ice zone, occurring in greatest abundance at the fringes, but the leopard seal is also found on sub-Antarctic islands and even in lower latitudes. The Ross seal is characteristically found in areas of denser pack ice, as is the Weddell; however, the Weddell seal has a much more southerly distribution, nearer continental shores, and also breeds around Antarctic islands.

The Weddell seal has evolved to fill a specific niche, the fast-ice zone close to shore, where it establishes pupping colonies and winters under the ice, keeping breathing holes open with its teeth. This location enables it to exploit inshore prey year round and also greatly reduces predation on its young by the killer whale and the leopard seal. A probable penalty of occupying this niche is higher adult mortality caused by tooth wear. At the other extreme is the crabeater seal (Fig. 10); its young are subject to very heavy predation, mainly by leopard seals, but live longer than Weddell seals if they survive their first year.

The female crabeater seal gives birth in the spring, and each cow and pup pair is joined by an adult male, forming a characteristic breeding group usually separated from other crabeaters by a kilometer or more of pack ice. The pup is weaned when it is about a month old and takes to the water, where it is preyed on by leopard seals. The leopard seals breed later; the very widely dispersed cow and pup pairs are not joined by a male, and mating occurs after the pup is weaned rather than before. Very little is known about the breeding behavior of the Ross seal. The stable fast-ice platform of the Weddell seal has made possible gregarious colonies and the development of a male hierarchical system, with dominant males defending three-dimensional underwater territories beneath the pupping colonies, where they mate with entering females. As in the case of the other seals that breed on ice, the female is larger than the male. The polygynous breeding behavior of the Weddell seal is intermediate between the highly polygynous mating system of the land-breeding elephant and fur seals, where the male is much larger than the female, and the monogamous pattern of the crabeater, leopard, and Ross seals, which breed on shifting pack ice.

The elephant seal (Fig. 11) gives birth in September or November, the fur seal in November or December. The two species have different preferences in breeding sites—the elephant seal favoring larger open beaches, the fur seal rocky shores and small, sheltered beaches—which also minimizes competition. Both have a highly developed polygynous social organization, in which dominant bulls prevent subordinate bulls from approaching the breeding females. The elephant seal pup is weaned at 23 days, having quadrupled its birth weight; during lactation the cow fasts. The cow mates 19 days after parturition, and leaves the beaches after weaning the pup. The female fur seal mates 5 days after giving birth; she makes frequent feeding trips to sea during lactation, which lasts 110 to 120 days.

Annual growth layers in the teeth can be used to estimate age, a technique developed in studies of the elephant seal (Laws 1952) and now widely applied to other marine and terrestrial mammals. Longevity ranges from 20 years in the elephant seal up to 40 years in the crabeater seal, with an annual adult mortality rate of about 5 to 10%. The pattern of the layers also indicates reproductive history. There is a transitional zone between immature and mature years, coinciding with the onset of sexual maturity, and lactation periods can be identified in the structure of elephant and fur-seal teeth. Sexual maturity is attained at 3 to 5 years in the Antarctic seals, but social maturity—that is, participation in breeding—is deferred in male elephant and fur seals until an age of 8 to 12 years.

The seals of Antarctica also show an adaptive radiation based on their feeding habits (Laws 1977, 1984). Of the two seals characteristic of ice-free waters, the elephant seal appears to consume a diet consisting of approximately 75% squid and 25% fish, whereas the fur seal at South Georgia feeds almost exclusively on krill near

Figure 12. At 9 m long, the minke whale is one of the smallest baleen whales. An exception to the rule that the largest species and individuals migrate to the highest polar latitudes, the minke whale is found as far south as the largest blue whale in the Southern Ocean, where it feeds, like the blue, on small krill; elsewhere it also feeds on fish. The minke is anomalous in other respects: its layer of blubber is relatively much thinner than that of the larger baleen whales, which may mean that it must feed more continuously, and it has a much higher pregnancy rate. It is generally believed that the minke whale has experienced a release from competition as populations of the blue whale have declined, so that it is now the most abundant Antarctic species of baleen whale and the only one currently hunted. (Photograph courtesy of D. R. Gipps.)

the breeding grounds. Fish and on rare occasions squid have also been found in the stomachs of fur seals at South Georgia; overall they probably eat equal amounts of krill, squid, and fish. Of the others, Weddell seals consume on average 53% fish, 11% cephalopods, and 36% other invertebrates; crabeater seals, 94% krill, 4% fish, and 2% squid; leopard seals, 37% krill, 13% fish, 8% squid, 3% other invertebrates, and 39% birds, seals, and carrion; and Ross seals, 22% fish, 64% squid, 9% krill, and 5% other invertebrates. The fur seal (and probably the crabeater and leopard seals) feed in surface waters, mainly at night, in the top 40 m; this is correlated with the diurnal vertical migrations of krill, which rise to the surface at night. The other seals are deep-diving, with the Weddell known to reach depths of at least 600 m.

The fur and elephant seals, which are mainly sub-Antarctic in distribution, were brought near extinction by sealers in the nineteenth century, and their populations are still recovering. The elephant seal was the object of a nationally controlled industry at South Georgia from 1910 to 1964. Based on a philosophy of sustained yield, this enterprise was a classic example of good management, and ceased for economic reasons when whaling ended there, rather than because of overharvesting (Laws 1979). The population of Antarctic fur seals at South Georgia has increased with protection from a few hundred in the 1950s to probably over a million today, with the numbers rising by 16.8% annually from 1958 to 1972 and by 14.5% from 1972 on—the fastest rate of population growth of any marine mammal. At the current rate the Antarctic fur seal will soon overtake the northern fur seal as the world's most abundant eared seal and should regain or exceed its original numbers of several million before the end of the century.

The crabeater seal, now probably numbering about 30 million, is the most abundant seal in the world, comprising two-thirds of the total world stock of seals and more than three-quarters of its biomass. The species appears to have experienced a decline in the age at which it reaches sexual maturity (Laws 1984; Bengtson and Laws, in press), a phenomenon correlated with an increase in the reproductive rate and thus an expansion of population. This expansion is in line with the drastic decline in whale stocks, which presumably led to an increase in the amount of krill available for other predators. The stocks of Antarctic seals now probably total at least 33 million, with a biomass of about 7 million tons; these seal populations annually consume over 130 million tons of krill (two or three times the current consumption by whales), at least 10 million tons of squid, and at least 8 million tons of fish (Laws 1977, 1984; Clarke 1983, recalculated).

Southern whale stocks

Whales are highly mobile, ranging over much of the world's oceans, with many species occurring in both northern and southern hemispheres from the equator to the polar seas (Mackintosh 1965; Brown and Lockyer 1984). Most baleen, or filter-feeding, whales migrate seasonally between tropical or subtropical breeding areas and polar feeding grounds, as do adult male sperm whales. The large baleen whales of the Southern Ocean comprise six species and one subspecies: in addition to the blue whale and its subspecies the pygmy blue, there are the fin, sei, minke, humpback, and southern right whales, the last of which rarely enters waters south of the convergence. There is one large toothed whale, the sperm whale. Eleven small toothed whales found south of the Antarctic Convergence include the killer whale, an important predator on seals, as well as the southern white-sided dolphin and two beaked whales (Ziphiidae); twenty species are recorded in the sub-Antarctic zone. Knowledge of the small toothed whales is still very slight (Brown and Lockyer 1984).

Before exploitation took effect, baleen whales were probably about four times more abundant, and had a biomass five times greater, in the southern hemisphere than in the northern, but all stocks have been greatly reduced by whaling. However, there is less diversity in the southern whales, and some types that are represented in the Arctic are lacking in the Antarctic (Laws 1977).

The reproductive cycle of baleen whales is closely tied to the extreme seasonality of the Antarctic environment and their long-range migrations between breeding and feeding areas. Mating and births occur in the tropics and subtropics, and gestation periods are less than a year. The characteristic fetal growth curve of the baleen whales is unique among mammals, not only because of its magnitude (a blue whale calf weighs 2½ tons at birth), but because of its unusual configuration. If size is expressed as body length or the cube root of body weight, there is an initial linear phase followed by an exponential phase; this exponential growth does not occur in other mammals (Laws 1959). Interestingly, the shift takes place at the time the whales begin to feed actively on krill in the Antarctic. The lactation period of most baleen whales is six to seven months; the calf is weaned shortly after entering Antarctic waters. Female sperm whales do not migrate to the Antarctic—only sexually mature males are found there—and are not so rigidly tied to an annual cycle as the baleen whales. Fetal growth is linear, with large birth size achieved by a longer gestation period of 16 months; there is a correspondingly long suckling period of several years.

The age of toothed whales can be determined by examining annual growth layers in the teeth, as in seals; in baleen whales annual layers in the wax ear plug are analyzed. A transitional zone in the ear plug analogous to that in seal teeth has been used to estimate the age at sexual maturity (Lockyer 1972). The interpretation of this zone is still controversial, but sexual maturity can also be established more directly by relating age to ovary and testis development. The age at sexual maturity ranges from 6 to 14 years and has apparently decreased in a density-dependent way as commercial whaling has reduced the whale stocks. In the baleen whales females reach larger sizes than males, but the sperm whale shows a more marked sexual dimorphism in favor of the male, achieved by a major spurt in growth at and just after puberty, when males begin to feed in Antarctic waters. Unlike the baleen whales, which are monogamous, the sperm whale has a breeding system based on the harem. Male social maturity is deferred until the age of 26 years, when growth is nearly complete. The great whales attain ages of 80 years or more, and adult mortality rates are about 3 to 5% (Allen 1980). Very little is known about the biology of toothed whales other than the sperm whale.

The migrations of the baleen whales are staggered, with the largest species tending to arrive earliest in Antarctic waters—the blue whale first, the fin and the humpback next, followed by the sei. The extent of penetration into colder waters is also correlated with body size; in general the larger species travel farthest south, the small minke whale (Fig. 12) being a notable exception. As a result, although the Antarctic distributions overlap, each occupies a characteristic latitudinal zone. There are also longitudinal variations in abundance both within and between species, and the longitudinal segregation of fin whales by size suggests the presence of intraspecific competition. Within each species, too, the migrations of different classes of individuals are staggered in relation to their energy needs. Larger and older whales tend to reach higher latitudes than smaller and younger ones; pregnant females arrive early and leave late, and females with calves postpone their entry into the cold polar waters. The average annual feeding period south of the Antarctic Convergence is about 120 days (Laws 1977; Brown and Lockyer 1984).

The major food of the baleen whales is krill, but the more northerly species feed mainly on copepods. The mouth and baleen are functionally adapted to the size of the food organisms eaten, notably in terms of the spacing and degree of fineness of the filtration mechanism. Three different types of feeding behaviors are found among baleen whale species: swallowing or gulping, in which a mouthful is taken and filtered; skimming, a more continuous process of filtering, with the jaws open 10 to 15° (typical of the right whale); and a combination of the two. As a result of all these mechanisms there is a considerable degree of inter- and intraspecific separation on the feeding grounds, although this separation is by no means complete (Laws 1977; Brown and Lockyer 1984). The relative amount of krill available increases enormously in the summer months; indeed, the baleen whales would be under nutritional stress were they present in Antarctic waters in winter at their summer densities. The boundary of the zone in which krill are found in quantities sufficient for baleen whales to feed on moves southward, reflecting progressive grazing by waves of immigrant baleen whales, as the krill are exposed by the retreat of the pack ice. In the years before 1940 the whales entered the East Wind Drift by March, but they now tend to remain in the West Wind Drift, presumably because food is more abundant than formerly. The sperm whale feeds predominantly on cephalopods ranging from 0.5 to 4 m in size, but also takes pelagic fish such as *Micromesistius* and *Dissostichus.*

The Antarctic baleen whales and the sperm whale have been the object of a fishery since 1904, when their numbers are estimated to have been about 1.1 million. At that time their biomass was about 45 million tons, and they consumed an estimated 190 million tons of krill, 14 million tons of squid, and nearly 5 million tons of fish annually. Competition for the available food probably limited sizes and numbers. By 1973 their numbers had declined through overhunting to about half a million, a biomass of 9 million tons, and they ate about 43 million tons of krill, 6 million tons of squid, and 130,000 tons of fish (Laws 1977).

Ecological interactions

The enormous reduction of the whale population means that some 150 million tons of krill formerly eaten by whales have probably become available to the remaining stocks of whales and to other predators. The importance of this resource is clear from a comparison with the total annual yield of all the world's fisheries, which is about 70 million tons. It has been suggested that some of the krill "surplus" could be safely harvested by man, while allowing the depleted stocks of whales to increase to a reasonable level. However, the "surplus" may have been taken up already by increases in the populations of other consumers (see Table 1).

There has been an increase in the pregnancy rates of fin and sei whales and an apparent but controversial decrease in their age at maturity (Brown and Lockyer 1984); in the case of the sei these changes preceded large-scale exploitation of the species, which began in the 1960s. If these shifts are real they would strengthen the conclusion that an indirect effect of whaling is involved, presumably through greater availability of food to the survivors. The age at which the minke whale reaches sexual maturity also appears to have declined, from 14 to 6 years (Brown and Lockyer 1984)—although this finding is also controversial—and the species may have expanded to partially fill the vacuum created by the removal of the blue whale. The present minke population levels may be double those existing before whaling began, although estimation of abundance is exceedingly difficult, so that such trends can be determined only over long periods of time.

The crabeater seal populations off the west coast of the Antarctic Peninsula, like those of the baleen whales, appear to have experienced a decrease in the age of sexual maturity, from about 4 years in the 1950s to 2½ years by the early 1960s. As in the case of the whales, this is likely to be the result of increased growth rates and to be associated with an increased reproductive rate, leading to an increase in numbers; the age at maturity has subsequently reverted to the earlier level, with an estimated annual rate of population increase of about 7.5% (Laws 1984; Bengtson and Laws, in press). In contrast, the 14 to 17% annual increase in the population of the South Georgia fur seals is unusually high—6 to 10% is the rule for marine mammals recovering from overhunting—and is probably also related to the increased abundance of krill accompanying the reduction of the whale stocks. We do not know the response of the fish and squid stocks to the increased availability of food, but it is likely to have been similar; commercial fishing activities complicate analysis.

Turning now to the birds, three of the most abundant species—the chinstrap, Adelie, and macaroni penguins—have also shown increases in population (Croxall and Prince 1979). Together these birds account for 90% of the Antarctic avian biomass, with breeding populations of 20 to 30 million pairs and total populations of about twice that. (Totals are estimated by counting members of the distinctive colonies and then making an allowance for immature age groups.) The chinstrap, a northerly species most abundant in the Scotia Sea, has shown average annual increases there of 6 to 10% over a period of 30 years in the colonies monitored. The even more northerly macaroni penguin at South Georgia has also shown large increases of up to 9% annually in the two decades from 1957 to 1977. Adelie penguins, which account for two-thirds of the Antarctic avian biomass, have increased at sites in the Scotia Sea, where they feed on *E. superba*; at Signy Island, one of the South Orkney Islands, the annual increase averaged 2 to 3% over the 31 years from 1947 to 1978.

Conversely, there is no indication of an increase in Adelie penguins in the Ross Sea, but little competition between whales and other groups for food is to be expected in this area. The whale and seal stocks there have always been relatively small, and there is a large area of the continental shelf where *E. superba* is replaced by *E. crystallorophias*, which does not congregate in the huge swarms characteristic of the Antarctic krill. The king penguin at South Georgia has increased nearly sixfold over 33 years, from 12,000 birds in 1946 to 57,000 in 1979, an average annual rate of increase of nearly 5%. Unlike the three other species, however, the king penguin feeds mainly on squid. The black-browed and gray-headed albatrosses at South Georgia (respectively mainly krill- and squid-feeders) have shown no significant change in numbers since the 1960s. The wandering albatross, which feeds chiefly on squid and fish, appears to have decreased by 19% since the 1960s. New colonies of blue petrels, a krill-feeder, are developing at South Georgia.

It seems reasonable to conclude that there have been increases over the past thirty years or more in most species, particularly the minke whale, the crabeater seal, and penguins dependent on *E. superba*, and larger than expected increases in the Antarctic fur seal. These appear to be due to the increased availability of food, especially krill. Increases in king penguins may be secondarily due to increases in krill-feeding squid, which are in any case available in larger quantities because of decreases in the

Table 1. Possible changes in patterns of consumption of Antarctic krill by the major groups of predators

	Annual consumption (tons × 10⁶)	
Predator	*1900*	*1984*
Whales	**190**[a]	**40**
Seals	50	**130**
Birds	50	**130**
Fish	100	70
Cephalopods	80	100
Total	470	470

SOURCES: Clarke 1983; Croxall 1984; Everson 1984b; Laws 1977, 1984

[a] Figures shown in boldface represent published estimates, adjusted in the case of birds to allow for population increases between 1979 and 1984. Other figures are tentative estimates based on indications from sources above and on back-projections of increases in seal and bird populations.

sperm whale, a principal consumer. Finally, growth in the number of fur seals at South Georgia has caused an increase in giant petrel populations there, particularly *Macronectes halli*, by increasing the availability of carrion. Potential rises in the population of krill-feeding fish would be expected to be nullified or reversed by the intensive commercial fishing activities. It is in the Atlantic sector that the responses to a "surplus" of krill are likely to have been greatest, for this is the region where there is the most overlap in the distributions of the whale, seal, and bird populations (Fig. 6).

Allowing for the very great uncertainty about the actual amounts consumed, these trends suggest that the total quantity of krill taken by predators may be about 500 ± 100 million tons, but apportioned differently now. This admittedly very crude estimate is not too different from the lower range of estimates that place annual krill production at 750 to 1,350 million tons, because to it should be added annual krill mortality from causes other than predation; this figure is, however, as yet unknown.

If krill harvesting is to increase, management must be soundly based and the rate of growth of the krill fishery must be slow enough to allow the ecosystem to come gradually to a new balance, as the numbers of krill-consumers other than man move toward their former levels of abundance. This requires a great expansion in our knowledge of the interactions within the food web, both in qualitative and quantitative terms.

References

Allen, K. R. 1980. *Conservation and Management of Whales.* Univ. of Washington Press and Butterworths, London.

Bengtson, J., and R. M. Laws. In press. Trends in crabeater seal age at maturity: An insight into Antarctic marine interactions. In *Antarctic Nutrient Cycles and Food Webs*, ed. W. R. Siegfried, P. R. Condy, and R. M. Laws. Springer-Verlag.

Brown, S. G., and C. H. Lockyer. 1984. Whales. In *Antarctic Ecology*, ed. R. M. Laws, pp. 717–81.

Clarke, A., and D. J. Morris. 1983. Towards an energy budget for krill: The physiology and biochemistry of *Euphausia superba* Dana. *Polar Biol.* 2:69–86.

Clarke, M. R. 1983. Cephalopod biomass—Estimation from predation. *Mem. Nat. Mus. Victoria*, no. 44:95–107.

Croxall, J. P. 1984. Seabirds. In *Antarctic Ecology*, ed. Laws, pp. 533–616.

Croxall, J. P., and P. A. Prince. 1979. Antarctic seabird and seal monitoring studies. *Polar Rec.* 19, no. 123:573–95.

———. 1983. Antarctic penguins and albatrosses. *Oceanus* 26:18–27.

Eastman, J. T. In press. The evolution of neutrally buoyant Nototheniid fishes: Their specializations and potential interactions in the Antarctic marine food web. In *Antarctic Nutrient Cycles and Food Webs*, ed. Siegfried et al. Springer-Verlag.

Ettershank, G. 1983. Age structure and cyclical annual size change in the Antarctic krill, *Euphausia superba* Dana. *Polar Biol.* 2:189–93.

Everson, I. 1977. *The Living Resources of the Southern Ocean.* Rome: UN Development Program, Food and Agriculture Organization.

———. 1984a. Zooplankton. In *Antarctic Ecology*, ed. Laws, pp. 463–90.

———. 1984b. Fish. In *Antarctic Ecology*, ed. Laws, pp. 491–532.

Foster, T. D. 1984. The marine environment. In *Antarctic Ecology*, ed. Laws, pp. 345–71.

Hewes, C. D., O. Holm-Hansen, and E. Sakshaug. In press. Alternate carbon pathways at lower trophic levels in the Antarctic food web. In *Antarctic Nutrient Cycles and Food Webs*, ed. Siegfried et al. Springer-Verlag.

Heywood, R. B., and T. M. Whitaker. 1984. The Antarctic marine flora. In *Antarctic Ecology*, ed. Laws, pp. 373–419.

Holm-Hansen, O. In press. Nutrient cycles in Antarctic marine ecosystems. In *Antarctic Nutrient Cycles and Food Webs*, ed. Siegfried et al. Springer-Verlag.

Ikeda, T., and P. Dixon. 1982. Observations on moulting in Antarctic krill (*Euphausia superba* Dana). *Aust. J. Mar. Freshw. Res.* 33:71–6.

Kock, K.-H. In press. Krill consumption by Antarctic Notothenioid fish. In *Antarctic Nutrient Cycles and Food Webs*, ed. Siegfried et al. Springer-Verlag.

Kooyman, G. L., R. W. Davis, J. P. Croxall, and D. P. Costa. 1982. Diving depths and energy requirements of king penguins. *Science* 217: 726–27.

Laws, R. M. 1952. A new method of age determination for mammals. *Nature* (London) 169:972.

———. 1959. The foetal growth rates of whales, with special reference to the fin whale, *Balaenoptera physalus* Linn. *"Discovery" Rep.* 29: 281–308.

———. 1977. Seals and whales of the Southern Ocean. *Phil. Trans. Roy. Soc. B* 279:81–96.

———. 1979. Monitoring whale and seal populations. In *Monitoring the Marine Environment*, ed. D. Nichols. Symp. Inst. of Biol. no. 24, 115–40.

Laws, R. M., ed. 1984. *Antarctic Ecology.* Academic Press.

Le Maho, Y. 1977. The emperor penguin: A strategy to live and breed in the cold. *Am. Sci.* 65:680–93.

Lockyer, C. 1972. The age at sexual maturity of the southern fin whale (*Balaenoptera physalus*) using annual layer counts in the ear plug. *J. du Conseil. Cons. Perm. Int. Explor. Mer* 34: 276–94.

Lutjeharms, J. R. E., N. M. Walters, and B. R. Allanson. In press. Oceanic frontal systems and biological enhancement. In *Antarctic Nutrient Cycles and Food Webs*, ed. Siegfried et al. Springer-Verlag.

Mackintosh, N. A. 1965. *The Stocks of Whales.* London: Fishing News Ltd.

———. 1973. Distribution of post-larval krill in the Antarctic. *"Discovery" Rep.* 36:95–156.

Marr, J. 1962. The natural history and geography of the Antarctic krill (*Euphausia superba* Dana). *"Discovery" Rep.* 32:33–464.

Nemoto, T., M. Okiyama, and M. Takahashi. In press. Aspects of the role of squids in food chains of the marine Antarctic ecosystem. In *Antarctic Nutrient Cycles and Food Webs*, ed. Siegfried et al. Springer-Verlag.

Payne, M. R., and P. A. Prince. 1979. Identification and breeding biology of the diving petrels *Pelecanoides georgicus* and *P. urinatrix exsul* at South Georgia. *New Zealand J. Zool.* 6:299–318.

Siegfried, W. R., P. R. Condy, and R. M. Laws, eds. In press. *Antarctic Nutrient Cycles and Food Webs. Proceedings of the Fourth SCAR Symposium on Antarctic Biology.* Springer-Verlag.

Tranter, D. J. 1982. Interlinking of physical and biological processes in the Antarctic Ocean. *Oceanogr. Mar. Biol. Ann. Rev.* 20:11–35.

Wilson, G. J. 1983. Distribution and abundance of Antarctic and sub-Antarctic penguins: A synthesis of current knowledge. Cambridge, England: SCAR and SCOR.

PART IV
Conservation and Ecosystem Management

Our understanding of ecology, as touched on in sections I, II and III of this collection, is being severely tested as we attempt to deal with earth's many environmental problems, most of which we brought on ourselves. In this final section, scientists explore the application of ecological principles to questions of sustainable agriculture, the maintenance of biodiversity, the management of entire ecosystems, and the preservation of endangered species.

The section leads off with a disturbing evolutionary analysis of our approach to pest management in agriculture, with Gould pointing out that, thanks to their rapid evolution, pests have repeatedly won the war we wage against them and that even the fantastic advances possible through the genetic engineering of crops may be no match for pest evolution. Pests in agriculture are not the only signs of ecological disruption. Biological invasions by exotic organisms may threaten human health and cause economic havoc throughout the world, as pointed out by Vitousek and colleagues. The key to minimizing the damage caused by exotic species is a solid understanding of the factors determining the success and impact of the invasive species that humans unwittingly introduce around the world.

Unfortunately, not all species are thriving as well as the agricultural pests and exotic invaders. Some species have required intensive captive breeding programs to prevent their local extinction. Simons and colleagues show how a combination of traditional natural history studies, captive breeding programs, and radio telemetry have been necessary to help rescue the U.S. national symbol—the bald eagle—in the southeastern United States.

Much of the biological disruption we see as pest outbreaks, invasions of exotics, or species extinctions can be traced to subtle alterations in the physical and chemical attributes of our environment. For example, the crown-of-thorns starfish that is destroying coral reefs throughout the southern Pacific Ocean seems to owe its success to excessive nutrient inputs associated with runoff from terrestrial human activities. Birkeland's article suggests that trying to control this pestilent starfish directly is misguided; instead, the root causes for its recent success must be clearly understood and addressed.

A far larger and less tractable problem is the disruption of the world global carbon cycle due to fossil fuel emissions. Post and a large interdisciplinary team of colleagues report that no scientific accounting has yet been able to balance the world's carbon fluxes, cautioning that this gap in our knowledge greatly hinders our ability to predict the trajectory for atmospheric carbon dioxide, and hence to predict climate change.

Although the speed and magnitude of global warming trends is in dispute, most scientists accept that some major climate alteration is underway. Economists and agronomists typically anticipate that global climate change will result in bands of vegetation moving toward the poles in ways that track the climate change. Ecologists know better. Pitelka and colleagues ask whether vegetation is likely to be able to keep up with climate shifts, or whether poor dispersal and habitat fragmentation will hinder the migration of vegetation, further exacerbating the consequences of global change.

Finally, no serious discussion of global environmental problems can leave out human population growth. Malone's commentary on the "Earth Summit" sponsored by the United Nations in 1992 starkly reviews the data—a fourfold increase in the world's human population in the last century alone, with an even greater explosion in energy and natural resource consumption. Whether sustainable development is a feasible goal, or whether the planet is inexorably headed for tragic deterioration of its life-support functions is a question whose answer depends largely on the insights and work of our next generation of ecologists and their ability to convey scientific findings to the public and world leaders.

The Evolutionary Potential of Crop Pests

Weeds, plant pathogens and insects are masters at surviving the farmer's assaults. New control strategies must anticipate pests' evolutionary responses

Fred Gould

One would be hard pressed to find a biology graduate who has not heard the story of the peppered moth, *Biston betularia*. In pre-industrial Britain the moths' light, peppered wings blended with the color of the lichen-covered tree bark on which they rested during the day. When industrial pollution killed the lichens and darkened the tree bark, the moths became more conspicuous targets for preying birds. Under intense selection pressure from these predators, moth populations in some polluted areas evolved dark wings and regained their cryptic status within a few decades (Kettlewell 1973).

This story of rapid adaptation is deserving of fame. Yet many equally or more spectacular cases exist in the scientific literature on crop pests and have considerable social relevance. Over the history of agriculture, farmers and plant breeders have applied selection pressure to plants to domesticate them. In doing so, and in attempting to protect their crops, they have forced the simultaneous adaptation of competing plants and of the fungi, bacteria, viruses and insects that feed on crops.

The pattern for the evolutionary battle between farmers and pests was set during the early days of agriculture, when farmers first separated weed seeds from desirable grain and when they selected the healthiest plants from one season as the source of seed for the next, thereby assuring that each generation would produce more of the repellents and toxins needed to fend off pests. In response, weeds evolved seeds that mimicked the crop seeds the farmer saved for the next year's planting; pathogens and insects often developed resistance to toxins or modified their habits to avoid them.

This conflict has escalated sharply since World War II. As modern agriculture attempts to share less and less of the crop yield with pests, the intensity of the selection pressure for pest adaptation increases. New lessons about adaptive response seem to emerge every time a new agricultural technology is applied. The lessons have often been unpleasant ones: Many of the short-term triumphs of pest control have carried within them the seeds of longer-term failure.

During the past 10 to 15 years, evolutionary biologists and agricultural scientists have joined forces in examining patterns in the evolutionary history of crop pests. The question that brings them together is an essential one for the future of agriculture: Are there strategies for protecting the world's food supply from pests that can anticipate and slow down the pests' evolutionary responses?

Although collaboration between the two fields is in its infancy, it has already yielded several promising ideas and much new understanding. For instance, we have begun to understand how resistance develops, or fails to, when insects are challenged by multiple toxins or by a combination of a toxin and other plant defenses. We are investigating how insects adapt to toxins when they are expressed only in certain plant tissues, or only in some plants, or at different times in the season. The genetic basis for the adaptation of some fungi to plant toxins is being explained and may help us develop strategies for coexisting with these genetically flexible pests.

Even if teams of academic scientists find solutions to the problems of pest adaptation, implementation of these solutions may be difficult. In the economically competitive environment of agriculture, the general orientation is toward short-term profits, whereas the concept of resistance management emphasizes long-term paybacks. In some cases, according to this approach, less is more: The lower the overall selective pressure challenging a pest, the longer the time for the pest to adapt. As a result of this approach, some pests may be left in the field. Even though these low densities of pests do not generally cause significant yield reduction, farmers may not be willing to take even small short-term risks for the promise of long-term stability of yield.

Over the next few decades, agriculture will continue its struggle with the problems of pests and resistance—a struggle that is likely to take some surprising turns as we experiment with the new strategies that genetic engineering offers the farmer. The connections between evolution and agricultural pest management are important both to agriculture's future and to science. Each crop-pest system has a unique mix of ecology and genetics, and experiments that elucidate the evolutionary pathways available in these systems might expand the range of potential solutions to the problems of pest management at a time when new directions are sorely needed. And agricultural systems may turn out to be the perfect testing ground for new evolutionary theories.

Later in this article I shall describe some of the ways in which evolutionary biology is being used to delay pest adaptation to control tactics. But it is important to set these developments against the backdrop of our past experience with pest evolution. The history of the war between farmer and pest teaches us not to underestimate the capacity of any pest species to resist attempts to

Fred Gould received his Ph.D. in ecology and evolutionary biology from the State University of New York at Stony Brook in 1977. He now teaches insect ecology at North Carolina State University and conducts research on the coevolution of plants and their herbivores, pest resistance and ecological aspects of pest management. Address: Department of Entomology, Box 7634, North Carolina State University, Raleigh, NC 27695-7634.

destroy it, whether it be a weed, a pathogen or an insect.

Weeds: Masters of Mimicry

Of all the crop pests, weeds boast the longest recorded history of adapting to agricultural practices. It is a history dotted with examples of one of nature's most interesting adaptive strategies: mimicry.

Within traditional agricultural systems there were two basic ways for weed seeds to survive the interval between cropping seasons. They could remain in the field, where they had to withstand the effects of weather, pathogens, seed feeders, burning and plowing, or they could hide among the crop seed that the farmer carefully stored for the next season's planting.

Success at the latter method requires that the weeds possess a number of important characteristics. They must ripen by harvest time. They must be held tight to their stems so that they do not fall to the ground on the way to threshing. And, finally, they must have a shape and density similar to that of the crop seed, so that they are not discarded during the winnowing operation, which separates crop seeds from anything that the wind blows less or more than the crop seeds. A surprising number of weeds have evolved all the characteristics required to become crop-seed mimics and lead a life of luxury between cropping seasons.

A great deal of morphological change lies within the evolutionary reach of a weed. One striking example comes from the mimicry of lentil seeds, *Lens culinaris*, by the common vetch, *Vicia sativa*. The lentil seed has a gently convex shape and, in fact, was the source of the word *lens*. Normal seeds of the common vetch are much more rounded than lentil seeds (*Figure 2*).

Early in this century botanists began speculating about a variant of the vetch seed that was causing trouble in the fields of central Europe. The variant had seeds that so closely resembled the lentil and looked so little like vetch that its origin was a puzzle. The change in seed shape was of tremendous practical importance: By mimicking the lentil seed the vetch could dominate a field planted for a lentil crop.

The prevailing hypotheses about the origin of the vetch variant ranged from hybridization of the taxonomically distinct vetch and lentil species to the view of V. S. Dmitriev (1952), who felt that the case offered evidence supporting

Figure 1. Winnowing, in which a farmer uses the wind to sift desirable seeds from weed seeds and debris, is one of the oldest of agricultural practices. But farmers' efforts to keep their fields from being overrun by weeds have often been defeated by the ability of some weeds to mimic the seed characteristics of crop plants. Seed mimicry is one of many evolutionary responses that have enabled crop pests—weed, plant pathogens and insects—to resist attempts to destroy them. The pest-control tactics used by modern farmers have increased the pressure on pests, and the pests' often-surprising adaptive responses suggest that agricultural scientists still have much to learn about the evolutionary biology of crop pests. (Reproduced from *The Grain Harvesters* with permission of the American Society of Agricultural Engineers.)

Trofim D. Lysenko's theory of species conversion. Breeding experiments in Wales in the late 1950s finally ended the speculation. By crossing normal vetch and the lentil-like variant, D. G. Rowlands (1959) demonstrated that the major change in seed shape could be attributed to a single recessive mutation. Farmers still occasionally struggle with vetch invasions of their lentil fields, especially in areas where traditional winnowing practices have not been supplanted by mechanized farming.

The Russian literature on species of the weed *Camelina* documents the most thoroughly studied case of mimicry in a weed (Sinskaia and Beztuzheva 1930). A number of *Camelina* subspecies are

Figure 2. Success at seed mimicry has given the common vetch the ability to contaminate commercial lentil fields. At left is shown the typical seed shape of the common vetch, *Vicia sativa*. In a lentil field near Albion, Washington, a U.S. Department of Agriculture plant pathologist recently found vetch seeds that had a distinctly different shape *(center)* that was quite similar to the flatter shape of the lentil, *Lens culinaris (right)*. Breeding experiments conducted in the 1950s after a long debate over the origin of a vetch variant that had been infesting European lentil fields established that a single recessive mutation could cause the vetch to vary its seed shape to mimic the lentil, allowing it to make it through the winnowing process to the farmer's next planting. The seeds illustrated above were supplied by Richard M. Hannan of the USDA's Regional Plant Introduction Station in Pullman, Washington.

found in flax crops; they appear to have diverged from their nonweedy relatives by developing traits that resemble those of flax *(Figure 3)*. One variety has become so dependent on flax culture that it is found only in flax fields.

In the 1920s two Soviet botanists, E. N. Sinskaia and A. A. Beztuzheva, gathered *Camelina* seeds from a large area of central and eastern Europe and sowed them near Leningrad. They found a great deal of variation in characteristics such as height, branching pattern, seed size and the seeds' propensity for shattering, or dropping from the plant when ripe. Similarly, many varieties of flax were found in the region surveyed.

Sinskaia and Beztuzheva concluded not only that *Camelina* had evolved traits that mimicked flax in general, but also that some characteristics of local weed populations had evolved specifically either to fit into the flax culture of an area or to adapt to local climate, or a compromise between the two. Interestingly, the weedy subspecies had almost uniformly developed a nonshattering trait, matching the general habit of flax, but in certain small areas a shattering subspecies was found. In these areas farmers had planted a rare, primitive oil-flax cultivar that lacked the nonshattering trait.

Vetch and *Camelina* have been adept at getting their seed into the farmer's furrow. But surviving from one season to the next is only half the battle for a weed; the other half is avoiding the woman or man who wields the hoe. The distinction between a crop plant and a weed is often but not always clear-cut in the field. For example, one of rice's most serious rivals, barnyard grass, is a skillful mimic. Spencer C. H. Barrett, now of the University of Toronto, and his colleagues discovered in weedy forms of barnyard grass so many ricelike traits—such as stem color, midrib size and leaf angle—that they found it harder to tell barnyard grass from rice than to distinguish two variants of barnyard grass from each other (*Figure 4*). Barrett found only one telltale visual distinction: The rice plants have ligules (projections at the base of the leaf blade), but the weeds do not. This phenotypic trait is apparently not available within the taxonomic lineage of barnyard grass—but then, farmers are unlikely to notice the absence of a ligule in the field (Barrett 1983).

Vetch, *Camelina* and barnyard grass are not closely related to the crops they mimic, and they appear to have acquired their useful traits by mutation. Some weeds, however, are so closely related to crops that hybridization between them is often noted. Wild rices are major weeds of rice in India and Africa, where they have sometimes reduced yields by 50 percent. Their normal resemblance to rice makes hand weeding a difficult chore. Their ability to respond to changes in rice culture, furthermore, has allowed them to resist eradication efforts.

Some Indian plant breeders decided that they could make weeding easier by breeding rice plants with reddish-purple coloration(Dave 1943). This approach worked well until the wild rice plants also developed reddish-purple color, presumably by exchanging the color-coding gene with cultivated rice through hybridization. According to Keith Moody of the International Rice Research Institute, farmers in Orissa, India, have taken a further step: alternating over time the planting of red and green rice seedlings. Some of the weeds will always make it through this temporally changing selection regime, but theoretically and practically the alternation with intensive weeding limits the population size of the weed. On a recent trip to India, Moody found a farmer planting rice that had a red leaf-collar and a generally green stem. The farmer told Moody that now he could eliminate both the red and green weeds. Moody wonders how much time will pass before red-collared wild rice appears.

In the mechanized farming dominant in the U.S., hand weeding (along with traditional methods of winnowing seed for the next season) may be a thing of the past, but the battle between farmers and weeds goes on. Currently, there is no premium on looking like rice if you are a weed in a California rice field; the chemical herbicides used to control weeds do not discriminate on the basis of appearance. The nature of the game has switched to biochemical mimicry. Agricultural industries spend millions of dollars each year inventing chemical agents that kill weeds without harming crops. They have succeeded marvelously. But this success has put enormous selection pressure on weeds to biochemically mimic crops, so that any agent that kills them will also kill the crop. The first cases of herbicide resistance were reported in the 1960s, and today many weeds appear to be starting on their way toward such a mimetic state. At his last count, Homer LeBaron estimated that there were 84 cases of weeds with resistance to at least one chemical herbicide, and some weeds of wheat in Australia have become resistant to a broad array of herbicides (Green, LeBaron and Moberg 1990).

Genetic engineers have recently succeeded in putting new genes into cotton plants that make them resistant to a previously deadly herbicide (Stalker, McBride and Malyj 1988). If genes for herbicide resistance are engineered into Indian rice, agronomists will not be surprised to find wild rices "borrowing" these engineered genes in the same way they borrowed genes for red plant color.

Pathogens: Moving Targets

Plant-pathogenic fungi and bacteria are old hands at dealing with toxic chemicals. Long before the arrival of human beings, these pathogens and naturally reproducing plants entered a protracted battle for survival that prominently featured chemical weapons. The mortality caused by pathogens in ancient times is thought to have selected for our modern flora, which produce hundreds of microbial toxins and other physiological defenses that thwart the attack of pathogens. Those who enjoy spicy food can be grateful that plants have experimented with a flavorful array of biochemical defenses against pathogens. If the taste of ginger was a gift of Dionysus, it was also a useful tool for the plant's early survival.

Consciously or unconsciously, farmers and plant breeders have entered this biochemical war. Just as the first farmers are thought to have collected seed for the next season from their best-yielding, healthiest plants, modern plant breeders collect superior seed, but in a more organized and precise manner. Sometimes unknowingly, ancient and modern breeders have often selected for plants that produce toxins that defend against specific pathogens, which in turn have adapted to defend themselves against the toxins.

Today, plant breeders know that pathogens are moving targets. When they breed a plant for cold tolerance, they expect that trait to be maintained in the plant's progeny and assume that their job is done. When, on the other hand, they breed a plant that is resistant to a pathogen, they know that the plant's offspring will be genetically similar, but they can never be sure that the pathogen's response to the resistance trait will not change. Indeed, it would be an enormous job to catalogue all the cases in which a pathogen evolved a means of undoing the work of a plant breeder. It is quite common for such adaptation to occur in less than three years.

Like weed resistance to herbicides, the resistance of plant-pathogenic fungi to synthetic fungicides is a recent phenomenon, but a significant one. It was first observed in the early 1960s after topically applied fungicides that did not penetrate the plant were replaced with systemic fungicides. Although the resistance to the new compounds was a surprise to plant pathologists at the time,

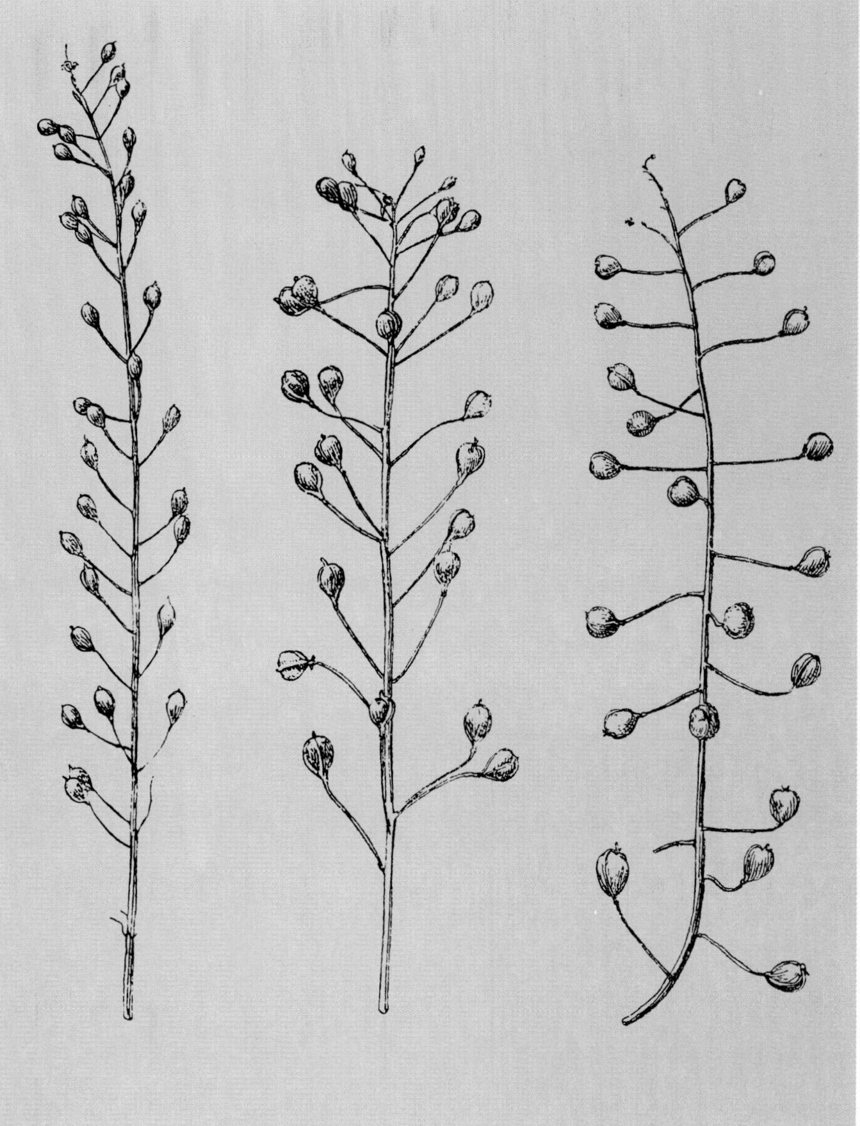

Figure 3. ***Camelina*, a weed that is a skillful crop mimic, has adapted to many variations in agricultural practices. Soviet botanists found that *Camelina* had evolved different forms in order to survive in the flax fields of eastern Europe. The plant mimics many characteristics of flax and has been known to adapt to local farming practices, to local flax varieties and to variations in climate. Above are drawings of three forms of *Camelina*. On the left is the wild form of *Camelina sylvestris* subspecies microcarpa, which has small fruits that "shatter"—that is, they drop their seeds when ripe. In the center is a weedy form of this subspecies with larger fruits that hold their seeds when ripe, as do most flax varieties. At right is a rare weedy *Camelina* that has large fruits but has a shattering characteristic similar to that of an old local variety of oil-flax. The botanists found that the characteristics of various *Camelina* forms often matched the characteristics of flax growing in the same area. (From Sinskaia and Beztuzheva 1930.)**

Figure 4. Survival in a hand-weeded field is easier for a weed that looks like a crop plant. A barnyard-grass seedling is easily mistaken for a cultivated-rice seedling, making barnyard grass *(Echinochloa crus-galli)* a serious nuisance in rice fields. In these renderings of a young rice plant *(Oryza sativa)* and two barnyard-grass seedlings, one can see how a rice-mimicking barnyard-grass plant can look more like a rice plant than like another variety of its own species. Left to right, the plants shown are cultivated rice, the *oryzicola* variety of barnyard grass and another barnyard-grass seedling, *Echinochloa crus-galli* var. *crus-galli*. In his research Spencer C. H. Barrett has found only one major visual distinction between rice and its mimic: the presence of a ligule on the rice.

some retrospective studies show that there was sufficient genetic variation in the original fungal populations to have allowed the rapid adaptation to have been predicted. By 1984 more than 100 species of fungi were known to be resistant to at least one fungicide (Green, LeBaron and Moberg 1990).

Plant pathologists and breeders have been working to devise evolutionary hurdles that will be more difficult for pathogens to jump. The conventional wisdom is that pathogens have more difficulty overcoming pathogen resistance that is based on many genes and, presumably, many diverse resistance factors. Other approaches are also being considered and will be discussed in the last section of this article.

Molecular biologists have begun to throw their expertise into the battle. They have found that by inserting into a plant's genome a gene that causes the plant to produce an excess of a pathogenic virus's coat protein, they can make the plant resistant to the virus (Beachy 1990). The mechanism of this resistance is the subject of continuing investigation, and there is currently no way to predict whether the pathogenic virus will evolve to circumvent the effect of this gene.

Molecular biology has made a very different but important contribution to this war by elucidating details of the biochemical battle between plants and pathogens. Willi Schäfer and his colleagues reported recently in *Science* that they had moved from one species of pathogenic fungus to another a gene that coded for a special detoxifying enzyme (Schäfer et al. 1989). The recipient fungus, which previously could attack corn plants but not peas, became capable of surviving within pea plants because it could now metabolize a toxin produced by the pea plants. This work proved that the limits of host range in some pathogens may involve as little as a single gene, coding for the right detoxification enzyme. If such a small change could extend a pathogen's range across the great taxonomic divide of the plant world from monocotyledons to dicotyledons, it is no wonder that a fungus's adaptation to a new variant of a crop species is often so rapid.

Insects: Winners in Chemical Warfare

Fungi have adapted to synthetic pesticides with impressive speed, but the real experts at chewing up the synthetic chemical agents of postwar pest control are the insects. Resistance to DDT, detected shortly after its introduction as one of the first so-called modern insecticides, is frequently cited as a textbook case of rapid adaptation. Since the DDT case the insects have, as a group, never met a chemical they couldn't take to the mat. George P. Georghiou has carefully documented more than 500 instances of insect adaptation to insecticides. In some situations insects have adapted to insecticides within a single season, even when the insecticide featured a new chemical twist. And there are now a few severe pests that have adapted to all or almost all pesticides that can legally be used to kill them.

Today most pesticide chemists are modest when predicting the life expectancy of a new type of insecticide. In the late 1960s, however, a different attitude prevailed in some scientific circles. In a 1967 *Scientific American* article, the reknowned insect physiologist Carroll Williams proclaimed that investigators were on the verge of developing "resistance proof" insecticides. His optimism seemed warranted at the time, for the new insecticides he described were mimics of insect juvenile hormones. How could an insect possibly become immune to the effects of its own hormones without wreaking havoc on its developmental and reproductive systems?

Five years after Williams's pronouncement, bits and pieces of information started coming in that demonstrated that insects can indeed adapt to the new hormone mimics, which interrupt the normal pattern of development in susceptible insect strains. The most precise work in this area is the research of Thomas Wilson and his colleagues in Vermont, who worked with natural and induced genetic variation for tolerance of juvenile hormone and its synthetic equivalents in the fruit fly, *Drosophila* (Shemshedini and Wilson 1990). Wilson's group found that a strain exposed to the mutagen EMS (ethyl methane sulfonate) developed 100-fold resistance to juvenile hormone. The resistant strain showed little decrease in fitness when reared on food with or without juvenile hormone. Recent molecular work by Wilson's group has demonstrated that the resistance in *Drosophila* can be traced to a single genetic change, induced by the mutagen, that affects the binding characteristics of one cytosolic juvenile-hormone binding protein.

If a single gene change can neutralize a hormone mimic, can we expect any control tactics to hold their own in the face of evolutionarily flexible pests? Biological control has often been held out as the solution when new problems arise with pesticide resistance. We would not have these problems, the argument goes, if we used naturally occurring insect pathogens, parasites and predators to control pests. This may be true in some cases, but there is now ample evidence that insect pests can adapt to some of their natural control agents if these agents exert sufficient selection pressure on the pest population.

For example, a colony of cabbage moths maintained at Cambridge, England, for five years contracted a viral disease. Over 90 percent of the population was wiped out, but the colony finally recovered from the disease outbreak. Later studies showed that the colony had become significantly more resistant to the virus than were other moth colonies sampled in that general area of England (David and Gardiner 1960). Additional support comes from experiments in which specific selection regimes have been imposed on insect populations by bringing them into contact with natural control agents in the laboratory; these experiments have generally given rise to insects with elevated resistance to pathogens. There are, however, exceptions where the resistance level has not changed, even following repeated generations of selection.

A noteworthy case of virus resistance was documented by L. L. J. Ossowski in 1980. Ossowski demonstrated that the wattle bagworm, a caterpillar pest in South Africa, was more tolerant of local strains of a virus than it was of foreign strains of the virus. Ossowski concluded that the caterpillars had developed a defense against only the specific virus strains it had encountered over time. He suggested that the best biological control agents may be those collected far from the area where the pest problem exists.

Today, as genetic engineers attempt to use biological systems to control pest populations, there are still lessons to be learned about the nature of genetic variability. Ten years ago there was a belief among genetic engineers that a stable way to combat caterpillar pests would be to incorporate into crop plants insect-toxin genes derived from the bacterium *Bacillus thuringiensis*, or *B.t.* Since the bacteria and the caterpillars had been in contact for millions of years, it seemed logical that the caterpillars would have already adapted to the toxins if they had the genetic po-

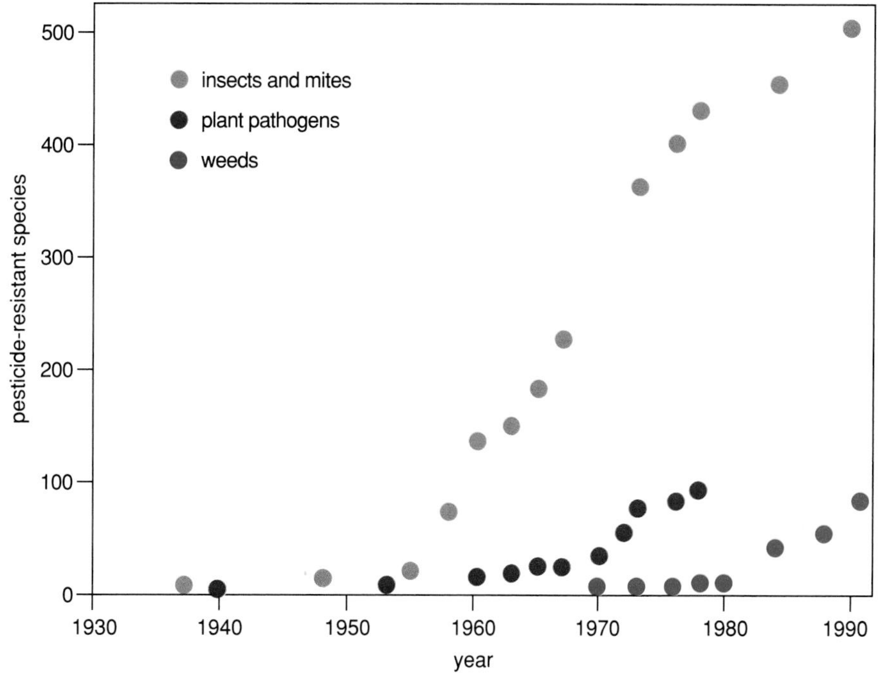

Figure 5. Hundreds of crop-pest species have become resistant to pesticides since synthetic chemicals were first used on a large scale for controlling agricultural pests in the 1940s. Insecticide resistance became a problem first, and today more than 500 insect species are known to be resistant to one or more insecticides. Resistance to pesticides has also developed in plant pathogens, primarily in fungi exposed to systemic fungicides. Genetic resistance to herbicides began to appear among weeds in the 1960s and has been reported in at least 84 species to date. The data are from George P. Georghiou and Homer LeBaron and from Green, LeBaron and Moberg 1990.

tential to do so. What was apparently forgotten in this argument was the fact that in natural environments outbreaks of these bacteria are extremely rare, and so the selection pressure for pest adaptation is low in nature. The caterpillars' genetic "potential" may never have had the occasion to prominently manifest itself. Proving this point are findings that some insect populations have recently developed over 100-fold resistance to these bacterial toxins as a result of unusually heavy reliance on *B.t.* for their control (Shelton and Wyman 1991). The stability of natural biological-control agents may reside not in their being immune to pest adaptation but in their reaching an evolutionary equilibrium with the pest, since both the pest and the biological-control agent can evolve.

If one were to dream of control measures to which insect pests truly could not adapt, food deprivation might be an obvious candidate. Indeed, the age-old practice of crop rotation is based on this approach: The farmer alternates planting of an insect's normal host plant with a plant that it cannot feed on, and the pest dies of starvation. This strategy has proved highly effective over hundreds of years, but insects are never completely static components in such a situation.

J. L. Krysan and his colleagues at Brookings, South Dakota, received calls during the early 1980s that indicated that some farmers who were rotating corn and soybean crops were still experiencing problems with the northern corn rootworm. Krysan's team determined that in some areas where corn and soybeans were rotated, the corn rootworm had a genetically altered diapause, or resting stage (Krysan et al. 1986). In large areas of the Midwest where corn is grown every year, almost all of the northern corn rootworms produce eggs that remain in the soil for one winter and then hatch and feed on the young corn roots in spring. In certain areas where farmers rotate corn and other crops that rootworms cannot feed on, about 40 percent of the eggs remain in diapause for a second winter. In this way they are synchronized to feed on the rotated corn crop. Krysan's colleagues in Illinois have recently reported a significant correlation ($p < 0.04$) between the percentage of rotated corn crops within a county and the percentage of rootworms that have an prolongeddiapause *(Figure 6)* (Levine, Oloumi-Sadeghi and Fisher, in press). Fortunately, this adaptation to crop rotation has not caused widespread problems, at least partially because eggs remaining in the soil for a long period of time are subject to mortality from both biotic and abiotic forces.

Evolutionary Biology on the Farm

Looked at from the farmer's viewpoint, the history of pest control is the saga of a long struggle to stay a step ahead of pest adaptation. Some of the techniques used to combat pests have proved relatively resistance-proof, but the successes have been limited. The results of society's recent experiment with synthetic chemical pesticides have been particularly disappointing.

Agricultural scientists now recognize that they need to maintain an arsenal of pest-control tools in anticipation of pests' evolutionary responses. That arsenal contains some potentially powerful weapons, among them the novel approaches offered by biotechnology and the promise of new developments in pesticide chemistry. But with or without the new weapons (whose use may be limited by economic, technological or social considerations and the high regulatory costs associated with their development), advances in managing pest resistance depend on improving our ability to predict the evolutionary future of pests.

Much of the discussion of resistance management over the past 10 to 15 years has centered on ways to reduce the rate at which pests adapt to conventional pesticides—a theme that dominated a major 1984 meeting on the management of pesticide resistance sponsored by the National Academy of Sciences (National Research Council 1984). Yet pests adapt not only to pesticides but also to other agricultural pressures, and they interact with other parts of their environment in important ways. The most interesting solutions may lie not only in more careful pesticide use but also in alternative strategies, such as manipulating plant-

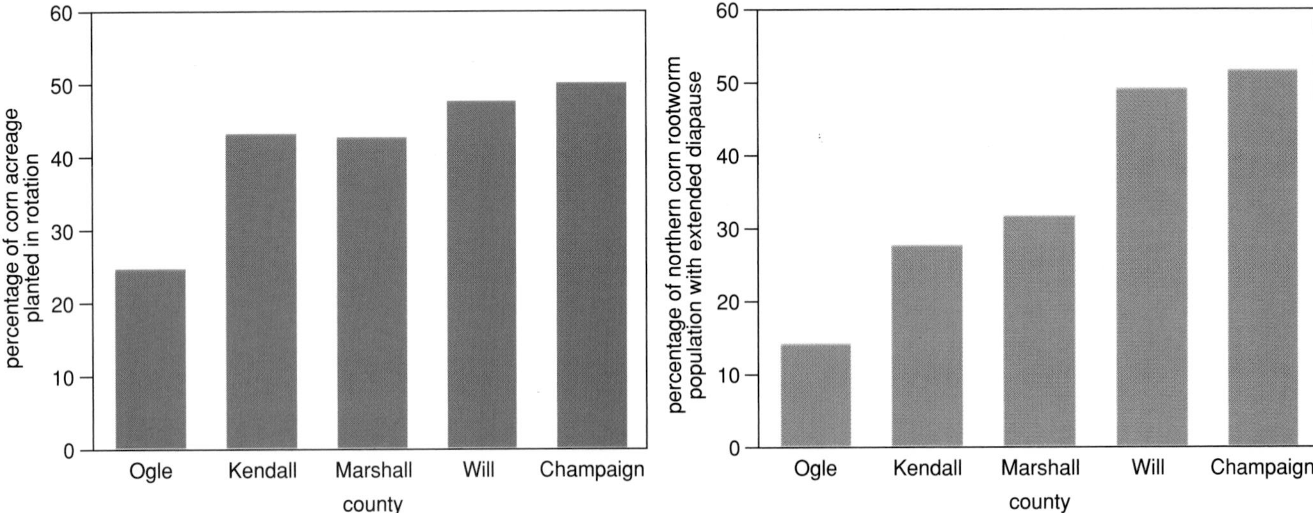

Figure 6. Eggs of the northern corn rootworm, a major pest in cornfields, typically remain in diapause, a resting stage, for one winter. The insect has recently evolved a longer diapause in areas where corn is rotated with soybeans. This genetic adaptation allows the insect egg to survive in the soil between the every-other-year planting cycles of its host plant instead of hatching too early among soybean roots, where it would starve. The bars show the relationship in five Illinois counties between the predominance of crop rotation as a practice and the proportion of the corn rootworm population with prolonged egg diapause. Crop rotation often succeeds in reducing pest populations by depriving the pests of their food, but some pests may adapt to the practice. Data are from Levine, Oloumi-Sadeghi and Fisher, in press.

pest interactions or combining the limited use of toxins with biological controls.

One idea, the multiple-toxin approach, provides a lesson in the complexity and importance of the population-genetic factors at work in pest adaptation. It has been suggested that one way to slow down a pest's adaptation to pesticides and to pest-resistant host plants is to challenge the pest with two or more different pesticides or plant-defense mechanisms. This notion is based on an intuitively appealing idea: The more hurdles placed in the path of a pest, the harder it is for the pest to adapt. When this view was explored mathematically through the use of population genetics–based models, it was found that in some cases the conventional wisdom was reasonable; in other cases, it could be very misleading (Gould 1986).

The success of the multiple-toxin approach depends on a number of specific attributes of the pests and of the toxins being used to kill them. If the inheritance of pest resistance to each of the toxins is recessive, and if high enough doses of persistent toxins are used so that most pests will die even if they are resistant to one of the toxins—and if, furthermore, there are some untreated pest habitats offering refuge in or near the crop—then a multiple-toxin approach is likely to work extremely well. On the other hand, if resistance to the toxins is inherited as an additive or dominant trait, and if the toxins degrade over time or are used at low doses, large populations of sexually reproducing pests can adapt to two toxins used in combination as fast as or faster than they would adapt to the two when used separately. And if, instead of reproducing sexually every generation, the pest goes through a number of parthenogenic generations between bouts of sexuality (as is the case with many aphids), recessive inheritance does not hinder adaptation as much. A further caveat is that the toxins must be distinct enough biochemically that the insect is unlikely to adapt to multiple toxins with a single genetic change.

These results have obvious utility. Even though most pesticides are too dangerous or expensive to apply as mixtures at high doses, plant breeders and genetic engineers can apply multiple-toxin approaches to their work. The same dynamics of pest resistance that would be at work in a field sprayed with pesticides will apply if a crop plant is given pest resistance by

midgut of susceptible insect

midgut of insect resistant to CryIA(b)

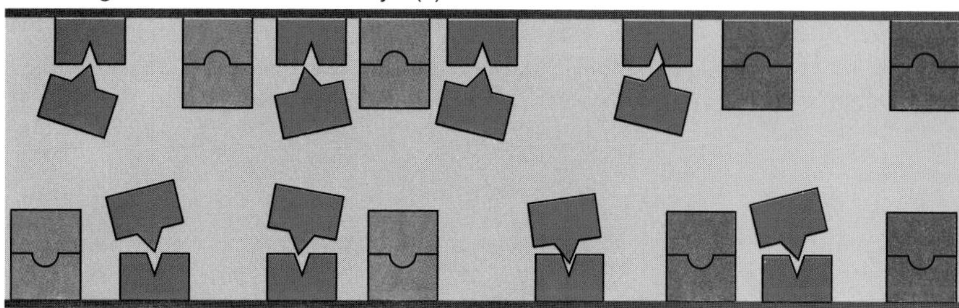

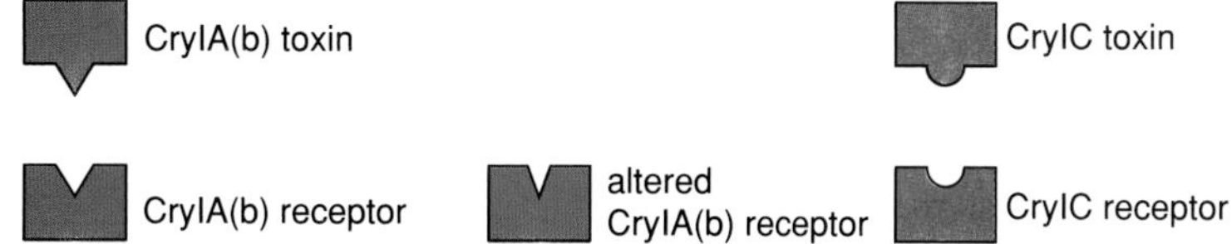

Figure 7. Insect resistance to toxins produced by the bacterium *Bacillus thuringiensis*, or *B.t.*, seems to be mediated by heritable changes in receptor proteins on the inner surface of the insect's midgut. Two of the *B.t.* toxins are insecticidal crystal proteins designated *CryIA(b)* and *CryIC*. In *B.t*-susceptible strains of the Indian meal moth, receptor proteins in the lining of the midgut bind to both toxins. Insects resistant to *CryIA(b)* have been found to have an altered *CryIA(b)* receptor in the midgut; the alteration greatly reduces the receptor's affinity for the toxin. However, the resistant insects have a slightly greater than normal abundance of *CryIC* receptors, and so they remain susceptible to this second toxin. In principle the Indian meal moth might evolve resistance to both *B.t.* toxins, but no such doubly resistant strain has been observed. Work aimed at creating transgenic crop plants that express *B.t.* toxins is now under way; it may be possible to control some insects for long periods of time by engineering such plants to express multiple *B.t.* toxins. (Adapted from Van Rie 1991.)

the insertion of two or more genes that express toxins at high levels.

The multiple-toxin approach is currently receiving detailed attention as such crops as cotton, corn and tomatoes are engineered to produce the *Bacillus thuringiensis* toxins discussed earlier. Different strains of the bacteria produce distinct toxins injurious to insects, and in many strains each bacterium produces a number of toxins. Some of the toxins are quite similar in their amino acid sequences; a single genetic change in an insect may produce resistance to both. On the other hand, studies with resistant colonies of the diamondback moth and the Indian meal moth have shown that adaptation to one type of toxin can be independent of adaptation to other distinct toxin types. And cross-breeding of resistant and susceptible insects has shown that *B.t.* toxin resistance is often inherited as a recessive or semi-recessive trait.

Working with the meal moth and the diamondback moth, molecular biologists have developed a reasonably good understanding of why resistance to one type of *B.t.* toxin has not led to cross-resistance to other *B.t.* toxins (Van Rie 1991). Their experiments have demonstrated that the *B.t.* toxins they worked with are effective only if the toxin binds to a receptor protein in an insect's midgut. Distinct types of toxins turn out to bind to different receptors. Insects with resistance to one type of toxin were found to have a single altered receptor protein, the one responsible for binding the specific toxin to which they were resistant *(Figure 7)*.

The alteration that led to resistance generally involved a decrease in the affinity of the receptor for that toxin, so that it took a lot more toxin to cause a fatal lesion in the gut. Because the other receptors in the gut were specified by different genes, a change in one receptor did not alter the receptors that bound other *B.t.* toxins, and so there was no cross-resistance. The critical experiments of simultaneously selecting these insect species with two toxins to be more certain that they will not come up with a single genetic change that makes them immune to both toxins have not been done, but they are in the planning stage. If such experiments indicate that cross-resistance is truly unlikely, and confirm the recessive nature of the resistance traits, we will be in a position to recommend testing the multiple-toxin strategy at a small-scale field level. Such tests could be conducted on an isolated island, just in case of some unexpected evolutionary event.

My laboratory has recently collaborated with scientists in Alabama, Belgium and Spain to investigate the question of whether the multiple-toxin approach is as likely to be effective with other insect species. We are aiming our efforts at one species in the *Heliothis* complex, a four-species group of caterpillars that attack cotton, corn, tomatoes and several other crops. *Heliothis* is one of the world's most severe insect pests and is a major target for control with genetically engineered plants. We selected *Heliothis virescens* for resistance to combinations of toxins and to single toxins, and exposed the insects to the toxins at different doses and for different portions of their larval stage.

With the *Heliothis* complex, or at least with *Heliothis virescens*, we are not sure that multiple *B.t.* toxins will be useful because preliminary results indicate that strains resistant to one toxin may be developing resistance to other toxins. If these preliminary findings hold true, it means that if we are to use multiple toxins for *Heliothis* control, we need to find and combine classes of toxins that are biochemically more distinct from each other than are pairs of *B.t.* toxins. Some possible candidates are insect-specific toxins produced by mites and scorpions, which sound horrendous but may be safer for human beings than the chemicals that give basil leaves their delightful flavor. Although genetic engineers are currently limited to working with protein-based toxins, it is hoped that in the future they will be able to manipulate the levels of expression of safe, natural plant-defense compounds that are generally more difficult to manipulate than single proteins.

When Is Less More?

Another idea for delaying pest adaptation is that of partial resistance. Plant pathologists and entomologists have long thought that if you could choose between a crop variety that had very strong resistance to a pest and one that had partial resistance to a pest, the partially resistant variety would probably last longer in the field than the variety that initially conferred strong resistance (Lamberti, Walker and van der Graaff 1981). Plant pathologists have some historical evidence that supports this view, and there are a number of potential explanations. One is that the selection pressure on the pest population is less in the case of partial resistance than it is when there is strong resistance.

Although partial resistance of crops to insect pests may not provide sufficient control in some cases, entomologists have found that combining partial resistance with the action of natural biological-control agents could offer useful protection. It was once generally assumed that the action of pest-resistance factors in plants differed so greatly from the action of biological-control agents that these natural enemies would not influence the rate at which pests adapted to partial plant resistance. It was therefore somewhat surprising when ecologically based population-genetic models showed that natural enemies of pests could significantly increase or decrease the rate of pest adaptation to the plant, depending on the characteristics of the natural enemy, including its hunting behavior (Gould et al. 1991).

Behavioral interactions between a pest and a crop with resistance to it can also be important, and an understanding of these interactions offers another intriguing prospect for managing adaptation: the expression of toxins in specific parts of crop plants. In a number of natural insect-plant associations, it has been shown that the plant does not produce an equal amount of toxin all season long or in all of its parts. In many cases it seems as if evolution has selected for plants that produce more of the protective toxins in their most vulnerable tissues at times of the year when insects can do the most harm. Some insects appear to respond to this heterogeneous defense strategy by avoiding the heavily defended tissues.

Biotechnology may offer a way to ex-

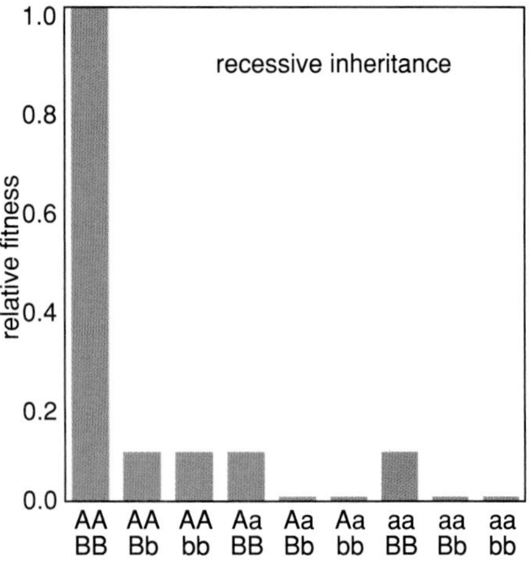

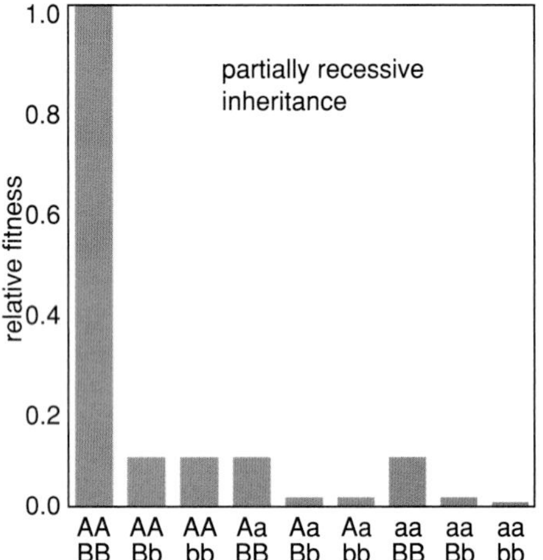

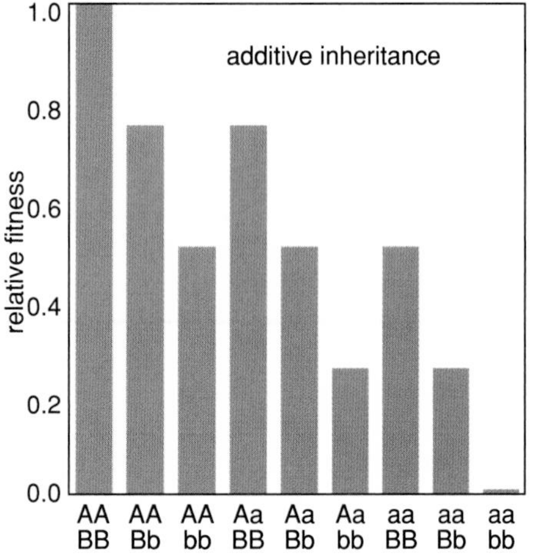

Figure 8. Onset of resistance to plant-produced toxins in a pest population was studied in a series of computer simulations, which showed that the useful lifetime of a plant's defensive strategy depends strongly on the intensity of the selection pressure applied to the pest and on how the resistance traits are inherited. The simulations examined a pest species with two genetic loci for resistance to plant-produced toxins; the *A* and *B* alleles confer resistance, whereas the *a* and *b* alleles do not. The effect of selection pressure is most

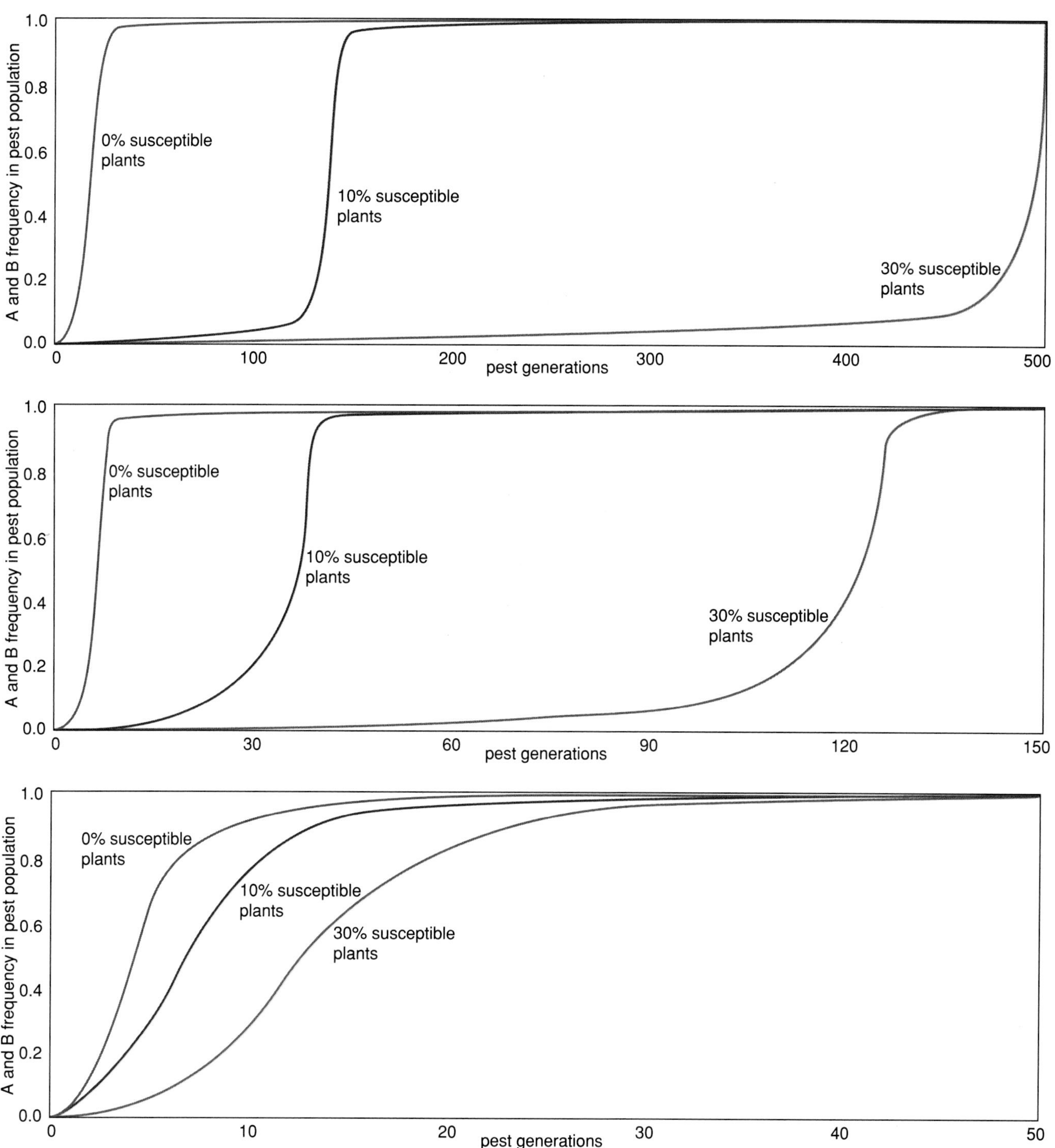

dramatic when resistance is inherited as a recessive trait *(top)*. In this case only the *AABB* genotype has high fitness when the pest must live on a host plant that produces both toxins. (All the genotypes are equally fit when the host plant has no defensive toxins.) If the simulated pests are exposed exclusively to toxin-producing plants, resistance emerges quickly; within about 30 pest generations the frequency of the *A* and *B* alleles goes from near 0 to near 1. Adding just 10 percent susceptible plants delays the development of resistance for almost 150 generations, and with 30 percent susceptible plants the resistant pests do not dominate the population until after 500 generations. The benefits of reduced selection pressure are smaller but still significant when resistance in the pest is inherited as a partially recessive trait *(middle)*. In this case, when individuals with a single *A* or a single *B* allele feed on plants expressing both toxins, their fitness is slightly higher than the fitness of *aabb* individuals (0.02 versus 0.01). If the pests feed exclusively on toxin-producing plants, the population becomes resistant in fewer than 10 generations. A 30-percent admixture of susceptible plants delays the onset of resistance until about generation 120. If resistance to the toxins is an additive trait, so that each *A* or *B* allele contributes incrementally to fitness on toxin-producing plants, the advantage of adding susceptible plants decreases further but is not completely lost *(bottom)*. Note that the three graphs have different horizontal scales. In all cases the initial frequencies of alleles *A* and *B* are set at 0.01. The effects would be more dramatic at lower initial frequencies, but the true initial frequencies are as yet unknown.

Figure 9. Tissue-specific expression of toxins by host plants, a concept being explored in experiments involving the caterpillar *Heliothis virescens (above)*, may be one way to delay the development of resistance in crop pests. (Photograph by Karl Suiter.)

ploit this interaction to protect important parts of plants while delaying the evolutionary response of insects to controls. It may become possible to restrict pests to nibbling on less-important, nontoxic parts of a plant, thus relieving the pressure on the pest for physiological adaptation while preserving tissues essential to a good harvest. Laboratory experiments with *Heliothis virescens* have shown that larvae, when given a choice of synthetic food with or without *B.t.* toxin in it, eat much more of the unadulterated food. Experimental results on growth and survival of *H. virescens* strains in situations where there is either choice or no choice indicate that having a choice of foods could slow adaptation to less than one-fifth the rate found in the situation in which *B.t.* toxin is present in all of the food (Gould and Anderson 1991).

My colleagues and I are currently conducting experiments that are similar to these, with the exception that we are using genetically engineered plants that produce *B.t.* toxin. We have chosen tobacco as our model plant because it is a major host of *H. virescens* and has become the "white rat" of plant genetic engineering. We have developed a technique with tobacco that enables us to mimic a crop plant that produces *B.t.* toxin solely in the apical bud area, usually the favorite part of the plant for *Heliothis virescens* larvae. From an agricultural perspective this is useful because when *H. virescens* feeds on the bud, it eliminates apical dominance (the controlling effect of the terminal bud on development of lateral buds), and the plant becomes deformed. If the larvae abandon the toxin-laden bud tissue to feed on older, larger leaves that lack the toxin, just as they abandon the toxin-laden synthetic food, their activity will cause much less reduction in yield. Such a system could have real advantages in many crops if the insect survived but there was little yield lost. In our experiments we have to perform an operation akin to grafting the top of a *B.t.*-toxin-producing plant to the lower part of a normal tobacco plant because plants expressing *B.t.* toxin only in specific tissues are not available. Although the technology for tissue-specific expression exists, it will take a number of convincing ecological and behavioral experiments to get genetic engineers to construct such plants. Our next step with this system will be to determine how natural enemies fit into such tissue-specific defense systems.

The utility of tissue-specific expression of toxins or other resistance factors is not limited to insects. Mike Bonman and his co-workers at the International Rice Research Institute have found that the most severe losses from rice blast disease occur when the blast-disease organism (*Pyricularia grisea*) attacks the panicle, or flower cluster, of the plant (Bonman et al. 1991). The growth of the blast organism on leaf tissue is less detrimental to yield. Bonman has found some rice genotypes that are more resistant to panicle infection than to leaf infection, and he has suggested that molecular genetics could be used to develop even more specific panicle resistance. This type of tissue-specific resistance would protect the rice against major losses while limiting selection pressure to the brief part of the season during which the panicle is exposed to infection.

Up to this point, I have been assuming that insecticides or resistant crops will start out being very effective and will sooner or later lose their effectiveness. But evolutionary theory holds out the tantalizing possibility that resistance-management approaches might be applied in ways that allow evolution to enhance rather than undo the effects of control tactics. I used a genetic model to explore a system in which an insect can feed on a crop or on other vegetation, but prefers the crop (Gould 1984). I plugged into the model, at low frequency, a single mutation that made the insect resistant to a new pesticide being sprayed on the crop. I also introduced at low frequency a mutation that programmed the insect to lay more of its eggs on the alternative vegetation. In many cases the insect population became resistant to the pesticide. However, at certain initial frequencies of the two novel traits the insect population evolved to a state in which it abandoned the crop in favor of the alternative food, thereby indirectly increasing the effectiveness of the pesticide over time.

These theoretical results suggest that there should be cases in which insects adapt to pesticides by avoiding them or by avoiding the types of plants that are often coated with them. Empirical work is being conducted in this area, and although some cases of such behavioral avoidance of pesticides can be found, they are rare compared with cases of physiological resistance. It may be that, contrary to the conventional biological wisdom that suggests that behavior evolves faster than physiology, it is easier for these pests to adapt physiologically than behaviorally when confronted with toxic substances.

Most of the work on developing evolutionarily sustainable pest-control strategies has focused on the manipulation of natural and synthetic toxins. As noted earlier, toxins are only one component of an agricultural system to which pests adapt. In the future, the evolutionary approach may help us to design better cultural and biological control strategies. For example, we have information indicating that some weed species can be divided into specific strains that are best adapted to distinct crops; however, we have yet to examine whether the rotation of certain types of crops makes it difficult for a weed population to adapt to all of the crops in the rotation. Similarly, it is likely that a weed population that is adapted to survive in fields where the soil is turned over with a moldboard plow may not be as adapted to survive in fields farmed with new no-tillage techniques. Temporal rotation of these farming practices could keep weed populations from reaching an optimal genetic solution to either practice. Recent thesis research by Heather Henter has shown that within a single 10-acre field, some lineages of pea aphids are far more resistant to their parasite, *Aphidius ervi*, than are others. (Parasite eggs laid within the bodies of resistant aphids usually degenerate before hatching.) Will studies of the genetic interaction between other parasites and pests reveal such striking variation at this fine scale? If so, understanding how such variation is maintained may help us to develop more efficient and sustainable biological-control programs. And if it is more difficult for insect pests to adapt behaviorally than physiologically, then instead of engineering plants to kill insects, it may be desirable to engineer plants that do not attract insects in the first place.

The need to understand more about pest adaptation to all components of agricultural systems may become more acute as replacing evolutionarily outmoded pesticides with new ones becomes more difficult for technological and social reasons. We are just starting to explore some of the ways in which we could hold pest evolution in check for at least a few more years. It will be a great challenge to agricultural scientists and evolutionary biologists to see just how far they can push the limits of their research fields to predict and direct the evolutionary future of pests.

It is sometimes said that necessity is the mother of invention. The needs of agriculture require more accuracy and precision from evolutionary theory than has previously been demanded. Perhaps this need will lead to a reexamination of old techniques of prediction and the development of new ones. As young evolutionary biologists sit and ponder what organism should be the subject of their thesis research, I hope they will consider the benefits that could accrue should they choose one of the hundreds of intriguing pests that plague the world's harvest.

Acknowledgments

This article had its origins in a previous paper (Gould 1990) and in a Sigma Xi lecture delivered at Kansas State University in April 1991. I would like to thank C. R. Carroll, D. J. Futuyma, R. Prokopy and R. L. Rabb for inspiring and fostering my early work in this area. M. T. Johnson and C. Nalepa offered helpful comments on the manuscript.

Bibliography

Barrett, Spencer C. H. 1983. Crop mimicry in weeds. *Economic Botany* 37:255–282.

Beachy, R. N., P. P. Abel, R. S. Nelson, J. Register, N. Tumer and R. T. Fraley. 1990. Genetic engineering of plants for protection against virus diseases. In *Plant Resistance to Viruses*, ed. D. Evered and S. Harnett, pp. 151–158. Ciba Foundation Symposium 133. New York: Wiley & Sons.

Bonman, J. M., B. A. Estrada and J. M. Bandong. 1989. Leaf and neck blast resistance in tropical lowland rice cultivars. *Plant Disease* 73:388–390.

Dave, B. B. 1943. The wild rice problem in the central provinces and its solution. *Indian Journal of Agriculture* 13:46–53.

David, W. A. L., and B. O. C. Gardiner. 1960. A *Pieris brassicae* (Linnaeus) culture resistant to a granulosis. *Journal of Insect Pathology* 2:106–114.

Dmitriev, V. S. 1952. Questions of the development of species and the control of weeds. *Soviet Agronomy* 4:17–27.

Georghiou, George P. 1986. The magnitude of the resistance problem. In *Pesticide Resistance: Strategies and Tactics for Management*, National Research Council. Washington: National Academy Press.

Gould, F. 1984. Role of behavior in the evolution of insect adaptation to insecticides and resistant host plants. *Bulletin of the Entomological Society of America* 30:33–41.

Gould, F. 1986. Simulation models for predicting the durability of insect-resistant germ plasm: a deterministic, two locus model. *Environmental Entomology* 15:1–10.

Gould, F. 1990. Ecological genetics and integrated pest management. In *Agroecology*, ed. C. R. Carroll, J. H. Vandermeer and P. M. Rosset, pp. 441–458. New York: McGraw-Hill.

Gould, F., and A. Anderson. 1991. Effects of *Bacillus thuringiensis* and HD-73 delta-endotoxin on growth, behavior, and fitness of susceptible and toxin-adapted *Heliothis virescens* (Lepidoptera: Noctuidae) strains. *Environmental Entomology* 20:30–38.

Gould, F., G. G. Kennedy and M. T. Johnson. 1991. Effects of natural enemies on the rate of herbivore adaptation to resistant host plants. *Entomologia Experimentalis et Applicata* 58:1–14.

Green, N. G., H. M. LeBaron and W. K. Moberg. 1990. *Managing Resistance to Agrochemicals: From Fundamental Research to Practical Strategies*. Washington: American Chemical Society.

Kettlewell, B. 1973. *The Evolution of Melanism*. Oxford: Clarendon Press.

Krysan, J. L., D. E. Foster, T. F. Branson, K. R. Ostlie and W. S. Cranshaw. 1986. Two years before the hatch: Rootworms adapt to crop rotation. *Bulletin of the Entomological Society of America* 32:250–253.

Lamberti, F., J. M. Walker and N. A. van der Graaff. 1981. *Durable Resistance in Crops*. New York: Plenum.

Levine, E., H. Oloumi-Sadeghi and J. R. Fisher. In press. Discovery of multiyear diapause in Illinois and South Dakota Northern Corn Rootworm (Coleoptera: Chysomelidae) eggs and incidence of the prolonged diapause trait in Illinois. *Journal of Economic Entomology*.

National Research Council. 1986. *Pesticide Resistance: Strategies and Tactics for Management*. Committee on Strategies for the Management of Pesticide Resistant Pest Populations. Washington: National Academy Press.

Ossowski, L. L. J. 1960. Variation in virulence of a wattle bagworm virus. *Journal of Insect Pathology* 2:35–43.

Rowlands, D. G. 1959. A case of mimicry in plants—*Vicia sativa* L. in lentil crops. *Genetica* 30:435–446.

Schäfer, W., D. Straney, L. Ciuffetti, H. D. Van Etten and O. C. Yoder. 1989. One enzyme makes a fungal pathogen, but not a saprophyte, virulent on a new host plant. *Science* 246:247–249.

Shelton, A. M., and J. A. Wyman. 1991. Insecticide resistance of diamondback moth (Lepidoptera: Plutellidae) in North America. *Proceedings of the Second International Diamondback Moth Workshop*. Taiwan.

Shemshedini, L., and T. G. Wilson. 1990. Resistance to juvenile hormone and an insect growth regulator in *Drosophila* is associated with an altered cytosolic juvenile hormone-binding protein. *Proceedings of the National Academy of Sciences* 87: 2072–2076.

Sinskaia, E. N., and A. A. Beztuzheva. 1930. The forms of *Camelina sativa* in connection with climate, flax and man. *Trudy Po Prikladnoi Botanike (Genetike I Selektsii)* 25:98–200.

Stalker, D. M., K. E. McBride and L. D. Malyj. 1988. Herbicide resistance in transgenic plants expressing a bacterial detoxification gene. *Science* 242:419–423.

Van Rie, Jeroen. 1991. Insect control with transgenic plants: resistance proof? *Trends in Biotechnology* 9:177–179.

Williams, C. M. 1967. Third-generation pesticides. *Scientific American* 217:13–17.

Biological Invasions as Global Environmental Change

Our mobile society is redistributing the species on the earth at a pace that challenges ecosystems, threatens human health and strains economies

Peter M. Vitousek, Carla M. D'Antonio, Lloyd L. Loope and Randy Westbrooks

The human species is noteworthy for its ability to forge into new environments and drastically alter them. No other species in the earth's history has spread throughout every continent and explored every remote island. We recognize some of the environmental consequences of the expansion and movement of the human population—chemical pollution, ravaged landscapes and the subsequent effects on natural ecosystems—but we fail to fully appreciate the extent to which we manipulate the distribution of life on the earth. In the course of our global travels, we don't merely bring the material trappings of our cultures, we also carry other species with us, frequently with a benevolent purpose in mind, but often unwittingly and without any particular intent. Unfortunately, the redistribution of the earth's species is proving to be ecologically and economically damaging, and the costs will continue to worsen.

Peter M. Vitousek is professor of biological sciences at Stanford University. Carla D'Antonio is assistant professor of integrative biology at the University of California, Berkeley. She has worked on the ecology of plant invasions for the past 10 years. Lloyd Loope assesses the extent of biological invasions and devises strategies to reduce their effects for the U.S. Department of the Interior at Haleakala National Park. Randy Westbrooks works to keep new noxious weeds out of the United States for the U.S. Department of Agriculture, Animal and Plant Health Inspection Service. Vitousek's address: Department of Biological Sciences, Stanford University, Stanford, California 94305.

One recent and notorious example is the Eurasian zebra mussel. Like many other aquatic organisms, the zebra mussel entered North America in the ballast water of ships. And, like many other introduced species, it spread rapidly once it arrived, covering the bottoms of rivers and lakes and venturing into the waterworks of municipalities and industries. According to a 1993 report from the Office of Technology Assessment, the cost of clearing blocked intake pipes will reach about $3.1 billion over a 10-year period. The economic consequences are enormous, but the ecological consequences may dwarf them: Zebra mussels reduce natural algae populations and biological productivity and increase the concentrations of nutrients in entire ecosystems. They continue to spread into rivers, lakes and canals throughout North America.

We suggest that biological invasions by notorious species such as the zebra mussel and by its many less-famous counterparts have become so widespread as to constitute a significant component of global environmental change. This point is not widely appreciated, even by the global-change research community and by those who work to control biological invasions. (See Elton 1958 for one important exception.) In part, this lack of appreciation reflects the natural limitations of an individual's spatial perception. Individuals can observe biological invasions almost anywhere, but it is much more difficult for an indi-

Figure 1. Biological invasions can be deceptively beautiful, as exemplified by these purple loosestrife (*Lythrum salicaria*) that flourish

Jim Steinberg/Photo Researchers, Inc.

in a Minnesota marsh. Since its introduction to North America the purple loosestrife has spread throughout the northeastern and midwestern United States, where it has become the dominant plant in many wetland areas. It is estimated that from 5 to 25 percent of the vascular plants in the United States reserves are non-native species.

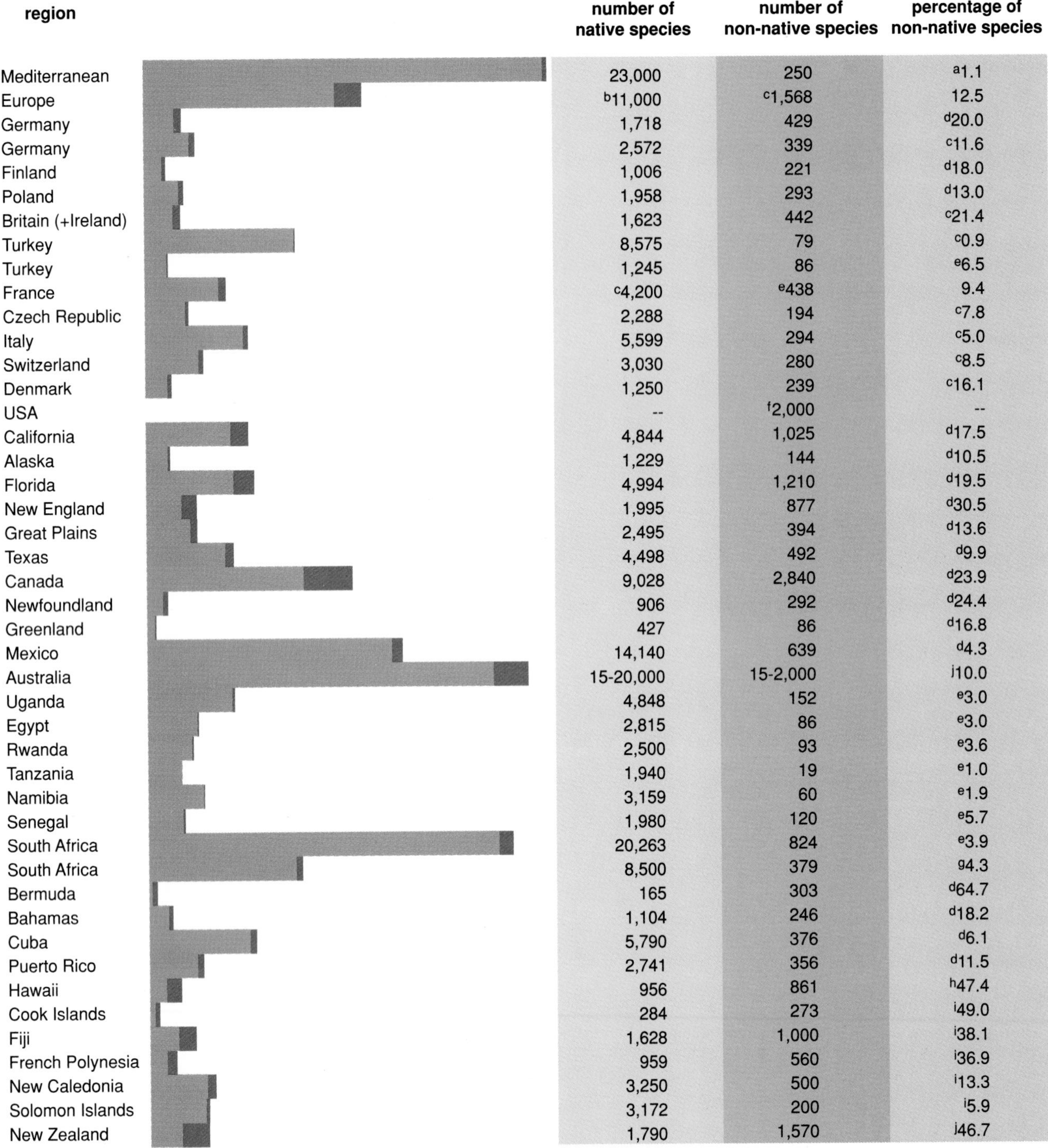

region	number of native species	number of non-native species	percentage of non-native species
Mediterranean	23,000	250	[a]1.1
Europe	[b]11,000	[c]1,568	12.5
Germany	1,718	429	[d]20.0
Germany	2,572	339	[c]11.6
Finland	1,006	221	[d]18.0
Poland	1,958	293	[d]13.0
Britain (+Ireland)	1,623	442	[c]21.4
Turkey	8,575	79	[c]0.9
Turkey	1,245	86	[e]6.5
France	[c]4,200	[e]438	9.4
Czech Republic	2,288	194	[c]7.8
Italy	5,599	294	[c]5.0
Switzerland	3,030	280	[c]8.5
Denmark	1,250	239	[c]16.1
USA	--	[f]2,000	--
California	4,844	1,025	[d]17.5
Alaska	1,229	144	[d]10.5
Florida	4,994	1,210	[d]19.5
New England	1,995	877	[d]30.5
Great Plains	2,495	394	[d]13.6
Texas	4,498	492	[d]9.9
Canada	9,028	2,840	[d]23.9
Newfoundland	906	292	[d]24.4
Greenland	427	86	[d]16.8
Mexico	14,140	639	[d]4.3
Australia	15-20,000	15-2,000	[j]10.0
Uganda	4,848	152	[e]3.0
Egypt	2,815	86	[e]3.0
Rwanda	2,500	93	[e]3.6
Tanzania	1,940	19	[e]1.0
Namibia	3,159	60	[e]1.9
Senegal	1,980	120	[e]5.7
South Africa	20,263	824	[e]3.9
South Africa	8,500	379	[g]4.3
Bermuda	165	303	[d]64.7
Bahamas	1,104	246	[d]18.2
Cuba	5,790	376	[d]6.1
Puerto Rico	2,741	356	[d]11.5
Hawaii	956	861	[h]47.4
Cook Islands	284	273	[i]49.0
Fiji	1,628	1,000	[i]38.1
French Polynesia	959	560	[i]36.9
New Caledonia	3,250	500	[i]13.3
Solomon Islands	3,172	200	[i]5.9
New Zealand	1,790	1,570	[i]46.7

Figure 2. Non-native species of vascular plants have invaded many parts of the globe. The total number (*bar length*) of native (*tan*) and non-native (*red*) species in a region varies from about 500 in Bermuda to more than 23,000 in the countries bordering the Mediterranean Sea. The percentage of non-native species in a particular region tends to be greater on island habitats, but many species have invaded the continents as well. Sources: a - Quezal, Burbero, Bonini and Loisel 1990; b - Heywood 1985, estimate of native species; c - Weber (in preparation); d - Rejmanek and Randall 1994; e - Rejmanek (unpublished); f - U.S. Congress, Office of Technology Assessment 1993; g - Kruger, Breytenbach, Macdonald and Richardson 1989; h - Wagner, Herbst and Sohmer 1990; i - Given 1992; j - Heywood 1989.

vidual to perceive that the invasions are almost everywhere. In part, it may also reflect a narrow view of global environmental change, one that emphasizes climatic change to the exclusion of other, equally significant components of human-caused global change.

There are several components to global change that are widely recognized as being caused by the explosive growth of industry and agriculture in the past two centuries. These include the increasing concentration of carbon dioxide in the atmosphere, alterations to the global biogeochemical cycle of nitrogen and other elements, the production and release of persistent organic compounds such as the chlorofluorocarbons, widespread changes in land use and land cover, and the hunting and harvesting of natural animal populations. To this list we add

Figure 3. Invasive saltcedar (tamarisk) trees have displaced the native flora in many southwestern river courses in the United States. The species is believed to increase soil salinity and to lower the water table.

another component—the introduction of non-native species to habitats and ecosystems that were previously isolated from each other. To varying degrees each of these components plays a role in enhancing the greenhouse effect or reducing the earth's biological diversity (through the extinction of genetically distinct populations or species).

Invasions Are Everywhere

Until recently the worst effects of biological invasions were largely considered to be restricted to oceanic islands. But it is becoming increasingly apparent that non-native species are also abundant on the continents. Attaching a number to the magnitude of these continental invasions is not always an easy task, however. Two different investigators may cite disparate numbers of introduced species for a particular region. This may be due to differences in the definition of "introduced species," the geographic boundaries used for the assessment or the year in which the study took place. Even so, the results of different studies almost always suggest that the scale of the continental invasions is much larger than is widely supposed.

Consider the extent of the invasions on the basis of some countries where reasonable data exist. Australia, Canada and the United States each harbor more than 1,500 species of invasive non-native plants. (California and Florida lead the continental U.S., each with close to 1,000 invasive introduced species of plants.) Flora of many European countries have several hundred introduced species, and South Africa alone supports nearly 800 species of non-native plant invaders.

As a proportion of the total flora and fauna in a particular country, the non-native species may comprise anywhere from a few percent to more than 20 percent of the total number. Remote islands tend to have a relatively greater proportion of non-native species, often as much as 50 percent of the total number.

Another general tendency is that the total numbers of invading introduced species are greater for plants than for animals. This is partly because the introduction of animal species (with the exception of insects) typically requires a greater degree of purposeful human intervention. This is especially true for birds and fishes. Nonetheless, isolated islands often have more non-native species of fish than native fishes. Even many continental areas, such as California, Europe and Brazil, have relatively large numbers of non-native fish species. Similarly, non-native birds have established wild populations in many parts of the world. Here again, the proportion of non-native species tends to be greater on islands than on the continents. Nevertheless, non-native avian species can be quite abundant on the continents—consider the widespread and abundant European house sparrow and the European starling in North America.

Parks as Canaries

One way to assess the extent of the biological invasions in a particular region is to take a closer look at parks and biological preserves. These habitats generally represent the least disturbed areas of land in a country. A former director of the U.S. National Park Service championed the concept that the park lands are an environmental analogue of miners' canaries—relatively pristine sites where the extent of large-scale environmental changes might be evaluated.

region	breeding birds		freshwater fish	
	native species	non-native species	native species	non-native species
Europe	[k]514	[l]27	--	[a]74
California			[b]76	[c]42
Alaska			55	[d]1
Canada			177	[e]9
Mexico			275	[e]26
Australia	--	[l]32	145	[e]22
South Africa	[m]900	[l]14	107	[e]20
Peru			--	[f]12
Brazil	1,635	[n]2	517	[g]76
Bermuda	--	[l]6		
Bahamas	[o]288	[l]4		
Cuba	--	[l]3	--	[f]10
Puerto Rico	105	[p]31	3	[h]32
Hawaii	57	[i]38	6	[i]19
New Zealand	155	[q]36	27	[j]30
Japan	[r]248	[l]4	--	[f]13

Figure 4. Numbers of native and non-native species of breeding birds and freshwater fish are shown here for selected regions of the world. (Sources: a - Holcik 1991; b - McGinnis 1984; c - Courtenay, Hensley, Taylor and McCann 1984; d - Moyle 1986; e - Macdonald, Kruger and Ferrar 1986; f - Welcomme 1981; g - Nomura 1984; h - Erdman 1984; i - Stone and Stone 1989; j - McDowall 1984; k - Jonsson 1993; l - Long 1981; m - Roberts 1985; n - Sick 1993; o - Paterson 1972; p - Raffaele 1989; q - Kinsky 1980; r - Higuchi, Minton and Katsura 1995.)

An examination of floristic lists reveals that anywhere from 5 to 25 percent of the vascular plants in the United States reserves are non-native species. For the most part the introduced species pose little or no threat to native species or ecosystems. There are some important exceptions, however. Fire-promoting grasses (such as cheatgrass) have invaded several semi-arid areas, tamarisk is prevalent in many riparian habitats of the Southwest, *Melaleuca* is rampant in the wetlands of the Florida Everglades, and purple loosestrife dominates waterways in the Northeast and Midwest. Plant invasions also threaten native biota in Hawaiian reserves, where the percentage of non-native species may be as much as 50 to 70 percent of the flora.

There are some serious threats to U.S. parks from animal invasions as well, especially ungulates. Feral pigs may be the single most damaging introduction in the national parks and reserves of the United States. The effects of the pigs on otherwise undisturbed areas are severe and pervasive in the Great Smoky Mountains National Park and in Haleakala and Hawaii Volcanoes National Parks. In Hawaii the pigs make a significant contribution to the dispersal of invading plants. Other harmful invaders include feral goats in Hawaii (now largely removed), feral burros in the Grand Canyon and other southwestern parks and mountain goats in Olympic National Park.

The consequences of invasion by several fish species into the aquatic and wetland ecosystems on the continental U.S. are as severe as many island invasions. In the streams of Sequoia-Kings Canyon National Park, for example, intentionally introduced brook trout and brown trout have displaced the native rainbow trout. In other parks, the introduction of brook trout and rainbow trout in waters that were originally barren of fish has greatly reduced the numbers of native invertebrates and amphibians. In the Great Smoky Mountains National Park the introduction of rainbow trout threatens the existence of native brook trout populations. Even the relatively pristine waters of Glacier National Park have been largely compromised by the introduction of fish in this century.

The introduction of non-native insects and microorganisms has also worked havoc on the forests of the U.S. parks and reserves. The white pine blister rust and the balsam woolly adelgid, which were costly to commercial forestry for many years, are now devastating the parks. Both were brought to the United States nearly 100 years ago on nursery stock from Europe. White pine blister rust attacks five-needled pines and is now killing sugar pines in the forests of Yosemite and Sequoia-Kings Canyon National Parks. Whitebark pine trees are also being hit hard throughout their entire range—fewer than 10 trees in 100,000 are rust resistant. Since the seeds of the whitebark pine are an important source of food for the grizzly bear and other animals, the decline of the tree may have severe consequences for the wildlife in Glacier, Yellowstone and Grand Teton National Parks. The balsam woolly adelgid attacks true firs of the genus *Abies*, causing death within 2 to 7 years by chemical damage and by feeding on the plant's vascular tissue. This small cottony insect is particularly damaging to the Fraser fir, which is found only in the southern Appalachian Mountains (primarily within the high elevations of the Great Smoky Mountains National Park). Since 1963 the adelgid has killed nearly every adult (cone-bearing) fir tree in the park.

Consequences of Invasions

Many invasions are reflections of other changes rather than agents of change themselves. For example, invading plants that only occupy roadsides thrive in these circumscribed habitats. These plants cannot be regarded as serious threats to native biological diversity except where ecological restoration is attempted. Moreover, some introduced species are beneficial to humanity—it would be impossible to support the present population of the United States entirely on native plants and animals. However, many invading species degrade human health and wealth, whereas others affect the structure of ecosystems or the maintenance of native biological diversity. We shall discuss an example of each of these to illustrate some of the consequences of recent invasions. For every example we discuss here, there are many others that are equally well documented and at least as harmful.

Newly introduced species can act as vectors of disease. A recent example of an introduced disease vector is the Asian tiger mosquito, larvae of which were brought into the United States in used automobile tires that were imported for retreading and resale. Two earlier introductions of the tiger mosquito in shipments of military tires had failed to become established. With the recent growth of commercial tire imports, however, the importation of mosquitoes has also increased. (In 1986, 6.8 of every 10,000 tires were infested with mosquito larvae.) The Asian tiger mosquito first became established in the United States in the 1980s, and by 1992 the mosquito had spread throughout 25 states. In its natural range, the mosqui-

Figure 5. Zebra mussels clog the municipal and commercial waterworks in the Great Lakes region. Since their introduction to North America in the late 1980s, the zebra mussels have spread throughout the waters in the Great Lakes. Here a water jet is used to clear zebra mussels from the walls of a pump room in a Detroit power station.

Peter Yates/Photo Researchers, Inc.

to is a vector of dengue fever and other human arboviruses. In the United States the mosquito can feed on most mammals and birds and is a vector for eastern equine encephalitis, an often fatal viral infection of people as well as horses (Craig 1993).

Biological invasions can be expensive. Non-native species can affect crops, rangelands and commercial forests, costing millions of dollars annually in lost yields and expensive efforts to control the invasions. Biological invasions can also be especially costly to developing economies, which typically have smaller margins for dealing with additional costs.

One prime example is the golden apple snail (*Pomacea canaliculata*) in Asian rice ecosystems. The snail was originally brought from South America to Taiwan to provide a supplemental source of protein and to increase the export income of small rice farms. Its benefits proved to be illusory. The local people find the snail distasteful—a recipe calling for "washing in a vinegar solution repeatedly to remove mucus and slime" may explain why. Moreover, the snail export market was closed because of health concerns. At the same time, both the environmental and economic costs of importing the golden apple snail have been very high. Snail populations grow rapidly, consuming young rice plants as they spread throughout irrigation canals. Despite the outcome of snail introduction in Taiwan, the entrepreneurs who originally imported the snail simply exported it to other countries. As of today the snail has spread throughout most of the Far East and southeastern Asia.

The life history of the snail invasion in the Philippines can be traced in the titles of publications over the span of a few years. A 1986 article, "Golden Kuhol propagation good source of income," promises economic opportunity, whereas a 1988 report, "The distribution and control of the introduced golden snail in the Philippines," hints at the need for restraint, and a 1989 article, "The golden apple snail: A serious pest of lowland rice in the Philippines," documents the costly turn of events. Rosamond Naylor of Stanford University has recently evaluated the economic costs of the snail invasion in the Philippines. In 1990, the total cost to farmers was between $27.8 million and $45.3 million, including the costs of picking the rice by hand, replanting destroyed crops and controlling the snail with molluscides and the loss of crop yields. This amounted to 25 to 40 percent of what the Philippines spent on rice imports in 1990. It represents just one year's damage in one of many infested countries.

Invaders alter ecosystem processes. Invaders don't simply consume or compete with native species—they can change the rules of existence for all species by altering ecosystem processes such as primary productivity, decomposition, hydrology, geomorphology, nutrient cycling or natural disturbance regimes. Invaders that affect each of these processes are known. One dramatic example is the invasion of the nitrogen-fixing tree *Myrica faya* into Hawaii Volcanoes National Park. Because the seeds of this tree are dispersed by a variety of birds, the tree can easily spread to new sites created by volcanic eruptions. The consequences for the native plant life are

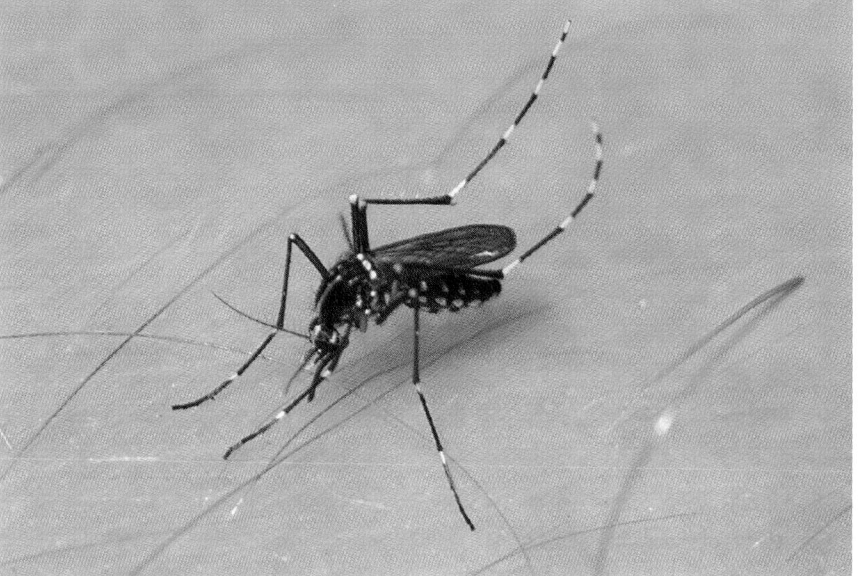

Figure 6. Asian tiger mosquito entered the United States by way of imported automobile tires in the 1980s. By 1992 the species had spread throughout 25 states. Here a mosquito rests on a human arm.

David T. Roberts/Photo Researchers, Inc.

profound. Usually the growth of native plants in young volcanic sites is limited by the poor availability of nitrogen in the soil. With the introduction of *Myrica*, however, there may be a rapid, fourfold increase in the amount of biologically available nitrogen added to the soil. This alters the plants and soil organisms that can thrive in the newly formed volcanic habitats. As it happens the species that do well in these altered habitats are non-native organisms. In essence, an invasion by one species changes the composition and the dynamics of an entire ecosystem.

Invasions reduce biological diversity. The eastern deciduous forests of North America are a diverse ecosystem that appear to be as resistant to biological invasions as any. Although these forests were cleared extensively in the 1800s, they have recovered substantially in this century. The scientific community has put a great deal of effort into understanding various assaults on these forests, including the effects of climatic change, the increased concentrations of atmospheric CO_2, acid rain and oxidant air pollution. However, by far the greatest perturbations in this century have involved repeated waves of invading pests and diseases (Sinclair, Lyon and Johnson 1987). Some pests, such as the gypsy moth, consume a variety of species, and their full effects on forest diversity are not yet known. Other more specialized pathogens in the eastern forest have virtually eliminated the once dominant American chestnut and the American elm. Other species that are declining because of newly introduced insects and diseases include the American beech, mountain ash, butternut, eastern hemlock, flowering dogwood and sugar maple (Langdon and Johnson 1992, Campbell and Schlarbaum 1994). We predict that the invasion of non-native species will continue to be the greatest threat to the diversity of these forests in the foreseeable future.

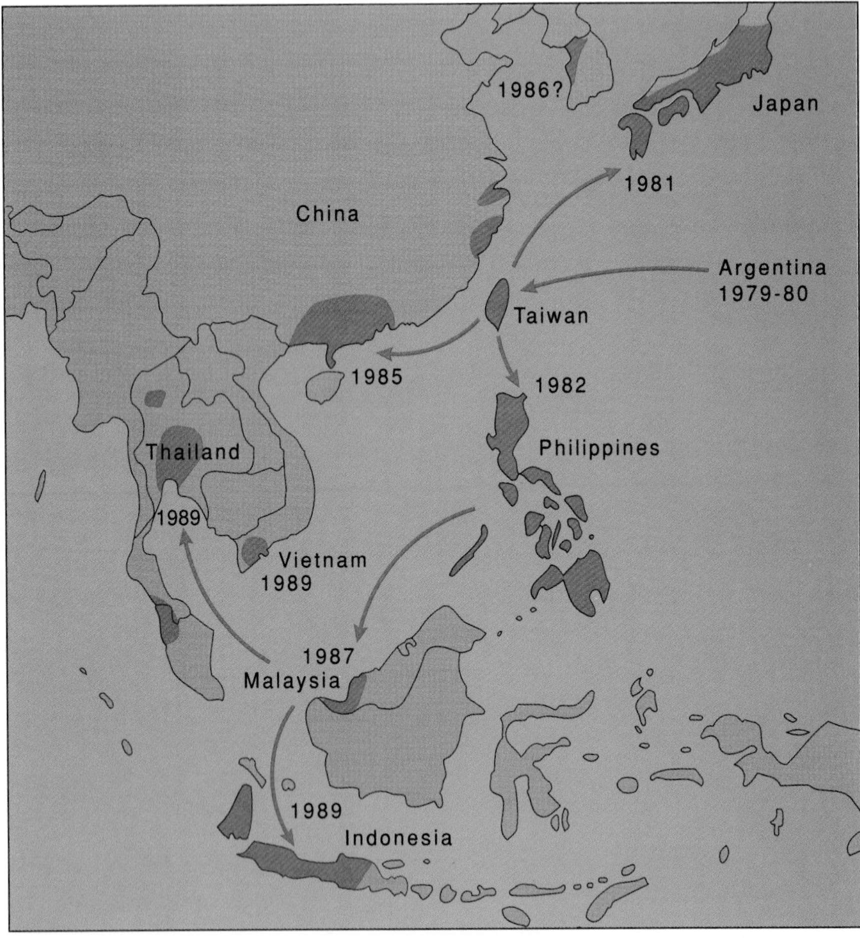

Figure 7. Invasive spread of the golden apple snail (*red*) threatens the rice crops of several countries in Southeast Asia. The snail was initially introduced to Taiwan from Argentina in 1980 as a potential source of protein. Within 10 years it had spread north to the Korean peninsula and south to Indonesia. Ironically, the snail not only consumes a significant proportion of the annual rice crop, it has failed to be a palatable source of protein in the Asian diet. The snail has proved to be a costly experiment in the redistribution of a species. (Adapted from R. L. Naylor, in press.)

Global Changes Interact

The recognition that biological invasions are a component of global change leads one to ask how the movement of species might interact with other changes taking place on a worldwide scale. It seems likely that each component of change interacts with others to varying degrees. We are far from understanding the dynamic among all of these components, but the ultimate consequences may go beyond simple additive effects. Here we introduce some examples of how two of these components might interact with changes in the distribution of plants and animals.

Changes in land use can promote invasions. The human species is now the premier agent of ecological disturbance on the planet. We have not merely increased the frequency and intensity of disturbances; we have also created landscapes that are unlike anything in the evolutionary history of many species. The alteration of natural disturbance regimes often promotes the invasion of species that would otherwise not be found in a region.

Consider the interaction between land use and the invasion of fire-promoting grasses. Many grasses from Eurasia and Africa have spread throughout the arid and semiarid ecosystems of the Americas, Australia and Oceania, where they increase the frequency and the intensity of fires. They also threaten tropical dry forests, and they are a major impediment to the restoration and reforestation of cleared lands (D'Antonio and Vitousek 1992).

Initial disturbances such as land clearing (often with fire) allow the invasion of these grasses. The invading grasses not only provide more fuel for the fire; they also create a microclimate that favors an increase in the frequency of fires. In turn, fires select against many native species and further promote the spread of fire-adapted grasses. The result is a positive-feedback system that perpetuates low-diversity shrublands and savanna.

External disturbances are not always required to set the cycle in motion. In some cases the invasion of grass alone can increase the probability of fire by increasing the load of fuel in an area. (This happened, for example, in Hawaiian woodlands.) It is also possible that the ready availability of forage grasses that withstand grazing and drought conditions can promote land-use change. In Mexico the conversion of millions of hectares of the Sonoran woodland desert to monocultures of African buffel grass has occurred in less than a decade. Similarly the dry and mesic forests of Central and South America have been replaced by grazing-tolerant (and fire-responsive) African pasture grasses (Parsons 1972).

Perhaps the most dramatic and well-documented example of a cyclic relationship between an introduced grass and the increased incidence of fire is the invasion of European cheatgrass into North America. This annual species invaded shrub and steppe habitat in the Great Basin, which was previously dominated by native shrubs and perennial grasses. After the invasion of cheatgrass the frequency of fires has increased from about once every 80 years to every 4 years or so. Almost 5 million hectares of land in Idaho and Utah are now nearly monospecific stands of cheatgrass (Whisenant 1990).

The suppression of natural disturbances by human activity can also promote the invasion of non-native species. This is especially true in aquatic ecosystems, where reproduction is often synchronized with natural disturbance cycles. The damming and impoundment of most of the rivers in the United States is correlated with the invasion of non-native species into rivers, streambanks and floodplains. It is also associated with the rapid conversion of species-rich riparian forests to species-poor stands of non-native species. For example, prior to the construction of the large network of dams that control the Colorado River, its floodplain forests were dominated by native cottonwood and willow. With the construction of dams, the groundwater tables have dropped and the intermittent floods that scoured the river banks have ceased. As a result the cottonwoods and the willows have been largely replaced by introduced saltcedar (tamarisk).

The fragmentation of natural habitats with the encroachment of farm lands and urban development has also encouraged the spread of non-native species. Urban forests and parks represent an increasing percentage of our remaining near-natural habitat and are often prime areas for newly introduced plants and animals. This is partly because the international traffic in cities is often the route by which a non-native species is introduced to a country. Having established themselves in these environments, the non-native species can then spread into less-urban habitats. The gypsy moth first became established in an urban forest and subsequently spread throughout the eastern United States, where it is now a major forest pest (Liebold, MacDonald, Bergdahl and Mastio 1995). Outbreaks of non-native fungal pathogens are also more common in fragmented forests close to urban areas.

Invasions promote extinction. The extinction of genetically distinct populations may be the least reversible of all global changes, and there is clear evidence that biological invasions contribute substantially to an increasing rate of extinction. As of 1991, 44 species of fishes in the continental United States were threatened or endangered by the introduction of non-native fishes. Of the 40 fish species known to have gone extinct since 1890, 27 were negatively affected by the introduction of non-native fishes (Wilcove and Bean 1994).

Although the native species on islands and in aquatic ecosystems have historically suffered the most from the introduction of non-native species (witness the rapid extirpation of the entire native forest-bird population on Guam due to the introduced brown tree snake), the potential for invasion-driven extinctions on the continents is substantial. One way to estimate the potential for extinction on the continents is to appreciate the relationship between species diversity and habitat area. For example, on many islands and isolated habitats the number of avian species present is linearly related to the logarithm of the land area that supports them (Preston 1960). If this relationship is extrapolated to include the total area of land on the earth, the projected number of avian species is substantially less than the actual number. The difference comes about because isolated habitats can support entirely different groups of birds. If these

Holt Studios International/Photo Researchers, Inc.

Figure 8. Eggs of the golden apple snail (*pink*) cluster on rice plants in Southeast Asia.

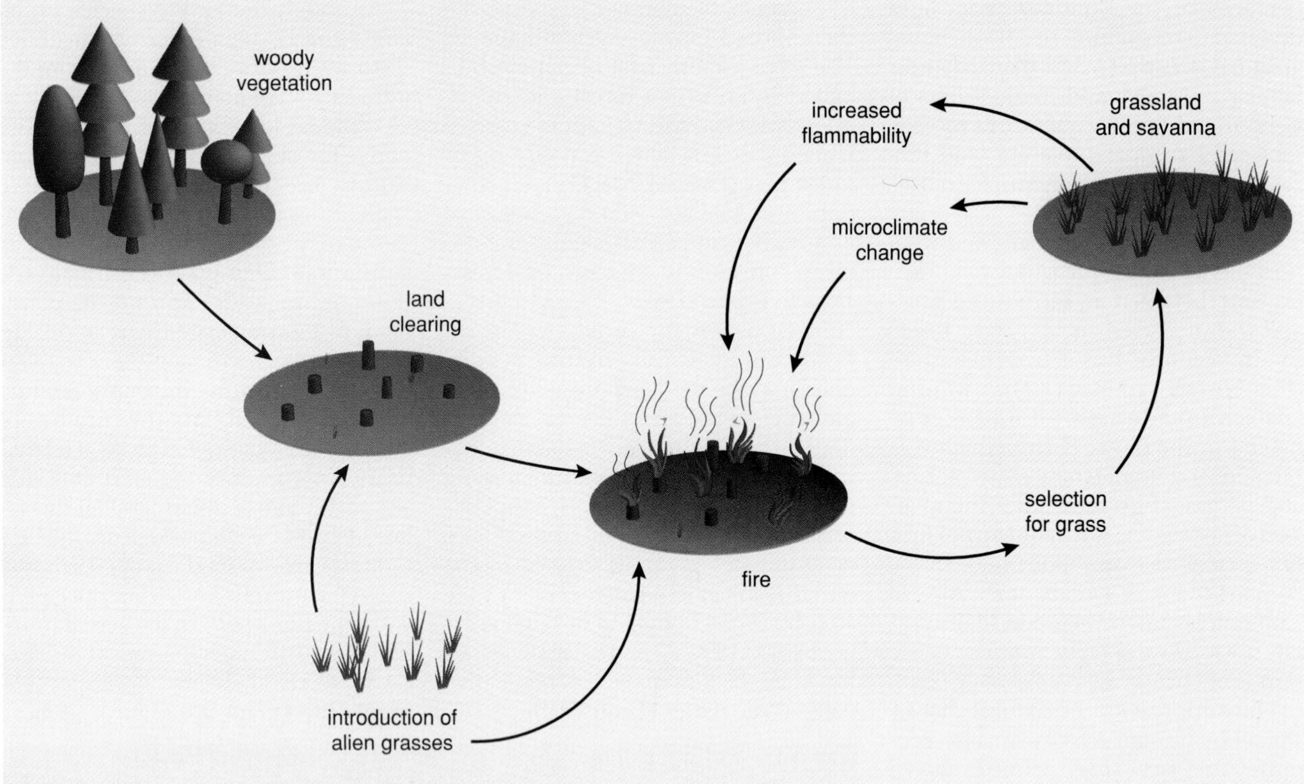

Figure 9. Grass invasions can initiate and maintain a grass-fire feedback system that prevents the regeneration of native woody species over large areas of the planet. The cycle can begin after the purposeful land clearing or after "natural" invasions of woodland or shrubland by introduced grasses. (Adapted from D'Antonio and Vitousek 1992.)

isolated areas were to be brought into contact with each other, some species would eventually be lost through competition, predation or disease. Regional distinctiveness begets global diversity.

One of us (Westbrooks) and colleagues applied this approach to calculate the potential for extinctions resulting from biological invasions. Based on known species–area relationships and the number of mammalian species present on each continent, they calculated that a single supercontinent comprising all the land area on the earth would support about 2,000 mammalian species. In fact the earth supports about 4,200 species of mammals. This analysis implies that the complete breakdown of biogeographic barriers might result in the eventual extinction of more than half of the earth's mammalian species.

We think that estimating the potential for extinction with species-area relationships is as valid as estimates that are based on the loss and fragmentation of habitats (Wilson 1992). This approach is also supported by paleobiological evidence. About 3 million years ago North and South America became connected by the Isthmus of Panama, which allowed a massive exchange of plants and animals across the two continents. The result was asymmetrical: Although some South American mammals (notably the opossum) crossed the isthmus, many more North American mammals spread into the Southern Hemisphere. The invasion of the North American mammals was associated with a significant increase in the extinction rate for the native mammals of South America (Marshall, Webb, Sepkowski and Raup 1982).

What Can Be Done?

In our attempts to convince others of the significance of biological invasions we have run into two major obstacles. First, there is a belief that invasions represent a natural process that has always been a part of evolutionary history. The second is a feeling that the ease of travel and the continued expansion of the global economy will make it impossible to prevent invasions.

It is of course true that invasions (like extinctions) have always been with us. What differs now is the increased *rate* of the invasions, which is so large as to represent a difference in kind rather than degree. For example, until recently all of the insects on the Hawaiian Islands arose from a successful colonization by one group or another every 50,000 years or so. Recently, however, about 15 to 20 insect species become established on the islands every year (Beardsley 1979). Similarly, in eastern North America, there is only one known instance in the postglacial past (within about 10,000 years) in which a species of tree (the eastern hemlock) declined over much of its range in a way that is consistent with an attack by a pathogen (Davis 1981). Yet several species of American trees have been devastated by introduced pathogens in the past century alone. Both examples suggest that recent invasions are very different from those of the past.

It is true that stemming the tide of biological invasions poses a huge challenge to the ingenuity of humankind. But it is not hopeless. A large part of the task is simply convincing our colleagues, students and the public at large that it is a problem worthy of our best efforts. A greater awareness of the issues increases the likelihood that citizens will respond in a positive way. In some respects this is already being achieved at a local level in California, Hawaii and Florida. Here the prevention of biological invasions is becoming popular as county councils consider the creation of emergency funds for the control of invasive species to assure the protection of biodiversity, the local lifestyles and the tourist industry. Ultimately, reducing the extent of biological invasions may prove to be as rewarding as reducing the rate of fossil-fuel consumption—while being less of a threat to economic growth and lifestyles.

The legal framework concerning the control and quarantine of plants and animals also needs improvement. Existing laws and policies such as the Federal Noxious Weed Act of 1974, the Nonindigenous Aquatic Nuisance Prevention and Control Act of 1990 and the Lacey Act can be enforced and strengthened. Given a reasonable amount of public support, intelligent new approaches can be devised. Moreover, concerned and informed citizens can participate personally by recognizing incipient invaders and preventing them from spreading. The concept of thinking globally and acting locally applies well to stopping biological invasions. No other form of global change offers educated and dedicated individuals such an opportunity to make a lasting difference.

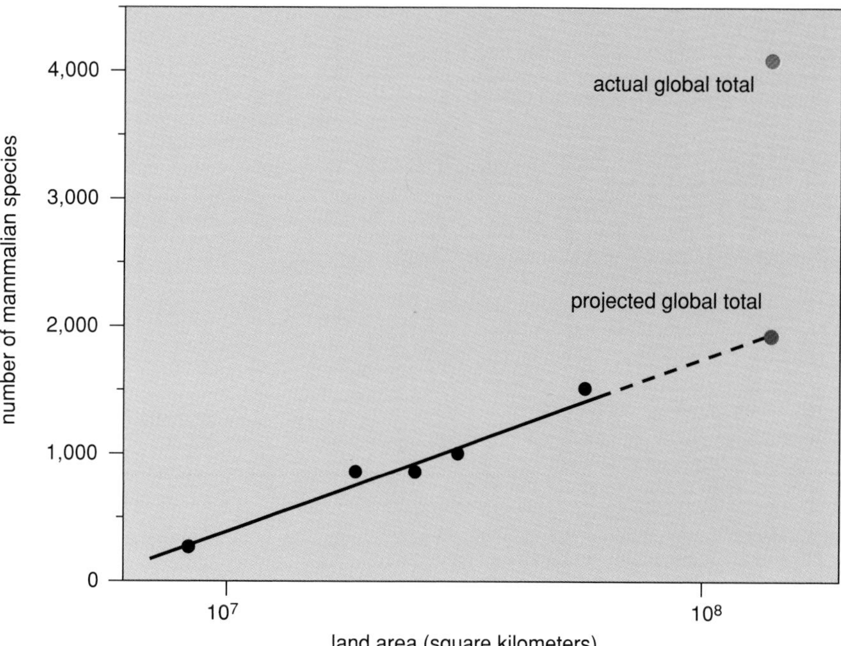

Figure 10. Number of mammalian species in a given area of land is dependent on the isolation of these species from each other. If the earth's land masses were united into a single supercontinent, the total area of land (more than 10^8 square kilometers) is projected to support about 2,000 species of mammals (*blue dot*). However, there are about 4,200 mammalian species on earth (*red dot*). The authors suggest that the current geographic isolation of the earth's species promotes and maintains biological diversity. By transporting other species across natural geographic boundaries, human beings threaten to reduce the diversity of life on the earth. The projected total number of mammalian species is based on surveys (*black dots*) showing a linear relationship between the number of species in a region and the logarithm of the area of land in the region. (Adapted from work by A. Launer of the Center for Conservation Biology, Stanford University.)

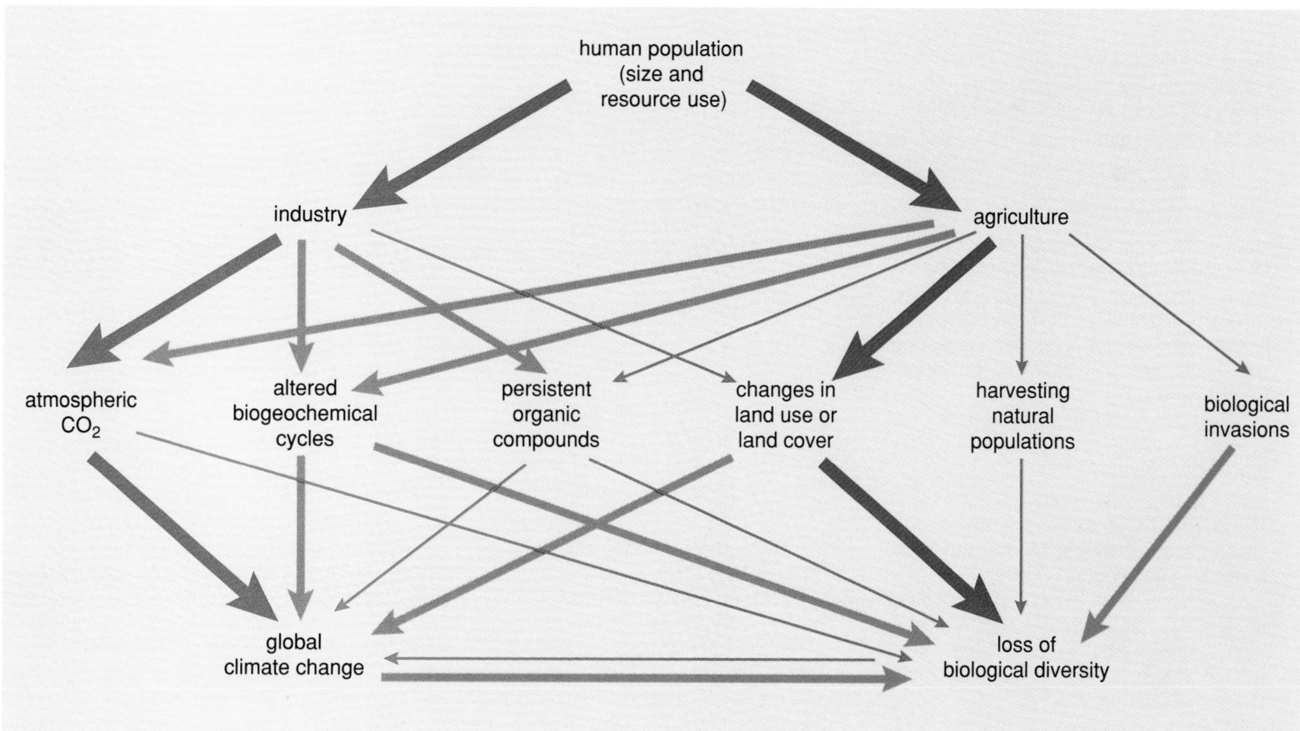

Figure 11. The growth of industry and agriculture in the past 200 years has promoted at least six identifiable components (*middle row*) of global environmental change. To varying degrees these components alter the earth's climate and reduce the planet's biological diversity. A general notion of the magnitude of these effects (*arrow thicknesses*) can be estimated, but the interrelationships and the synergistic effects of all six components have yet to be fully appreciated. (After Vitousek 1994.)

Bibliography

Beardsley, J. W. 1979. New immigrant insects in Hawaii: 1962 through 1976. *Proceedings of the Hawaiian Entomological Society* 23:35–44.

Campbell, F . T., and S. E. Schlarbaum. 1994. Fading forests: North American trees and the threat of exotic pests. New York: *Natural Resources Defense Council.*

Caraco, N. F., J. J. Cole, P. A. Raymond, D. L. Strayer, M. L. Pace, S. E. G. Findlay and D. T. Fischer. In press. Zebra mussel invasion in a large, turbid river: Phytoplankton response to increased grazing. *Ecology.*

Castello, J., D. Leopold, and P. Smallidge. 1995. Pathogens, patterns, and processes in forest ecosystems. *Bioscience* 45:16–24.

Courtenay, W. R., D. A.. Hensley, J. N. Taylor, and J. A. McCann. 1984. Distribution of exotic fishes in the continental United States. Pp. 41–77. In *Distribution, Biology, and Management of Exotic Fishees*, ed. Courtenay, W. R. and J. R. Stauffer. Baltimore: Johns Hopkins University Press.

Craig, G. B.. Jr. 1993. The diaspora of the Asian tiger mosquito. In *Biological Pollution: The Control and Impact of Invasive Exotic Species,* ed. McKnight, B. Indianapolis: Indiana Academy of Sciences.

Craven, R. B., D. A. Eliason, D. B. Fancy, P. Reiter, E. G. Campos, W. L. Jakob, G. C. Smith, C. J. Bozzi, C. G. Moore, G. O. Maupia and T. P. Monath. 1988. Importation of *Aedes albopictus* and other exotic mosquito species into the United States in used tires from Asia. *Journal of the American Mosquito Control Association* 4:138–142.

Crosby, A. W. 1986. *Ecological Imperialism: The Biological Expansion of Europe 900–1900.* Cambridge: Cambridge University Press.

D'Antonio, C. M., and P. M. Vitousek. 1992. Biological invasions by exotic grasses, the grass-fire cycle, and global change. *Annual Review of Ecology and Systematics* 23:63–87.

Davis, M. B. 1981. Quaternary history and the stability of plant communities. Pp. 132–153. In *Forest Succession: Concepts and Applications.* Berlin: Springer-Verlag.

Elton, C. S. 1958. *The Ecology of Invasions by Animals and Plants.* London: Methuen and Co.

Erdman, D. S. 1984. Exotic fishes in Puerto Rico. Pp. 162–176. In *Distribution, Biology, and Management of Exotic Fishes*, ed. Courtenay, W. R. and J. R. Stauffer. Baltimore: Johns Hopkins University Press.

FAO. 1989. Integrated "Golden" Kuhol Management. Food and Agricultural Organization, United Nations.

Given. D. R. 1992. An overview of the terrestrial biodiversity of Pacific Islands. Report, South Pacific Regional Environment Programme, Apia, Western Samoa.

Heywood, V. H. 1989. Patterns, extents and modes of invasions by terrestrial plants In *Biological Invasions: A Global Perspective*, ed. J. A. Drake et al. SCOPE 37. New York: John Wiley and Sons.

Higuchi, H., J. Minton, and C. Katsura. 1995. Distribution and ecology of birds in Japan. *Pacific Science* 49(1):69–86.

Holcik, J. 1991. Fish introductions in Europe with particular references to its central and eastern part. California Journal of Fisheries and Aquatic Sciences 48 (supplement 1):13–23.

Huenneke, L. F. In press. Outlook for plant invasions. Interactions with other agents of global change. In *Assessment and Management of Plant Invasions*, ed. J. O. Luken, and J. W. Thiereth. New York: Springer-Verlag.

Kendall, K. C. 1995. Whitebark pine: ecosystem in peril. Pp. 228–230. In *Our Living Resources: A Report to the Nation on the Distribution, Abundance, and Health of U. S. Plants, Animals, and Ecosystems*, ed. E. T. LaRoe, G. S. Farris, C. E. Puckett, P. D. Doran, and M. J. Mac. Washington, D.C.: U. S. Department of the Interior, National Biological Service.

Kinsky, F. C. 1980. Amendments and additions to the 1970 annotated checklist of the birds of New Zealand. *Notornis* 27 (supplement): 1–3.

Kruger, F. J., G. J. Breytenbach, I. A. W. Macdonald, and D. M. Richardson. 1989. The characteristics of invaded Mediterranean-climate regions In *Biological Invasions: A Global Perspective,* ed. Drake, J. A. et al. SCOPE 37. New York: John Wiley and Sons.

Langdon, K. R., and K. D. Johnson. 1992. Alien forest insects and diseases in eastern USNPS units: Impacts and interventions. *The George Wright Forum* 9(1):2–14.

Long, J. L. 1981. *Introduced Birds of the World.* New York: Universe Books.

Macdonald, I. A. W., F. J. Kruger and A. A. Ferrar (eds.) 1986. The ecology and management of biological invasions in southern Africa: Proceedings of the National Synthesis Symposium on the ecology of biological invasions. Cape Town: Oxford University Press.

Marshall, L. G., S. D. Webb, J. J. Sepkowski and D. M. Raup. 1982. Mammalian evolution and the great American interchange. *Science* 215: 1351–1357.

McDowall, R. M. 1984. Exotic fishes: the New Zealand experience In *Distribution, Biology, and Management of Exotic Fishes*, ed. Courtenay, W. R. and J. R. Stauffer. Baltimore: Johns Hopkins University Press.

McGinnis, S. M. 1984. *Freshwater fishes of California.* Berkeley: University of California Press.

Moyle, P. B. 1986. Fish introductions into North America: patterns and ecological impact. In *Ecology of Biological Invasions of North America and Hawaii*, ed. Mooney, H. A. and J. A. Drake. *Ecological Studies 58.* New York: Springer-Verlag.

Naylor, R. L. In press. Invasions in agriculture: assessing the cost of the golden apple snail in Asia. *Ambio.*

Nomura, H. 1984. *Dicionario dos Peixes do Brasil.* Brasilia: Editerra.

Office of Technology Assessment. 1993. Harmful Non-Indigenous Species in the United States. U. S. Government Printing Office: Washington, D.C.

Parsons, J. 1972. Spread of African pasture grasses to the American tropics. *Journal of Range Management* 25:12–17.

Paterson, A. 1972. *Birds of the Bahamas.* Vermont: Darrel Publications.

Preston, F. W. 1960. Time and space and the variation of species. *Ecology* 41:611–627.

Quezal, P., M. Burbero, G. Bonini, and R. Loisel. 1990. Recent plant invasions in the circum-Mediterranean region. Pp. 51–60 In *Biological Invasions in Europe and the Mediterranean Basin*, ed. DiCastri, F., A. J. Hansen, and M. Debusche. The Netherlands: Kluwer Academic Publishers.

Raffaele, H. A. 1989. *A Guide to the Birds of Puerto Rico and the Virgin Islands.* Princeton: Princeton University Press.

Rejmanek, M., and J. Randall. 1994. Invasive alien plants in California: 1993 summary and comparison with other areas in North America. *Madrono* 41(3):161–177.

Roberts, A. 1985. *Roberts' Birds of Southern Africa.* Cape Town: Trustees of the John Voelcker Bird Bood Fund.

Sick, H. 1993. *Birds in Brazil: A Natural History.* Princeton: Princeton University Press.

Simpson, G. G. 1980. *Splendid Isolation: The Curious History of South American Mammals.* New Haven: Yale University Press.

Sinclair, W.A., H. H. Lyon, and W. T. Johnson. 1987. *Disease of Trees and Shrubs.* Ithaca: Cornell University Press.

Stone, C. P. 1985. Alien animals in Hawaii's native ecosystems: toward controlling the adverse effects of introduced vertebrates. Pp. 251–297, In *Hawaii's Terrestrial Ecosystems: Preservation and Management*, ed. C. P. Stone and J. M. Scott. Honolulu: Cooperative National Park Resources Studies Unit, University of Hawaii.

Stone, C. P., and D. B. Stone. 1989. *Conservation Biology in Hawaii.* Honolulu: University of Hawaii Press.

U. S. Congress, Office of Technology Assessment. 1993. Harmful non-indigenous species in the United States. OTA-F-565. Washington D. C.: U.S. Congress Government Printing Office.

Vitousek, P. M. 1994. Beyond global warming: ecology and global change. *Ecology* 75:1861–1876.

Vitousek, P. M., and L. R. Walker. 1989. Biological invasion by *Myrica faya* in Hawaii: Plant demography, nitrogen fixation, ecosystem effects. *Ecological Monographs* 59:247–265.

Wagner, W. L., D. R. Herbst, and S. H. Sohmer. 1990. *Manual of the Flowering Plants of Hawaii, Volume 1.* Honolulu: University of Hawaii Press and Bishop Museum.

Weber, E. The introduced flora of Europe, a taxonomic and biogeographic analysis. Manuscript.

Welcomme, R. L. (compiler). 1981. Register of international transfers of inland fish species. United Nations, Food and Agricultural Organizations, Fisheries Technical Paper 213.

Whisenant, S. 1990. Changing fire frequencies on Idaho's Snake River plains: Ecological and management implications. Pp. 4–10 In *Proceedings from the Symposium on Cheatgrass Invasion, Shrub Dieoff, and Other Aspects of Shrub Biology and Management.* U.S. Forest Service General Technical Report INT–276.

Wilcove, D. S., and M. J. Bean. 1994. *The Big Kill: Declining Biodiversity in America's Lakes and Rivers.* Washington, D.C.: Environmental Defense Fund.

Wilson, E. O. 1992. *The Diversity of Life.* New York: Norton and Company.

Restoring the Bald Eagle

Ted Simons, Steve K. Sherrod,
Michael W. Collopy, M. Alan Jenkins

Ancient falconry techniques, animal husbandry, and modern ecological theory are aiding the recovery of the bald eagle

Despite Benjamin Franklin's persistent lobbying on behalf of the wild turkey, our founding fathers chose the bald eagle (*Haliaeetus leucocephalus*) as our national symbol. A common resident throughout much of North America in the eighteenth century, the bald eagle was viewed as a symbol of strength, courage, beauty, and freedom. Ironically, populations have plummeted over the past two centuries, so that today the bald eagle also symbolizes the effects of environmental contamination, habitat loss, human persecution of wildlife, and the impending free-fall of biological diversity throughout the world (Lewin 1986). Since the banning of DDT in 1972, populations have shown encouraging signs of recovery (Grier 1982); nevertheless, today over 90% of the remaining nesting pairs are confined to relict populations centered in Florida, the Chesapeake Bay area, Maine, the Great Lakes, and the Pacific Northwest (Green 1985).

Over the last 15 years, nationwide conservation efforts have been focused on restoring the bald eagle to portions of its former range. For the most part, these efforts have involved the reintroduction of birds into the few remaining fragments of suitable habitat. Often these habitat islands are imbedded in a landscape highly modified by man's activities, where, without direct intervention, there would be little likelihood of natural recolonization.

The foundation of most projects to reintroduce birds of prey is the ancient falconry technique known as hacking (Sherrod et al. 1981). The term comes from the *hack*, the board on which the hawk's meat was laid and to which the hawk returned. Hacking, formerly used with great success for restoring populations of the peregrine falcon (*Falco peregrinus*), has more recently been applied to bald eagles (Cade and Temple 1977; Nye, in press). Centuries ago, falconers discovered that most birds of prey are philopatric—that is, they form an attachment to the place where they are raised and tend to return to that location when they are ready to breed. The irony of this philopatric tendency is that it makes wild birds unlikely to recolonize vacant habitats. For example, although over 2,500 bald eagles migrate from the northern United States and Canada to winter in southeastern states, few, if any, stay to breed in what appears to be suitable and vacant habitat. Thus despite a large yearly influx of birds, there currently are only about 120 active nests of bald eagles in the southeastern United States outside of the state of Florida (Bagley 1987). Current populations of the southern bald eagle are estimated to be about one-third of their historic size (Fig. 2). Young bald eagles released into suitable but unoccupied habitat will tend to return to that habitat to nest when they reach adulthood at four to six years of age. Thus, hacking has proved to be an effective tool of wildlife management, because it establishes birds in the scattered islands of remaining suitable habitat, and it overcomes the population inertia that results from philopatry.

The goals for the restoration of the species are determined by the US Fish and Wildlife Service and outlined in a document entitled the Recovery Plan (Murphy et al. 1984). This plan has established a goal of 90 new nests in the Southeast, increasing the regional nesting population to approximately 40% of its estimated historic level. At that point the population would no longer be viewed as in danger of extinction, and consideration would be given to changing its status from endangered to threatened. Although most of the 14 eagle hacking projects under way in the United States are in their early stages, the results have been encouraging, and to date, at least seven new nesting territories have been established by hacked birds (Nye, in press).

Finding a suitable source of birds for reintroduction is an obstacle for all hacking projects. This problem is particularly acute in the case of the southern bald eagle. While hacking projects in the northern United States have used chicks removed from healthy populations in Canada and Alaska, southern bald eagles are considered by many to be a distinct subspecies (King 1981). They show several unique adaptations to their environment, all of which are believed to have some genetic basis.

Figure 1. (Next page) A program to restore the southern bald eagle (*Haliaeetus leucocephalus*) to its historic range attempts to reestablish the eagles in an area by placing fledglings that have been hatched in captivity into artificial nests. The eagles will return to these nests to be fed until about 6 months of age, when they can hunt for themselves entirely. The eagle shown here on the Mississippi River is a juvenile about a year old. The restoration program relies on the fact that when eagles reach adulthood and are ready to breed—at 4 to 6 years of age—they return to the area where they were raised. (Photogrsaph by Frank Oberle.)

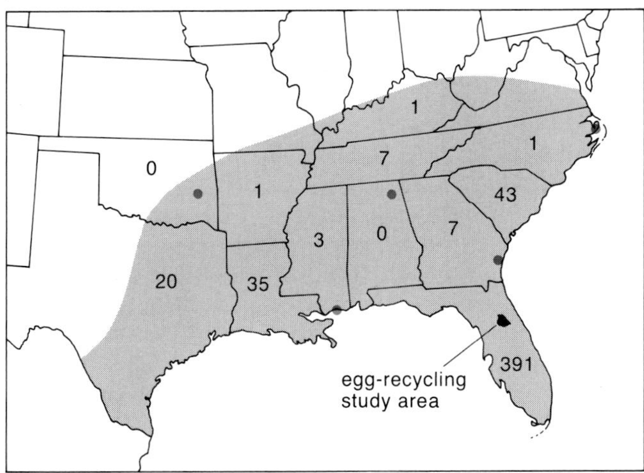

Figure 2. The population of the southern bald eagle has declined from an estimated 1,500 breeding pairs, historically distributed from eastern Texas to the Carolinas, to about 500 pairs today. Over 80% of the remaining birds are concentrated in central and southern Florida. The restoration program seeks to reestablish breeding birds across the Southeast through a combination of egg-recycling, captive propagation, and hacking at strategic sites (*red dots*) in five states.

Southern birds are smaller (presumably an adaptation to the warmer climate), less migratory as adults, and, in contrast to their northern counterparts, winter breeders. Our work, a large cooperative project involving the Sutton Avian Research Center, the states of Florida, Georgia, Alabama, Mississippi, Oklahoma, and North Carolina, the University of Florida, the US Fish and Wildlife Service, and the National Park Service, is an effort to develop a restoration program that takes into account the unique characteristics of southern bald eagles. The first and most important stage of this project was an attempt to determine whether we could use the relict Florida population, which contains over 80% of the birds remaining in the Southeast, as a renewable source of eagles by employing a technique called egg recycling.

Egg recycling

The technique of egg recycling relies on a female bird's ability to lay a replacement clutch of eggs—to recycle. This ability, presumably an adaptation to the loss of eggs to predators, storms, and other hazards, had been demonstrated in related species, such as ospreys (*Pandion haliaetus*) and falcons, but not in wild bald eagles (Kennedy 1977; Morrison and Walton 1980). However, if Florida bald eagles could be induced to recycle, and if they could do it without a significant reduction in their breeding success, our plan was to use that surplus production as a source of birds for hacking projects throughout the Southeast.

We set out to examine this question in 1985 with four objectives: to determine whether recycling occurs in southern bald eagles; to determine how egg removal affected subsequent nesting success and productivity; to determine how the timing of clutch removal influenced recycling; and to determine whether there were any differences in behavior or survival between late-fledging eagles from donor nests and fledglings from undisturbed control nests.

Donor and control nests were located in two areas of north-central Florida (in Alachua and Marion counties, and in the Ocala National Forest) where there are large numbers of nesting bald eagles (Fig. 2). Aerial surveys during the breeding seasons from 1985 to 1987 were initiated prior to egg-laying (in October and November) and repeated approximately every week until nearly all eggs hatched (mid-March). From mid-March until the eaglets fledged, surveys were conducted approximately every two weeks to monitor chronology and productivity at all nests.

A substantial amount of time and persuasion was required to locate accessible nests for egg collecting and to obtain permission from landowners to visit the nests. A total of 42 suitable donor nests were eventually located, and 87 eggs were removed over three breeding seasons. One egg-collecting trip was made in 1985, and two in 1986 and 1987. Eggs were removed from the nests by climbers and quickly dispatched to the Sutton Center in Oklahoma (Fig. 3).

We found that the rates at which eggs were recycled were high, ranging from 70.6% in 1987 to 100% in 1985, and averaging 78.6% (Table 1). Although none of the adult birds we studied was marked or banded, we are confident about our estimates of recycling, which were based on the chronology of egg laying, on the proximity of alternate nests, and on the history of eagle nesting in the area. Of the 33 birds that recycled over the three years, 21 did so in their original nests, and 12 recycled in nearby alternate nests.

Table 1. Recycling data for nests of the southern bald eagle

Year	Nest type	*n*	*n* recycled	% recycled	Mean recycling interval (days)	% nests fledging young	Fledglings per active nest
1985	Donor	9	9	100.0	32.4	77.8	1.22
	Control	31	—	—	—	71.0	1.19
1986	Donor	16	12	75.0	31.0	56.2	1.00
	Control	47	—	—	—	66.0	0.98
1987	Donor	17	12	70.6	26.2	70.6	1.24
	Control	54	—	—	—	68.5	1.17
1985-7	Donor	42	33	78.6	29.9	66.7	1.14
	Control	132	—	—	—	68.2	1.11

The age of an egg at the time of collection was estimated by subtracting the 35-day incubation period from the hatching date (Bent 1937). During the three years of the study, the ages of the eggs when collected varied widely, averaging 15.9, 16.2, and 15.8 days old, respectively. Although our sample sizes are small, it does appear that the probability of recycling decreases as the nesting season progresses—recycling did not occur readily at nests from which eggs were taken late in the egg-laying period (mid-January).

Recycling intervals (the number

of days between the removal of eggs and subsequent laying of replacement eggs) averaged nearly 30 days, ranging from 20 days to 57 days. Overall, there was no relationship between the age of the clutch when removed and the recycling interval.

The ultimate impact of removing eggs was judged by comparing the productivity of donor and control (i.e., unmanipulated) nests. We found no significant differences between these two groups in the percentage of nests fledging young or in the number of young fledged per active nest. We believe that if precautions are taken to collect eggs early in the season, Florida bald eagles will recycle readily and produce young at normal rates.

Figure 3. Climbers who collect eggs wear surgical gloves and masks to protect the eggs from contamination. The overhanging nest edge can make access to nests precarious, but most clutches are removed in less than fifteen minutes. (Photograph by M. A. Jenkins.)

Captive propagation

Preparations for the captive propagation phase of the project began at the Sutton Avian Research Center a full year before any eggs were collected. A chick-raising laboratory with facilities for the production, preparation, and storage of food was built. Specialized equipment, such as cannisters for holding eggs in the field, portable field incubators, and a motor home to be used as a field laboratory also had to be built and tested. Redundancy, backup, and monitoring capabilities were incorporated into each phase of the project to ensure against the inevitable problems caused by bad weather, power outages, equipment malfunction, and human error.

Once eggs are removed from a nest, they are put into a protective cannister, lowered to the ground, and placed in a portable field incubator. This incubator, powered by a portable generator, is fitted inside with netting to cradle the eggs and protect them from vibration and shock. It is designed to maintain internal temperatures within 0.25°C in the face of ambient temperatures that range between freezing and 27°C. The eggs are then taken to the motor-home field laboratory and placed in a larger incubator that also has been specially cushioned against road vibration and equipped with backup power and temperature alarms. About two days are required to obtain the approximately twenty eggs normally taken during a collecting trip. During that period, and the nonstop 33-hour ride back to the Sutton Center, the eggs are monitored carefully and turned by hand every three hours around the clock.

At the Sutton Center, the eggs are placed under Cochin sitting hens. These hens, which are kept "broody" by exposure to long photoperiods, make excellent surrogate parents, and their attention greatly increases hatching success. Prior to hatching, at about 35 days of age, the eggs are transferred back to an incubator to minimize the possibility of transmitting diseases to the newly hatched chicks.

After hatching, chicks are brooded on thermostatically controlled hot water bottles until they are capable of thermoregulation, at about 3½ weeks of age. For the first several weeks chicks are fed with a latex eagle-head puppet to ensure the proper stimulus for imprinting (Fig. 4). By three weeks they are able to feed themselves from trays of ground food left in their individual artificial nests, and by six weeks they are capable of tearing up whole food on their own. As soon as the chicks' vision begins to sharpen, at about one week, all feeding is done from

Figure 4. A latex bald eagle puppet is used to feed a three-week-old eaglet. Birds are observed through one-way mirrors and are kept isolated from human contact to prevent imprinting. (Photograph courtesy of the Oklahoma Department of Wildlife.)

Figure 5. These four-week-old eagle chicks are kept separated from their siblings to prevent them from attacking one another—a source of mortality in wild chicks. (Photograph by G. McKee.)

behind one-way mirror dividers. This is one of the precautions taken to minimize the chicks' awareness of their human caretakers, so that the birds do not imprint on their foster parents and lose the instincts necessary for survival in the wild.

The staple of the young eagles' diet consists of Coturnix quail, raised year-round at the center. Eagle chicks, which weigh about 85 g when they hatch, weigh 3.5 to 5 kg some eight weeks later and consume the equivalent of 800 quail (125 kg) in that interval. Only about half that number are fed to each bird, however, with the balance of their diet made up of venison, rabbits, chickens, rats, and a supplement of multiple vitamins.

Sibling aggression, a common behavior in birds of prey, must be controlled in birds reared in captivity. During the first month of development, although chicks must be kept within sight of their nest mates to permit proper imprinting, they must also be kept physically separated to prevent them from attacking and killing each other (Fig. 5). This aggressive behavior—called the "Cain and Abel" conflict (Stinson 1979)—appears to be an adaptation to eliminate competition in the nest during periods of food shortage, and it persists until chicks are about a month old. Until that time, the mere presence of a dominant sibling close by may inhibit a subordinate chick's feeding and development. In the wild this behavior often means that fewer chicks survive to fledging than actually hatch. By preventing sibling chicks from killing each other in the laboratory, and by providing optimum conditions for hatching, nutrition, and development, we have achieved productivities up to 60% greater than are normally attained by wild birds (Tables 1 and 2).

Table 2. Captive rearing and hacking of southern bald eagles

	Annual breeding season			*Total*	
	84–85	*85–86*	*86–87*	*84–86*	*84–87**
Eggs collected	18	34	35	52	87
Viable eggs**	17	33	32	50	82
% Eggs collected that are viable	94.4	97.1	91.4	96.2	94.3
Chicks hatched	17	30	24	47	71
% Viable eggs that hatched	100.0	90.9	75.0	94.0	86.6
Chicks reared to hacking age	13	28	20	41	61
% Hatched chicks reared to hacking age	76.5	93.3	83.3	87.2	85.9
Chicks that were hacked successfully	12	28	19	40	59
% Hatched chicks that were hacked	70.6	93.3	79.2	85.1	83.1
% Viable eggs resulting in hacked birds	70.5	84.4	59.4	80.0	72.0
% Collected eggs resulting in hacked birds	66.7	82.4	54.3	76.9	67.8

* 1987 was an atypical breeding season in Florida because of unusually warm, wet weather during the incubation period. These conditions apparently fostered the growth of bacteria in the birds' nests and the infection of many developing embryos (Sherrod et al., in press). The result was an abnormally low hatching success in eggs reared both in the wild and in captivity. Therefore, the 84–86 statistics are probably more typical of the results that can be obtained under average conditions.

** Viable eggs are fertile eggs that showed some sign of development.

At six weeks of age, the chicks, now nearly fully grown, are moved from the temperature-controlled laboratory to a "hardening yard," where they are acclimated to ambient temperatures. After about two weeks of acclimation, their sex is determined, they are banded, and they are then flown to the hack sites.

Hacking

There currently are four hack sites in the program, and a fifth, in North Carolina, will be added in 1988 (Fig. 2). The criteria for selecting a site are illustrated by the Horn Island hack site located 15 km off the Mississippi coast. Historic data indicate that the bald eagle was once a fairly common breeder on the barrier islands and adjacent coastal areas (Burleigh 1945). The species was extirpated by the early 1950s, and today only a single coastal nest can be found along the northern Gulf Coast. The Horn Island site will serve the important function of reconnecting the relict populations in Louisiana and Florida. This strategy, a basic tenet of conservation biology, will promote genetic exchange between the subpopulations and reduce the probability of local population extinctions (Wilcove et al. 1986). Additional evidence of the area's suitability for eagles is provided by the abundance of ospreys, an ecologically similar species. And finally, protected public lands, such as Gulf Islands National Seashore and nearby Bon Secour National Wildlife Refuge, ensure the long-term preservation of those habitats. These characteristics (evidence of historic nesting, high-quality protected habitat, and the potential for reconnection of relict populations) are shared by all the release sites.

The hack tower consists of a platform about 9 m high, with cages, artificial nests, and an adjacent room from which the birds can be observed and fed (Fig. 6). Here, as in the earlier stages of the program, great care is taken to ensure that the birds have no direct contact with people. The birds are nearly full grown when they are placed in the hacking tower, and they require about one kilogram of fresh whole fish, rabbits, or other meat a day.

Before release, each bird is fitted with a lightweight radio transmitter that allows us to monitor the birds for about six months. The device is attached with a backpack-style harness made of tubular Teflon ribbon and falls off within a year. Regular observations are stepped up when the birds reach ten weeks of age in order to pinpoint the best time for release. Nestlings become noticeably more restless just prior to fledging, and it is important that they not develop an aversion to the hack tower by being confined too long. If released properly, young eagles usually return to the tower to feed within 72 hours and continue to do so for up to three months as they sharpen their hunting and flying skills. This gradual transition to independence is probably crucial to their survival, especially in light of recent evidence that most fledglings embark on a long nonstop northward migration when they are about six months old. Hacked birds are not expected to establish breeding territories until they are four to six years old. Nevertheless, we already have two records (one in Oklahoma and the other in Alabama) of hacked birds returning to their release sites a year or more after release.

Figure 6. Hacking towers, this one on Horn Island, Mississippi, are placed in isolated patches of vacant habitat in an effort to reconnect the relict populations remaining in the Southeast. Cages containing artificial nests surround a central observation room from which the birds, unaware of their human caretakers, can be fed and observed. (Photograph by T. Simons.)

Management strategy

Once the viability of a restoration program based on egg recycling was demonstrated, the next step was the development of a management strategy that addressed the objectives of the recovery plan. This was not a simple matter, in large part because of a lack of good demographic data. The current state of our knowledge about bald eagles illustrates that wildlife management today is an imprecise science at best. Although they are one of the most conspicuous and intensively studied of all North American birds, and in spite of the fact that they have received special attention as our national symbol and as an endangered species, we still know very little about the population biology of bald eagles. The characteristics of a population are determined by both fecundity and survival. At present, almost all that we know about bald eagles comes from studies conducted during the breeding season. As a result, we have a fair understanding of fecundity in these birds, but we can only make rough estimates of juvenile and adult survival rates, and can only guess at the percentage of adult birds that attempt to breed each year (Newton 1979).

One approach to these shortcomings has been the use of stochastic population models to determine the most sensitive aspects of a species' life history. These models, which incorporate random fluctuations in life-history parameters, have been used recently to understand better the population dynamics of several endangered birds, including California condors (*Gymnogyps californianus*) (Mertz 1971); bald eagles (Grier 1980a, b); and dark-rumped petrels (*Pterodroma phaeopygia*) (Simons 1984). The patterns are similar for each of these long-lived, low-productivity ("k-selected") birds (MacArthur and Wilson 1967). Modeling has shown that populations of these and many other endangered species are extremely sensitive to changes in adult survival rates (Grier 1980b); their populations are less sensitive to changes in juvenile survival and are rather tolerant of variations in reproductive success. In addition, a low intrinsic rate of population increase subjects small populations of these species to high probabilities of extinction (Fig. 7). When applied to the conservation of bald eagles, these models indicate that even under optimum conditions, population recovery will require several decades; that founder populations established by hacking should be fairly large (at least 30 birds) to minimize the chances that random events will send a population to extinction; and that future conservation efforts must be based on a better understanding of survival rates, because unless adult survival rates are high (above 85%), efforts that focus on fecundity, such as hacking, may be futile.

Our plans for future work have been shaped by the results of this modeling, by the work on egg-recycling, captive propagation, and hacking, by the recovery plan objectives, and by the need for better information on survival rates.

The egg-recycling results indicate that the Florida eagle population can withstand a harvest of 100 eggs per year. About 20% of the eggs we collect in Florida will produce breeding adult birds. This estimate is derived from modeling, captive-propagation results, and the success of hacking efforts to date (Nye, in press). Population modeling predicts that for a given number of birds, hacking all of the birds in one year will, on the average, yield the same results as hacking smaller numbers of birds over many years. Other factors suggest that the optimal way to hack birds would be as one large group. First, the economics of hacking strongly favor larger releases. Second, it is reasonable to assume that birds released into vacant habitats as part of a single large cohort will be more likely to find a mate when they reach breeding age, than birds in small cohorts. On the other hand, caution argues that putting all of one's eagle eggs in one basket may be risky, given the uncontrollable effects on survival of weather and food supplies.

Our plan for the management phase of this project is an attempt to strike a balance between these biological and economic factors. It will run for at least five years and involve release sites in a minimum of five states (eventually with several sites per state). Each year, beginning in 1989, the sites in the target state will release a large cohort of up to 75 birds, while the remaining birds will be distributed among the sites in the other states. This effort should realize the recovery-plan goal of 50 to 60 new nests in the five target states, about two-thirds of the regional objective. Assuming the results are favorable, the program will then be shifted to the remaining southeastern states and will continue for another three to five years.

The expanded restoration program has been coupled with new field studies intended to broaden our understanding of the biology of bald eagles. This work

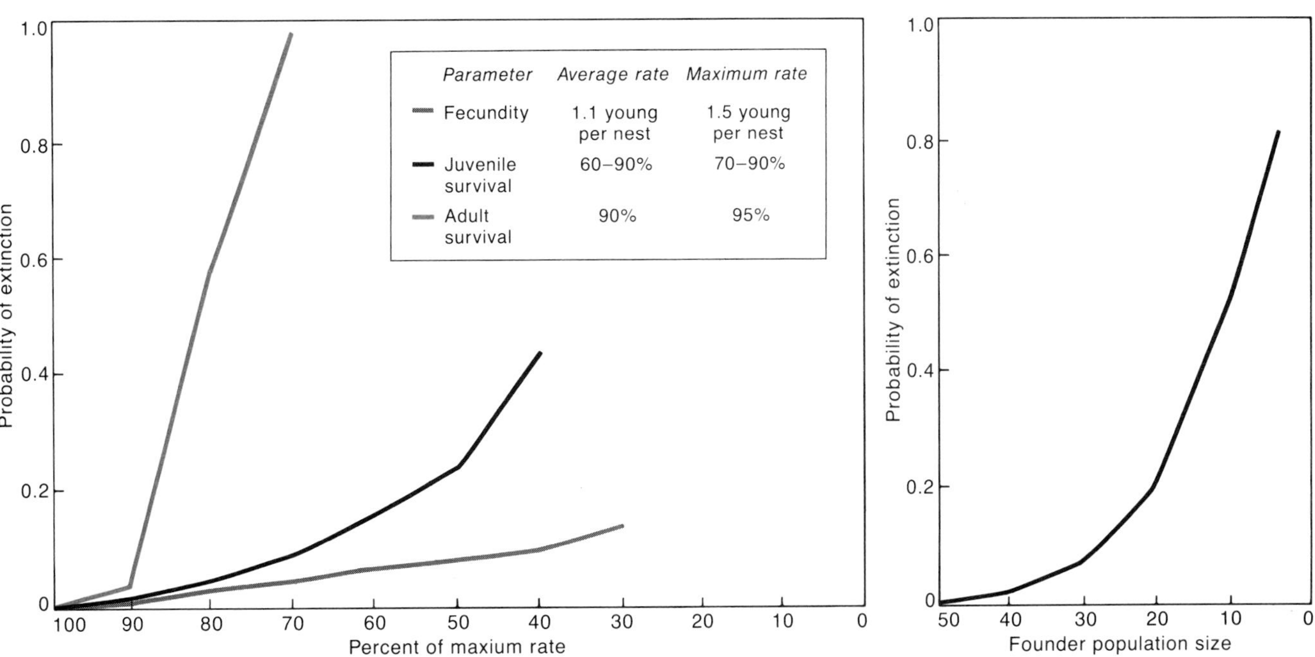

Figure 7. Mathematical modeling, based on data derived from the field studies of the southern bald eagle (summarized in Table 1 and Grier 1980b), shows that the probability of established populations becoming extinct increases as the rates of fecundity, of annual survival among juvenile birds (1–4 years old), and especially of annual survival among adult birds (5 years and older) decrease (*left*). The simulations depict each parameter as a percentage of the maximum rate, assuming the average rate for the other two parameters, and assuming a founder population of 50 birds. (See Grier 1980a, b and Simons 1984 for details of the model.) The modeling also shows that small populations of eagles have a high probability of becoming extinct (*right*), because of a low rate of fecundity combined with random fluctuations in the annual rates of fecundity and survival; this simulation assumes rates of fecundity and survival that fluctuate around average values.

includes the close monitoring of donor nests in Florida and the initiation of a long-term population study aimed at determining the dispersal patterns and survival rates of wild birds. Radio telemetry will be a valuable tool for much of this work.

Recent advances in the technology of batteries, miniaturization, and solar cells have made radio-telemetry studies of eagles both feasible and affordable (Kenward 1980). Miniature battery-powered transmitters weighing less than 30 g and capable of air-to-air ranges of up to 100 km are now available at a reasonable cost, as are slightly heavier solar-assisted transmitters with life spans of several years or more. Studies begun as part of this program have documented the dispersal and survival patterns of hacked eagles in Oklahoma and of fledglings from donor and control nests in Florida. Ten Florida chicks were radio-tagged in 1987 and tracked for several months. Preliminary data indicate that the age at which nestlings disperse does not differ between control (128 days) and donor (131 days) nests. Several Florida fledglings were located again after migrations to the Chesapeake Bay area, and Oklahoma birds also have been located after migrating to northern Wisconsin and Minnesota. In one instance, a combination of ground and aerial tracking was used to follow a bird flying from the Oklahoma hack site on a continuous 11-day migration to southern Canada in 1987. Surprisingly, the bird did not feed or follow rivers or other natural features during migration. Instead, it flew due north each day—even over downtown Omaha—only varying from that course as a result of the strong westerly winds it encountered en route. The bird was located again in early August on a lake in northern Wisconsin, and a second bird from the Oklahoma hack site was found on a lake in northern Minnesota about the same time. The Minnesota bird was resighted on the Fourche La Fave River in western Arkansas in mid-February, less than 100 km from the Oklahoma hack site. Sadly the bird was shot, and died on 28 February 1988. Unfortunately, shooting is still a major form of unnatural mortality in bald eagles and other large birds of prey.

The development of lightweight satellite telemetry transmitters holds the greatest promise for understanding the population biology of eagles. Employed for almost a decade on a variety of larger animals, including caribou, sea turtles, whales, and grizzly bears, the transmitters work with a Doppler positioning system carried aboard the National Oceanic and Atmospheric Administration's TIROS-N satellites. Researchers from the Applied Physics Laboratory at Johns Hopkins University were the first to refine the technology to the point where it could be applied to large free-flying birds (Strikwerda et al. 1985, 1986). They attached a prototype transmitter weighing 170 g to a young male bald eagle in July 1984 and tracked the bird for almost eight months over a distance of 4,554 km. The results, an unprecedented record of movement patterns and habitat selection in a free-flying eagle, provided a glimpse of the technology's potential. The weight of satellite transmitters will have to be reduced by about 50% before they become practical for large-scale applications, but such a reduction is thought to be feasible, and commercially produced transmitters are expected to become available within the next two years (Tomkiewicz and Beaty 1987). When they are, and we begin to understand the movement patterns and survival rates of bald eagles, we will for the first time be able to make the best possible use of the limited resources available for the conservation of this and other wide-ranging wildlife species.

Conservation programs targeted at species like the bald eagle can be extremely effective mechanisms for preserving biological diversity. The acquisition and preservation of breeding and wintering habitats for a particular species promote the conservation of untold other species that piggyback on this process. In addition to these direct benefits, there are many others that are less tangible. In some ways, this project is as symbolic as the birds it is attempting to conserve. Bald eagles are not on the verge of extinction, and when viewed in the context of global conservation needs and of other critically endangered species, the attention may seem misplaced. In fact, it is precisely the symbolic nature of widespread species like the bald eagle—with their ability to capture the imagination of the public—that makes them such worthwhile conservation investments. As symbols of wilderness and of the freedom wilderness represents, bald eagles have the unique capacity to inspire people and to foster a sympathetic attitude toward the needs of other threatened species and toward related environmental issues such as habitat destruction and water quality. Clearly, without that sympathy and the political will it engenders, the needs of more obscure species will go unmet. It may be trickle-down conservation, but in light of the ever-increasing pressure on global resources, it may prove to be one of the more fruitful conservation strategies available in the years ahead.

References

Bagley, F. M. 1987. Summary of 1986–1987 southern bald eagle nesting success. US Fish and Wildlife Service Field Office, Jackson, MS.

Bent, A. C. 1937. Life histories of North American birds of prey. Part I. Washington: US National Museum Bulletin No. 167.

Burleigh, T. D. 1945. The bird life of the Gulf Coast region of Mississippi. *Occ. Papers Mus. Zool.* 20:331–490. Louisiana State University.

Cade, T. J., and S. A. Temple. 1977. The Cornell University falcon programme. In *Report of Proceedings, World Conference on Birds of Prey, Vienna, 1975*, ed. R. D. Chancellor, pp. 353–69. London: International Council for Bird Preservation.

Green, N. 1985. The bald eagle. In *Audubon Wildlife Report 1985*, ed. R. L. Di Silvestro, pp. 508–31. New York: National Audubon Society.

Grier, J. W. 1980a. Ecology: A simulation model for small populations of animals. *Creative Computing* 6:116–21.

———. 1980b. Modeling approaches to bald eagle population dynamics. *Wildl. Soc. Bull.* 8:316–22.

———. 1982. Ban on DDT and subsequent recovery of reproduction in bald eagles. *Science* 218:1232–35.

Kennedy, R. S. 1977. A method for increasing osprey productivity. In *National Park Service Trans. and Proc. Ser.*, No. 2, *Transactions of the North American Osprey Research Conference*, ed. J. C. Ogden, pp. 35–42. Washington, DC.

Kenward, R. 1980. Radio-monitoring birds of prey. In *A Handbook on Biotelemetry and Radio Tracking*, ed. C. J. Amlaner and D. W. Macdonald, pp. 97–104. Pergamon.

King, W. B. 1981. *Endangered Birds of the World*. Smithsonian Institution Press.

Lewin, R. 1986. A mass extinction without asteroids. *Science* 234:14–15.

MacArthur, R. H., and E. O. Wilson. 1967. *The Theory of Island Biogeography*. Princeton Univ. Press.

Mertz, D. B. 1971. The mathematical demography of the California condor population. *Am. Nat.* 105:437–53.

Morrison, M. L., and B. J. Walton. 1980. The laying of replacement clutches by falconiforms and strigiforms in North America. *Raptor Res.* 14:79–85.

Murphy, T. M., et al. 1984. *Southeastern States Bald Eagle Recovery Plan.* Atlanta: US Fish and Wildlife Service.

Newton, I. 1979. *Population Ecology of Raptors.* Vermillion, SD: Buteo Books.

Nye, P. E. In press. A review of bald eagle hacking projects and early results in North America. In *Proceedings of the International Symposium on Raptor Reintroduction*, ed. D. Garcelon. Raptor Research Foundation.

Sherrod, S. K., W. R. Heinrich, W. A. Burnham, J. H. Barclay, and T. J. Cade. 1981. *Hacking: A Method for Releasing Peregrine Falcons and Other Birds of Prey.* Ithaca, NY: The Peregrine Fund.

Sherrod, S. K., M. A. Jenkins, G. McKee, S. Tatom, and D. Wolfe. In press. Using wild eggs for production of bald eagles for reintroduction into the southeastern United States. In *Proceedings of the 3rd Nongame and Endangered Wildlife Symposium*, ed. R. Odom. Social Circle, GA: Georgia Department of Natural Resources.

Simons, T. R. 1984. A population model of the endangered Hawaiian dark-rumped petrel. *J. Wildl. Manag.* 48:1065–76.

Stinson, C. H. 1979. On the selective advantage of fratricide in raptors. *Evolution* 33:1219–25.

Strikwerda, T. E., H. D. Black, N. Levanon, and P. E. Howey. 1985. The bird-borne transmitter. *Johns Hopkins APL Technical Digest* 6:60–67.

Strikwerda, T. E., M. R. Fuller, W. S. Seegar, P. W. Howey, and H. D. Black. 1986. Bird-borne satellite transmitter and location program. *Johns Hopkins APL Technical Digest* 7:203–07.

Tomkiewicz, S. M., and D. W. Beaty. 1987. Wildlife satellite telemetry —a progress report 1987. Paper presented at the 1987 Argos Users Conference, September 15–17, Washington.

Wilcove, D. S., C. H. McLellan, and A. P. Dobson. 1986. Habitat fragmentation in the temperate zone. In *Conservation Biology*, ed. M. E. Soule, pp. 237–56. Sinauer.

Ted Simons, a research biologist with the National Park Service, holds an M.S. (1979) and a Ph.D. (1983) in wildlife biology from the University of Washington. Steve K. Sherrod obtained his Ph.D. in biology from Cornell University (1982); he is the director of the Sutton Avian Research Center in Bartlesville, Oklahoma, a private nonprofit organization established in 1983 for the conservation of rare and endangered birds. Michael W. Collopy, with a Ph.D. in natural resources from the University of Michigan (1980), is an associate professor and chairman of the Department of Wildlife and Range Sciences, University of Florida. M. Alan Jenkins has an M.S. in zoology from Brigham Young University (1974) and is the assistant director of the Sutton Avian Research Center. Address for Dr. Simons: Gulf Islands National Seashore, Ocean Springs, MS 39564.

The Faustian Traits of the Crown-of-Thorns Starfish

Charles Birkeland

The crown-of-thorns starfish, *Acanthaster planci,* is one of the most influential species in the diverse biotic communities that make up tropical reefs. Although nine or ten species of starfish in the Pacific occasionally feed on coral or coral mucus, the crown-of-thorns starfish affects the reef community to a far greater extent than do any of these. In a short period of time, *A. planci* can alter the composition of a coral reef over large areas. An individual crown-of-thorns starfish digests an average of about 5 to 6 m^2 of coral surface per year (Pearson and Endean 1969; Dana and Wolfson 1970). An outbreak of *A. planci* may consist of tens of thousands to millions of individuals, capable of devouring from 0.5 to 5 or 6 km^2 of living coral tissue in a single year (Fig. 1).

Figure 1. A phalanx of crown-of-thorns starfish grazes on a coral reef off the coast of Palau. *Acanthaster planci* feeds by extruding its stomach over the surface of the reef and digesting the coral externally. Rapid growth and a high ratio of stomach surface area to total biomass make *A. planci* a formidable predator in the coral-reef community. Crown-of-thorns starfish tend to feed in tight aggregations, often, as here, piled several layers deep; chemicals released when corals are damaged attract more *A. planci* to the scene. (Unless otherwise noted, all photos are by the author.)

The impact of the crown-of-thorns starfish has been dramatic (Fig. 2). Between late 1968 and early 1969 it killed over 90% of the coral along 38 km of coastline in northwestern Guam (Chesher 1969). Extensive mortality also occurred at 21 of 45 islands surveyed in Micronesia in the period from 1969 to 1972 (Marsh and Tsuda 1973). Commenting on the situation in the northwestern Pacific, Yamaguchi has stated that "it is safe to assume that most reefs in the Ryukyus have been devastated in the past 15 years" (1986, p. 24). In the southwestern Pacific, an outbreak on the Great Barrier Reef that apparently began in the vicinity of Green Island before 1962 spread southward in a series of subsequent outbreaks over the next ten years, traveling hundreds of kilometers (Kenchington 1977). In recent surveys covering hundreds of reefs in the Great Barrier system, about 28% were found to be affected to varying degrees by *A. planci* (Zann and Eager 1987).

The ecological effects of these incursions are complex. When corals are eaten by the crown-of-thorns starfish, they are replaced initially by algae (Fig. 3). This change is occasionally followed by an increase in sessile animals other than reef-building corals, such as alcyonacean soft corals or *Terpios* sponges (Pearson and Endean 1969; Bryan 1973; Nishihira and Yamazato 1974). In some cases, coral communities composed predominately of branching *Acropora* corals have collapsed into rubble several months after predation by *A. planci,* demonstrating the major effect of the crown-of-thorns starfish on reef topography (Sano et al. 1987).

Structural traits otherwise found only in temperate starfish have produced a fast-growing, voracious, but relatively fragile tropical predator

As algae invade the space vacated by living coral tissue killed by *A. planci,* the food supply of herbivorous fishes increases and that of coral-eating fishes decreases. Thus a decrease in the density of populations of coral-eating fishes usually follows an outbreak of *A. planci* (Bouchon-Navaro et al. 1985; Williams 1986; Sano et al. 1987; Wass 1987). An increase in the population densities of herbivorous fishes occurs sometimes, but not always. At American Samoa, Wass (1987) found that herbivorous fishes increased from about 9 to 44% of the reef-fish population when the corals were killed by *A. planci.* At Moorea in French Polynesia the butterfly fish *Chaetodon*

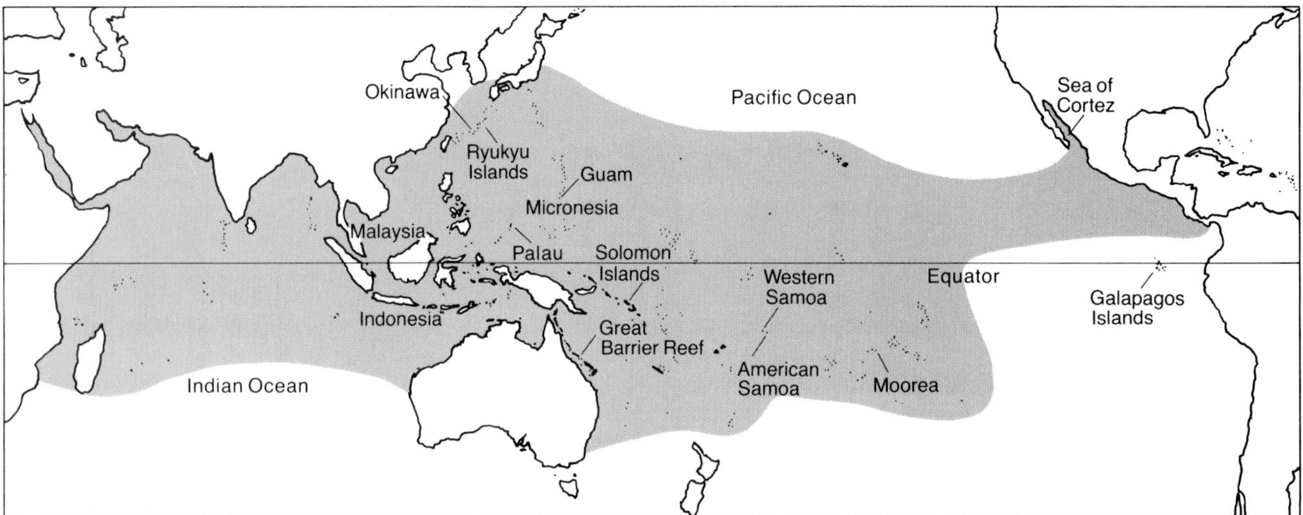

Figure 2. The crown-of-thorns starfish is found throughout the tropical Pacific and Indian oceans in the shaded area; it is not present in the Galapagos Islands or in the waters south of Panama. *A. planci* is thought to have evolved in the area of Malaysia and Indonesia, perhaps when the archipelago was partitioned into isolated seas during the recent glacial periods. It may then have spread rapidly across the Indo-Pacific, taking advantage of the vast food resource offered by the coral reefs. The ability of *A. planci* to extend its range is seen in modern outbreaks, which are sometimes followed by a series of secondary outbreaks hundreds of kilometers downstream along major water currents. (After Zann and Eager 1987.)

Figure 3. The ecological effects of the crown-of-thorns starfish are visible in this view of two starfish feeding on tabletop *Acropora*. The green area at the right indicates living coral tissue. Coral skeleton recently cleaned of living tissue by *A. planci* is white. Increasingly darker shades of gray at the left show algal succession on the coral skeleton.

citrinellus, an algae-eater, also increased in abundance, replacing two coral-eating species (Bouchon-Navaro et al. 1985).

However, increases in herbivorous fish populations were not observed following outbreaks of *A. planci* on the Great Barrier Reef and at Okinawa (Williams 1986; Sano et al. 1987). Although the removal of a food resource often leads to a decrease in a population that consumes it, an increase in the carrying capacity of a community does not necessarily result in an increase in population. Such increases depend also on the increased survival of larval recruits. It is likely that other factors independent of carrying capacity interfered with the recruitment of herbivorous fishes during the years studied.

The crown-of-thorns starfish can have a major influence on the structure of coral communities even at densities lower than those classified as an outbreak (Glynn 1976). Coral communities exposed to predation in the eastern tropical Pacific are generally dominated by branching pocilloporid corals (Glynn 1976). In similar areas where *A. planci* is absent, massive agariciid and poritid corals are most prominent (Birkeland et al. 1975; Glynn and Wellington 1983). Certain species of crabs and shrimps that live among the branches of pocilloporid corals drive away the crown-of-thorns starfish by nipping at its tube feet and spines. *A. planci* readily accepts pocilloporid corals as prey when they are unprotected, but larger colonies are usually defended by these crustacean symbionts.

The repulsion of the crown-of-thorns starfish by crustacean defenders effectively shifts its feeding preference from pocilloporid to agariciid corals. However, dense stands of *Pocillopora* can act as refuges for massive agariciid corals (Glynn 1976). *A. planci* will not move across a continuous band of *Pocillopora* both because of the crustacean symbionts and because nematocysts in the pocilloporid coral tissue can sting the starfish's tube feet. Agariciid colonies have been known to survive for as long as 190 years if they are protected by a continuous surrounding band of *Pocillopora* colonies (Glynn 1985).

When many of the pocilloporid colonies on the Pacific coast of Panama were killed by the elevated water temperatures produced by the El Niño of 1982–83, this barrier to predation by *A. planci* was eliminated. The agariciid corals were not as strongly affected by the rise in water temperature, but they were eaten by the crown-of-thorns starfish after the high temperatures had killed the *Pocillopora*. The extent of agariciid mortality on this occasion indicates that *A. planci* is a powerful factor in shaping reef structure even at relatively low densities.

One modern approach to ecology is to concentrate on ecological processes and to avoid referring to individual species as much as possible (Mann 1982). However, some species have such distinct, controlling effects on communities that they cannot be averaged into trophic levels or rate processes. Their natural histories must be taken into account if the ecological system is to be understood.

Why outbreaks occur

Population increases of 5 to 6 orders of magnitude have been observed in the crown-of-thorns starfish within a single year (Birkeland 1982). No mechanism directly affecting adult starfish has yet been discovered that could account for increases of such magnitude and abruptness. Factors affecting larval *A. planci,* however, are plausible causes. An individual can produce as many as 65 million eggs each spawning season (Kettle and Lucas 1987). Thus a small improvement in the rate of larval survival would result in a tremendous increase in numbers. The distribution of outbreaks following an initial event indicates that *A. planci* spreads over hundreds of kilometers downstream along major water currents (Kenchington 1977; Yamaguchi 1986). This pattern suggests that larval survival and distribution are probably key factors in determining the time and place of outbreaks.

Evidence from a variety of sources indicates that nutrients—phytoplankton and perhaps bacteria and organic detritus—are an important element in larval survival. Records of previous outbreaks indicate a strong correlation with heavy rainfall, which facilitates terrestrial runoff (Birkeland 1982). Outbreaks usually occur in the coastal waters of high islands or along continental shelves, where runoff from the land can release a pulse of nutrients; they rarely occur at atolls in intermediate locations. Such nutrient pulses are often associated with phytoplankton blooms. The spawning season of *A. planci* coincides with the beginning of the rainy season on both sides of the equator—the time phytoplankton blooms are most likely to occur. In controlled laboratory experiments, Lucas (1982) has shown that *A. planci* larvae survived on particular species of phytoplankton only when these nutrients occurred at densities characteristic of a phytoplankton bloom.

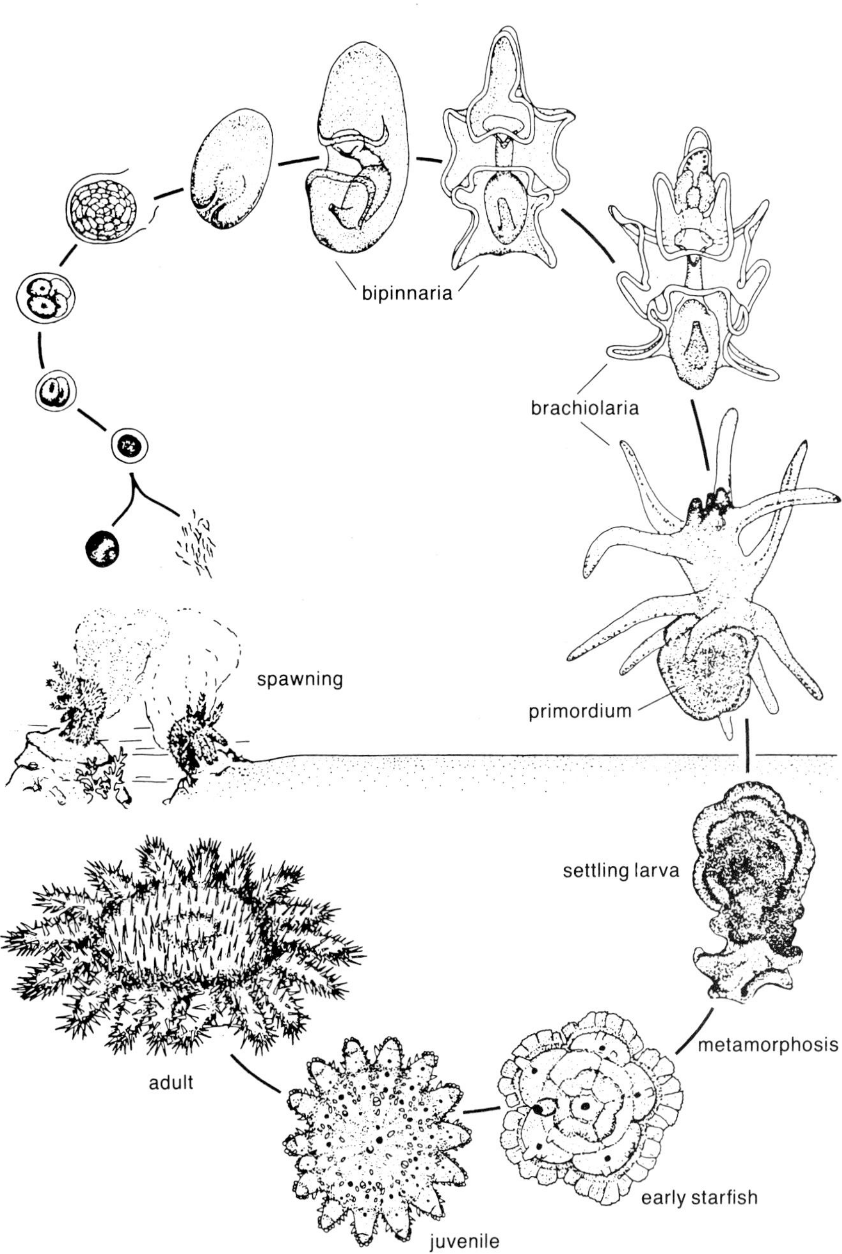

Figure 4. The life cycle of the crown-of-thorns starfish is distinguished by a larval stage lasting two to three weeks followed by rapid growth through the vulnerable juvenile stage. The primordium of the larva in the late brachiolaria phase is the early form of the small starfish. The settling larva attaches itself to the ocean floor by means of the brachiolar arms and a small adhesive disk located between the arms. The primordium then absorbs the larval body and grows into a small starfish, which crawls away. The newly metamorphosed starfish has five arms, but over the next few months the juvenile acquires a new arm every nine or ten days. The average adult *A. planci* has from 14 to 18 arms, although some individuals may have as few as 8 or as many as 21. (Drawing © Australian Institute of Marine Science.)

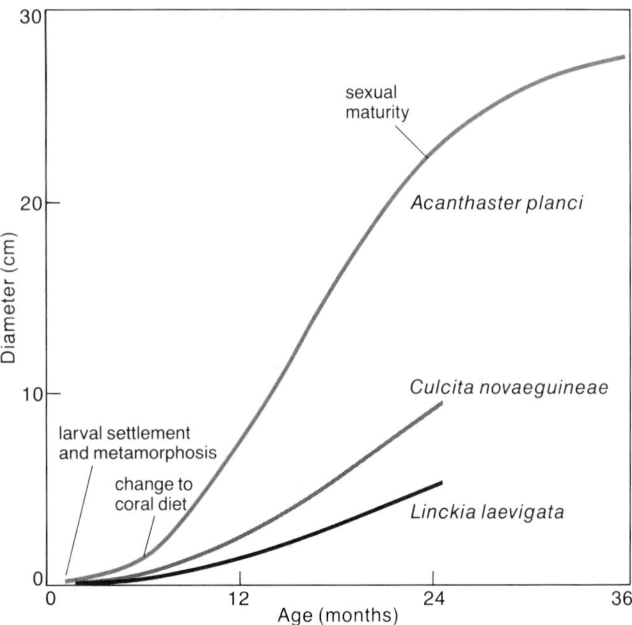

Figure 5. In this graph comparing the growth rates of three species of coral-reef starfish, the early shift of *A. planci* from a juvenile diet of algae to an adult diet of coral coincides with a conspicuous spurt in growth. *Culcita novaeguineae* and *Linckia laevigata* do not change to adult form until they are about two years old, and their growth is much slower. Once *A. planci* reaches sexual maturity, some energy is diverted from body growth to reproductive products, and thus the growth rate slows considerably.

By devising a way to raise larvae in chambers near a coral reef both in ambient seawater and in seawater enriched with phytoplankton, Olson (1987) demonstrated that they could survive in concentrations of nutrients normal for coral reefs—that is, in the absence of phytoplankton blooms. This suggests that the diversity of nutrients in natural seawater may enhance the development of larvae, as opposed to the controlled monocultures of food used in the laboratory. However, the mean length of time required for larvae to reach a late stage of development was about 9% less for enriched seawater than for natural seawater—12.6 as compared to 13.8 days. Although this difference was statistically significant, Olson concluded that it was not large enough to explain the abrupt escalation of population that occurs in an outbreak.

A large variation in recruitment from year to year is characteristic of coral-reef animals whose larvae feed on phytoplankton. The larval biology of some tropical starfish is similar to that of *A. planci* (Fig. 4; Yamaguchi 1973a). But starfish with nearly identical larval histories can have very different rates of recruitment due to characteristics of the juveniles and events following metamorphosis (Birkeland 1974).

Crown-of-thorns starfish as small as 8 to 10 mm in diameter can change from the juvenile mode of feeding on encrusting algae to the adult mode of preying upon corals. This transition begins from 4½ to 7 months after metamorphosis and takes about a month (Yamaguchi

Figure 6. The crown-of-thorns starfish contrasts sharply with other coral-reef asteroids not only in growth rate but in structure. With its disk-shaped body and multiple prehensile arms, *A. planci (upper left)* has a large surface area compared to other tropical forms. Adults range from 25 to 40 cm in diameter, although they can be as large as 60 to 70 cm. The stomach covers an area about the size of the central disk—approximately half the total diameter. *C. novaeguineae (lower left)* is typical of the heavy, robust reef starfish of the family Oreasteridae. Its thick, calcareous skeleton is probably a good defense against predation by most fish, but limits the prey available to it. The brightly colored *L. laevigata (above)* is a typical member of the Ophidiasteridae, a family of reef starfish with finger-like arms and a small central disk. *L. laevigata* is not preyed on by most fishes, perhaps because of the high ratio of extremely thick skeleton to digestible tissue. (Photos of *C. novaeguineae* and *L. laevigata* courtesy of R. F. Myers.)

1973a, 1974; Lucas and Jones 1976; Lucas 1984). Once the juvenile switches to an adult diet of corals, its rate of growth increases significantly (Lucas 1984; Zann et al. 1987). An individual can grow to a diameter of 20 to 25 cm in about two years (Lucas 1984; Yamaguchi 1974). At the time it reaches sexual maturity, *A. planci* is several times larger than most other starfish species at the same stage of development (Fig. 5; Lawrence 1987).

By contrast, *Culcita novaeguineae,* another starfish that feeds on corals, changes from a flat, pentagonal juvenile to a cushion-form adult at a diameter of about 9 cm—ten times the width and about eighty times the surface area of *A. planci* when it makes the transition from juvenile to adult diet (Fig. 6; Yamaguchi 1973b). *C. novaeguineae* makes the change to adult form at about two years of age, or about four times the age of *A. planci* at its changeover (Yamaguchi 1977a). Another common asteroid of coral reefs, *Linckia laevigata,* may make the shift to adult form at a diameter of about 5 cm, when it is approximately two years old (Yamaguchi 1973b, 1977b).

The growth of juvenile *A. planci* may be slower if corals are not readily available (Lucas 1984; Zann et al. 1987). If a dense larval population occurs where corals are scarce, growth and survival rates might not be much better than those of other coral reef starfish. In areas where corals are common, however, dense recruitment of *A. planci* larvae might result in an outbreak because of the rapid switch to a diet of corals. In other asteroid species, an individual that first obtains a good meal has been observed to grow several times larger than other individuals in the same group that obtained meals later, and its chances of survival are greater (Mead 1900). Nauen (1978) found that starfish may not grow for several months until they come upon a good meal, after which they begin to grow.

The crown-of-thorns starfish has an exceptionally large area of food intake—the extruded stomach—in relation to the biomass it supports (Fig. 7; Birkeland, in press). This high ratio between the ability to take in food and total biomass is probably an important factor in its relatively rapid growth. Other large coral reef starfish have a much greater biomass to support in relation to the size of their feeding apparatus—what might be called a "brontosaurus handicap." All juvenile asteroids remain concealed until they reach adult size, whether or not they have attained reproductive maturity. The rapid growth of *A. planci* through the especially vulnerable juvenile stage increases its chances of survival after metamorphosis.

Structure and success

In the shallow waters of temperate regions starfish have a controlling influence on benthic communities (Menge 1982), but most of the diverse array of asteroids on coral reefs have a negligible effect on their communities. The large stomach surface of *A. planci* allows it to consume coral tissue at five or six times the rate of *C. novaeguineae,* the next most important asteroid predator of corals (Glynn and Krupp 1986). But a more crucial characteristic that sets *A. planci* apart from other coral-reef starfish is its thin, pliable form. Other large starfish that evolved on coral reefs developed thick, calcareous bodies. *A. planci* has a body form that is large, multi-armed, elastic, and prehensile—traits that are common in influential species in temperate regions but which are found in no other starfish on coral reefs. The flaccid, disklike structure of *A. planci* not only makes it the most efficient of the coral-reef asteroids in rapid growth, allowing it to pass quickly through the hazardous juvenile stage, but also permits it to prey upon large coral colonies that are inaccessible to other starfish (Fig. 8).

Starfish with a thick, calcareous structure are unable to climb onto branching coral colonies and are also less able to adhere to large hemispherical corals. Asteroids such as *C. novaeguineae* can kill only small colonies 3 to 8 cm in diameter, and can only partially eat larger colonies (Glynn and Krupp 1986). Both the recovery of a coral reef and the structure of the future community are profoundly affected by whether all large corals are killed or only nonbranching or small branching colonies. Thus even if other coral-reef starfish occurred in the enormous numbers found in *A. planci* outbreaks, they would not affect the reef communities to the same extent.

The evolution of *A. planci* and its affinity with temperate forms may provide some insights into the appearance of the structural traits that make it such a successful predator of coral. On the basis of cross-fertilization, rearing, and electrophoretic studies, Lucas and his co-workers have concluded that *A. planci* and *A. brevispinus* are recently separated sibling species, with the mollusc-eating *A. brevispinus* being the archetype (Lucas and Jones 1976; Lucas et al. 1985). They hypothesize that the ancestor of *A. planci* found a vast food resource in the coral reefs and spread rapidly across the Indo-Pacific.

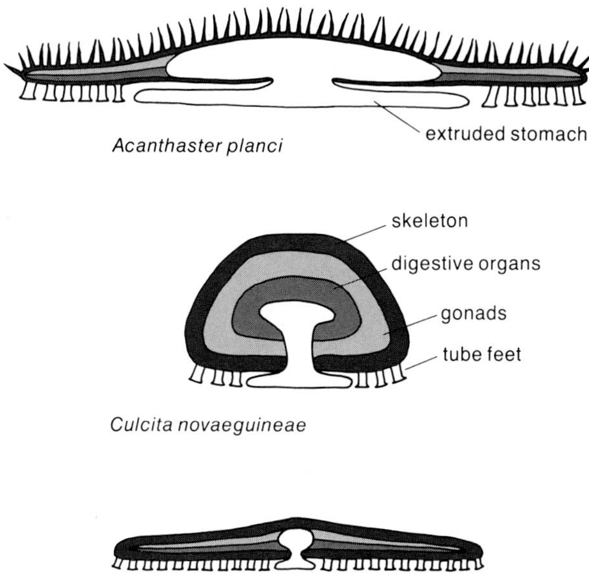

Figure 7. A schematic diagram shows the relative stomach area and skeletal thickness of three coral-reef starfish. The ratio of extruded stomach to biomass is over seven times greater in *A. planci* than in *C. novaeguineae*; the stomach area of *C. novaeguineae* in turn exceeds the small intake area of *L. laevigata,* which corresponds to its small central disk. The thickness of the calcareous skeleton—greater in *C. novaeguineae* and *L. laevigata,* least in *A. planci*—enhances resistance to predators but limits mobility and thus access to some coral prey.

The ability of *A. planci* to extend its distribution in this way can be seen in the spreading of secondary outbreaks over hundreds of kilometers within a few years (Kenchington 1977; Yamaguchi 1986). Since *A. planci* is found from the Sea of Cortez to the Pacific coast of Panama, and since no evidence of acanthasterids has been found in the Caribbean, this sequence of events probably took place within the last three million years, after the rise of the Isthmus of Panama.

Lucas and his colleagues further suggest that the separation of *A. planci* and *A. brevispinus* might have occurred in the region of Malaysia and Indonesia, because this archipelago was partitioned into isolated seas during the recent glacial periods (Potts 1983). Both of the sibling species exist in this region, and the split could have occurred when their ranges were partitioned into isolated pockets during the Pleistocene epoch.

The family Acanthasteridae, which consists of a single genus, bears a strong resemblance to the Solasteridae, which are similar in ecology, behavior, feeding structure, and early transition from juvenile to adult feeding patterns. The two families have traditionally been placed next to each other in taxonomic arrangements on the basis of morphology. Immunological studies have supported this association, although one researcher has recently placed Acanthasteridae near the Oreasteridae because of certain skeletal characteristics (Mochizuki and Hori 1980; Blake 1979).

Figure 8. The pliable, prehensile form of the crown-of-thorns starfish allows it to feed on branching corals and adhere to large massive corals that are unavailable to rigid, thick-bodied oreasterids and ophidiasterids. Here a single *A. planci* clings to the branches of an *Acropora nobilis* coral in the waters off American Samoa. The crown-of-thorns can feed on almost any species of coral, while other tropical starfish are limited to small or encrusting coral colonies.

Like *A. planci,* solasterids are magnificent predators because of their large stomachs, pliable bodies, and multiple prehensile arms. They have been a major influence in subtidal habitats in the temperate Pacific (Mauzey et al. 1968; Birkeland 1974; Sloan 1980; Menge 1982; Birkeland et al. 1982). Solasterids are cosmopolitan in distribution and are generally found on both sandy and rocky substrata, although they occur in deeper waters in the tropics. Species in the genus *Crossaster* resemble *A. brevispinus* and are found in deeper waters in the Philippines and Indonesia, within the geographic ranges of *A. brevispinus* and *A. planci.* Both acanthasterids and *Crossaster* species are absent from the tropical western Atlantic.

Similarities of diet and feeding methods are also suggestive. *Crossaster papposus,* a northern species, feeds on molluscs and anthozoans—a group including sea pens, anemones, and soft corals—in addition to a variety of other taxa (Sloan 1980). When it feeds on molluscs it assumes the arched position used by *A. brevispinus.* Half the diet of *C. papposus* in the Puget Sound region consists of an octocoral, *Ptilosarcus gurneyi,* showing a propensity for this kind of prey (Mauzey et al. 1968; Birkeland 1974).

Since a thick, calcareous skeleton restricts prey to small or encrusting coral colonies, and since the thicker skeleton and smaller ratio of stomach to biomass lead to slower growth and a longer period at a vulnerable size, why do most large asteroids on coral reefs have these apparently disadvantageous traits? Blake has theorized (1983) that coral-reef starfish evolved under intense pressure from strong-jawed fishes that prey on shelled invertebrates, a factor much less important for starfish in temperate benthic communities. Juvenile oreasterids and adults of other coral-reef starfish are under continuous pressure from predators and generally remain hidden. But in spite of the fact that adult oreasterids often display bite marks, they appear to be protected by their thick calcareous skeletons and no kills have been observed. Unsuccessful predation is a stronger selective force for a defensive trait than either successful predation or absence of predation (Vermeij 1982). Perhaps because of the large ratio of calcareous to digestible matter, *L. laevigata, L. guildingii,* and other relatively large ophidiasterids do not appear to be the favored prey of fish.

By contrast, both *A. brevispinus* and *A. planci* depend on spines for defense. As compared to its sibling species, *A. planci* has developed longer spines which are toxic; it is the only known venomous starfish in the world (Halstead 1978). Nevertheless, the toxic spines appear to be a less effective deterrent than the thick calcareous skeletons of other coral-reef starfish that live in exposed locations. *A. planci* is eaten by a variety of predators, although it is not known to be the preferred prey of any fish (Endean 1973; Moran 1986). During the daylight hours individuals are often found on the undersides of overhanging coral plates or in other concealed locations, especially when the population is not abundant; they are probably hiding from visual predators such as fish. When sparse, *A. planci* forages in the open primarily at night, but when it is abundant enough to overwhelm its predators it forages both day and night. Predation by the shrimp *Hymenocera picta* and the marine worm *Pherecardia striata* appears to have kept the crown-of-thorns starfish at relatively low levels of density on the Pacific coast of Panama (Glynn 1984).

The traits that provide for quick success may also increase the likelihood of an early demise

The structural basis of the exceptional influence of *A. planci*—an elastic, disklike form with multiple prehensile arms combined with a large ratio of stomach to biomass—has developed often in temperate Pacific regions. It is found not only in *Crossaster* but in such genera as *Pycnopodia, Solaster,* and the multi-armed *Heliaster,* which is most commonly found in the Galapagos Islands and the Sea of Cortez, where the pressure of predators on echinoderms is less than in other regions of the eastern Pacific (Glynn and Wellington 1983; Lucas et al. 1985). Intense predation by fish has probably been a major factor in preventing these traits from developing in coral-reef starfish (Blake 1983). Thus the general form from which *A. planci* originated must have evolved in other habitats. The rapid dissemination of the crown-of-thorns starfish across the Indo-Pacific appears to be analogous to the expansion of an introduced species—except that in this case the species originated in the center of the region. *A. planci* is an "introduced" form in that it came equipped with a structure that would not have evolved under the intense pressure of predation by fishes on a coral reef.

Faustian traits

As *A. planci* grows larger, achieving a thousand-fold increase in linear dimension and a ten-million-fold increase in biomass (Kettle and Lucas 1987), the morphological traits which were beneficial for a small asteroid can become detrimental. Lucas (1984) has determined that the increase in biomass with growth exceeds the rate of food intake by an exponential factor of 0.32. As the individual increases in size, more of the caloric content and dry-weight biomass of the body is apportioned to the gonads, less to the body wall (Kettle and Lucas 1987). Larger individuals have less ossification of body wall and become more flattened, pliable, and fragile in construction. Thus the elastic, disklike structure may represent a Faustian bargain for *A. planci,* allowing it quick growth in the early years but resulting in a loss of integrity and strength in later life.

In exchange for access to a broad range of coral prey unavailable to other starfish, a large ratio of food intake to biomass, and rapid growth through the juvenile stage with a better chance of survival and greater population density, *A. planci* is relatively susceptible to predation in comparison to other large coral-reef starfish such as oreasterids. As the crown-of-thorns starfish grows larger it becomes more fragile and thus more susceptible to damage by wave action or exposure to air. Large *A. planci* tend to lose structural integrity when lifted from the water. Dense populations may also overexploit their food supply, which may contribute to the abrupt disappearance of some outbreak populations. Thus the traits that provide for quick success may also increase the likelihood of an early demise. The thousands of *A. planci* involved in an outbreak generally disappear within a year or so.

Scientists and managers sometimes wish to determine the "normal" density of *A. planci* in a given region in order to establish a scale for monitoring or a target for management plans. In some areas affected by terrestrial runoff—for example, high islands in the Caroline Islands or the region of the Great Barrier Reef influenced by the Burdekin River—the abundance of *A. planci* fluctuates widely. The factors of high fecundity, transport of larvae by water currents, rapid growth to adult size, and relatively short persistence of dense populations would seem to make it meaningless to attempt to define normal density. It is probably more meaningful to determine the sequence of events that lead to outbreaks than to obtain a quantitative estimate of the normal population for the coast of a high island or a continental shelf.

In view of the extensive mortality of corals when the crown-of-thorns starfish reaches high population densi-

Charles Birkeland is a professor at the Marine Laboratory of the University of Guam. He has served on advisory committees on the crown-of-thorns starfish for the Great Barrier Reef Marine Park Authority of Australia since 1984. Address: Marine Laboratory, University of Guam Station, Mangilao, Guam 96923.

ties, the public has become understandably worried about outbreaks. The Australian government has allocated some three million Australian dollars (about $2,460,000 in American dollars) to the study of the ecology of the starfish and the development of techniques for its management. In 1969 the US Department of the Interior provided $215,000 for a survey of *A. planci* populations around Micronesia. American Samoa invested $75,000 in the removal of nearly half a million *A. planci* from the shallow coastal waters of the main island in 1978, and Japan spent more than 600 million yen (about $2,400,000) to remove over 13 million *A. planci* from the Ryukyu Islands between 1970 and 1983 (Yamaguchi 1986). No other tropical starfish has caused such widespread concern.

The question of how far back in time outbreaks have occurred is germane, but it is difficult to obtain solid information. Core sampling and radiocarbon dating of sediments on the Great Barrier Reef, along with the distribution and density of *A. planci* skeletal elements in the sediments, suggest that outbreaks have been taking place there for thousands of years (Walbran and Henderson 1988). Older fisherman on such Pacific islands as Samoa, the Solomons, New Ireland, Pohnpei, and Palau remember various brief periods decades ago when the crown-of-thorns starfish was so abundant that it was unsafe to walk on the reefs at night; at such times they had to alter their fishing schedule temporarily (Birkeland 1981; Flanigan and Lamberts 1981).

However, the outbreaks recalled by the fishermen occurred at intervals of several decades—two or three times a century. Outbreaks appear to be more frequent now, and they are sometimes chronic. For example, no outbreaks are remembered on Guam before the 1950s, but there was a major outbreak in late 1967 and a smaller one in 1979. The second outbreak has perpetuated itself for a decade and is still with us. Similarly, *A. planci* has become a chronic problem in the Ryukyu Islands, beginning in 1969 and continuing to the present (Yamaguchi 1986).

Crown-of-thorn outbreaks are not the only tropical marine phenomenon that has increased in frequency during the past two decades. Red tides, dinoflagellate blooms, and paralytic shellfish poisoning have been increasing at a geometric rate in the western Pacific since 1975 (Maclean 1984; Holmes and Catherine 1985). The clearing of land for agricultural and urban use has also accelerated in the past twenty years, increasing the runoff of nutrients and sediments in coastal regions. If phytoplankton blooms increase the development rate and hence the larval survival of *A. planci,* then it is plausible that *A. planci* outbreaks might be related in part to increased coastal development in southeast Asia and the western Pacific.

Large-scale efforts to control outbreaks, like attempts to control red tides, the warming of the climate, and the rising sea level, are probably not feasible or at least exorbitantly expensive. The problem will recur unless the causes are acted on, and thus the greatest effort at this time should be put into probing causes. We may discover that reducing the frequency and magnitude of outbreaks is a more complex task than we now realize, requiring careful management of developing coastal lands.

References

Birkeland, C. 1974. Interactions between a sea pen and seven of its predators. *Ecol. Monogr.* 44:211–32.

———. 1981. *Acanthaster* in the cultures of high islands. *Atoll Res. Bull.* 255:55–58.

———. 1982. Terrestrial runoff as a cause of outbreaks of *Acanthaster planci* (Echinodermata: Asteroidea). *Mar. Biol.* 69:175–85.

———. In press. The influence of echinoderms on coral-reef communities. In *Echinoderm Studies III*, ed. M. Jangoux and J. M. Lawrence. Rotterdam: Balkema Press.

Birkeland, C., P. K. Dayton, and N. A. Engstrom. 1982. A stable system of predation on a holothurian by four asteroids and their top predator. *Austr. Mus. Mem.* 16:175–89.

Birkeland, C., D. L. Meyer, J. P. Stames, and C. L. Buford. 1975. The subtidal communities of Malpelo. In *The Biological Investigation of Malpelo Island, Colombia,* ed. J. B. Graham, pp. 55–68. Smithsonian Contributions to Zoology, no. 176.

Blake, D. B. 1979. The affinities and origins of the crown-of-thorns sea star *Acanthaster* Gervais. *J. Nat. Hist.* 13:303–14.

———. 1983. Some biological controls on the distribution of shallow water sea stars (Asteroidea: Echinodermata). *Bull. Mar. Sci.* 33: 703–12.

Bouchon-Navaro, Y., C. Bouchon, and M. Harmelin-Vivien. 1985. Impact of coral degradation on a chaetodontid fish assemblage (Moorea, French Polynesia). *Proc. Fifth Internat. Coral Reef Congress, Tahiti* 5:427–32.

Bryan, P. G. 1973. Growth rate, toxicity, and distribution of the encrusting sponge *Terpios* sp. (Hadromerida: Suberitidae) in Guam, Mariana Islands. *Micronesica* 9:237–42.

Chesher, R. H. 1969. Destruction of Pacific corals by the sea star *Acanthaster planci. Science* 165:280–83.

Dana, T., and A. Wolfson. 1970. Eastern Pacific crown-of-thorns starfish populations in the lower Gulf of California. *Trans. San Diego Soc. Nat. Hist.* 16:83–90.

Endean, R. 1973. Population explosions of *Acanthaster planci* and associated destruction of hermatypic corals in the Indo-West Pacific region. In *Biology and Geology of Coral Reefs*, III, ed. O. A. Jones and R. Endean, pp. 389–438. Academic Press.

Flanigan, J. M., and A. E. Lamberts. 1981. *Acanthaster* as a recurring phenomenon in Samoan history. *Atoll Res. Bull.* 255:59–62.

Glynn, P. W. 1976. Some physical and biological determinants of coral community structure in the eastern Pacific. *Ecol. Monogr.* 46:431–56.

———. 1984. An amphinomid worm predator of the crown-of-thorns sea star and general predation on asteroids in eastern and western Pacific coral reefs. *Bull. Mar. Sci.* 35:54–71.

———. 1985. El Niño-associated disturbance to coral reefs and post-disturbance mortality by *Acanthaster planci. Mar. Ecol. Progr. Ser.* 26: 295–300.

Glynn, P. W., and D. A. Krupp. 1986. Feeding biology of a Hawaiian sea star corallivore *Culcita novaeguineae* Muller and Troschel. *J. Exp. Mar. Biol. Ecol.* 96:75–96.

Glynn, P. W., and G. M. Wellington. 1983. *Corals and Coral Reefs of the Galapagos Islands.* Univ. of California Press.

Halstead, B. W. 1978. *Poisonous and Venomous Marine Animals of the World.* Darwin Press.

Holmes, P. R., and W. Y. L. Catherine. 1985. Red tides in Hong Kong waters: Response to a growing problem. *Asian Mar. Biol.* 2:1–10.

Kenchington, R. 1977. Growth and recruitment of *Acanthaster planci* (L.) on the Great Barrier Reef. *Biol. Conserv.* 11:103–18.

Kettle, B. T., and J. S. Lucas. 1987. Biometric relationships between organ indices, fecundity, oxygen consumption and body size in *Acanthaster planci* (L.) (Echinodermata; Asteroidea). *Bull. Mar. Sci.* 41: 541–51.

Lawrence, J. M. 1987. *A Functional Biology of Echinoderms.* Johns Hopkins Univ. Press.

Lucas, J. S. 1982. Quantitative studies of feeding and nutrition during larval development of the coral reef asteroid *Acanthaster planci* (L.). *J. Exp. Mar. Biol. Ecol.* 65:173–93.

———. 1984. Growth, maturation, and effects of diet in *Acanthaster planci* (L.) (Asteroidea) and hybrids reared in the laboratory. *J. Exp. Mar. Biol. Ecol.* 79:129–47.

Lucas, J. S., and M. M. Jones. 1976. Hybrid crown-of-thorns starfish *(Acanthaster planci* x *A. brevispinus)* reared to maturity in the laboratory. *Nature* 263:409–12.

Lucas, J. S., W. J. Nash, and M. Nishida. 1985. Aspects of the evolution of *Acanthaster planci* (L.) (Echinodermata, Asteroidea). *Proc. Fifth Internat. Coral Reef Congress, Tahiti* 5:327–32.

Maclean, J. L. 1984. Red tide: A growing problem in the Indo-Pacific region. ICLARM *Newsletter* 7:20.

Mann, K. H. 1982. *Ecology of Coastal Waters: A Systems Approach.* Univ. of California Press.

Marsh, J. A., Jr., and R. T. Tsuda. 1973. Population levels of *Acanthaster planci* in the Mariana and Caroline Islands 1969–1972. *Atoll Res. Bull.* 170:1–16.

Mauzey, K. P., C. Birkeland, and P. K. Dayton. 1968. Feeding behavior of asteroids and escape responses of their prey in the Puget Sound region. *Ecology* 49:603–19.

Mead, A. D. 1900. On the correlation between growth and food supply in starfish. *Am. Nat.* 34:17–23.

Menge, B. A. 1982. Effects of feeding on the environment: Asteroidea. In *Echinoderm Nutrition,* ed. M. Jangoux and J. M. Lawrence, pp. 521–51. Rotterdam: Balkema Press.

Mochizuki, Y., and S. H. Hori. 1980. Immunological relationships of starfish hexokinases: Phylogenetic implication. *Comp. Biochem. Physiol.* 65B:119–26.

Moran, P. J. 1986. The *Acanthaster* phenomenon. *Oceanogr. Mar. Biol. Ann. Rev.* 24:379–480.

Nauen, C. E. 1978. The growth of the sea star, *Asterias rubens,* and its role as benthic predator in Kiel Bay. *Kieler Meeresforsch* 4:68–81.

Nishihira, M., and K. Yamazato. 1974. Human interference with the coral reef community and *Acanthaster* infestation of Okinawa. *Proc. Second Internat. Coral Reef Symp.* 1:577–90.

Olson, R. R. 1987. In situ culturing as a test of the larval starvation hypothesis for the crown-of-thorns starfish, *Acanthaster planci. Limnol. and Oceanogr.* 32:895–904.

Pearson, R. G., and R. Endean. 1969. A preliminary study of the coral predator *Acanthaster planci* (L.) (Asteroidea) on the Great Barrier Reef. *Queensland Fisheries Branch, Fisheries Notes* 3:27–55.

Potts, D. C. 1983. Evolutionary disequilibrium among Indo-Pacific corals. *Bull. Mar. Sci.* 33:619–32.

Sano, M., M. Shimizu, and Y. Nose. 1987. Long-term effects of destruction of hermatypic coral by *Acanthaster planci* infestation on reef fish communities of Iriomote Island, Japan. *Mar. Ecol. Progr. Ser.* 37:191–99.

Sloan, N. A. 1980. Aspects of the feeding biology of asteroids. *Oceanogr. Mar. Biol. Ann. Rev.* 18:57–124.

Vermeij, G. J. 1982. Unsuccessful predation and evolution. *Am. Nat.* 120:701–20.

Walbran, P. D., and R. A. Henderson. 1988. Modern and ancient perspectives on the crown-of-thorns starfish *(Acanthaster planci* L.) in the Great Barrier Reef Province, Australia, assessed from the sediment record of John Brewer, Green Island and Heron Island Reefs. *Abstracts of the Sixth Internat. Coral Reef Symp.*: 104.

Wass, R. C. 1987. Influence of *Acanthaster*-induced coral kills on fish communities at Fagatele Bay and at Cape Larsen. In C. Birkeland, R. H. Randall, R. C. Wass, B. D. Smith, and S. Wilkins, 1987, *Biological Resource Assessment of the Fagatele Bay National Marine Sanctuary,* pp. 193–209. NOAA Technical Memoranda, Marine and Estuarine Management Division 3.

Williams, D. M. 1986. Temporal variation in the structure of reef slope fish communities (central Great Barrier Reef): Short-term effects of *Acanthaster planci* infestation. *Mar. Ecol. Progr. Ser.* 28:157–64.

Yamaguchi, M. 1973a. Early life histories of coral reef asteroids, with special reference to *Acanthaster planci* (L.). In *Biology and Geology of Coral Reefs,* II, ed. R. Endean and O.A. Jones, pp. 369–87. Academic Press.

———. 1973b. Recruitment of coral reef asteroids, with emphasis on *Acanthaster planci* (L.). *Micronesica* 9:207–12.

———. 1974. Growth of juvenile *Acanthaster planci* (L.) in the laboratory. *Pac. Sci.* 28:123–38.

———. 1977a. Estimating the length of the exponential growth phase: Growth increment observations on the coral-reef asteroid *Culcita novaeguineae. Mar. Biol.* 39:57–59.

———. 1977b. Population structure, spawning, and growth of the coral reef asteroid *Linckia laevigata* (Linnaeus). *Pac. Sci.* 31:13–30.

———. 1986. *Acanthaster planci* infestations of reefs and coral assemblages in Japan: A retrospective analysis of control efforts. *Coral Reefs* 5:277–88.

Zann, L., J. Brodie, C. Berryman, and M. Naqasima. 1987. Recruitment, ecology, growth and behavior of juvenile *Acanthaster planci* (L.) (Echinodermata: Asteroidea). *Bull. Mar. Sci.* 41:561–75.

Zann, L., and E. Eager. 1987. The crown of thorns starfish. *Australian Science Mag* 3:14–55.

The Global Carbon Cycle

Wilfred M. Post, Tsung-Hung Peng, William R. Emanuel, Anthony W. King, Virginia H. Dale and Donald L. DeAngelis

The dynamic responses of natural systems to CO_2 remain a puzzle—and the earth's climate may hang in the balance.

More than three decades have passed since Roger Revelle and Hans Suess (1957) drew scientific attention to a planetary-scale "experiment" in which mankind is "returning to the atmosphere and oceans the concentrated organic carbon stored in sedimentary rocks over hundreds of millions of years." In the years since that warning, we have begun to grasp the significance of this unplanned and uncontrolled experiment. We are witnessing a dramatic increase in carbon dioxide in the atmosphere, and we now recognize the potential climatic impact of that increase. What began as speculation among scientists about the interactions between CO_2 and climate is today a popular subject for government reports, Congressional hearings, newspaper articles and international policy debates.

The debates are set against a backdrop of new findings about CO_2 in the atmosphere. In recent years we have gained information on the contribution of land clearing to atmospheric CO_2 levels. We have found that long-term records of atmospheric CO_2 concentrations can be obtained from ice cores. And we have better estimates of how carbon is stored and exchanged by oceanic, atmospheric and terrestrial reservoirs.

But the research of the past few years has uncovered more complexities than were previously appreciated. We are unable to balance all the fluxes of the global carbon cycle over the period from 1800 to the present, and different mathematical models give results that are hard to reconcile. Moreover, recent studies have added to our awareness of the sensitive feedback relationships between the concentration of CO_2 in the atmosphere and the terrestrial and oceanic processes that regulate exchanges with the atmosphere. The popular phrase "greenhouse effect" describes one part of the interaction between CO_2 and climate, in which a higher concentration of CO_2 is expected to bring about a global warming. Feedback processes, however, could either moderate the increase in CO_2—and thereby stabilize the global system—or turn a gradual rise into an even more rapid climb. Hence, while we can document the growing human contribution of carbon dioxide to the atmosphere, and the potential for additional increases, we are in a poor position to predict how continued increases will affect the global carbon cycle.

Writing in *American Scientist* in 1977, Charles F. Baes, Jr., and his colleagues accurately articulated what was then known about CO_2 and climate. In this article we present what we believe are the most important advances in carbon-cycle research since that article appeared. We take inventory of the world's carbon reservoirs, and we discuss what is known about the role of oceanic and terrestrial systems in exchanging CO_2 with the atmosphere. Finally, we describe a new global systems approach, which shows promise in resolving current difficulties.

Wilfred M. Post, Tsung-Hung Peng, William R. Emanuel, Anthony W. King, Virginia H. Dale and Donald L. DeAngelis are staff scientists with the Environmental Sciences Division of Oak Ridge National Laboratory. Post, King and Dale are ecosystem ecologists, Peng is a geochemist, Emanuel is an electrical engineer and DeAngelis is a theoretical ecologist. Address: Environmental Sciences Division, Oak Ridge National Laboratory, Oak Ridge, TN 37831–6335.

Carbon Reservoirs and Fluxes

The carbon dioxide that makes up a small (but vitally important) constituent of the atmosphere is part of a vast planetary cycle, in which carbon cir-

H. Silvester (Photo Researchers, Inc.)

culates among three active reservoirs and undergoes several changes of chemical form. The reservoirs are the atmosphere, the oceans and a terrestrial system that includes a variety of stocks, such as forests and the organic carbon found in soil (Figure 6). Of the three reservoirs the oceanic one contains by far the largest amount of carbon. The atmosphere is the smallest in terms of carbon storage, but it plays a significant role in the cycle as a conduit between the other two reservoirs. The flux of carbon among the reservoirs is influenced by the current inventory of carbon in each reservoir and by the turnover rates, which vary as functions of environmental factors.

The size of the atmospheric carbon reservoir has been accurately known since 1958, when Charles Keeling began continuous measurements of the atmospheric concentration of CO_2 at the Mauna Loa Observatory in Hawaii. In 1958 the average annual concentration in the atmosphere was 315 microliters of CO_2 per liter of air, which works out to a concentration of about 0.03 percent and a total of 671 gigatons (billions of metric tons) of carbon in the atmosphere. Since then the amount of carbon in the atmosphere has grown exponentially (Figure 3b). In 1988 the concentration was 351 microliters per liter, or 748 gigatons of carbon. In contrast, analysis of air trapped in polar ice shows that over the past 160,000 years, atmospheric CO_2 has varied from 200 microliters per liter at the height of the last glaciation to between 260 and 300 microliters per liter during interglacial periods. Ice-core measurements for recent times agree well with the Mauna Loa data and suggest that concentrations during the period from 1750 to 1800 were

Figure 1. Burned Amazonian rainforest, crossed by an unpaved road, suggests the impact of human activities on the global carbon cycle. As a result of deforestation, as much as 2.6 gigatons of carbon stored in vegetation returned to the atmosphere as carbon dioxide in 1980. Estimates of this flow, the size of the carbon stocks in tropical vegetation and the impact of past land-use activities on the global carbon cycle are the subjects of considerable debate.

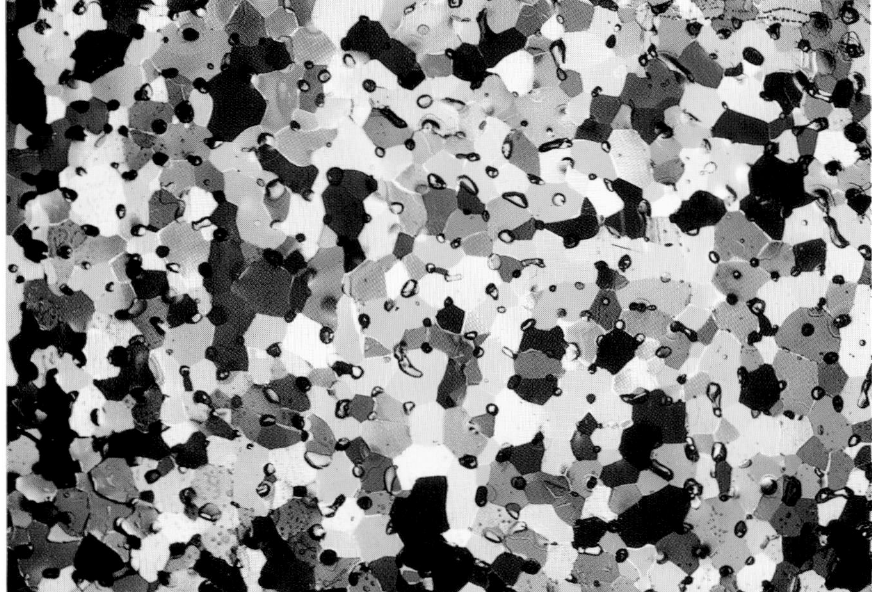

Figure 2. Trapped air bubbles, analyzed to provide a record of atmospheric CO_2 concentrations, are visible in samples of recently formed ice. This photograph, made between crossed polarizers, shows a section of a Byrd Station Antarctic ice core, half a millimeter thick and 61 millimeters across in this picture. The section was taken from a depth of 56 meters, where the trapped air is estimated to be 450 years old. The bubbles appear amber-colored and are located primarily at crystal grain boundaries. Ice-core measurements have replaced the less-reliable technique of analyzing ^{13}C in tree rings to compare current atmospheric CO_2 levels with those of the recent past. (Photograph courtesy of A. J. Gow, U.S. Army Cold Regions Research and Engineering Laboratory.)

approximately 279 microliters per liter, an important benchmark for estimating the impact of recent human activity.

The ocean stores carbon in three forms: dissolved inorganic carbon (consisting of dissolved CO_2 and the bicarbonate and carbonate ions HCO_3^- and CO_3^{2-}), dissolved organic carbon (consisting of both small and large organic molecules) and particulate organic carbon (consisting of live organisms or fragments of dead plants and animals). Based on data from the Geochemical Ocean Sections Study, about 37,000 gigatons of dissolved inorganic carbon is found in the oceans. In 1979 Kenneth Mopper and Egon Degens estimated that the oceans contain an additional 1,000 gigatons of dissolved organic carbon and 30 gigatons of particulate organic carbon. New measurement techniques, however, may substantially increase the estimates of organic carbon.

There is considerable uncertainty about how much carbon is stored on land. Estimates of the car-

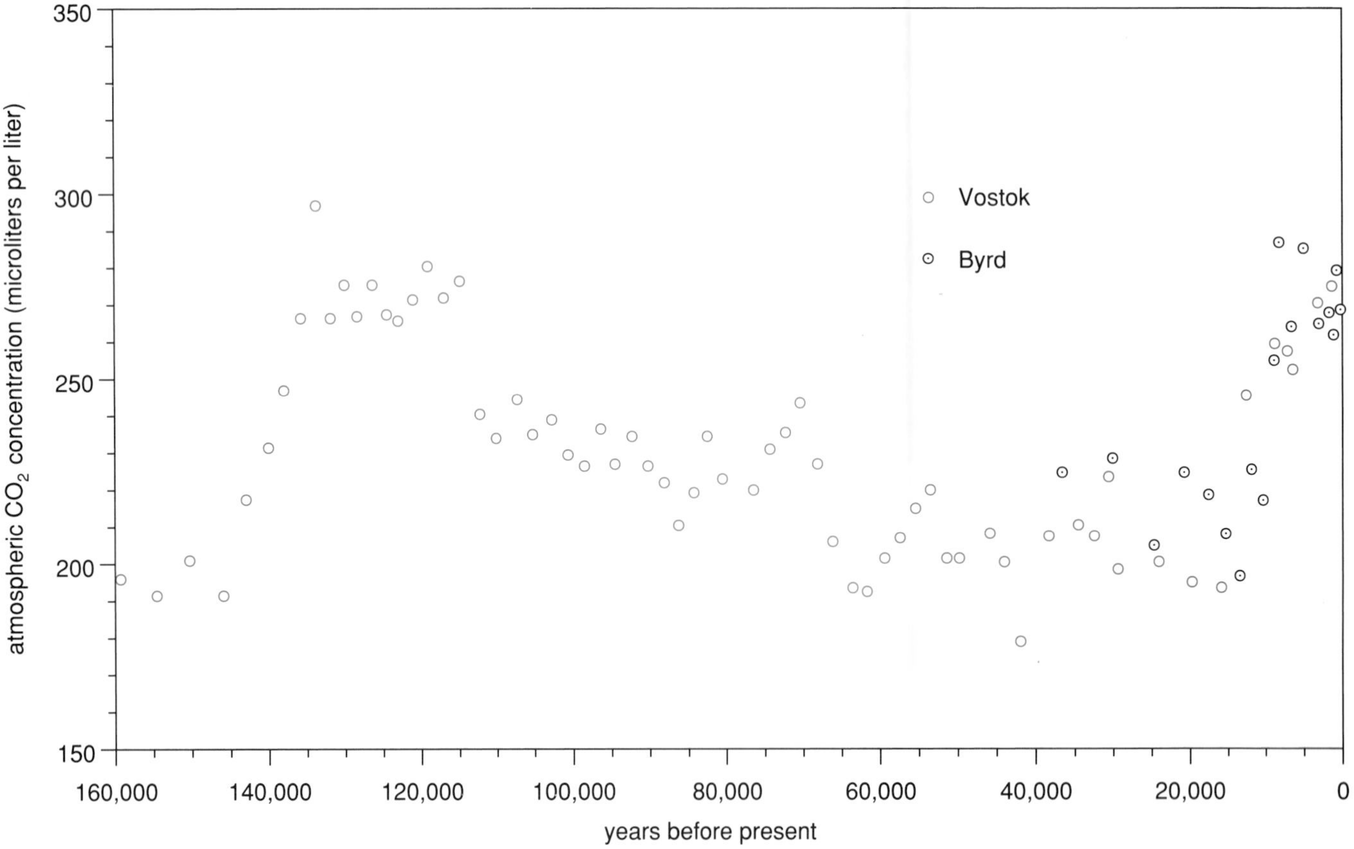

atmospheric CO_2 concentration (microliters per liter)

350

300

250

200

150

Vostok

Byrd

160,000 140,000 120,000 100,000 80,000 60,000 40,000 20,000 0

years before present

Figure 3a. Carbon dioxide concentrations in the atmosphere have varied over the glacial cycles of the earth's history, peaking at just under 300 microliters per liter of air during the interglacial period approximately 130,000 years ago and reaching that level again at the end of the last glaciation 10,000 years ago. This graph shows CO_2 measurements from air bubbles trapped in Antarctic ice sampled at Vostok and Byrd stations (Barnola et al. 1987, Neftel et al. 1982).

bon in plants range widely, from 420 to 830 gigatons, depending on the methods used to classify ecosystems into types, to determine the area of each type, and to measure the carbon stocks of each type. The same problems arise in estimating the storage of carbon in litter and soil organic matter; the most probable range for soil carbon is between 1,200 and 1,600 gigatons.

Since the Industrial Revolution, people have contributed carbon to the atmosphere primarily through the burning of fossil fuels such as coal, petroleum and natural gas. This activity injects into the cycle annually a substantial amount of carbon—equivalent to 0.8 percent of the current carbon content of the atmosphere—from the earth's geological reservoir, which otherwise would not play a role in the global carbon cycle in the short term. Carbon emissions from fossil-fuel burning are estimated to have increased at a rate near 4.3 percent per year from 1860 until 1973, except for brief periods during the Great Depression and the world

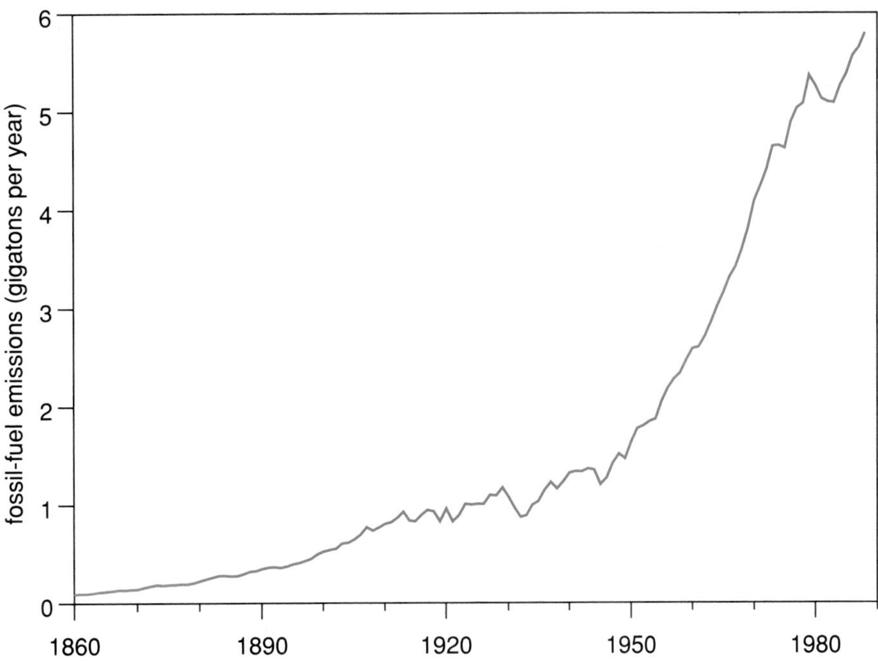

Figure 4. Fossil-fuel burning, cement production and natural-gas flaring have released increasing amounts of carbon into the atmosphere since 1860. With the exception of short periods during the Great Depression and world wars, emissions grew about 4.3 percent annually until 1973. Following the 1973 oil embargo and a decline induced by sharp oil price increases in the early 1980s, the amount of carbon entering the atmosphere in these ways resumed its steady increase in the mid-1980s and reached 5.9 gigatons in 1988 (Marland et al. 1989).

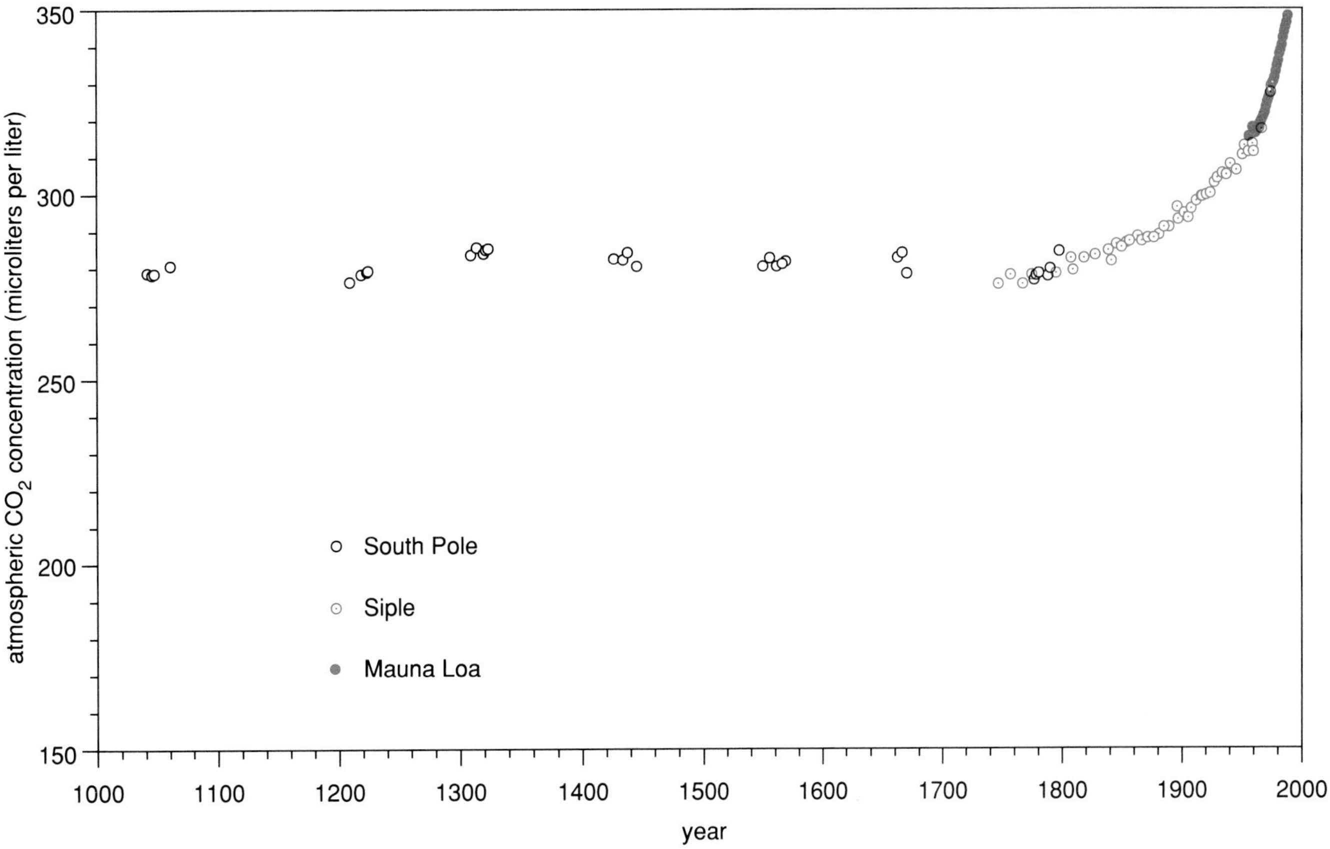

Figure 3b. Atmospheric CO_2 began increasing in the 18th century, and direct measurements made at Mauna Loa Observatory in Hawaii since 1958 indicate that the increase has accelerated. In 1988 the atmospheric carbon reservoir was estimated at 748 gigatons, equivalent to a CO_2 concentration of 351 microliters per liter and larger than at any time during the past 160,000 years. The South Pole and Siple ice-core data are from Neftel et al. 1985, Friedli et al. 1986 and Siegenthaler et al. 1988.

wars. The 1973 oil embargo halted this growth, but emissions have been increasing again since the mid-1980s, and the amount of carbon contributed to the atmosphere from fossil-fuel burning was about 5.9 gigatons in 1988 (Figure 4). Remaining reserves of recoverable fossil fuels total more than 4,000 gigatons.

To evaluate the cumulative impact of fossil-fuel burning, we must place it in context with other contributions to the carbon cycle—including human land use, which transfers carbon from the terrestrial reservoir to the atmosphere. While fossil-fuel burning contributed about 5.9 gigatons of carbon to the atmosphere in 1988, the atmosphere annually exchanges more than 100 gigatons of carbon with terrestrial ecosystems and a similar amount with the world's oceans. Thus the overall flows of carbon into and out of the atmosphere amount to more than 25 percent of the total atmospheric reservoir (Figure 5).

This article will discuss these processes in terms of net fluxes—the difficult-to-measure balances and imbalances in exchanges of carbon between and within the reservoirs. Determining how human activities affect the concentration of atmospheric CO_2 requires understanding the effects of the natural fluxes and of feedbacks between increased atmospheric CO_2 and changes in these fluxes.

Ocean Mixing and Circulation

The largest pool of carbon in the world cycle, the oceanic reservoir, has a major part in determining the concentration of CO_2 in the atmo-

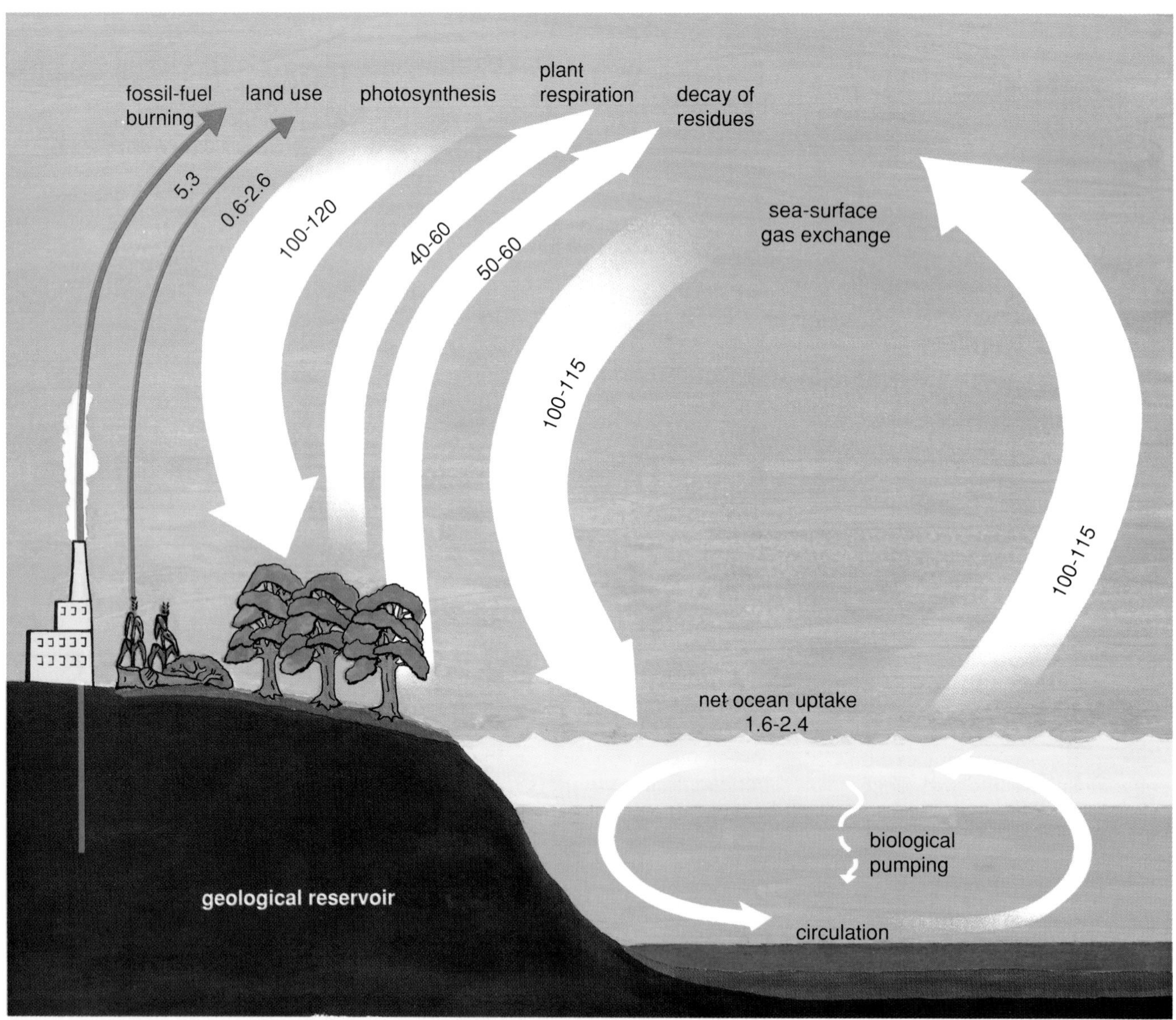

Figure 5. Massive flows between reservoirs make up the global carbon cycle. Plants on land take in carbon dioxide through photosynthesis. The terrestrial biosphere returns carbon to the atmosphere through plant respiration, the decomposition of plant residues and natural fires. The clearing of land for human activities also transfers carbon from the terrestrial system to the atmosphere. Meanwhile, gases are exchanged rapidly at the ocean surface; carbon is transferred to the deep ocean by circulation and by biological production (the use of carbon by surface organisms to produce compounds that enter the food chain of marine life). The ocean is believed to accumulate about two gigatons of carbon per year; most of this carbon is eventually dissolved in deep-ocean waters. Fossil-fuel burning transfers carbon from the geological reservoir to the atmosphere. All carbon fluxes are 1980 estimates, in gigatons.

sphere through physical processes (mixing and circulation), chemical processes (carbon chemistry and buffering effects) and biological processes (production and decomposition of organic matter and the formation and dissolution of carbonate shells). Biological processes maintain the carbon structure of the oceans: deep-ocean water is richer in dissolved inorganic carbon than surface water, in which the dissolved carbon is reduced by photosynthesis and the subsequent sinking of the organic matter produced. These vertical gradients help to stabilize the atmospheric CO_2 concentration, as does ocean alkalinity, which regulates carbonate chemistry.

Because of the effectiveness of winds over the oceans' vast surface area, CO_2 is exchanged rapidly across the sea-air interface, resulting in an approximate equilibrium between the partial pressures of CO_2 in the atmosphere and in the surface ocean water. As a result, little CO_2 can be taken up by surface seawater without a process that transfers carbon to the deeper water, lowering the concentration of CO_2 at the surface.

The rate at which carbon in ocean surface water is mixed into deeper layers was poorly known until both radioactive and stable chemical tracers could be used for making estimates. Since the 1970s natural ^{14}C and both tritium (^3H) and ^{14}C produced by atmospheric tests of nuclear weapons have served as tracers in studies of ocean mixing and circulation. These tracer data permit calibration of models of carbon turnover in the oceans. Such models typically divide the oceans into a well-mixed surface layer exchanging carbon with the atmosphere and with deeper waters, and a deep-water reservoir that is further subdivided to represent mixing and circulation effects. The most widely applied model based on such layers, or boxes, describes vertical transfer in terms of diffusion; it is therefore called a box-diffusion model. The calibrated model suggests that the net carbon uptake by the oceans lies in the range from 23 to 30 gigatons for the years between 1958 and 1980—or 26 to 34 percent of the fossil-fuel carbon put into the atmosphere during that period. This considerable range between the upper and lower estimates indicates how inadequately we understand mixing processes in the ocean.

In addition to the vertical mixing emphasized by the box-diffusion model, larger-scale advective flows (currents) transport carbon in the ocean. This transport often follows contours of equal density, called isopycnals. At low and middle latitudes the densest water is at the greatest depths; at high latitudes, however, such dense waters occur at shallow depths. They can even be exposed to the atmosphere under polar low temperatures. Direct contact of excess atmospheric CO_2 with dense, cool water in polar outcrop areas such as the North Atlantic creates a shortcut for significant amounts of carbon to enter waters that sink to form the deep waters for much of the world's oceans. It is not known what fraction of excess CO_2 has entered the ocean through this process, however, because very little is known about high-latitude oceanography and deep-water formation.

Biological Pumping

In a box-diffusion model, carbon uptake is calculated by deriving a coefficient of vertical diffusivity, which is actually a surrogate for several important water-mixing effects: upwelling, downwelling, vertical diffusion, advection and the gravitational drift of biogenic materials. Understanding the ocean carbon cycle lies in determining the effects of such controlling processes, which also include biological production and destruction.

Marine life flourishes near the ocean surface. Through photosynthesis, organisms take up dissolved inorganic carbon and manufacture both inorganic compounds (such as the carbonate of foraminifera shells) and the organic matter that provides energy to the marine food chain. Many of the substances created through this process, which is called primary production, sink to the deeper ocean, often in the form of fecal pellets and dead organisms. The sinking materials undergo remineralization and bacterial decomposition, and a minor fraction is deposited on the floor of the open ocean. The transport mechanism that carries carbon from the upper ocean to deep waters is called biological pumping.

The rate at which biological pumping transfers organic material from the surface to the deeper ocean is called new production. New production is hard to measure directly, and therefore it is estimated as 15 to 20 percent of net primary production (carbon assimilated through photosynthesis less what is released by respiration of photosynthetic organisms). Until recently, the magnitude

Figure 6. Major active reservoirs in the natural global carbon cycle are the oceans, terrestrial system and atmosphere. The oceans are the largest active reservoir; the atmosphere, the smallest. Geological stores of recoverable fossil fuels form a reservoir that was relatively inactive in the carbon cycle before people began mining and burning fossil fuels. Reservoir sizes are expressed in gigatons of carbon.

of new production was thought to be approximately 3.4 gigatons per year (Koblentz-Mishke, Volkovinsky and Kabanova 1970). But many investigators now believe that sampling and incubation procedures used in estimating ocean production have inhibited the growth of phytoplankton, resulting in a systematic underestimate of oceanic production. Recent measurements that eliminate this inhibition indicate that new production may be as much as 8.3 gigatons per year, or 2.4 times as large as the estimates derived from the work of Koblentz-Mishke and her colleagues. The lack of agreement among oceanographers about these revisions adds uncertainty to our understanding of this important flux in the global carbon cycle.

Special attention has been directed to the important role of polar oceans in biological pumping; moreover, the polar seas offer us a glimpse of ocean activity under extreme climatic conditions. In upwelling areas of the Antarctic Ocean, nutrient supplies—with the possible exception of iron—do not limit primary production, as they do in most of the world's oceans. Instead, there are other limiting factors such as reduced incident solar radiation during seasonal extensions of sea ice. This suggests that by reducing the extension of sea ice, warming at high latitudes might enhance biological production and increase the downward flux of carbon.

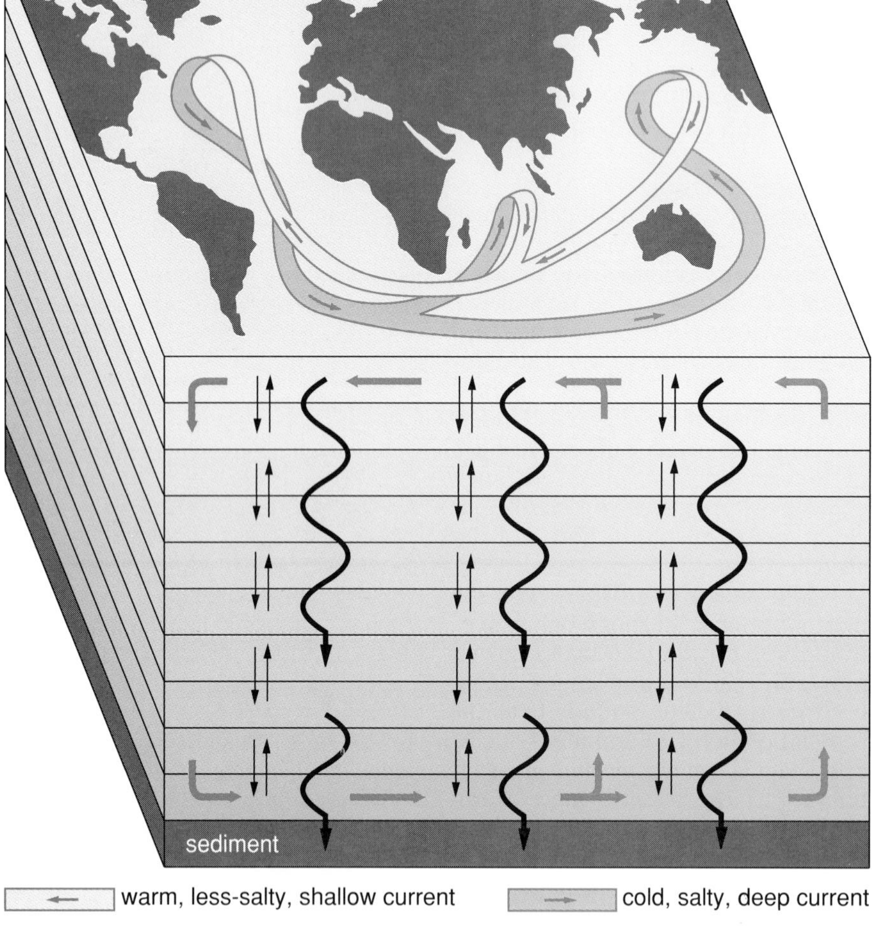

Figure 7. Flows of carbon in the world's oceans can be depicted as movements within and between box-shaped ocean compartments. The blue arrows in this vertical cross section indicate global circulation: upwelling of deep waters in the northern Indian Ocean and equatorial Pacific, downwelling in the North Atlantic. Horizontal transport occurs in both the deep, cold ocean layers and the warm surface waters. Wavy arrows represent biological pumping, which transfers carbon to the deep waters. Diffusion also moves carbon between layers. Model calculations considering these flows, calibrated with measurements of the movements of chemical tracers, suggest that the oceans took up between 26 and 34 percent of the fossil-fuel carbon put into the atmosphere between 1958 and 1980. (After Broecker 1985.)

Recent research has also shaken long-held assumptions about another aspect of the biological transfer of carbon in the oceans. Recently several geochemists using a new technique for the oxidation of organic compounds in seawater (Sugimura and Suzuki 1988) have found a large quantity of dissolved organic compounds, largely carbohydrates and proteins, in the world's oceans. The new estimates of dissolved organic carbon in both surface and deep water are at least twice as large as previously accepted values. These compounds, which are likely produced by photosynthetic organisms at the ocean surface, are transported by advection or mixing rather than by sinking. The presence of such large quantities of newly detected dissolved and actively decomposing organic carbon complicates the interpretation of depth profiles of oxygen and carbon, which are used for making ocean carbon-flux estimates.

There have been some difficulties in reproducing the results of Sugimura and Suzuki, but the measurements made using this method remain striking. Their apparatus, a complete combustion column, can extract large amounts of dissolved organic carbon from water samples from which it was thought all dissolved organic carbon had been removed. If a much larger pool of this carbon exists, and if it has a mean lifetime of 100 years as they suggest, then the mean production rate of dissolved organic carbon would be about 4.3 gigatons per year—comparable to the conventional estimate of new production. The estimated pool of dissolved organic carbon based on the new measurements is probably at least twice the size of either the atmospheric CO_2 pool or the carbon pool represented by terrestrial plants. Its production and dispersal is a potentially significant biologically induced flow of organic matter from the ocean surface to deep water. Small changes in such a pool would appreciably alter the ocean-atmosphere CO_2 exchange rate.

The Terrestrial Carbon Cycle

Terrestrial carbon dynamics have also presented daunting challenges to those seeking to understand the global carbon cycle. As in marine

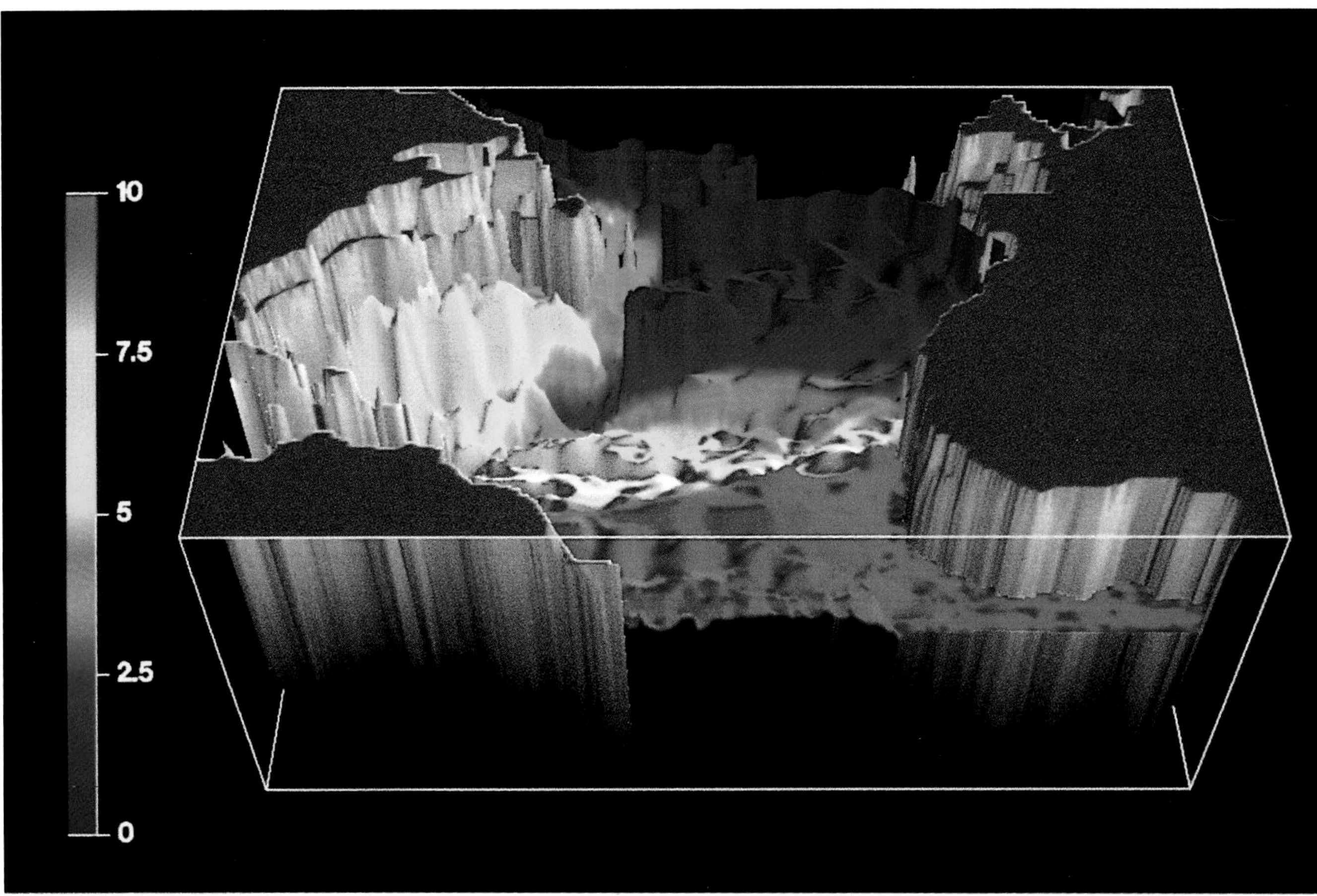

Figure 8. Outcrop of cold, dense waters in the North Atlantic provides a direct path for atmospheric carbon invading deep ocean water. The outcrop is seen in purple near the top of this three-dimensional image of an isopycnal surface, or equal-density horizon, in the Atlantic. The computer-generated image, which shows the area between the equator and 55 degrees north, is produced by a model based on principles of ocean dynamics and forced with seasonally varying wind stress, heating and fresh-water fluxes at the sea surface. Colors show the "age" of water on the isopycnal surface—the time since it was last exposed to the atmosphere. The vertical dimension of the box indicates the depth—descending from the surface to 500 meters—at which water of this density is found in the ocean. Near the equator, water on this isopycnal has not been exposed to the atmosphere for a decade. The model was constructed by Frank O. Bryan and William R. Holland of the National Center for Atmospheric Research.

biological pumping, the cycle on land begins with primary production by photosynthetic plants that take up inorganic carbon as CO_2 and make organic compounds, which serve as the source of chemical energy in the food chain. Terrestrial ecosystems return carbon to the atmosphere by respiration, decay and fires. The terrestrial carbon reservoir is actually a collection of carbon pools with a wide range of net primary production rates, respiration rates and carbon turnover times.

Data from the International Biological Program, an effort to assess biological productivity worldwide, became available in the late 1970s. Jerry Olson and his colleagues used this information in 1983 in constructing a new global estimate of terrestrial carbon pools. Their estimate considered both human influences and natural factors and accounted for periodic disturbances that might cause regional variations from pristine conditions in which ecosystem carbon fluxes are thought to be in balance. In calculating the net uptake of carbon by terrestrial vegetation, Olson distinguished three components of live vegetation: low vegetation, woody parts of trees and nonwoody parts of trees. While the net primary production amounts for the three pools are similar, tree wood and low vegetation are larger pools with slower turnover rates than tree leaves.

The total net primary production of terrestrial vegetation has been estimated at 62 gigatons per year. This is assumed to be approximately balanced, over a period of several years, by an equivalent return of carbon to the atmosphere from decomposition of litter and soil organic matter. The return flow comes from two pools of "dead" organic matter: the detritus/decomposer pool, made up of litter and decomposers at the soil surface, and the active soil carbon pool, which consists of that fraction of the carbon in soils, and the associated decomposer organisms, that are in relatively active exchange with the atmosphere. Figure 10 shows the relationship of these pools to other components of the terrestrial ecosystem.

The global totals do not illustrate the wide variations in the activity of these pools. The rates of input for different ecosystem types, such as temperate forest and tundra, vary over several orders of magnitude. The turnover time of carbon in the detritus/decomposer pool can range from less than a year in moist tropical forests to decades in cold, dry boreal forests. Soil contains both active carbon pools and relatively inactive ones; the activity of soil organic matter varies with depth, soil texture, climate and the chemistry of the organic matter. For example, the closer the organic matter is to the surface, the more readily decomposed it is, and, therefore, the greater its impact on annual or seasonal CO_2 cycles. Pools that are larger but have lower exchange rates may regulate long-term trends. In addition, some soils, even though undisturbed by human activities, are not currently in a steady state. Globally, peatland and wetland soils may be accumulating 0.1 to 0.3 gigaton of carbon per year, and desert soils may store 0.01 gigaton of carbon per year in the form of carbonates.

The balance between these processes—assimilation by photosynthesis and the release of carbon from both living and dead material—determines the magnitude of the net exchange of carbon between the atmosphere and the world's terrestrial systems. Over time periods shorter than a decade, carbon fluxes may be out of balance at specific locations, shifting with the availability of nutrients, changes in the weather, and sporadic disturbances. But over long periods, for reasonably undisturbed ecosystems, uptake and loss are generally assumed to be in balance, so

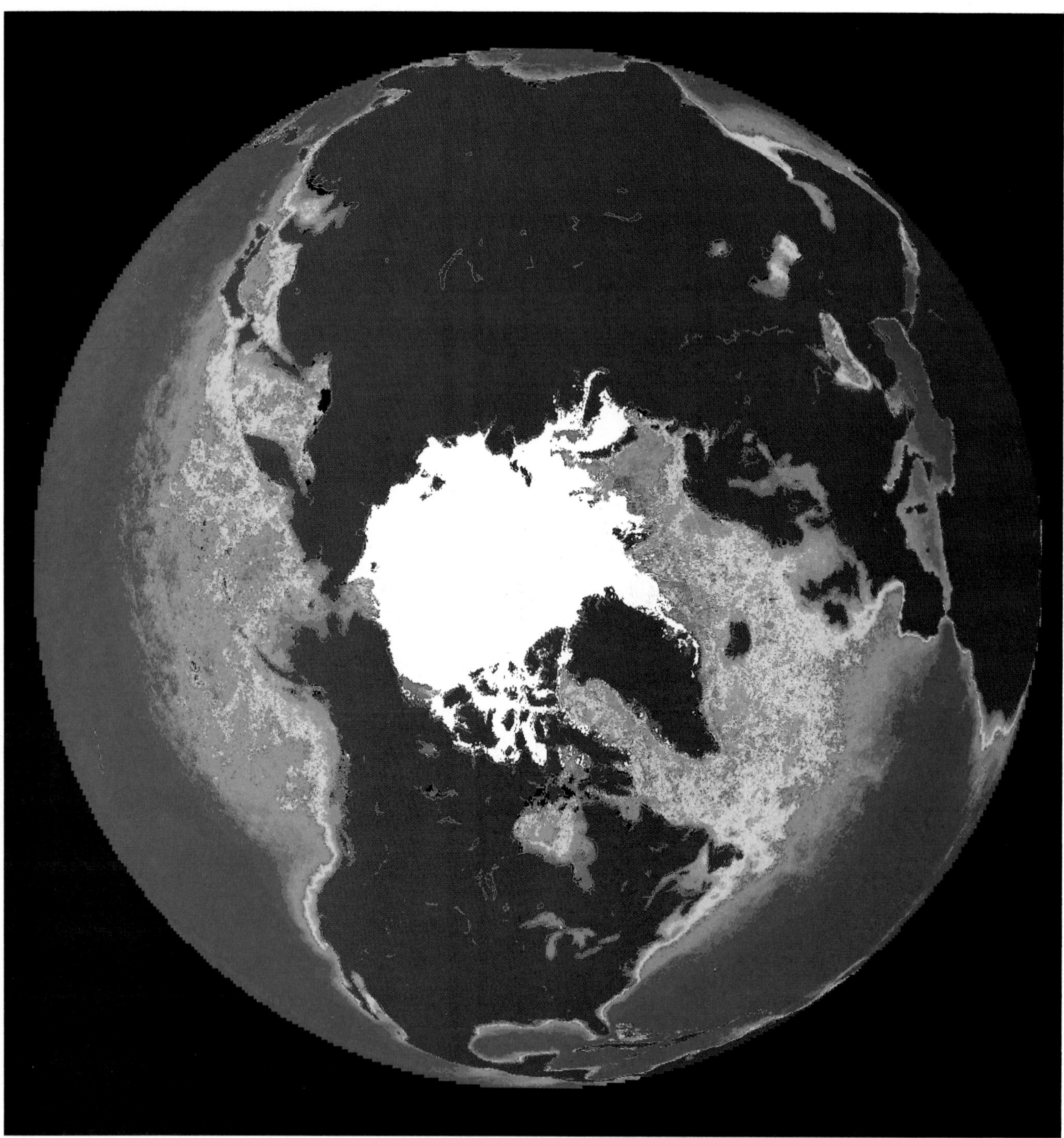

Figure 9. Polar oceans are the sites of abundant biological activity, as shown in these composite images produced by the Coastal Zone Color Scanner, which operated on the *Nimbus*-7 satellite from 1978 to 1986. Phytoplankton blooms in both the Arctic and Antarctic oceans are highly seasonal, limited by the availability of sunlight. The images are based on cumulative radiometric measurements of phytoplankton

that the average standing crop of carbon reaches a steady-state level.

We shall assume, as do the authors of nearly all studies of human disturbance of vegetation, that the exchange of carbon between the atmosphere and undisturbed terrestrial ecosystems is more or less in balance over annual or decade time scales. Some investigators, notably Ariel Lugo and Sandra Brown (1986), argue that we cannot assume there are enough reasonably undisturbed regions of the terrestrial world for such an assumption to be usefully applied to carbon-cycle calculations. Nevertheless, if care is taken to incorporate some natural disturbances into what we consider to be equilibrium carbon-pool sizes, this assumption provides a basis for estimating the net flux of carbon between the atmosphere and terrestrial systems due to human activities—a flow that has been impossible to measure directly.

Impact of Human Land Use

Several methods have been developed to estimate the recent effect of land use, such as forestry and agriculture, on the net flux of carbon between the atmosphere and the terrestrial ecosystem. One approach,

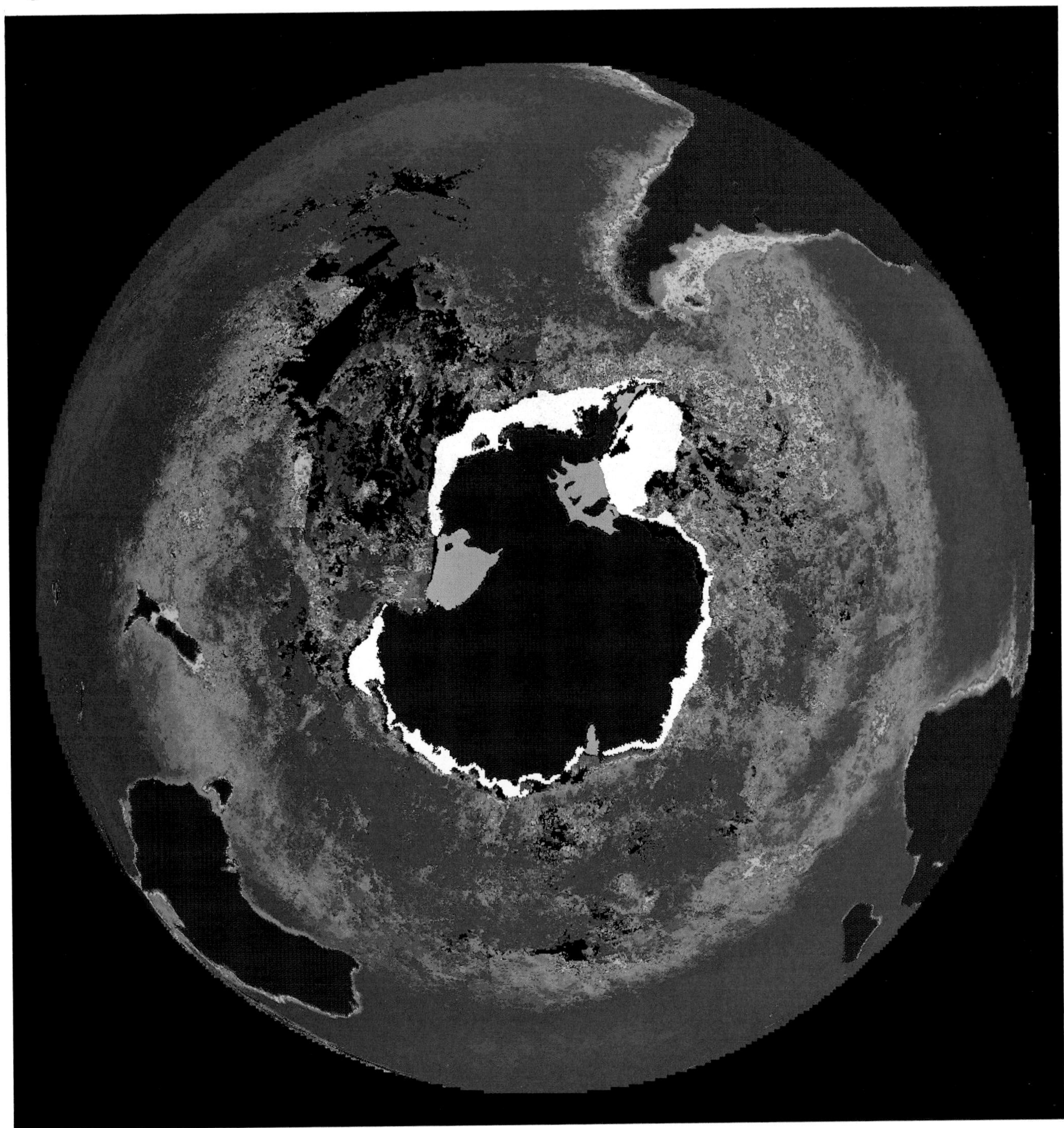

pigment concentrations, an index of photosynthetic activity near the surface. Yellow to red hues indicate the highest concentrations, blues and purples the lowest. Regions where no measurements were taken or where the sensor's view was obscured by clouds are colored gray. (Images by Gene Carl Feldman, National Aeronautics and Space Administration Goddard Space Flight Center.)

called reconstruction, relies on data recording changes in land use, from which one can estimate changes in the amount of carbon stored in vegetation and soil. A second approach attempts to deduce the flux from carbon pools on land into the atmosphere through the technique called deconvolution.

During the 1980s several studies attempted to reconstruct the impact of land-use changes on the net release of carbon into the atmosphere. Studies focusing on land clearing for cropping in specific regions produced estimates of the flow of carbon to the atmosphere in the range from 0.4 to 2.6 gigatons for 1980. Almost all of this amount came from the tropics; the analyses lacked comprehensive treatments of the impacts of fire suppression in the Northern Hemisphere (which results in increased carbon storage within extant forests and forest expansion into sparsely wooded areas) or of logging, harvesting of fuel wood, deliberate burning, and grazing in tropical savannas.

Other studies have attempted to reconstruct a time series of the net biotic flux of carbon since 1800 by using land-use data. The most recent estimate (Houghton 1989) reconstructs yearly changes in the amount of carbon in terrestrial systems by considering 10 geographic regions, each with up to 14 types of ecosystems and seven types of land-use changes. The analysis tracks the area, age and carbon content of each disturbed ecosystem, using response curves to describe the change in carbon stocks after a disturbance. It also computes the oxidation of fuel wood and wood products. The reconstruction yields an estimate that the total net flow of carbon to the atmosphere as a result of changes in land use

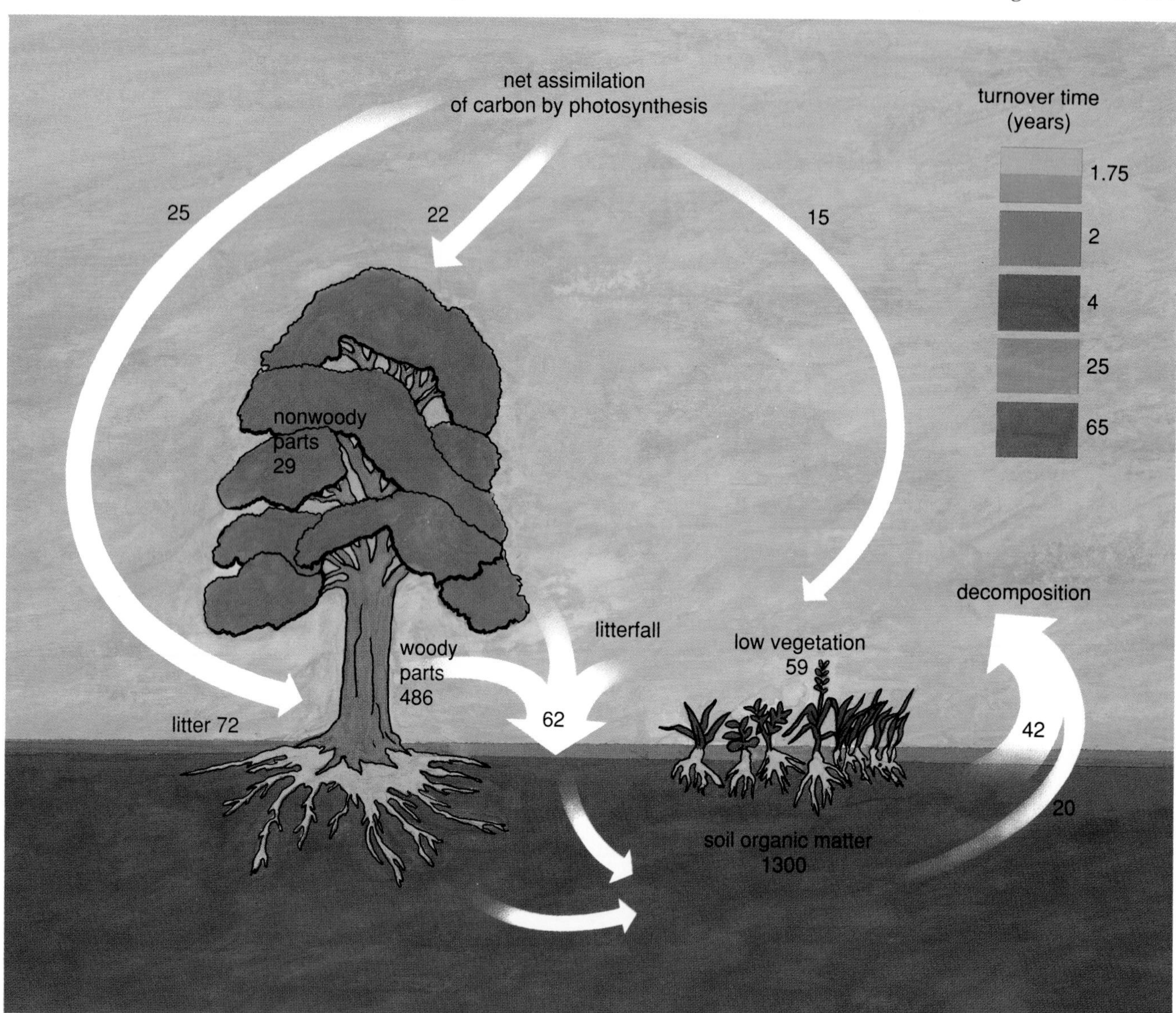

Figure 10. Terrestrial flows of carbon are determined by varying rates of photosynthesis, respiration and decay, and by the turnover times of the carbon reservoirs in the biosphere. Carbon is assimilated from the atmosphere by the woody and nonwoody parts of trees and by ground vegetation such as grasses and low bushes. An estimated 62 gigatons of carbon in plant material falls to the ground as litter or enters the soil by root mortality each year. The carbon contained in litter and soil is returned to the atmosphere through decomposition. While the woody parts of trees represent the largest active reservoir of terrestrial carbon, they exhibit very slow turnover; nonwoody tree parts store less carbon but turn over rapidly. As a result, these reservoirs assimilate similar amounts of carbon over time. The illustration represents the flows of carbon within the system and sizes of terrestrial carbon compartments in gigatons.

between 1800 and 1980 amounted to between 90 and 120 gigatons.

But the estimates derived by the reconstruction approach are inconsistent with the observed increase in atmospheric carbon. If the estimated release of carbon due to changes in land use (90 to 120 gigatons) is added to recorded releases from burning fossil fuels (150 to 190 gigatons), and if estimated ocean uptake (40 to 78 gigatons) is subtracted, the increase in atmospheric carbon for the period from 1800 to 1980 should, using the extremes, lie in the range from 162 to 270 gigatons. The observed increase is approximately 150 gigatons—below the minimum calculated using the reconstruction estimates. Since the ranges of predicted and observed increases in atmospheric carbon do not even overlap, many scientists remain skeptical that we can analyze the impact of fossil-fuel burning on the global carbon cycle.

Because changes in terrestrial carbon pools are difficult to measure, it is often assumed that reconstruction of the terrestrial flux is in error. This may or may not be correct, but recent work suggests several possible sources of error in the reconstructions. The amount of carbon in initial, undisturbed ecosystems may have been overestimated, so that projections of the amount released by changes in land use were too high. An even larger problem may lie in the simplifying assumption that human influences on many ecosystems were negligible before land-use conversion. Carbon stocks may have been gradually lowered over long periods before land areas were completely converted to crops or shifting cultivation.

An alternative method for estimating the land-use flux of carbon is deconvolution. The essential idea is to subtract fossil-fuel emissions from measured changes in atmospheric carbon, making allowances for the uptake of carbon by the oceans; the difference should be the contribution of the terrestrial system. Given our assumption of carbon balance in natural systems, this flux is equivalent to that from land-use changes.

Deconvolution studies have produced a wide spectrum of results. Studies done in the early 1980s used measurements of ^{13}C in tree rings to estimate atmospheric carbon levels in the period before the Mauna Loa record begins. However, local environmental conditions influence the relationship between ^{13}C in tree rings and atmospheric CO_2 levels, complicating this method. Analysis of bubbles trapped in ancient glacial ice now provides a direct means of measuring historical levels of ^{13}C and CO_2 partial pressure. Depending on which ocean model is employed, deconvolution based on the ice-core record gives a cumulative release of from 90 to 150 gigatons of carbon from 1800 to 1980.

The deconvolution estimates suggest a historical pattern that does not agree with the pattern derived from historical reconstruction. Figure 13 compares the results of studies done by the two techniques. The reconstruction estimate shows an exponential increase in carbon release since 1900, due largely to the increased rate of tropical-forest clearing over the past 50 years. The deconvolution estimate suggests a steadily declining land-use flux, making the terrestrial system a CO_2 source in the 19th century and a sink in the latter part of the 20th century. This lack of apparent increase demands some explanation, given the vast scale on which tropical forests are being destroyed today. One line of speculation suggests there was substantial land clearing in the northern temperate zones in the 19th century, and much of this area may now be serving as a CO_2 sink. Analyses of key regions indicate, however, that vegetation regrowth is unlikely to be large enough to account for most of the discrepancies.

Figure 11. Soil carbonate, or caliche, serves as a carbon reservoir in dry ecosystems, forming at a global rate of about 0.01 gigaton per year. Caliche in the desert landscape of La Mesa, near Las Cruces, New Mexico, is clearly visible as a white layer in the soil profile. (Photograph by William H. Schlesinger, Duke University.)

If the reconstruction method turns out to be correct, attention will focus on adjustments to the ocean models as a means of achieving consistency. Could any conventional ocean model reconcile the estimated rate of land-use release with the Mauna Loa and ice-core measurements? Ian Enting and J. V. Mansbridge (1987), using a linear-programming technique to answer this question, concluded that the discrepancy is too large. They mention four possible remedies, one of which is particularly worthy of notice here: the possibility of nonlinear effects in the uptake of CO_2 by the oceans. This notion is strengthened by other lines of evidence suggesting that the complex dynamics of the global circula-

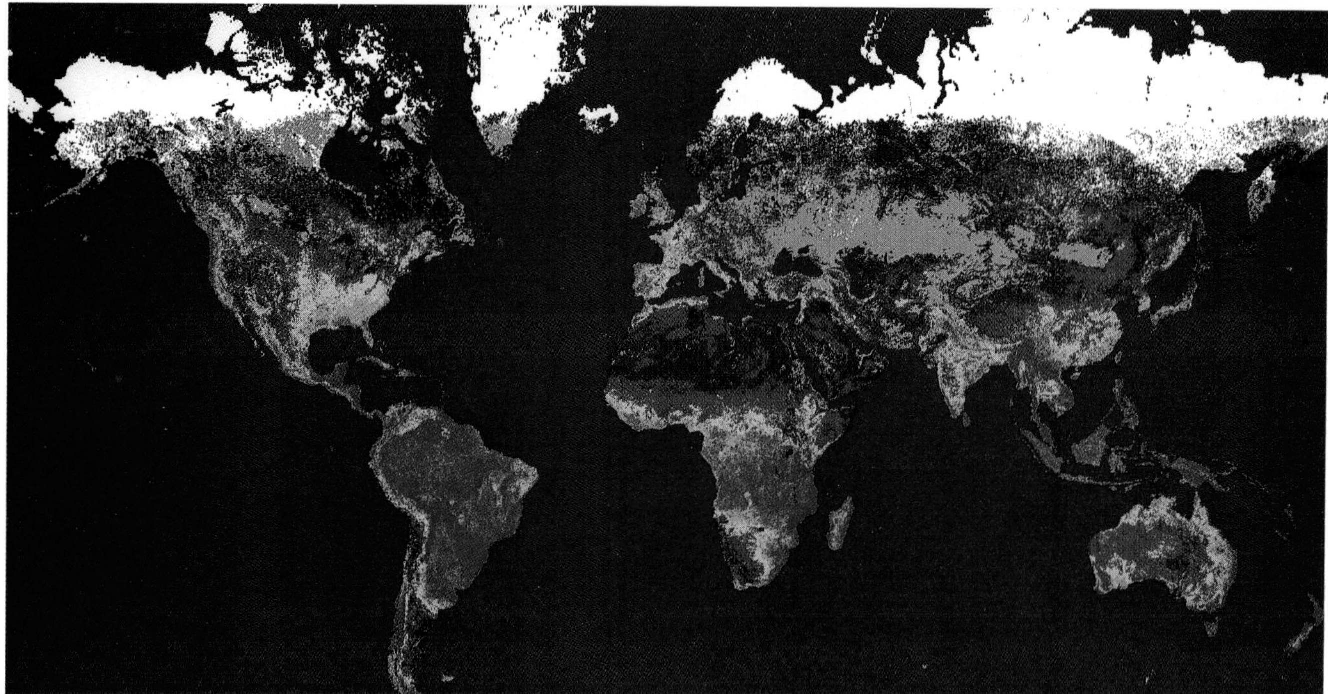

January–February

Figure 12. Carbon stores in the terrestrial ecosystem are dispersed unevenly around the globe. These images show plant activity during two-month periods in 1987 using the Global Vegetation Index, which is compiled with data from sensors mounted on National Oceanic and Atmospheric Administration satellites. The images show the cycles of photosynthetic activity in the Northern and Southern hemispheres.

tion of ocean waters are important in describing CO_2 uptake. Such considerations are leading to the development of three-dimensional general-circulation models of the oceans, which can accommodate nonlinear effects.

Much of the progress in understanding the global carbon cycle has been accomplished by a divide-and-conquer strategy, in which scientists from many disciplines work separately on separate pieces of the problem. Although this approach will continue to refine our knowledge of the global carbon cycle, the pieces do not always fit together. Over the past decade, a new perspective has emerged. This approach recognizes that the oceans, terrestrial ecosystems, the atmosphere and climate form an interconnected system that can be studied in its entirety. Components of this global system interact through the hydrologic cycle of evaporation and precipitation, through the flow of carbon and nutrients in food chains, and through biological and geochemical reactions that result in trace-gas exchanges. All of these phenomena in turn have an influence on climatic conditions at the earth's surface. It is just such feedback relations—where the output of a system affects its own input—that give rise to complex responses.

Three kinds of research are important to this approach: global modeling, which couples system components that were previously considered separately; spatial studies, which explore regional differences in carbon exchanges; and the analysis of temporal patterns of CO_2 variation in the atmosphere.

Global System Modeling

Mathematical models of carbon flow between the oceans and the atmosphere and between terrestrial ecosystems and the atmosphere have long been useful tools in carbon-cycle research. In the area of ocean-atmosphere dynamics, recent models have improved our understanding by incorporating three spatial dimensions and by allowing for the simultaneous transfer of heat and carbon, thus coupling the biogeochemical system with the climate system.

One step in this direction is the lateral-transport model of the global oceans proposed by Wallace Broecker and his colleagues in 1985. The model divides the Atlantic and Pacific oceans into five latitudinal zones and the Indian Ocean into three latitudinal zones. It incorporates upwelling coupled with a divergence of surface waters in the tropics, the Antarctic regions and the North Pacific. A corresponding convergence of surface-water flow coupled with downwelling is necessary in the temperate regions of all the oceans and in the North Atlantic.

In Broecker's model the oceans can take up 35 percent of the CO_2 released from fossil-fuel burning during the period from 1958 to 1980—slightly more than traditional globally averaged models suggest. Oceanographers have hoped to build more complex ocean models that would help balance the global carbon cycle, but preliminary results do not show significant increases in carbon uptake from these models. Still more detail could be included in models based on real-world measurements of temperature, salinity and currents, and on data describing the distribution of tracers in the sea. More work is needed to ascertain that the ocean-circulation models describe the ocean adequately and to include biological processes.

General-circulation models of the atmosphere have been coupled with similar models of fluid and heat transport in the oceans to understand the exchange of heat and moisture

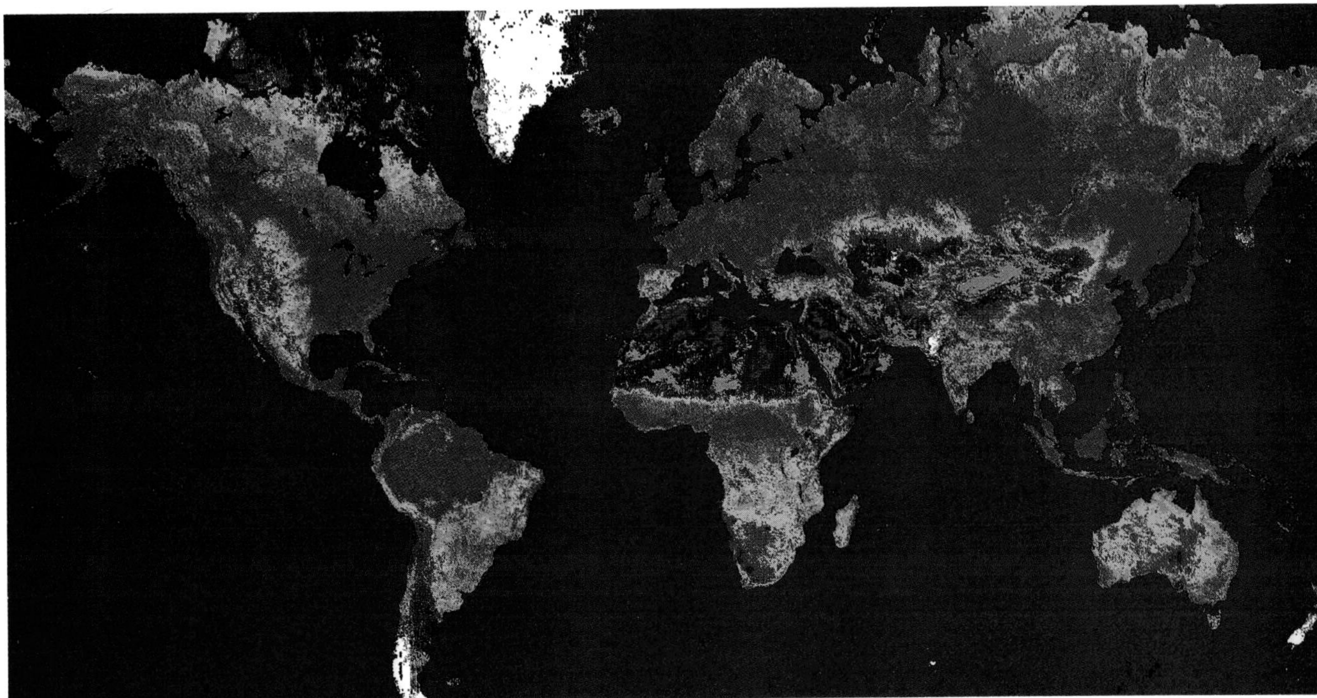

Deep green represents the highest "greenness index." Yellow indicates moderate vegetation, and browns and grays show minimal activity. Data values were omitted for the areas in white because of cloud cover and other technical problems. (Images courtesy of Kevin P. Gallo, NOAA/National Environmental Satellite, Data, and Information Service, and the U.S. Geological Survey/EROS Data Center.)

between the oceans and atmosphere. And a convergence of efforts is linking models of atmospheric and oceanic circulation, and incorporating carbon exchanges and flows into the ocean models. This combined approach promises to be a powerful tool in developing an understanding of the complex relationship between ocean biogeochemistry and climate dynamics.

Carbon-cycle models that attempt to describe terrestrial processes currently do not include a central feature of the dynamics of the system—the dependence of its major processes on environmental conditions such as temperature, moisture and CO_2 concentration. All ecological processes are sensitive to temperature and moisture. These environmental conditions control the faster-responding biological processes of the carbon cycle: photosynthesis, respiration, translocation and transpiration in plants, and the turnover of microbial decomposers. Photosynthesis and respiration also can be affected by increases in atmospheric CO_2. Such fast responses are constrained by slow ecosystem dynamics that allocate carbon to various compartments (leaves and fine roots, wood, litter and soil organic matter), decompose dead organic matter and alter ecosystem composition through replacement of plant species.

Feedback mechanisms (in which CO_2 emissions lead indirectly to greater carbon uptake) could produce changes in terrestrial production large enough to compensate, at least in part, for decreased production and storage caused by human land use. This could, in turn, account for some or all of the inconsistencies in historical reconstructions of carbon flows. But in order to describe realistically how terrestrial carbon dynamics respond to varying environmental factors, two classes of terrestrial models must be merged: physiological models of the fast carbon dynamics and ecosystem models of the slow carbon dynamics. This is not a simple matter. Models of fast processes are designed for small time intervals and small spatial extents—hours and centimeters. Because of nonlinearities and complex spatial variations, the models cannot simply be integrated to take in a larger scope. Furthermore, computational errors grow unacceptably when small deviations accumulate over long periods.

Another significant challenge is the wide variation of environmental conditions across the earth's surface. Global-scale analyses of carbon dynamics that take into account the spatial distribution of terrestrial biological and environmental factors have only recently been attempted. These models offer provocative results. For example, Gerd Esser (1987) incorporates a terrestrial CO_2 "fertilization" response function that, in his simulation, increases net primary production by about 5 gigatons per year by 1980. In the model, the additional production stimulated by excess atmospheric CO_2 over the period from 1860 to 1980 is responsible for an additional terrestrial uptake of 73 gigatons of carbon, offsetting a significant portion of the impact of land clearing during the same period.

It is not yet possible to independently evaluate the results of such models by direct observations. A promising direction in global modeling is offered by model formulations that would simulate certain observable variations in the global system, such as the seasonal and latitudinal variation of CO_2 and ^{13}C in the atmosphere.

Spatial and Seasonal Patterns

The spatial distribution of CO_2 sources and sinks is an important area of study. Annual mean atmospheric concentrations of CO_2 vary continuously with latitude and are higher at the North Pole than at the South Pole. It is thought that this gradient is maintained by the geographical distribution of sources and sinks in the oceans and on land, coupled with atmospheric mixing. The fact is, however, very little is known about this distribution.

Recently, some information on the distribution of sea-surface CO_2 concentrations has been compiled. These studies locate sources of CO_2 in the equatorial Pacific and Atlantic, the northwestern Pacific and the northwestern Indian oceans, where there is upwelling of deep waters. The subantarctic Southern Ocean and the northern North Atlantic Ocean are sinks of CO_2, with low surface concentration. To incorporate these geographic variations into a model of the global carbon cycle, we need to know the exchange rate across the air-water interface, which depends not only on CO_2 concentration but also on temperature and wind speed. Given data on these quantities, the spatial information yields an estimate of the global uptake of CO_2 by the ocean. An estimate made by Pieter Tans, Inez Fung and Taro Takahashi (1990) using this method is 1.6 gigatons per year, considerably lower than estimates obtained by other methods. Sensors aboard future satellites should improve these estimates by mapping sea-surface roughness (an indicator of wind speed) and color (an indicator of chlorophyll or phytoplankton concentration). The satellite observations in turn will need to be calibrated with ground-truth experiments.

During the past decade it has become clear that estimates of regional and global terrestrial carbon dynamics must also take into account their variability across space and time. Currently, our knowledge of carbon dynamics involving vegetation is limited to fairly simple extrapolations of measurements made in relatively small plots. Even if such measurements could be made exactly, they would fail to account for year-to-year variations in climate, or the effects of longer-term disturbances and successional changes.

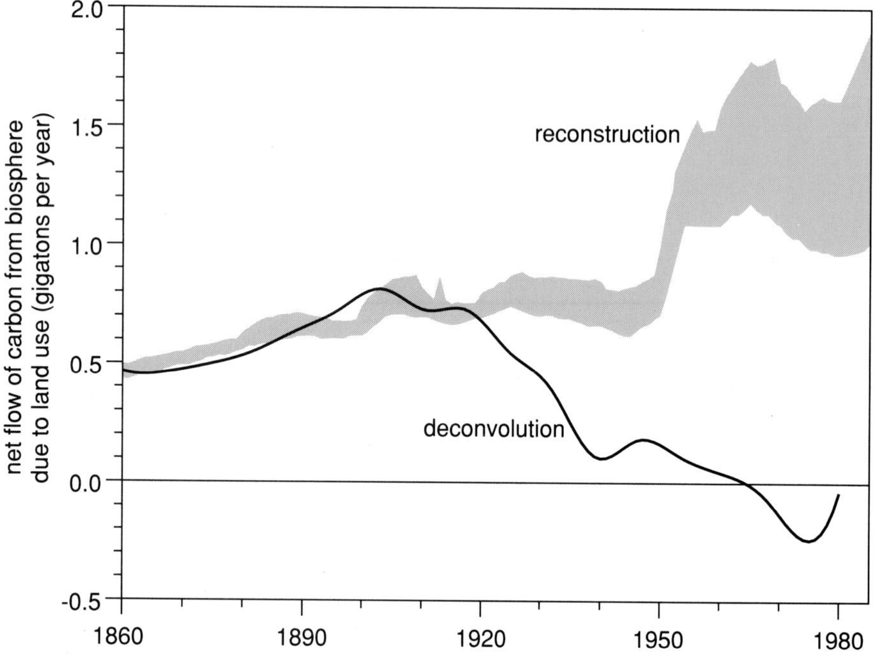

Figure 13. Impact of human land use on the global carbon cycle has proved difficult to estimate. This graph compares the results of recent studies using the techniques of historical reconstruction and deconvolution. The reconstruction study (Houghton 1989) used historical records to calculate changes produced by disturbances of vegetation and soil since 1860. The results are shown as a range of values between high and low estimates. The dramatic increase since 1950 is caused primarily by tropical deforestation. The deconvolution method works differently: it estimates the effect of fossil-fuel emissions and ocean uptake on changes in atmospheric CO_2, then infers that any carbon flows not accounted for must represent the effects of human land use. This method produces a different historical profile, as illustrated by results from a deconvolution study based on a box-diffusion model of ocean uptake and on Siple ice-core and Mauna Loa CO_2 measurements (Siegenthaler and Oeschger 1987).

Moreover, extrapolating from a few acres to an entire continent has obvious hazards. In tropical regions, where most attention has been focused lately, it is now clear that even the most recent estimates of the standing stock of carbon in vegetation—and perhaps estimates of net primary production as well—are too large and should be revised. This may also be true for other regions of the world.

Several global maps of contemporary terrestrial ecosystems have been constructed, an exercise that can serve as a basis for integrating detailed regional information to make global estimates. There is also much promise in using satellite imagery for assessing current patterns of vegetation and land-use, and for determining exchanges of energy, water and CO_2 between the terrestrial surface and the atmosphere. Problems exist, however, in interpreting remotely sensed images in terms of carbon concentrations and land-use change. So far, the interpretation has been done only for selected areas. Possible approaches include the use of remote sensing to classify vegetation cover, to relate observed seasonal photosynthetic activity to atmospheric CO_2 concentration and to monitor changes in productivity and, therefore, in terrestrial carbon storage.

Another important modeling issue is raised by seasonal and year-to-year fluctuations in atmospheric CO_2 levels. Annual variations, which are very small at the South Pole, are strong and regular in the Northern Hemisphere, reflecting the seasonal exchange between the atmosphere and the terrestrial ecosystems (Figure 14). Atmospheric CO_2 measurements show significant growth in the amplitude of this seasonal cycle. At Mauna Loa the annual fluctuation in atmospheric CO_2 grew from 5.5 to 6.4 microliters per liter from 1958 to 1981, a mean rate of increase of 0.66 percent per year. Other records over the past decade exhibit growth rates of from 1 to 2 percent per year.

An increasing amplitude in the seasonal oscillation in CO_2 concentration indicates increased plant activity, but not necessarily increased net carbon storage. The seasonality of fossil-fuel use is insufficient to account for the increase; a more likely explanation lies in the strong temperature dependence of CO_2 respiration from

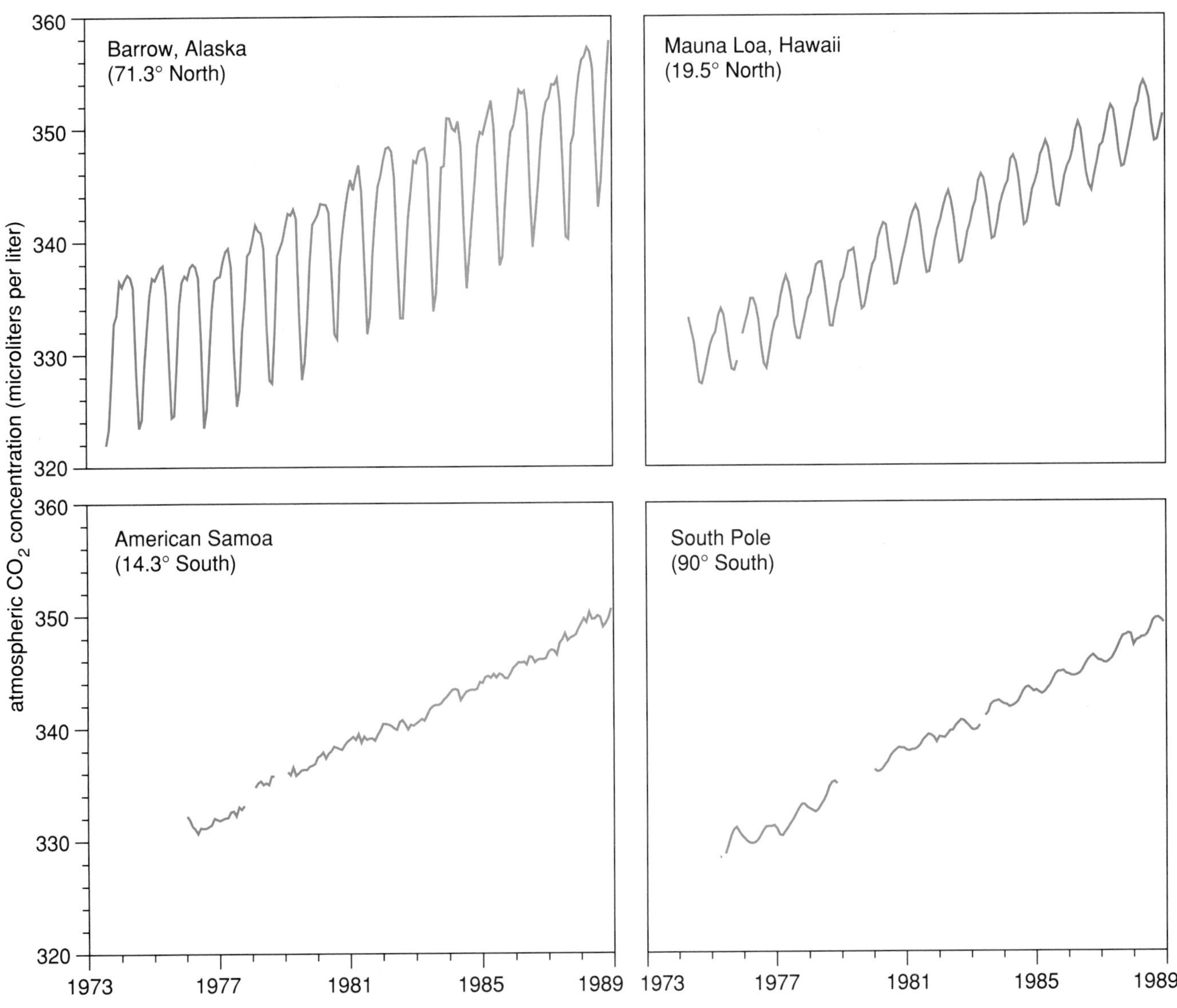

Figure 14. Seasonal cycles of atmospheric CO_2 concentrations are illustrated by measurements from four monitoring stations where the National Oceanic and Atmospheric Administration takes continuous readings. Analysis of the measurements shows that seasonal cycles in the Northern Hemisphere—which reflect the dominance of photosynthesis in the summer and plant respiration in the winter—seem to be increasing in amplitude. This may be a result of global warming. The dots represent monthly averages. The data are from Peterson et al. 1986; Komhyr et al. 1989; Thoning, Tans and Komhyr 1989; Waterman et al. 1989; and Gillette et al. 1987.

plants and soils and the observed warming trend over the past decade in the Northern Hemisphere.

Understanding Carbon Dynamics

The challenge of future carbon-cycle research is to understand relationships among the components of the global biogeochemical-climate system. Our inability to balance all of the carbon fluxes over the period from 1800 to the present may result from overlooking a dynamic response of terrestrial vegetation and ocean processes to changes in environmental conditions. The unexplained response—whose magnitude is between +0.5 and –2 gigatons annually, depending on various assumptions—represents about 4 percent of net primary production in the terrestrial system, or about 3 percent of the exchange between the ocean and the atmosphere. Current methods do not provide a way to detect the changes in carbon storage that would accommodate such small net fluxes. For the purposes of understanding the carbon cycle and predicting future atmospheric levels of CO_2, therefore, it is essential that we understand how terrestrial vegetation and ocean processes respond to changes in CO_2 and climate. Integrative research tools are now being developed to directly determine the responses of these natural systems over short time scales. In the area of ocean dynamics, additional data and more comprehensive models are needed to link the climatic effects of the atmosphere-ocean system to geochemical events.

In years to come there will likely be shifts in carbon storage by terrestrial ecosystems as small shifts in climate cause imbalances from year to year between production and decomposition-respiration. Observing these shifts may help to determine the magnitude of the terrestrial response. Useful techniques will include remote-sensing measurements of productivity changes as well as carbon-isotope and CO_2 measurements. These, too, need further development and must be interpreted by coupled and geographically explicit process models. Geographically oriented tools also will be important to understanding the spatial distribution of terrestrial ecosystems in relation to the heterogeneity of climate changes and geological constraints.

Much has been learned in the past decade about the global carbon cycle and how its complexities control CO_2 levels in the atmosphere. The greatest lesson is how dynamic and interactive are the components of the global biogeochemical-climate system.

References

Baes, C. F., H. E. Goeller, J. S. Olson and R. M. Rotty. 1977. Carbon dioxide and climate: The uncontrolled experiment. *American Scientist* 65:310–320.

Barnola, J. M., D. Raynaud, Y. S. Korotkevich and C. Lorius. 1987. Vostok ice core provides 160,000-year record of atmospheric CO_2. *Nature* 329:408–414.

Bolin, B. 1986. How much CO_2 will remain in the atmosphere? In *The Greenhouse Effect, Climatic Change, and Ecosystems,* ed. B. Bolin, B. R. Döös and J. Jäger, pp. 93–155. SCOPE 29. John Wiley, Chichester.

Broecker, W. S., T.-H. Peng, G. Östlund and M. Stuiver. 1985. The distribution of bomb radiocarbon in the ocean. *Journal of Geophysical Research* 90:6953–6970.

Emanuel, W. R., G. G. Killough, W. M. Post and H. H. Shugart. 1984. Modeling terrestrial ecosystems in the global carbon cycle with shifts in carbon storage capacity by land-use change. *Ecology* 65:970–983.

Enting, I. G., and J. V. Mansbridge. 1987. The incompatibility of ice-core CO_2 data with reconstructions of biotic CO_2 sources. *Tellus* 39B:318–325.

Esser, G. 1987. Sensitivity of global carbon pools and fluxes to human and potential climatic impacts. *Tellus* 39B:245–260.

Friedli, H., H. Lötscher, H. Oeschger, U. Siegenthaler and B. Stauffer. 1986. Ice core record of the $^{13}C/^{12}C$ ratio of atmospheric CO_2 in the past two centuries. *Nature* 324: 237–238.

Fung, I. Y, C. J. Tucker and K. C. Prentice. 1987. Application of advanced very high resolution radiometer vegetation index to study atmosphere-biosphere exchange of CO_2. *Journal of Geophysical Research* 92:2999–3015.

Gillette, D. A., W. D. Komhyr, L. S. Waterman, L. P. Steele and R. H. Gammon. 1987. The NOAA/GMCC continuous record at the South Pole, 1975–1982. *Journal of Geophysical Research* 92:4231–4240.

Houghton, R. A. 1989. The long-term flux of carbon to the atmosphere from changes in land use. Extended Abstract from the Third International Conference on Analysis and Evaluation of Atmospheric CO_2 Data Present and Past. Hinterzarten, 16–20 October 1989. *Environmental Pollution Monitoring and Research Programme No. 59.* World Meteorological Organization.

Houghton, R. A., J. E. Hobbie, J. M. Melillo, B. Moore, B. J. Peterson, G. R. Shaver and G. M. Woodwell. 1983. Changes in the carbon content of terrestrial biota and soils between 1860 and 1980: Net release of CO_2 to the atmosphere. *Ecological Monographs* 53: 235–262.

Keeling, C. D., R. B. Bacastow, A. F. Carter, S. C. Piper, T. P. Whorf, M. Heimann, W. G. Mook and H. Roeloffzen. 1989. A three-dimensional model of atmospheric CO_2 transport based on observed winds: 1. Analysis of observational data. In *Geophysical Monograph 55,* ed. D. H. Peterson. American Geophysical Union, Washington, D.C.

Koblentz-Mishke, O. I., V. V. Volkovinsky and J. G. Kabanova. 1970. Plankton primary production of the world oceans. In *Scientific Exploration of the South Pacific,* ed. W. Wooster, pp. 183–193. National Academy of Sciences, Washington, D.C.

Komhyr, W. D., T. B. Harris, L. S. Waterman, J. F. S. Chin and K. W. Thoning. 1989. Atmospheric carbon dioxide at Mauna Loa Observatory: 1. NOAA/GMCC measurements with a nondispersive infrared analyzer, 1974–1985. *Journal of Geophysical Research* 94:8533–8547.

Lugo, A.E., and S. Brown. 1986. Steady state terrestrial ecosystems and the global carbon cycle. *Vegetatio* 68:83–90.

Marland, G., T. A. Boden, R. C. Griffin, S. F. Huang, P. Kanciruk and T. R. Nelson. 1989. *Estimates of CO_2 Emissions from Fossil Fuel Burning and Cement Manufacturing, Based on the United Nations Energy Statistics and the U. S. Bureau of Mines Cement Manufacturing Data.* ORNL/CDIAC-25. Oak Ridge National Laboratory, Oak Ridge, Tennessee.

Mopper, K., and E.T. Degens. 1979. Organic carbon in the ocean: Nature and cycling. In *The Global Carbon Cycle,* ed. B. Bolin, E. T. Degens, S. Kempe and P. Ketner, pp. 293–316. SCOPE 13. John Wiley and Sons, New York.

Neftel, A., H. Oeschger, J. Schwander, B. Stauffer and R. Zumbrunn. 1982. Ice core sample measurements give atmospheric CO_2 content during the past 40,000 yr. *Nature* 295:220–223.

Neftel, A., E. Moor, H. Oeschger and B. Stauffer. 1985. Evidence from polar ice cores for the increase in atmospheric CO_2 in the past two centuries. *Nature* 315: 45–47.

Olson, J. S., J. A. Watts and L. J. Allison. 1983. *Carbon in live vegetation of major world ecosystems* ORNL-5862. Oak Ridge National Laboratory, Oak Ridge, Tennessee.

Peterson, J. P., W. D. Komhyr, L. S. Waterman, R. H. Gammon, K. W. Thoning and T. J. Conway. 1986. Atmospheric CO_2 variations at Barrow, Alaska, 1973–1982. *Journal of Geophysical Research* 88:3599–3608.

Revelle, R., and H. Suess. 1957. Carbon dioxide exchange between atmosphere and ocean and the question of an increase of atmospheric CO_2 during the past decades. *Tellus* 9:18–27.

Siegenthaler, U., and H. Oeschger. 1987. Biospheric CO_2 emissions during the past 200 years reconstructed by deconvolution of ice core data. *Tellus* 39B:140–154.

Siegenthaler, U., H. Friedli, H. Loetscher, E. Moor, A. Neftel, H. Oeschger and B. Stauffer. 1988. Stable-isotope ratios and concentrations of CO_2 in air from polar ice cores. *Annals of Glaciology* 10:1–6.

Sugimura, Y., and Y. Suzuki. 1988. A high temperature catalytic oxidation method for determination of non-volatile dissolved organic carbon in seawater by direct injection of liquid sample. *Marine Chemistry* 24:105–131.

Takahashi, T., W. S. Broecker, S. R. Werner and A. E. Bainbridge. 1980. Carbonate chemistry of the surface waters of the world oceans. In *Isotope Marine Chemistry.*

Tans, P. P., I. Y. Fung and T. Takahashi. 1990. Observational constraints on the global atmospheric CO_2 budget. *Science* 247:1431–1438.

Thoning, K. W., P. P. Tans and W. D. Komhyr. 1989. Atmospheric carbon dioxide at Mauna Loa Observatory: 2. Analysis of the NOAA/GMCC data 1974–1985. *Journal of Geophysical Research* 94:8549–8565.

Waterman, L. S., D. W. Nelson, W. D. Komhyr, T. B. Harris, K. W. Thoning and P. P. Tans. 1989. Atmospheric carbon dioxide measurements at Cape Matatula, American Samoa, 1976–1987. *Journal of Geophysical Research* 94:14817-14829.

Plant Migration and Climate Change

A more realistic portrait of plant migration is essential to predicting biological responses to global warming in a world drastically altered by human activity

Louis F. Pitelka and the Plant Migration Workshop Group

Terrestrial plants are notorious for their sedentary habits; indeed, that is one way our ambulatory species identifies them as members of the other kingdom. Of course, populations of plants do move, infiltrating new territory by creep of root and shower of seed. But how much does our self-absorbed species really know about these stately migrations? Do scientists know enough, for instance, to predict what would happen to plant communities if the earth's climate suddenly changed?

We seem to believe that we do. Forecasts of global warming are often illustrated with maps showing the poleward movement of plant species, as though the biosphere were made up of puzzle pieces that could be rearranged at will. In reality, it may take plant populations years or decades to move substantial distances. Moreover, today they must move through a landscape that human activity has rendered increasingly impassable. Under the circumstances, it is possible that many species might perish, caught in the double bind of climate change and habitat degradation. To reliably assess this risk, we need a better understanding of plant migration.

The interdisciplinary research required to predict how migration might constrain the response of the plant kingdom to climate shifts is just beginning. The evidence consists of diverse pieces. Some scientists have examined the fossil record of plant migrations following ancient climatic upheavals. Others have studied contemporary invasions of exotic species. Still others have analyzed the mathematics of dispersal mechanisms and the interaction of those mechanisms with contemporary landscape patterning.

These three lines of research provide disparate—even contradictory—insights. Past migrations have been much faster than simple calculations based on seed dispersal by wind or by vertebrate animals would predict. Plants appear to be capable of long-distance jumps and of rapidly spreading from pre-established outlier positions. But it is also evident that human activity has greatly altered patterns of plant migration. People disperse seeds farther and faster than the seeds' own dispersal mechanisms can take them. But people also fragment the landscape, creating habitat patchworks that are usually less able to support either plant species or their animal conveyances than are undisturbed landscapes.

Some of these factors would seem to retard plant migration and others to accelerate it. Which factor or combination of factors will predominate for which species in the event of future climate change? And what will happen if the climate changes faster than it has in the past? Perhaps the only way to explore questions this complex is by means of computer models.

The Fossil Record

The paleorecord provides compelling evidence that plant migration can be rapid enough to track climate change and that migrations can take place by means of sporadic long-distance leaps. For example, by mapping the accumulation of pollen in lake sediments at sites widely dispersed in time and space, paleoecologists have produced maps that indicate some tree species advanced rapidly during the early Holocene epoch (our own epoch, beginning about 10,000 years ago), following the retreat of the North American glaciers (see Figure 3). The high migration rates and the ability of species to jump large bodies of water reveal an underlying potential for relatively rapid response to climate change.

How many kilometers can a tree species traverse in a year? A plausible set of velocities can be calculated from analyses of past migrations. Indeed, these velocities might be especially pertinent because they describe movement across the northern temperate latitudes, exactly those regions where greenhouse-warming scenarios predict the most severe temperature change. In eastern North America, the migration velocities were as high as a kilometer a year. Moreover, these velocities seem to have been unaffected by dispersal barriers, even ones the size of the Great Lakes and the North and Baltic seas (Woods and Davis 1989, Kullman 1996).

Dispersal this rapid around seemingly significant barriers is a puzzle theoretical biologists have tried to solve for nearly half a century. How could trees move so fast and leap obstacles so broad? Simple calculations based on the ranges of birds or mammals that disperse seeds and on patterns of wind-borne seed movement do not predict such rapid migration. This outcome has led modelers to suspect that migration is accomplished mainly by rare long jumps that escape our observation (Clark *et al.*, in press). In addition, contemporary empirical studies (for example, measuring the amount of seed caught in traps placed near trees) do not pick up seed caught in vigorous uplift that would then inject it into high-level circulatory patterns. Long jumps might create dispersed outlier populations, too sparse to be detected in the pollen record, that might serve as foci for rapid invasion when conditions became more favorable.

Louis F. Pitelka is director of the Appalachian Laboratory of the University of Maryland Center for Environmental Science. He prepared the article collaboratively with the participants in a workshop held in fall 1996. The research fields of Pitelka and his coauthors (see "Authors and Acknowledgments," p. 473) encompass paleoecology, modeling, biogeography and plant ecology. Address for Pitelka: Appalachian Laboratory, UMCES, Gunter Hall, Frostburg, MD 21532. Internet: pitelka@al.umces.edu.

Paul Souders/© Corbis

Figure 1. Northern Europe was not always blanketed with Norway spruce, but over several thousand years *Picea albans* mounted an impressive migration that produced the dark-green forests that now fill images of Scandinavia such as this morning scene near Ingdal, Norway. The biological record of the large spruce migration provides evidence that plant populations can advance rapidly by establishing distant outlier populations. But predicting the plant migrations that might accompany a global warming is difficult, partly because of the complex effects of human alterations of the environment. For instance, *Picea* populations in plantations (such as one in Nordmoen, Norway, *right*, photographed during acid-rain research) sometimes exist south of the tree's natural range; they might be left stranded and less productive if global warming moves the range northward. (Photograph at right courtesy of Louis Pitelka.)

One rare documented example of the long-jump-and-outlier model of spread is the western migration of Norway spruce (*Picea abies*) across northern Europe during much of the Holocene and especially the spruce migration across western Sweden about 3,000 years ago (Bradshaw and Zackrisson 1990). Wood fragments found underneath living groups of *krummholz* (stunted, multi-stemmed) spruce in the Scandes Mountains demonstrate that spruce grew there as early as 8,000 years ago, more than 5,000 years before its presence could be inferred from pollen data *(see Figure 4)*. Isolated trees hundreds of kilometers in advance of the migrating front apparently served as foci for the later invasion of *Picea*. Although the long-jump-and-outlier model of spread might seem far-fetched as a general explanation for rapid migration, studies of contemporary plant invasions suggest it is not limited to Holocene spruce.

Some scientists think that, with suitable technique, it may be possible to distinguish local from windblown pollen and thus to detect outliers even in the pollen record. One team used a "geographic method" to look for evidence of outlying colonies of eastern hemlock (*Tsuga canadensis*) and American beech (*Fagus grandifolia*) in the pollen record preserved in lake sediments from periods before human settlement in Michigan, Wisconsin and Minnesota. They were confident that pollen percentages consistently significantly higher than those at surrounding sites identified outlying colonies (Davis, Schwartz and Woods 1991).

The outlier model of spread is consistent with results from other lines of inquiry. Recent mathematical analyses show that if a species' dispersal pattern (the spatial distribution of propagules around a source) is Gaussian, or bell-shaped, an invasion will exhibit constant-speed traveling waves. But if the pattern has "fat" tails (at least some seeds move especially long distances), invasion will accelerate (*see*, for example, Kot,

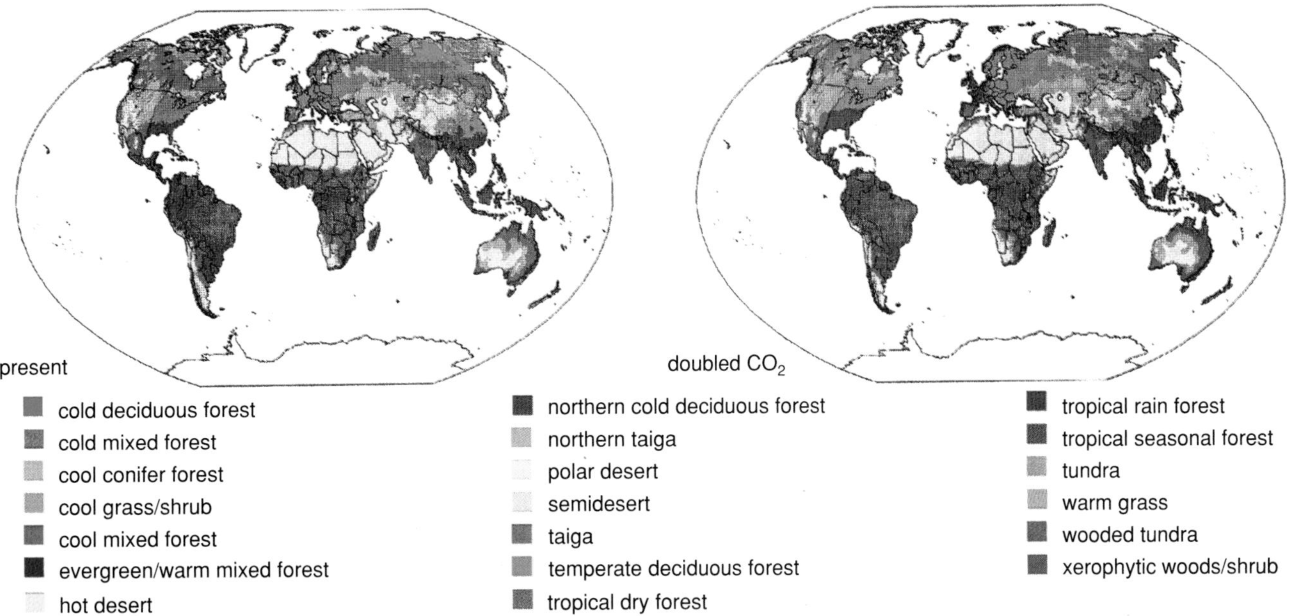

Figure 2. In many early simulations of climate change, vegetation ranges tended to shift around like puzzle pieces, ignoring the question of whether vegetation can migrate to match altered conditions. This map, produced by the BIOME1 model (Prentice *et al.* 1992), simulates the distribution of global natural vegetation for present climate conditions *(left)* and then under conditions brought on by a doubling of carbon dioxide levels in the atmosphere *(right)*, based on a widely used general-circulation model of the atmosphere called GFDL (Manabe and Wetherall 1987). Ecologists have found, however, that plant migration is a far more complex phenomenon than such models suggest, and that feedback responses from vegetation subjected to climate change might alter the course of that change.

Lewis and van den Driessche 1996). Outlier populations, then, may increase the rate of invasion, exerting a continual outward pull that causes the wave to move forward faster and faster. Converging lines of evidence suggest, therefore, that once an outlier tree population was established across a barrier, the population rapidly filled in the intervening space (Davis, Schwarz and Woods 1991). In this context, it is interesting that the paleorecord indicates that migration rates slowed rather than accelerated in the later Holocene. This suggests they may have been limited by some factor other than intrinsic dispersal rates, such as the rate of climate change (Clark *et al.*, in press). In other words, plants may not have reached their maximum potential rates of spread.

On the whole, the picture that emerges from the fossil record is one of relatively rapid migration: Plants have moved fast enough to track climate change, and may be capable of faster migration than is seen in the paleorecord. These migrations have come in response to change that in some cases has been as fast as the rapid warming predicted for the next few decades. During the last glacial period, there were repeated episodes of warming in which the mean annual global temperature rose by 5 degrees Celsius or more within a few decades (Nicholls *et al.* 1996).

Holocene patterns of spread, however, do not guarantee rapid plant migration in response to future climate change. First, people have greatly altered landscapes since the early Holocene. Second, early Holocene range shifts indicate that patterns of spread may depend on many variables, including local topography, and not just the plants' intrinsic dispersal potential. Finally, the tree species whose pollen records are clearest are not necessarily representative of the entire plant community. Many different patterns of migration are found in the fossil record, and gaps in the record and problems in interpretation make it difficult to characterize these patterns precisely. A study of the paleorecord in southern Italy suggests that several tree genera—oak, beech and lime—kept pace with climate fluctuations in the last 15,000 years, advancing, retreating and advancing again, but perhaps they were able to do so because they had previously established outlier populations in nearby refuges, which served as a kind of hedge against change (Watts, Allen and Huntley 1996). In other regions there is no evidence for such rapid adjustments, although this might be because the pollen record lacks the temporal resolution needed to identify them.

Contemporary Invasions

Clearly an exotic species invading a native plant community is not the same thing as a native species shifting its range in the wake of climate change. But in both cases plant populations come to occupy new territory via dispersal and reproduction. And since invasions have occurred with increasing frequency in the past 200 years, we have detailed accounts that, judiciously weighed, provide an opportunity to learn about range adjustments.

Studies of contemporary plant invasions also provide evidence for the long-jump-and-outlier mechanism of spread. A particularly dramatic example is the recent expansion of cheatgrass, the Eurasian grass *Bromus tectorum*, in western North America (Mack 1986). This aggressive alien grass occupied most of its current range of about 200,000 square kilometers within the last decade of a 40-year invasion process (see Figure 5). The evidence suggests that *B. tectorum* arrived in the intermountain West, the region bounded by the Rocky Mountains to the east and the Cascade–Sierra Nevada ranges to the west, in the 1880s, as a contaminant in agricultural seed. Before 1900, the species had established itself in at least five locations in the region. Over the next 20 years at least 50 additional foci appeared, but the total area occupied by the grass was still small. Then, within only about 10 years (1920–1930), the grass filled in between these foci, becoming a dominant species throughout the region.

B. tectorum is just one of many examples of invasions that have two distinct phases: a quiescent phase, during which ranges shift only slightly, followed by an active phase, during which something triggers explosive expansion (Forcella and Harvey 1981). The lag times of modern invasions range from decades to a century, numbers that provide an interesting commentary on the paleorecord of rapid migration. These periods are too brief to be picked up in fossil records, but they are long enough that plant populations might not persist if the climate changed as rapidly as some scenarios suggest it might.

It is difficult to see how the intrinsic migratory potential of an annual such as *B. tectorum* could limit its ability to adapt to future climate change, even if we assume it has a lag time of 40 years. But other species that have lower reproductive rates or take longer to reach reproductive maturity, such as trees and other perennials, could have longer lag times and might be stranded in unsuitable habitat by rapid climate change. This raises the unpleasant prospect that, in the event of rapid climate change, unwanted species—weeds—would be the species that would have little trouble shifting their ranges. Invaders by definition, most weeds are capable of rapid propagation, spread and often adaptability; they are plants that over the course of evolution may have traded off other traits more valued by people. Thus it is likely that species that are more desired, and perhaps cultivated, might be vulnerable to extinction.

A second important lesson of modern invasions is that it can be very difficult to establish new populations. A study of attempts to introduce parasites and predators into pest insect populations in Canada found that the best predictor of successful colonization was simply how many times colonization was attempted (Bierne 1975). Similarly, newly founded outlier populations of plants may fail again and again, and even when they succeed, it may be a while before the founding populations can send out enough propagules to fill in the intervening ground. Repeated opportunity may thus be crucial to a species' ability to shift its range. As we discuss below, people have modified landscapes in ways that tend to reduce such opportunities.

The record of modern invasions also allows scientists to examine the relation between dispersal modes and migration rates. It is tempting to attribute the rapid spread of some species in the Holocene to their dispersal modes, but there is little evidence for such a link. To the contrary, differences in dispersal modes did not seem to matter; virtually all species migrated into new areas with apparently equal efficiency. (Again this observation must be interpreted cautiously. The fossil record may lack the temporal resolution needed to distinguish dispersal modes, or some factor other than dispersal modes may have limited rates of migration.)

Although plants have many different strategies for dispersing seeds, only three are long-haul mechanisms: dispersal by water, by wind, and by birds and large mammals. Although the northward-flowing rivers of northern Europe are thought to have played a role in the extremely rapid postglacial migrations of some woody taxa, such as hazel (*Corylus avellana*) (Huntley and Birks 1983), only wind and vertebrate transport are likely to be important for predicting large-scale migrations of broad classes of plants.

The distance a propagule falls from its source depends on wind velocity and

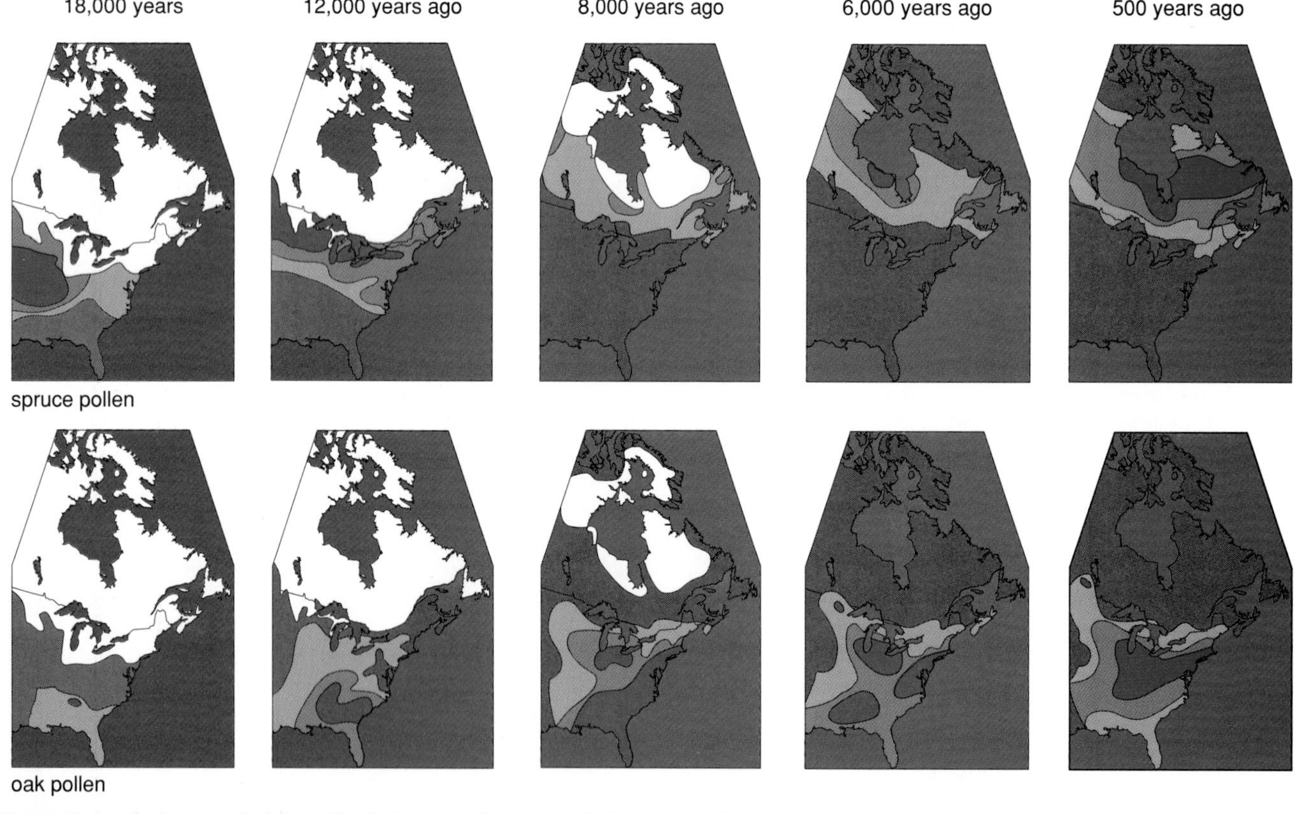

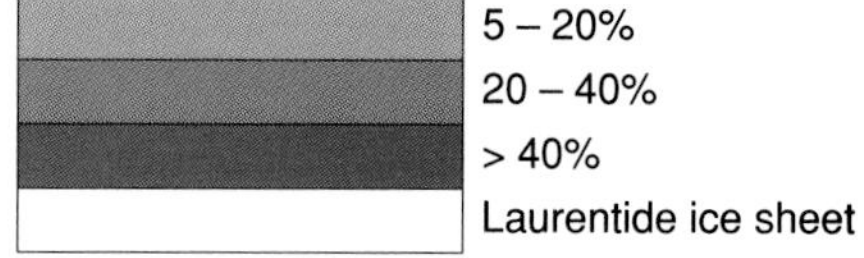

Figure 3. As glaciers receded from North America during the Holocene era, forests became established in their place. Pollen from sediments can be used to trace how tree species migrated northward as the climate warmed. These isopoll maps present density contours for spruce and oak pollen and show that both species moved rapidly at rates of up to one kilometer a year, their progress virtually unimpeded by large obstacles such as the Great Lakes and the Gulf of Saint Lawrence. (Adapted from Webb, Jacobson and Grimm 1987.)

"drag," or resistance to fall (which is enhanced by seed wings or plumes). Windstorms may carry propagules long distances, especially if vigorous uplift lofts them higher than their normal release height. Although these events are difficult to observe, there is circumstantial evidence that they take place; for example, plants have rapidly colonized Krakatau and other newly formed volcanic islands.

Many plant propagules are adapted for dispersal by vertebrates (Webb 1986). Large terrestrial mammals and birds of all sizes are important agents of long-distance dispersal and can transport seeds tens of kilometers. Moreover, some species preferentially cache nuts along forest edges, which encourages spread (Johnson and Adkisson 1985). Blue jays, which exhibit this behavior, may have enabled Fagaceous trees such as beech to adjust easily to postglacial warming in eastern North America (Johnson and Webb 1989).

Interesting as these natural dispersal mechanisms are, they are not necessarily relevant to future range shifts; for some species, the natural dispersal mechanism is likely to be superseded by human agency. People routinely introduce and cultivate plants thousands of kilometers outside their native ranges, thus serving as a kind of super dispersal mechanism. The Chinese tallow (*Sapium sebiferum*), for example, was intentionally introduced into South Carolina in the 1700s and has since spread throughout the southeastern United States. It is a serious threat to coastal prairie ecosystems, which are converted to woodland by this fast-growing tree. Remarkably, Chinese tallow is still sold in nurseries as an ornamental, even though it is known to be environmentally harmful (Stein and Flack 1996). Although human-assisted invasions have typically been considered ecological disasters, it is conceivable that people might cushion the effect of climate change for native species by moving seed longer distances than it could travel on its own.

Of course, dispersal is only the first step toward successful colonization. Plants must also successfully germinate, grow and reproduce in order to serve as the source of new propagules that can disperse in their turn. The likelihood of stray seed establishing itself depends on the existing vegetation and on the type and frequency of disturbances in the local environment. In the absence of disturbances, many plant communities are quite resistant to invasion. Shade from an established canopy or a thick leaf-litter layer can serve as a barrier to the establishment of both native and exotic plants. Plants also influence the chemistry of the litter and soil in ways that can render it inhospitable to invading plants. The unique soils in conifer forests, tundra regions and wetlands are all self-protective in this way.

In general, the disturbance of an established plant community by fire, flood, windstorms, and burrowing or grazing animals renders it more vulnerable to invasion. This suggests that future climate changes that kill existing plants or lead to more frequent fires or windstorms would increase the probability that dispersed propagules would be able to establish themselves. But people have also profoundly altered disturbance regimes. Cheatgrass, for example, was able to establish itself so easily in part because settlers introduced livestock, and the grasses native to the intermountain West were intolerant of trampling. Large mammals introduced by people, such as water buffalo and wild boar, have also contributed to rapid spread of exotic invasive plants in Australia and Hawaii (Russell-Smith and Bowman 1992, Stone 1985).

The net effect of human agency is difficult to predict. Although people disturb large tracts of land by tillage and grazing, they also suppress some natural disturbances, such as wildfires. Fire suppression prevents the invasion of forests with dense understory vegetation by shade-intolerant saplings. On the other hand, it has allowed Midwestern oak woodlands to be invaded by fire-sensitive species such as buckthorn and Australian sclerophyll (drought-tolerant) forests to be in-

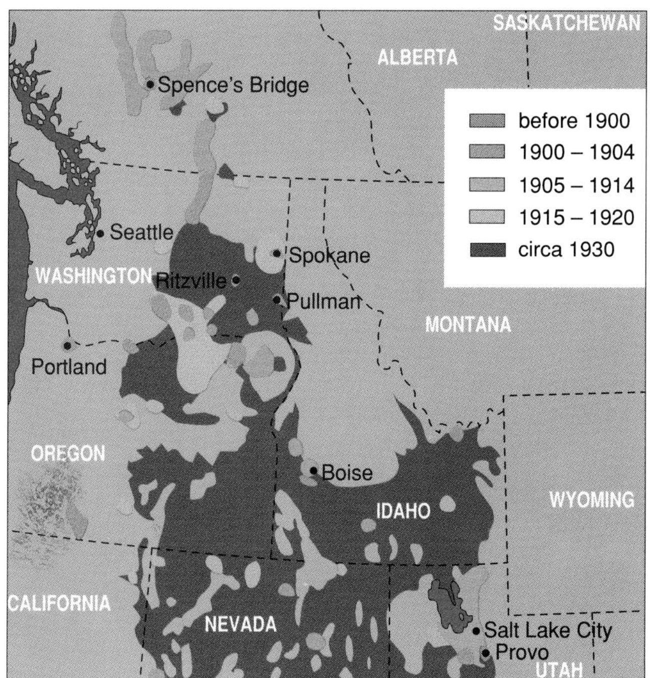

Figure 4. Cheatgrass, *Bromus tectorum (right)*, invaded western North America during the past century, but the invasion (mapped at left) did not proceed at a steady pace. It appears to have had a quiescent phase, during which the plant's range shifted only slightly, followed by an active phase, in the years 1920–1930. Other examples of migrations by aggressive alien species also exhibit two stages. (Map adapted from Mack 1986; photograph courtesy of Richard Mack.)

Farrell Grehan/Corbis

Figure 5. Long-distance plant dispersal requires the help of water, wind and birds or large mammals. Plant propagules are adapted for various kinds of dispersal. Seeds of the milkweed *(upper left)* ride on the wind, just as cockleburs ride on large, furry animals. Non-native plants often have human help. The Chinese tallow *(Sapium sebiferum)*, introduced into South Carolina as a landscaping plant and still sold at nurseries, now threatens coastal prairie ecosystems. The photograph at right was taken at Old Santee Canal State Park near Charleston, where berry-laden tallow branches stand out against the darker leaves of another invasive species, the Chinese privet *(Ligustrum sinense)*, imported for similar reasons and now also naturalized in the region. At lower left a cardinal enjoys tallow berries in winter, illustrating another common means of dispersal. (Photographs by: *lower left*, Lee Lowder, Old Santee Canal State Park; *right*, Lil Chappell.)

vaded by rain-forest species (Harrington *et al.* 1995).

Our species' flair for creating outlier populations of plants must also be considered. Because people have moved plants around the globe for horticultural and landscaping purposes, we may have set the stage for massive invasions by exotic species. Botanical gardens and similar large repositories of alien species have long been sources of plant invasions (Parker 1977), but such specialized facilities are not the only culprits. People routinely plant both native and alien species in their gardens and parks, far from the native ranges of these plants. Some of the most noxious plant invaders, such as the water hyacinth *(Eichhornia crassipes)*, are probably escaped cultivars. Global climate change could trigger a new round of escapes, if conditions changed in such a way that species long occurring as agricultural weeds or maintained by cultivation outside their natural ranges were able to grow in the wild.

What about Landscape Patterning?

The record of modern invasions suggests that human agency acts to accelerate the migration of some species, particularly those that are adept hitchhikers or whose usefulness has captured our attention. But this view fails to take into account what many scientists think is the most alarming effect of human activity: habitat fragmentation. Almost any aerial view of the earth reveals a patchwork landscape, in which undisturbed habitat exists as scattered islands in a sea of cultivated or developed land. Habitat destruction and fragmentation, by themselves, can drive species to extinction. Together with global climate change, they could lead to wholesale extinction, stranding plants in inhospitable landscapes without egresses. According to the Nature Conservancy, half of the endangered plant species in the U.S. are restricted to five or fewer populations (Schemske *et al.* 1994). If climate change renders these remnant sites inhospitable, there may not be enough plants left to launch migrations across largely hostile landscapes.

Landscape ecologists have developed several models that predict how habitat alteration might constrain plant migration. In general, the migratory rate is the product of the dispersal rate and the rate of population growth at the edge of the invasion. But if the plant population is spreading through a habitat mosaic, there is also a threshold effect: If the population is to expand at all, the fraction of the total habitat that is suitable for the species must exceed some minimum value (see Figure 9). Finally, because many species persist as metapopulations, with local extinctions being balanced by colonizations, there is also a second subtler threshold: If the species is to persist, let alone spread, the number of populations also must not fall below some minimum value. Indeed metapopulation dynamics may explain why the geographic distribution of some plants is more restricted than physiological limits allow (Schemske *et al.* 1994).

These three theoretical rules have very down-to-earth implications. By carving up landscapes with roads, buildings and agricultural land, people reduce the availability of suitable habitat, reduce the number of populations in a metapopulation, and even depress dispersal rates and rates of population growth. In some cases, all of these factors will conspire to inhibit migration. There are known to be cases where the opposite is true—blue jays move between habitat islands, and such birds may move farther in a fragmented environment—but these cases are not the rule. Decreases in bird populations likely will reduce long-distance seed dispersal.

On the whole, habitat fragmentation is likely to have the greatest impact on

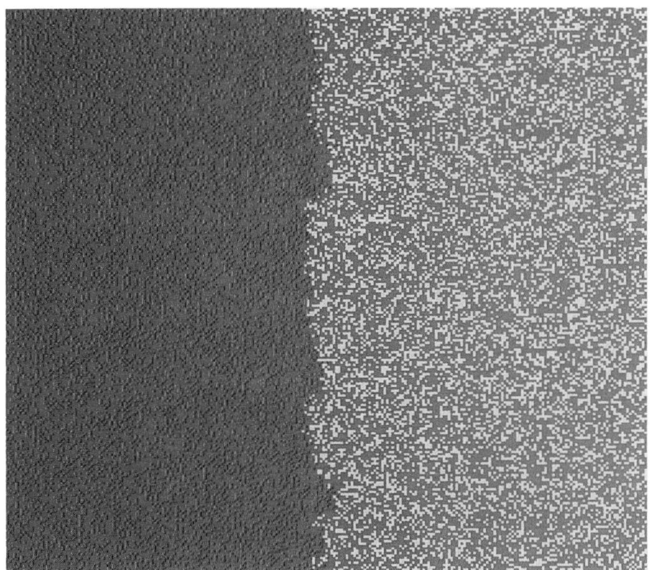

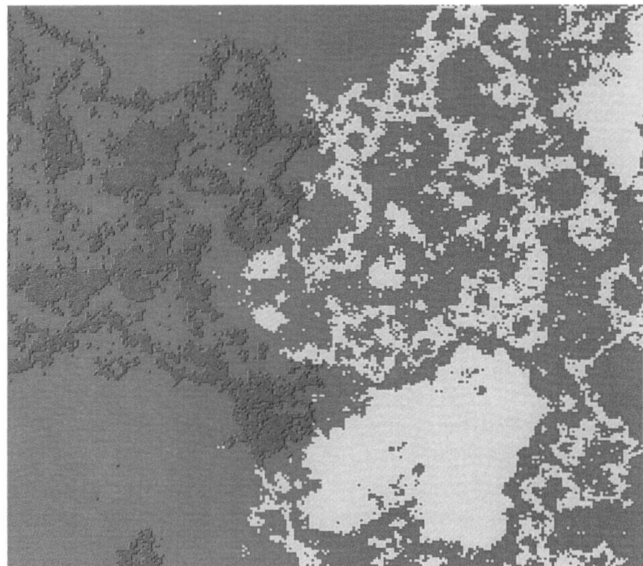

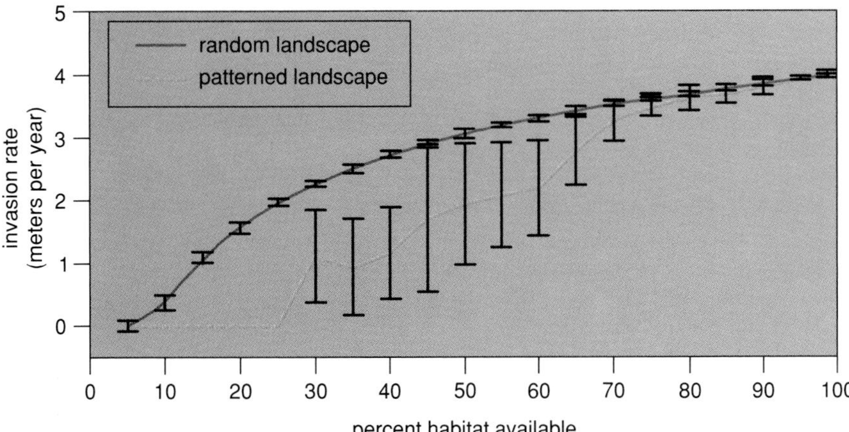

Figure 6. Landscape-ecology models predict how habitat alteration might constrain plant migration. If a plant population is spreading through a habitat mosaic, the models show there is a threshold effect: Some minimum fraction of the total habitat must be suitable if the population is to expand. These images show model predictions of the invasion of an invading annual plant species whose propagules are spread according to an exponential distribution with a mean distance of 1 meter and a maximum range of 6 meters. Each map has an area of 256 square meters, broken into one-meter sites. The invading species *(red)* is introduced onto the left side of the map and its asymptotic rate of invasion measured after 20 generations. In the random landscape *(top left)* the probability of any site being suitable is 0.4. A fractal algorithm was used to generate a patterned map *(top right)* in which habitat corridors for plant migration decrease in size and connectance as the fraction of suitable sites declines (but the overall probability remains 0.4). Above, the rate of invasion under each scenario is graphed (with bars representing 95 percent confidence intervals from 10 iterations) against the fraction of the map occupied by suitable habitat sites. Invasions slow significantly when the suitable habitat is less than 65 percent. (Images and data courtesy of Robert Gardner.)

species whose seeds are dispersed by large land animals. Landscapes are increasingly fragmented into parcels smaller than the home ranges of large vertebrates, and this refashioning is often accompanied by the hunting and local extirpation of species or the restriction of their movements. In parts of Africa, for example, even as introduced mammals contribute to the spread of some exotic plants, the substantial restriction of elephant populations may have far-reaching consequences for native trees such as *Balanites wilsoniana* (one of the species called torchwood), for which they seem to be the exclusive or nearly exclusive dispersal agents (Chapman, Chapman and Wrangham 1992).

What is the upshot of these pressures on plant communities during climate change? We know that even during the Holocene, when only natural dispersal mechanisms were operating, entire plant communities did not migrate intact to new ranges. Instead the compositions of biological communities were reshuffled by differences in the ability of species to re-establish themselves. Given that human activity accelerates the migration of some species and inhibits that of others, we can only expect that, in the future, this reshuffling would result in communities quite different from those we know today. Native species that rely on natural dispersal mechanisms would probably be slowed by habitat fragmentation. Those species that human beings actively disperse, such as timber or endangered species that we deliberately move to new sites, would be much less affected by fragmentation and might migrate faster than in the past. But so might species that take silent advantage of human mobility, such as weeds that are carried with harvested crops.

Regional Problems

Although climate change is a global process, some of its most important consequences will take place on a regional scale. Looking at plant communities on this scale, one notices some plants such as forest trees that are cultivated near the limits of their natural ranges and others that are maintained in relatively small, isolated reserves. These classes of plants are vulnerable to climate change.

Norway spruce is currently the major commercial timber tree in southern Sweden, where it grows in naturally regenerated forests and in plantations outside its natural range. Depending on the pattern of climate change, future warming might alter considerably spruce's role in natural forests. And it might strand plantations even farther outside the tree's natural range, threatening their productivity. Trees grown in heavily exploited agricultural landscapes have little opportunity for dynamic response to changing conditions. Replacing monoculture plantation systems with forestry systems that permit natural migration and regeneration, and that allow variation in species composition over time, will facilitate future species movements and might make the spruce a more viable species during warming.

Given the inexorable loss of habitat to human activity, conservationists often attempt to minimize species loss by arranging the remaining habitat in ways that increase its connectivity. Unfortunately many of these arrangements may be unlikely to function well through periods of climate change. For example, as a compromise between logging and conservation interests, a special system of habitat-conservation areas has been designed in the Pacific Northwest for the spotted owl. The expectation that this network of reserves on federal forest land will adequately protect the owl is based, however, on landscape-ecology models that assume old-growth forests in the targeted sites will regenerate at known rates (Lamberson *et al.* 1994). Because climate change is likely to alter the rates and patterns of regeneration of Douglas fir forests, it adds enormous uncertainty to these plans, yet it has not been factored into the calculations.

More generally we can think of our entire nature reserve and park system as a static network, with little flexibility in the face of climate change. Our reserves and parks are fixed in place; when the climate changes, what will become of the rare and unique species inhabiting these isolated sites? Even if other suitable sites exist, will these species be able to move to them?

One way to assess the risk to which climate change will expose endangered species is to calculate the average size of patches occupied by species and the mean distance between patches before and after different climate-change scenarios. Although we cannot say precisely what the effects of changes in these landscape features will be, we do know that marked reductions in either patch size or connectivity can threaten species. It is disheartening, for example, that the patch size and connectivity of old-growth forests in the Pacific Northwest have declined by more than an order of magnitude in the past 50 years (Groom and Schumaker 1993). This degree of landscape modification is certain to have severe consequences for plant and animal migration.

Global Consequences

At the global level we confront a difficult but potentially very important issue affecting not only plant migration but its significance: feedback between climate change and the biosphere. Changes in vegetation structure can affect the physical properties of the land surface, such as its *albedo* (the percentage of sunlight it reflects), surface roughness and canopy conductance to water vapor. Changes in vegetation structure and function can also influence the exchange of carbon dioxide and other radiatively active trace gases, such as methane and nitrous oxide, between the atmosphere and the biosphere. By these means plants can influence the percentage of incoming sunlight absorbed at the surface or the atmospheric concentration of greenhouse gases, and thus the rate and magnitude of climate change.

Positive feedback from the biosphere could exacerbate global warming. Models of earth's climate 6,000 years ago suggest that variations in the earth's orbit could not alone have produced the high annual average surface temperatures deduced from paleoclimate analyses. Instead orbital forcing must have been exaggerated by subsequent vegetation responses (Foley *et al.* 1994, TEMPO 1996). Because boreal forests absorb much more solar radiation than tundra does, poleward shifts in the location of the forest/tundra boundary during a period of warming can amplify climate changes by as much as 50 percent.

Moreover, under some circumstances, the carbon cycle might act as a second feedback loop. If during warming the rate of forest dieback on the southern edge of a forest range (in the Northern Hemisphere) were higher than the rate of forest expansion on its northern edge, there would be a net release of carbon to the atmosphere, exacerbating the greenhouse effect responsible for the climate change in the first place. What happens depends critically on the poleward migration rate. Calculations by one team suggest that transient carbon releases owing to this and related mechanisms might be substantial enough during predicted warming that the biosphere might no longer be the sink for carbon dioxide it is today (Smith and Shugart 1993).

How, then, do we make sense of the complex picture of plant migration described in this article? In order to accu-

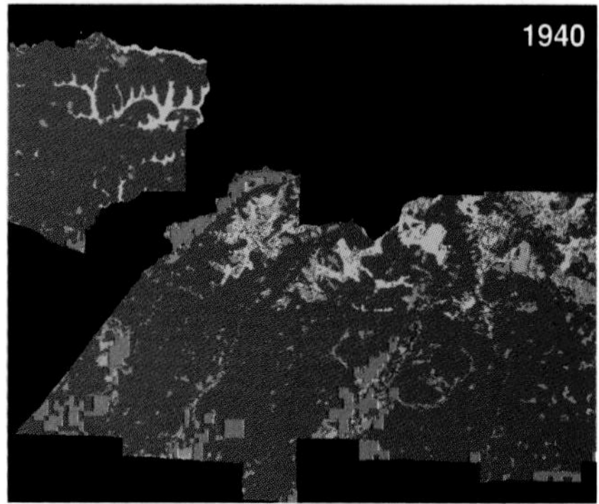

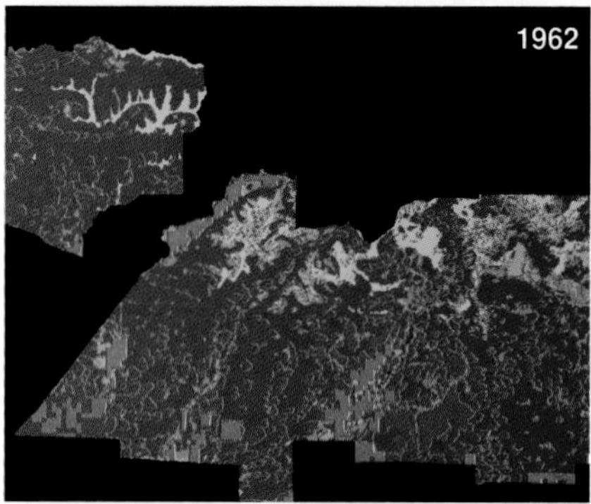

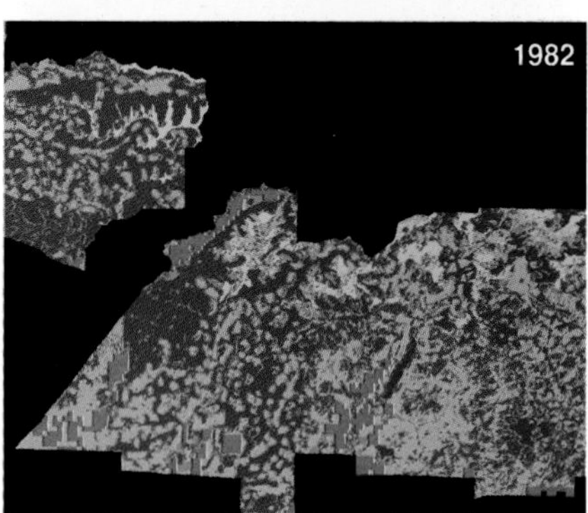

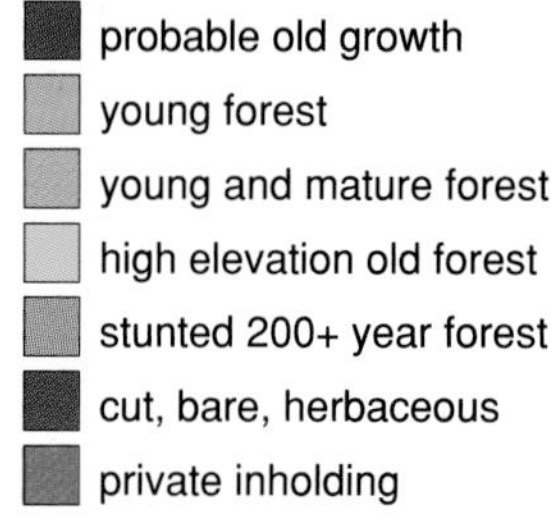

Figure 7. Ecological systems in old-growth forests in the Pacific Northwest may be particularly vulnerable to climate change because the size of old-growth patches and their connectivity have both declined dramatically as a result of cutting in recent years. These vegetation maps show a portion of the Olympic National Forest (approximately 38 x 46 kilometers) in 1940, 1962 and 1982. Each pixel represents a 57 x 57–meter patch of vegetation. Landscape modifications of this magnitude would be expected to have severe consequences for both plant and animal migration in a climate-change scenario. (Images courtesy of Nathan Schumaker, U.S. Environmental Protection Agency.)

rately predict future climate change, we must develop models that treat the biosphere as dynamic rather than static and that link changes in vegetation to global carbon and nitrogen cycles. The most comprehensive models of global vegetation, dynamic global vegetation models or DGVMs, can be used in combination with climate and carbon-cycle models to assess feedbacks between the biosphere and the general atmospheric circulation and the oceans (Foley *et al.* 1996).

DGVMs, like static models, include representations of photosynthesis, respiration, plant growth and decomposition, but they also include the processes of establishment, mortality and competition needed to simulate the waxing and waning of plant communities. Because they are global in scope, these models are necessarily coarse in some respects. For example, they typically describe the biosphere as a set of five to ten plant functional types, such as broad-leave deciduous trees or needle-leaved evergreen trees, and they have a grid resolution no finer than half a degree, or about 50 kilometers.

Although several DGVMs have been used for preliminary assessment of the response of the biosphere to simulated climate change, they do not fully capture biosphere dynamics. In particular, they lack any explicit representation of plant migration or of the interaction of dispersal processes and landscape patterning. Instead, they are typically executed in one of two "bracketing modes": Either dispersal is assumed to be fully effective, so that ranges that have become newly hospitable are invariably colonized, or dispersal is assumed to be always limiting, so that plant types are unable to shift their ranges (Sykes and Prentice 1996).

Neither of these approximations is satisfactory. The first mode would be a realistic approximation only if the climate changed very slowly, which is not the eventuality of concern. The second mode is also unrealistic, because it is more likely that rapid climate change will promote a few species to dominance than that no species will be able to keep up. Theoretically, even one vigorous migrator could set in motion the feedback loops described earlier.

One challenge of upgrading the models will be to incorporate dispersal modes in the plant functional types. As we have mentioned, only two of the many plant dispersal modes are likely to be important in large-scale migrations of functional types: wind and vertebrate dispersal. Wind dispersal is especially common among shade-intolerant species such as grasses and many forest trees, and these species often are highly clumped. Species dispersed by vertebrate animals—including perhaps 70 percent of tropical trees—may be either shade-tolerant or shade-intolerant but are often adapted to fill small gaps or patches of disturbed ground, growing as scattered individuals in highly diverse communities. These correlations might be exploited in the redefinition of functional types.

We still lack the knowledge to predict precisely how the interaction of dispersal modes and landscape patterning will affect the response to climate change. Given this predicament, the best approach might be to gradually narrow the range of possible outcomes by analyzing specific regions and vegetation types. For example, one could analyze the effect of climate change on Europe, which has many tree species with low dispersal rates and a highly fragmented landscape. This approach, together with that of evaluating various scenarios of climate change, offers our best hope of avoiding surprises in biosphere dynamics.

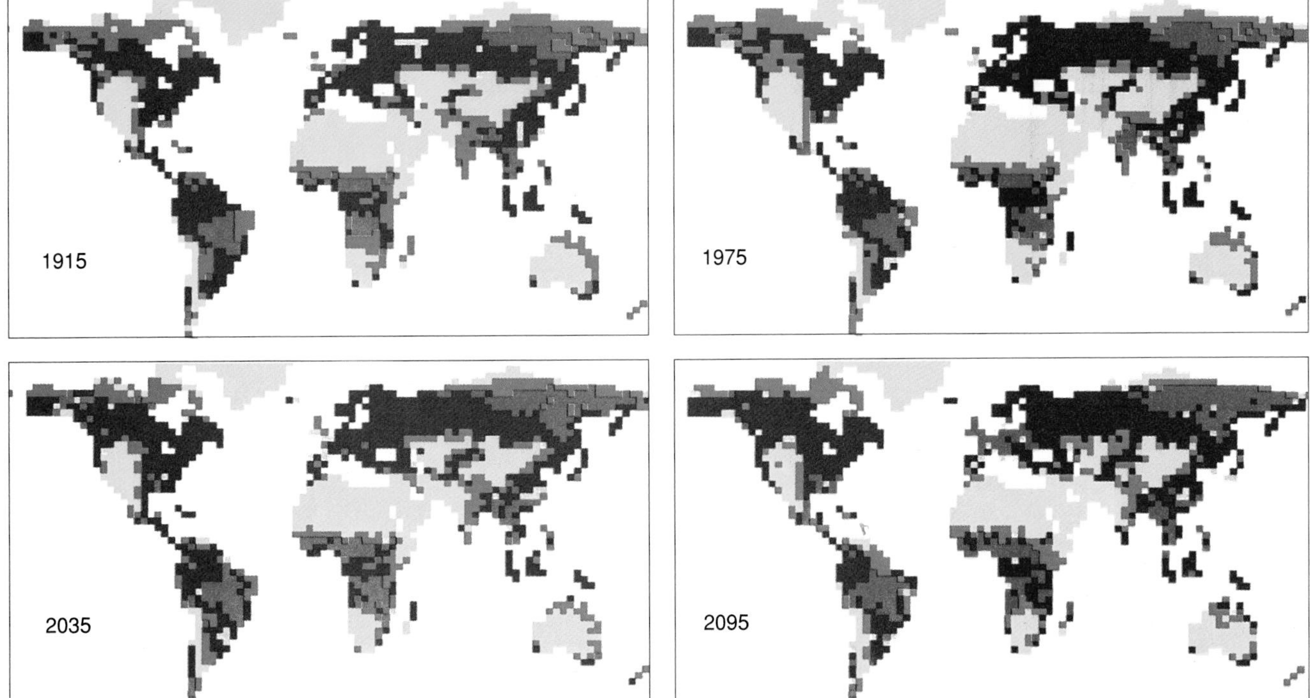

Figure 8. Linking climate models such as the one illustrated in Figure 2 with models of dynamic structural change in global vegetation could make it possible to predict biological responses to climate change and how these responses might affect the course of climate change. These preliminary simulations come from a DGVM, or dynamic global vegetation model, called IBIS (Foley 1996). In this simulation structural change in global vegetation is modeled based on population dynamics but no migration. The vegetation categories are desert *(yellow)*, grassland *(green)*, deciduous forest *(red)*, mixed forest *(pink)* and evergreen forest *(blue)*. The Hadley Centre general-circulation model of the atmosphere was used to predict climate change from known and predicted carbon dioxide concentrations from 1860 to 2099. (Preliminary unpublished model output courtesy of John Foley, University of Wisconsin; output maps by Alberta Fischer, Potsdam Institute for Climate Impact Research.)

It is not too soon to begin. In a recent issue of the journal *Nature* a team of scientists reported that measurements made from weather satellites between 1981 to 1991 had revealed a noticeable increase in the absorption of light by vegetation at northern latitudes. Not only has plant growth increased, the growing season has lengthened. Scientists cannot be sure what is causing the bloom, but it is certainly possible that the satellites have spotted one of the first signs of the effects of global warming on the biosphere.

Authors and Acknowledgments

This paper is the product of a workshop on plant migration and climate change held at Batemans Bay, Australia, in October 1996. In addition to Louis Pitelka the members of the Plant Migration Workshop Group are: Robert H. Gardner, also at the Appalachian Laboratory; Julian Ash of the Department of Botany and Zoology, Australian National University, Canberra; Sandra Berry, Habiba Gitay, Ian R. Noble and Alison Saunders of the Research School of Biological Sciences, Australian National University; Richard H. W. Bradshaw, Southern Swedish Forest Research Center, Alnarp, Sweden; Linda Brubaker, College of Forest Resources, University of Washington (Seattle); James S. Clark, Department of Botany, Duke University; Margaret B. Davis and Shinya Sugita, Department of Ecology, Evolution and Behavior, University of Minnesota, St. Paul; James M. Dyer, Department of Geography, Ohio University; Robert Hengeveld, Institute for Forestry and Nature Research, Wageningen, Netherlands; Geoff Hope, Research School of Pacific and Asian Studies, Australian National University; Brian Huntley, Department of Biological Sciences, University of Durham, U.K.; George A. King, Environmental Protection Agency Environmental Research Laboratory, Corvallis, OR; Sandra Lavorel, Centre d'Ecologie Fonctionnelle et Evolutive, Montpellier, France; Richard N. Mack, Department of Botany, Washington State University; George P. Malanson, Department of Geography, University of Iowa; Matthew McGlone, Landcare Research NZ, Lincoln, New Zealand; I. Colin Prentice, Department of Plant Ecology, Lund University, Sweden; and Marcel Rejmanek, Division of Biological Sciences, University of California, Davis. Funds to support the workshop were provided by the Electric Power Research Institute, the National Science Foundation and the Australian National University.

Bibliography

Bierne, B. P. 1975. Biological control attempts by introductions against pest insects in the field in Canada. *Canadian Entomologist* 107:225–236.

Bradshaw, R. H. W., and O. Zackrisson. 1990. A two-thousand-year history of a northern Swedish boreal forest stand. *Journal of Vegetation Science* 1:519–528.

Chapman, L. J., C. A. Chapman and R. W. Wrangham. 1992. *Balanites wilsoniana*: Elephant dependent dispersal. *Journal of Tropical Ecology* 8: 275–283.

Clark, J. S., C. Fastie, G. Hurtt, S. T. Jackson, C. Johnson, G. King, M. Lewis, J. Lynch, S. Pacala, I. C. Prentice, E. W. Schupp, T. Webb III and P. Wyckoff. In press. Dispersal theory offers solutions to Reid's Paradox of rapid plant migration. *BioScience.*

Davis, M. B., M. W. Schwartz and K. Woods. 1991. Detecting a species limit from pollen in sediments. *Journal of Biogeography* 18:653–668.

Foley, J. A., I. C. Prentice, N. Ramankutty, S. Levis, D. Pollard, S. Sitch and A. Haxeltine. 1996. An integrated biosphere model of land surface processes, terrestrial carbon balance and vegetation dynamics. *Global Biogeochemical Cycles* 10:603–628.

Foley, J. A., J. E. Kutzbach, M. T. Coe and S. Levis. 1994. Feedbacks between climate and boreal forests during the Holocene epoch. *Nature* 371:52–54.

Forcella, F., and S. J. Harvey. 1981. *New and exotic weeds of Montana, II. Migration and distribution of 100 alien weeds in northwestern USA, 1881–1980.* Helena, Montana: Montana Department of Agriculture.

Groom, M. J., and N. Schumaker. 1993. Evaluating landscape change: Patterns of worldwide deforestation and local fragmentation. In *Biotic Interactions and Global Change,* ed. P. M. Kareiva, J. G. Kingsolver and R. B. Huey. Sunderland, Mass.: Sinauer Associates, pp. 24–44.

Harrington, G. N., and K. D. Sanderson. 1994. Recent contraction of wet sclerophyll forest in the wet tropics of Queensland due to invasion by rainforest. *Pacific Conservation Biology* 1:319–327.

Huntley, B., and H. J. B. Birks. 1983. *An Atlas of Past and Present Pollen Maps for Europe: 0–13,000 Years Ago.* Cambridge, England: Cambridge University Press.

Johnson, W. C., and C. S. Adkisson. 1985. Dispersal of beech nuts by blue jays in fragmented landscapes. *American Midland Naturalist* 113:319–324.

Johnson, W. C. and T. Webb III. 1989. The role of blue jays (*Cyanocitta cristata* L.) in the post glacial dispersal of fagaceous trees in Eastern North America. *Journal of Biogeography* 16:561–571.

Kot, M., M. A. Lewis and P. van den Driessche. 1996. Dispersal data and the spread of invading organisms. *Ecology* 77:2027–2042.

Kullman, L. 1996. Norway spruce present in the Scandes Mountains, Sweden, at 8000 BP: New light on Holocene tree spread. *Global Ecology and Biogeography Letters* 5:94–101.

Lamberson, R., B. Noon, C. Voss and K. McKelvey. 1994. Reserve design for territorial species: The effects of patch size and spacing on the viability of the Northern Spotted Owl. *Conservation Biology* 8:185–195.

Lavorel, S., R. H. Gardner and R. V. O'Neill. 1995. Dispersal of annual plants in hierarchically structured landscapes. *Landscape Ecology* 10:277–289.

Mack, R. N. 1986. Alien plant invasion into the Intermountain West: A case history. In *Ecology of Biological Invasions of North America and Hawaii,* ed. H. A. Mooney and J. A. Drake. New York: Springer-Verlag, pp. 191–212.

Manabe, S., and R. T. Wetherall. 1987. Large scale changes in soil wetness induced by an increase in carbon dioxide. *Journal of Atmospheric Science* 44:1211–1235.

Nicholls, N., G. V. Gruza, J. Jouzel, T. R. Karl, L. A. Ogallo and D. E. Parker. 1996. Observed climate variability and change. In *Climate Change 1995: The Science of Climate Change,* ed. J. T. Houghton, L. G. Meira Filho, B. A. Callander, N. Harris, A. Kattenberg and K. Maskell. Cambridge: Cambridge University Press, pp. 133–192.

Parker, C. 1977. Prediction of new weed problems, especially in the developing world. In *Origins of Pest, Parasite, Disease, and Weed Problems,* ed. J. M. Cherrett and G. R. Sagar. Oxford: Blackwell Scientific Publications, pp. 249–264.

Prentice, I. C., W. Cramer, S. P. Harrison, R. Leemans, R. A. Monserud and A. M. Solomon. 1992. A global biome model based on plant physiology and dominance, soil properties, and climate. *Journal of Biogeography* 19:117–134.

Russell-Smith, J., and D. M. J. S. Bowman. 1992. Conservation of monsoon rainforest isolates in the Northern Territory, Australia. *Biological Conservation* 59:51–64.

Schemske, D. W., B. C. Husband, M. H. Ruckelshaus, C. Goodwillie, I. M. Parker and J. G. Bishop. 1994. Evaluating approaches to the conservation of rare and endangered plants. *Ecology* 75:584–606.

Schimel, David S. 1995. Terrestrial ecosystems and the carbon cycle. *Global Change Biology* 1:77–91.

Smith, T. M., and H. H. Shugart. 1993. The transient response of terrestrial carbon storage to a perturbed climate. *Nature* 361:523–526.

Stein, B. A., and S. R. Flack. 1996. *America's Least Wanted: Alien Species Invasions of U.S. Ecosystems.* Arlington, Virginia: The Nature Conservancy.

Stone, C. P. 1985. Alien animals in Hawai'i's native ecosystems: Towards controlling the adverse effects of introduced vertebrates. In *Hawai'i's Terrestrial Ecosystems, Preservation and Management,* ed. C. P. Stone and J. M. Scott. Honolulu: University of Hawai'i, pp. 251–297.

Sykes, M. T., and I. C. Prentice. 1996. Climate change, tree species distributions and forest dynamics: A case study in the mixed conifer/hardwoods zone of northern Europe. *Climatic Change* 34:161–177.

TEMPO (J. E. Kutzbach, P. J. Bartlein, J. A. Foley, S. P. Harrison, S. W. Hostetler, Z. Liu, I. C. Prentice and T. Webb III). 1996. The potential role of vegetation feedback in the climate sensitivity of high-latitude regions: A case study at 6000 years B.P. *Global Biogeochemical Cycles* 10:727–736.

Vitousek, P. M., C. M. D'Antonio, L. L. Loope and R. Westbrooks. 1996. Biological invasions as global environmental change. *American Scientist* 84:468–478.

Watts, W. A., J. R. M. Allen and B. Huntley. 1996. Vegetation history and climate of the last glacial at Laghi di Monticchio, southern Italy. *Quaternary Science Reviews* 15:113–132.

Webb, S. L. 1986. Potential role of passenger pigeons and other vertebrates in the rapid Holocene migrations of nut trees. *Quaternary Research* 26:367–375.

Webb, T., III, G. L. Jacobson, Jr., and E. C. Grimm. 1987. Plate 1. Isopoll Maps. In *North America and Adjacent Oceans During the Last Deglaciation,* ed. W. F. Ruddiman and H. E. Wright, Jr. Boulder, Colo.: Geological Society of America.

Woods, K. D., and M. B. Davis. 1989. Paleoecology of range limits: Beech in the Upper Peninsula of Michigan. *Ecology* 70:681–696.

The World after Rio

Thomas F. Malone

For two weeks last June, the population of the bustling city of Rio de Janeiro was swollen by more than 30,000 persons. Heads of state and leaders of private organizations around the globe had gathered to address profoundly complex issues central to the quality of life and to human development on the planet they share. I was among those who converged on Rio for the "Earth Summit," formally the United Nations Conference on Environment and Development (UNCED), and the parallel Global Forum sponsored by nongovernmental organizations.

I was keenly aware that UNCED was far from the first major international gathering on environmental issues; 20 years earlier I had been privileged to attend a U.N. conference on the environment in Stockholm. It was clear that the central environmental issues defined at the Stockholm conference remained salient two decades later. But the debate over the earth's future had been reshaped. The new ways of thinking and talking about our environmental dilemma that emerged from the Rio conference are important to understanding the problems, promise and opportunities facing scientists, educators and political leaders in the aftermath of UNCED.

Delegates to the Stockholm conference had a goal to which most citizens of the world would subscribe today—a vision of a physically attractive, biologically healthy and productive environment. It has become clear, however, that this goal cannot be discussed in isolation. Our vision of environmental quality has been broadened. Along with a healthy and productive natural environment, we must have human development. All people, in the present and in future generations, must have equitable access to the goods necessary to meet basic needs such as food, clothing, shelter and good health. And all must share in those legitimate aspirations that give meaning to sheer existence: culture, education, leisure, social interaction. Both the responsibilities and the opportunities for achieving human development while preserving environmental quality fall particularly on the makers of knowledge and the makers of policy. These new dimensions were explored at the seminal, interdisciplinary Forum on Global Change and the Human Prospect sponsored by Sigma Xi in November 1991.

Thomas F. Malone, former Foreign Secretary for the National Academy of Sciences, is director of The Sigma Xi Center in Research Triangle Park, North Carolina, Distinguished University Scholar at North Carolina State University and a former president of Sigma Xi. Address: P.O. Box 13975, Research Triangle Park, NC 27709.

Two Worlds, One Earth

The world population, and the global capacity to transform natural resources into the goods and services that meet human needs and wants, are both doubling every few decades. During the past century population has grown four-fold and economic production 15-fold. Both growth rates impinge upon the life-support capacity of the global environment.

But there are sharp asymmetries in demography and in economics between the two groups that constituted the major blocs at the Earth Summit: the seven industrialized nations in the North and the 122 still-industrializing nations in the South. The North, with about one-fifth of the world's total population of five billion people, produces and consumes four-fifths of all goods and services each year. The South, with four-fifths of the world's people, produces and consumes only one-fifth. For every person added to the population of the North, 20 individuals are added in the South. For every dollar of economic growth per person in the South, 20 dollars accrue to each individual in the North. In the North, the stress on the environment comes from an energy- and technology-powered economy and high consumption (hence the concerns about greenhouse-gas-induced warming and stratospheric ozone depletion). In the South, it comes from population pressure and low consumption (resulting in deforestation, desertification and soil deterioration). One-fourth of the people in the South exist in a condition of absolute and degrading poverty, inimical to human dignity.

The outlook for the 21st century is troubling. Even if population growth in the South and economic growth in the North are both constrained, the world population will more than double, and the global economy will increase nearly sixfold *(Figure 1)*. The world may be embarking on a grand experiment to ascertain the limits to the life-support capacity of the biosphere. Estimates indicate that human activity uses, diverts or wastes 40 percent of the photosynthetic productivity on Planet Earth.

These circumstances suggest the need for four actions: stabilizing the world population, transforming the energy- and technology-powered economic system into one that is environmentally benign, reducing the economic and demographic asymmetry between North and South and, in particular, reducing absolute poverty wherever it is found.

These objectives were implicit rather than explicit in the deliberations in preparation for UNCED and in the documentation that emerged from it. The North came with a principal concern for global environmental issues. The South was preoccupied with economic development and poverty. The results of the Earth Summit reflect this dichotomy. That some common ground was found represents a triumph of international diplomacy.

The Rio Agenda

World consciousness was raised at the Stockholm conference in 1972. Modest but effective institutional innovations were made within the U.N. system and among participating nations, and the following years witnessed a slowdown, or reversal, of environmental deterioration in many countries—especially the United States. The role of science and technology in economic development was addressed at another U.N. conference in Vienna in 1979. That one foundered on the controversial New International Economic Order and on the failure of the North and the South to reach agreement on the level of financing for proposed programs.

Impetus for the Earth Summit was provided by the report *Our Common Future*, by the Independent World Commission on Environment and Development, chaired by the Norwegian prime minister, Gro Brundtland. This report made a persuasive case that environmental quality and economic development are inextricably linked. *Our Common Future* held up the concept of "sustainable development" as a vision that would unite the North and South in a common endeavor. This concept envisions an interplay of population growth, increasing economic production and consumption, and evolving technology in a manner that meets the basic needs and legitimate aspirations of the present generation without foreclosing options for future generations. In simplified terms, it means living on natural-resource "interest," rather than drawing down natural-resource "capital." Although somewhat ambiguous, the term "sustainable development" caught on and emerged as a central theme at the Earth Summit.

A period of intensive planning, discussion, negotiation and compromise preceded UNCED, producing two major documents that constituted principal agenda items for the summit. They were the Rio Declaration on Environment and Development and Agenda 21.

The declaration constituted a set of 27 principles governing the rights, responsibilities and relationships of nations in the pursuit of sustainable development through "a new global partnership." Only muted reference was made to the reduction and elimination of "unsustainable patterns of production and consumption," the task of "eradicating poverty," the promotion of "appropriate demographic policies," and the "inherently destructive" nature of warfare. The declaration represented a compromise, fashioned mainly by Ambassador Tommy Koh of Singapore. A consensus could not be reached on a more ambitious and inspirational Earth Charter proposed during the planning period.

Agenda 21 is a document that may well serve as a blueprint for action. It is a several-hundred-page strategic plan divided into 40 chapters, embracing 115 program areas to be pursued by the global partnership during the 21st century in the interest of sustainable development. It addresses issues as diverse as demographic pressure, agriculture, forests, fresh water resources, the atmosphere, the oceans, toxic chemicals, hazardous and radioactive waste, poverty, trade and technology cooperation. It contains proposals for institutional arrangements to ensure implementation of the activities recommended for each program area, and financial arrangements to underwrite their costs.

Primary responsibility for implementation will rest with sovereign states, although UNCED recommended the creation of a high-level U.N. Commission on Sustainable Development to coordinate international efforts toward that goal. Since it is clear that the South will require financial assistance in taking action on Agenda 21, a funding mechanism—the Global Environmental Facility—was proposed that would operate under the auspices of the World Bank, the U.N. Development Program and the U.N. Environment Program. International funding for Agenda 21 was estimated at $125 billion annually. About half of this amount would represent new money that might be obtained if nations (mostly in the North) increased the portion of their gross national product that is now allocated to overseas development assistance to 0.7 percent by the year 2000. This was recognized as a desirable goal. The failure to obtain a firm commitment to this target was a distinct disappointment at Rio de Janeiro. The issue of financing is now under consideration at the 47th session of the U.N.

Various other topics on the UNCED agenda diverted public attention away from Agenda 21. The United States was at the center of controversy over conventions on climate change and biodiversity that were opened for signature at Rio. In both cases, the U.S.'s position isolated it from the other nations of the North. It is likely that residual reservations on both will be resolved in the negotiations that will transform conventions into treaties of historical significance. Deep concern by African countries over the issue of desertification led to a decision to recommend that a process be started that would lead to international negotiation of a convention for coping with this environmental phenomenon.

While the heads of state were working over tough policy issues, lively and informed discussions were under way at the Global Forum, which was attended by 20,000 represen-

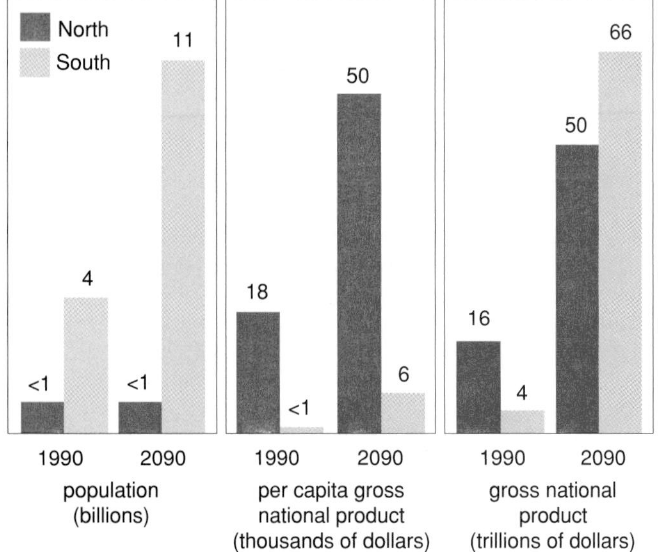

Figure 1. Current and projected population and economic contrasts between the industrialized nations of the North and industrializing nations of the South, based on assumed growth rates of 0 percent in population and 1 percent in per capita gross national product (GNP) in the North, and 1 percent in population and 2 percent in per capita GNP growth in the South. (Adapted from *World Development Report 1991*, World Bank and Oxford University Press.)

tatives of nongovernmental organizations. The debate at the forum included and went beyond the items on the UNCED agenda. Thirty-eight statements or treaties were drafted as a contribution to the deliberations initiated at the Earth Summit. Perhaps more important, informal networks were established that will provide a continuing mechanism for interaction within the nongovernmental community, and between that community and governments.

A Bend in the Road

The significance of the Earth Summit is not found in the specific details of the programmatic and institutional recommendations it produced, in the blizzard of documents generated in the preparatory phase, nor in the controversy over draft conventions at Rio. It is found in the nature of the particular juncture in the continuing saga of humankind on Planet Earth that brought so many individuals (including 110 heads of state or government) to Rio to deliberate on the problems and opportunities that lie ahead during the first century of the third millenium.

In expanding the common view of our global dilemma to include the question of equitable human development, the Earth Summit represented a major step toward maturity in the consideration of the environmental *problematique* that has attracted increasing attention for the past two decades. The issues are daunting. The task is formidable. Both individual and collective responsibilities are substantial. However, avenues for solving problems and seizing opportunities exist.

A major avenue is identifiable in the realization that knowledge is the driving force of human progress. It is knowledge about the world we inhabit and our place in that world that has brought us to this societal juncture. It is progress in the understanding of matter, energy, life processes and information, and their interactions, that has empowered individuals to transform the natural resources in the environment into the goods and services that satisfy basic needs and aspirations. The research community has been eminently successful in generating knowledge. Knowledge generation by individual investigators *must* continue; the need for new knowledge will only increase as societies advance. But greater emphasis, I believe, must now be given to integrating, disseminating and applying that knowledge.

Integration implies breaking new ground in interdisciplinary research and education at universities. Institutional reform and innovation to support these efforts will be necessary. Integration that crosses several domains of knowledge, such as the social sciences and the humanities, is mandatory.

Dissemination of knowledge touches deeply the role of education at all levels—informal as well as formal. A scientifically illiterate society is ill-prepared to make the kinds of decisions now required. The demographic and economic asymmetry between North and South is closely related to the knowledge asymmetry. The urgent need to address this aspect of "capacity building" is often overshadowed by a preoccupation with technology transfer and financial "aid." New technology for information handling and interactive communications provides attractive opportunities for the scientific community to make a major contribution to this dimension of the knowledge issue.

Application involves forging new modes of global partnership among business and industry, governments and the scholarly community. This is probably as crucial to the vitality of the research enterprise as it is to the fate of society. The coupling of knowledge generation and integration with policy formulation need not compromise the objectivity nor erode the standards of excellence that have been the hallmarks of the research community.

In addressing the central role of knowledge, it is useful to emphasize that the coupling of knowledge generation and integration with policy formulation will be most effective when pursued at the level where crucial decisions are made. In a highly diverse world society of sovereign nations, the optimal place is at a regional level. Proposals by the International Council of Scientific Unions and the Carnegie Commission on Science, Technology and Government for regional networks are now being actively pursued. In the United States, an elevated role for individual states is indicated.

Since knowledge is the domain of the universities, that community has a splendid opportunity to articulate a declaration of principles, set forth a realistic vision of the world that science and technology can help to achieve and make a commitment to the innovations in education and research that will be necessary to realize that vision. The biblical injunction, "where there is no vision, the people perish," underscores a challenge to which the university community might appropriately respond.

The concept of sustainable development has been extraordinarily useful, but it suffers as a societal goal, for example, by not distinguishing between economic development and the emerging concept of human development. The latter embraces economic considerations but incorporates other essential human aspirations. It merits elaboration.

The opportunities for leadership are limitless. But these opportunities will be seized only if the individual members of the research community become informed and involved. An important step would be taken, in my view, were scientists to commit themselves to the notion of "tithing," in which each individual dedicates a portion of research time to becoming informed on human-development issues and to participation in policy-relevant or strategic research.

The most important contribution of the Earth Summit was its function as a catalyst for stimulating governments, business and industry, nongovernmental organizations, and individual scientists and engineers to action. A framework now exists within which to begin addressing the issue of enhancing the quality of human life and human development everywhere on Planet Earth—now and in future centuries.

The outlines of a vision of that future are dimly visible: human development in an equitable society in a sustainable environment. The steps to achieve this vision are emerging. Many of them are found in Agenda 21. An urgent task is to focus on the overarching need to bring to bear on these societal problems the expanding storehouse of knowledge about the world in which we live and human values and behavior in that world. The research community—its individual members and organizations—deeply committed to knowledge and intrinsically international in outlook, is in a position to exercise leadership.

Bibliography

Sigma Xi, The Scientific Research Society. 1992. *Global Change and the Human Prospect: Issues in Population, Science, Technology and Equity*, Forum Proceedings, Sigma Xi, The Scientific Research Society, Research Triangle Park, NC.

International Environmental Research and Assessment: Proposals for Better Organization and Decision-Making. 1992. A Report of the Carnegie Commission on Science, Technology and Government, Carnegie Corporation.

Human Development Report, 1991. U.N.Development Programme, Oxford University Press.

The following *American Scientist* staff artists contributed illustrations to the articles in this collection:

Elyse Carter
Robin M. Gowen
Brian Hayes
Susan Hochgraf
Linda K. Huff
Rebecca Lehmann-Sprouse
Edward D. Roberts III
Linda Price Thompson

In addition, the work of the following freelance artists appears on the pages cited:

Beverly Benner Fig. 4, p. 113; Fig. 5, p. 115; Fig. 10, p. 258 (with L. Huff)

Sally Black Fig. 2, p. 144; Fig. 4, p. 146; Fig. 8, p. 150; Fig. 12, p. 153

Aaron Cox Fig. 5, p. 104; Figs. 8 and 9, p. 106; Fig. 11, p. 108; Fig. 2, p. 112; Figs. 7–11, pp. 116–120; Fig. 2, p. 220; Fig. 4, p. 222; Fig. 9, p. 226; Figs. 10 and 11, p. 227

Virge Kask Fig. 3, p. 92; Fig. 4, p. 93; Fig. 2, p. 208

D. W. Miller Fig. 2, p. 170